Histochemistry in Pathology

To Rachel and Rui without whose continued forbearance this book would not have been possible

Histochemistry in Pathology

EDITED BY

M. Isabel Filipe MB BS PhD FRCPath
Reader and Honorary Consultant, Department of Histopathology
UMDS, Guy's Hospital, London, UK

Brian D. Lake BSc PhD FRCPath
Professor of Histochemistry, Department of Histopathology, Hospital
for Sick Children & Institute of Child Health, Great Ormond Street,
London, UK

SECOND EDITION

CHURCHILL LIVINGSTONE
EDINBURGH LONDON MELBOURNE AND NEW YORK 1990

CHURCHILL LIVINGSTONE
Medical Division of Longman Group UK Limited

Distributed in the United States of America by Churchill
Livingstone Inc., 1560 Broadway, New York, N.Y. 10036,
and by associated companies, branches and representatives
throughout the world.

First edition 1983
Second edition 1990

ISBN 0 443 04019 2

British Library Cataloguing in Publication Data
Histochemistry in pathology.
   1. Medicine. Diagnosis. Histology
   I. Filipe, M. Isabel
   616.07583

Library of Congress Cataloging in Publication Data
Histochemistry in pathology/edited by M. Isabel Filipe,
   Brian D. Lake. — 2nd ed.
             p.   cm.
      Includes bibliographical references.
      Includes index.
      1. Histology, Pathological. 2. Histochemistry.
   I. Filipe, M. Isabel. II. Lake, Brian D.
      [DNLM: 1. Histocytochemistry. 2. Histological
   Technics. QS 525H673]
   RB27.H57 1990
   616.07'583 — dc20
   DNLM/DLC
   for Library of Congress   90-2038
                                                CIP

Printed and bound in Great Britain by
Butler & Tanner Ltd, Frome and London

Preface to the First Edition

Over the past few years we have become increasingly aware of the need for a practical book on the use of histochemistry in human pathology. The medical literature contains much information on histochemical methods and their interpretation in pathological material, but the references are widely scattered and there is no book which collects this information together. Professor A. G. E. Pearse has contributed extensively to the subject over the years and it was his suggestion that we should undertake the task of compiling a practical manual in which useful methods of known value in diagnosis could be described.

Histochemistry has now penetrated all fields of biological science, has contributed to the understanding of the relationship between structure and function in cells and tissues and has acquired an important role in diagnosis.

Its value in localising events at the histological level is unique and is an area not accessible by any other means. It provides a direct visual correlation between structure and function, so that individual cellular metabolic activities can be readily distinguished. In this and other aspects it has advantages over biochemical analysis. For example, a minimal amount of tissue is required; the functional heterogeneity at individual cell levels rather than the average activity can be visualised and a permanent record for interpretation is provided which can always be reviewed and often allows retrospective study.

Histopathology has moved in the last few decades from *macro* to *micro*, and the pathologist is now facing a new era with smaller and smaller biopsies to interpret. It is particularly in this

respect that the more specific techniques of histochemistry will greatly reduce the number of equivocal and inconclusive diagnoses with consequent benefit to the patient.

Histochemical techniques are now an essential part of the histopathological diagnosis of muscle disease and storage disorders, in the classification and diagnosis of leukaemias, lymphomas, and tumours of the neuroendocrine system, and in determining the site of origin of a variety of malignancies. Histochemistry increases the accuracy of the histological diagnosis in the assessment of malignant transformation and has a role in monitoring response to therapy, and in prognosis. Its use, however, has not yet gained wide acceptance.

This book is therefore aimed at the pathologist. It is not a pathology text. We assume that the reader is familiar with general pathological principles and has access to standard pathological texts. In the same way the book is not a histochemistry text. It includes histochemical techniques of established value in particular diagnostic problems. There is no pretence to cover every possible pathological condition, but we hope to have covered most disorders that are encountered in routine pathology.

We have excluded histological methods and subjects for which histochemistry as a means of diagnosis is well covered in pathology textbooks.

Finally, good co-operation between clinician, surgeon and pathologist should be encouraged, so that maximum information can be extracted from a biopsy specimen. However, histochemical techniques should not be used 'blind' as a sterile academic exercise; they should be used intelligently in relation to the clinical history. The methods included are simple and well-tried and all can be carried out in any routine laboratory.

We hope that many pathologists and technicians will find this book useful and thus be persuaded to adopt histochemistry as an integral part of histopathology.

London,   M. I. F.
1983   B. D. L.

Preface

In the Preface to the First Edition we stated that 'Histochemical techniques are now an essential part of the histopathological diagnosis' . . . 'in a restricted number of conditions' . . . 'their use however has not yet gained wide acceptance . . .' Seven years later, histochemistry is firmly established as an indispensable tool in routine diagnosis, covering a wide spectrum of pathology. Its role has now greatly expanded to encompass diagnosis, prognosis and monitoring response to therapy.

We were encouraged by the response of colleagues to the first edition and, seven years later, when techniques have changed or improved, others developed and new pathology areas become important, we felt that a new edition, incorporating in particular immunopathology, which was deliberately restricted to the well-established areas in the first edition, would be of value.

In this second edition, the layout has changed slightly, some chapters have been omitted as they became irrelevant, others adapted to the recent developments and new chapters introduced. A comprehensive guide to immunohistochemical methods has been included as a new chapter and an appendix. Fine needle aspiration cytology has its advocates and is a rapidly expanding diagnostic technique: methodology, interpretation and caveats are described and discussed in a new chapter.

Antibody studies in soft tissue tumours, in their infancy in the first edition, are now a major part of the diagnosis and have been integrated into a new chapter. Lymphomas are extensively covered in the second edition. This is one of the fields where the development of an increasing number of monoclonal antibodies has led to new concepts of disease and classification, with important implications in clinical management.

We have attempted to avoid duplication which has arisen in immunological discussion in many of the chapters, but have left the paragraphs where the content was particularly relevant.

Immunohistochemistry in routine diagnosis, in its early stages included in the first edition, is now a well-established method of diagnosis. Here also included in the second edition is information on in situ hybridization, flow cytometry and other areas which have just started having an impact in diagnostic pathology and are thought to be important in the future.

The book is intended to be of use both to pathologists and laboratory workers in the team approach to routine everyday diagnosis and to the diagnosis of rarer conditions.

London,   M. I. F.
1990   B. D. L.

Acknowledgments

The second edition of this book could not have been prepared but for the help, advice and encouragement of many friends and colleagues, and the co-operation of the contributors. To those whose chapters have disappeared, we thank you for your earlier support. We thank the contributors who speedily updated and rewrote their chapters and we welcome the new contributors to the second edition.

We acknowledge the expert technical assistance for many years of the staff at UMDS Guy's Hospital, particularly Mrs Angela Sandey. Expert and invaluable technical assistance at the Hospital for Sick Children has been provided in particular by Mrs Virpi Smith FIMLS and Mrs Dyanne Rampling FIMLS, and by many enthusiastic rotating MLSOs.

We would like to thank our Publishers, Mr Timothy Horne and his colleagues at Churchill Livingstone for their help in putting the book together and for its speedy publication.

Chapter 7 (Dr J. Lowe)

The basis for this work was partly funded by grants from the Special Trustees of the Nottingham University Hospitals and has been possible through the generous tolerance of Professor D. R. Turner. I would like to acknowledge the technical help of K. Morrell, D. Powe, D. McQuire, G. Thomas, and T. J. Palmer. Mr W. Brackenbury kindly performed the photomicrography and produced illustrations.

Contributors

Balbir Bhogal AIMLS
Chief Technician, Immunofluorescence
Laboratory, St. John's Hospital for Diseases of
the Skin, London, UK

Fred T. Bosman MD PhD
Professor and Chairman, Department of
Pathology, University of Limburg, Maastricht,
The Netherlands

Carlo Capella MD
Department of Human Pathology, University of
Pavia at Varese, Italy

A. C. Chu MRCP
Wellcome Senior Research Fellow, Senior
Lecturer and Consultant Dermatologist
Hammersmith Hospital, London, UK

Susan E. Daniel BSc MD MRCPath
Senior Lecturer in Neuropathology, Parkinson's
Disease Society Brain Bank, Institute of
Neurology, London, UK

C. E. H. du Boulay DM BM MRCPath
Department of Pathology, Southampton General
Hospital, Southampton, UK

S. J. Earley FIMLS
Department of Haematology, Hospital for Sick
Children, Great Ormond Street, London, UK

M. Isabel Filipe MB BS PhD FRCPath
Reader and Honorary Consultant, Department
of Histopathology, UMDS Guy's Hospital,
London, UK

I. M. Hann MD MRCPath
Department of Haematology, Hospital for Sick
Children, Great Ormond Street, London, UK

David Hopwood MD BSc PhD MRCPath
Senior Lecturer in Pathology, Dundee Medical
School; Honorary Consultant, Tayside Health
Board, Ninewells Hospital, Dundee, UK

P. G. Isaacson DM FRCPath
Professor of Morbid Anatomy, University
College and Middlesex Hospital, London, UK

A. C. Jöbsis MB BS MD
Department of Pathology, Bergweg Hospital,
Rotterdam, The Netherlands

Margaret A. Johnson PhD
Research Lecturer in Neuromuscular Disease,
Department of Experimental Neurology,
University of Newcastle, Newcastle-upon-Tyne,
UK

Brian D. Lake BSc PhD FRCPath
Professor of Histochemistry, Department of
Histopathology, Hospital for Sick Children and
Institute of Child Health, Great Ormond Street,
London, UK

S. Leibowitz MD BSc MB BCh FRCPath
Emeritus Professor of Histopathology, Guy's
Hospital Medical School, London, UK

Z. Lojda MD
Professor, Laboratory of Histochemistry,
Faculty of Medicine, Prague, Czechoslovakia

J. Lowe BMedSci BM MS MRCPath
Department of Pathology, University Hospital,
Queen's Medical Centre, Nottingham, UK.

Janet McLelland MB BS MRCP
Senior Registrar, Newcastle Royal Infirmary,
Newcastle-upon-Tyne, UK

A. J. Norton MB BSc MRCPath
Senior Lecturer in Histopathology, St.
Bartholomew's Hospital Medical College,
London, UK

J. R. O'Donnell MB BCH BAO FRCPI MRCPath
FRCP(Glasg.)
Consultant Haematologist in Administrative
Charge, Department of Haematology, Victoria
Infirmary, Glasgow, UK

Svante R. Orell ML(Stockholm) FRCPA FIAC
Director of Cytology, Department of
Histopathology, Flinders Medical Centre,
Adelaide, South Australia

M. G. Ormerod PhD
Senior Scientist, Institute of Cancer Research,
Royal Cancer Hospital, The Haddow
Laboratories, Sutton, Surrey, UK

T. J. Palmer BSc MRCP MRCPath
Senior Lecturer and Honorary Consultant in
Histopathology, UMDS, Guy's Hospital Medical
School, London, UK

Bernard C. Portmann MD FRCPath
Honorary Senior Lecturer and Consultant
Pathologist, Liver Unit and Department of
Morbid Anatomy, King's College Hospital and
School of Medicine and Dentistry, London, UK

R. N. Poston MD MRCPath
Senior Lecturer and Honorary Consultant,
Department of Experimental Pathology, UMDS,
Guy's Hospital, London, UK

Guido Rindi MD
Research Assistant, Department of Human
Pathology, University of Pavia at Varese, Italy

R. A. Risdon MD MB BS MRCS FRCPath
Professor of Histopathology, Hospital for Sick
Children, Great Ormond Street, London, UK

Maria Clara Sambade MD
Consultant, Department of Pathology, Medical
School, Hospital S. João, Porto, Portugal

Francesco Scaravilli MD PhD FRCPath
Reader, Institute of Neurology, Department of
Neuropathology, National Hospital for Nervous
Diseases, London UK

Andy E. Sherrod
Assistant Professor, Department of Pathology,
University of Southern California School of
Medicine, Los Angeles, USA

J. M. Skinner MB ChB FRCPath FRCPA
Associate Professor and Head of Department,
Department of Pathology, Flinders Medical
Centre, Adelaide, South Australia

J. P. Sloane MB BS FRCPath
Consultant Histopathologist, Royal Marsden
Hospital, Sutton, Surrey, UK

Manuel Sobrinho-Simões MD PhD
Professor of Pathology, Medical School,
Hospital S. João, Porto, Portugal

Enrico Solcia
Professor of Pathology and Director, Pathologic
Anatomy, Department of Human Pathology,
University of Pavia at Varese, Italy

Peter J. Stoward MA MSc DPhil FRSE
Professor and Head of Department of Anatomy,
University of Dundee, Dundee, UK

C. R. Taylor MA MD DPhil
Professor of Pathology, University of Southern
California; Chief of Immunopathology, LA
County, USC Medical Center, Los Angeles, USA

Susan Van Noorden MA(Oxon)
Senior Scientific Officer, Histopathology
Department, Royal Postgraduate Medical
School, London, UK

Nancy E. Warner SB MD
Professor and Chairman, Department of
Pathology, University of Southern California
School of Medicine, Los Angeles, USA

Moshe Wolman MD
Professor and Chairman, Department of
Pathology, Tel Aviv University Sackler School
of Medicine, Tel Aviv, Israel

Contents

1. General principles of fixation

D. Hopwood

For the pathologist there is a dilemma presented by fixation of tissues. On the one hand, with the development of more histochemical methods which have become useful diagnostically, a variety of fixative procedures from an increased armamentarium are found to be necessary. In some cases, no fixation is required. On the other hand, the clinician needs a simple uniform routine for dealing with biopsies and surgical specimens.

The problem is fundamentally related to two factors, namely the multiplicity of reaction, both chemical and physical, involved in tissue fixation and the specific needs of the techniques. Placing the specimen in formaldehyde solution of one kind or another immediately limits the number of investigations the pathologist can pursue unless there has been prior consultation with the clinician.

The processes involved in fixation from theoretical and practical aspects have been reviewed (Pearse 1980, Hopwood 1985).

At the most informed level, there is discussion between the clinician and the pathologist over handling of the biopsy specimen. Here, the diagnosis is usually suspected and the histochemical options available are maximal. Examples of this are seen in storage diseases and secreting APUD tumours. Other situations where the pathologist receives unfixed material regularly include muscle and kidney biopsies, where this routine has now become well-established. Also, material from theatre for frozen section is received fresh and, if some tissue remains, then appropriate histochemistry is possible. This also includes lymph nodes where culture may be initiated (Table 1.1).

A less favourable, although more common, receipt of tissue is at the cut-up. Here, the biopsy or surgical specimen will have been placed, usually

Table 1.1 Fixation and the clinician

1. Prior discussion, diagnosis suspected, pathologist to collect
   a. metabolic diseases, APUD tumours
   b. muscle, kidney, jejunum
   c. liver

2. Suspected by pathologist at cut-up (minus one)
   a. lipids, EM, some enzymes
   b. postfixation required, e.g. dichromate

3. After paraffin wax sections
   tumours — mucosubstances
   pigments, inorganic substances — X-ray
   dispersive microscopy
   immunoperoxidase

unopened, in the routine fixative of the department — most often formaldehyde. Fixation will have occurred at room temperature at varying periods up to 24 hours, or more if the specimen has come from another hospital. The histochemical possibilities are more limited here and depend on the diagnostic acumen of the pathologist, often in the light of clinical information given. At this stage, it is possible to perform lipid and mucin stains, investigate some of the enzymes and perform immunohistochemistry to detect many diagnostically useful antigens. At this point in processing, some specific form of postfixation may be indicated, such as chromaffin reaction. Electron microscopy is also possible from this material. Success depends a great deal on the time the tissue has already spent in formaldehyde. Some institutes do not process all their material, but retain a small portion in fixative whilst the remainder goes forward for routine sectioning and staining.

The most common way specimens are closely examined is as a paraffin section, stained with haematoxylin and eosin. At this stage there are limitations on the type of histochemical investigations possible. Commonly, this used to be limited to the nature of mucosubstances and very few enzyme histochemical studies were attempted. In more recent years advances in immunohistochemistry have allowed the detection of many antigens which are used in diagnosis, particularly of tumours. The application of lectins to diagnosis is possible in paraffin sections, and DNA flow cytometry is an expanding area. It is also possible to determine the nature of particulate and crystalline material in paraffin wax-embedded tissue by a combination of electron microscopy and X-ray energy dispersive microscopy (Crocker et al 1980).

Additives

A number of substances have been added to the fixative solution used in histochemical techniques. These are largely related to electron histochemistry, where they fall into two areas. In the ultrastructural demonstration of mucosubstances, both ruthenium red and alcian blue have been added to the primary and secondary fixative solutions (Behnke & Zelander 1970). The non-ionic detergent Triton X-100 has been used as an agent to facilitate the penetration of ruthenium red during fixation. It should also be mentioned that other non-ionic detergents (e.g. saponin) have been used to enhance the penetration of antibodies and lectins into cells (Laurila et al 1978).

A variety of substances have been used as cryoprotectants to prevent damage during freezing and thawing. These include polyvinylpyrrolidone, sucrose, glycerol and dimethylsulphoxide. It should be remembered that these may produce artefacts which may be relevant ultrastructurally. These additives may also inhibit enzyme activity.

The addition of phenol (2%) to the standard buffered formalin fixative eliminates the problem of formalin pigment formation without deleterious effects on staining properties or morphology (Slidders & Hopwood 1989).

Artefacts

There are a number of artefacts which are associated with fixation and the pathologist should be aware of these.

The physical loss of some tissue components occurs during fixation. These are, for obvious reasons, mostly small molecules. Important in the present context are various cofactors for enzymes (Hopwood 1969), and polypeptide hormones. Elsewhere in this chapter, the loss of mucosubstances, lipids and nucleic acids during fixation have been mentioned. Similarly, inorganic substances are lost and gradients they may have in life disappear.

The rearrangement of substances, e.g. glycogen and iron, within cells and tissues is the well-known phenomenon of false localization. The diffusion of enzymes may also occur, even in fixed material incubated with buffer at 37°C (Hardonk et al 1977).

Loss of enzyme activity usually follows fixation, although sometimes this may be reversed subsequently by washing the tissue in buffer, usually to a histochemically acceptable level. Different fixatives affect enzyme activity to a different extent (Lake & Ellis 1976). The same fixative reduces activity to varying levels when reacting with individual enzymes (Hopwood 1972). Arborgh and his colleagues (1976) showed that most enzyme fixation took place rapidly and that, from a histochemical viewpoint, any time between 5 minutes and 24 hours was satisfactory provided morphological preservation was adequate. It is important to differentiate between histochemical preservation of enzyme activity and enzyme activity measured in fixed tissues biochemically (Christie & Stoward 1974).

Besides the loss of enzyme activity, fixation also alters other protein-dependent reactions, e.g. antigen-antibody reactions. The dilemma is the same as with the enzymes, that is, morphological preservation versus biological activity.

Further points to bear in mind are:

1. Potential reaction between various components in a fixation mixture
2. False-negative effects on fixation
3. False-positive results.

Non-specific binding occurs with glutaraldehyde and with formaldehyde in line with their general reactivity towards tissues (Hopwood 1969). Glutaraldehyde also has the effect of producing aldehyde groups in tissues due to its bifunctional nature and will give false-positive PAS reactions unless the free aldehyde groups are reduced by sodium borohydride prior to staining.

Further, Pentilla et al (1975) have shown that injured cells behave differently to non-injured in respect to their volume in various fixatives. Four per cent formaldehyde should be used with buffers of 310 mmol and glutaraldehyde-formaldehyde mixtures with 100–150 mmol buffers. The osmolarity of two commonly used fixative solutions are: glutaraldehyde 480 mmol, paraformaldehyde 1280 mmol (Hayat 1973). The problems of cell volume change have been pursued further by Collins et al (1977), who showed changes which were more obvious with SEM and time-lapse cinematography than on electron microscopy in transmission mode.

Fixatives as histochemical reagents

Fixation may be used to differentiate various cellular features histochemically. Some years ago, we showed that glutaraldehyde fixation discriminated between adrenalin and noradrenalin-containing cells by light and electron microscopy in the adrenal medulla (Coupland & Hopwood 1966), the noradrenalin only forming an electron dense deposit with the glutaraldehyde.

Simionescu et al (1972) introduced a lead-containing fixative which demonstrates glycogen and dextrans in tissues by electron microscopy.

Karnovsky's osmium tetroxide potassium ferrocyanide fixative also enhances glycogen and components of the cell membranes and fuzzy coat (1971). Ainsworth (1977) described the use of osmium tetroxide-potassium ferrocyanide to demonstrate polyvinylpyrrolidone and dextrans used in tracers.

No fixation

In certain circumstances, it is essential that the tissue is not fixed before it is studied. From a diagnostic viewpoint, probably the most important are the metabolic disorders and most enzyme techniques, in particular dehydrogenases where fixation must be avoided, and the typing of lymphomas. It is often also useful to make dabs from fresh unfixed spleen, lymph nodes, lung and brain biopsies, tumours, etc. Porphyria cutanea tarda gives a brick-red autofluorescence in cryostat sections. Fixation may also affect lectin binding.

Mucosubstances

The pathologist's chief interests in mucosubstances are directed towards glycogen and acid mucosubstances as diagnostic aids. Glycogen is best demonstrated in unfixed cryostat sections covered with colloidin and stained by the PAS technique (Lake 1970). Rossman's fluid retains 70% of glycogen (Smitherman et al 1972), whilst formaldehyde retains about 12%. Nonetheless, in the majority of surgical specimens, this is enough to be demonstrated (but see Ch. 16). Polarization of glycogen within the cells is less after aqueous fixatives than after alcoholic fixatives. The mechanism whereby glycogen is fixed remains uncertain, but Smitherman and his colleagues (1972) think that physical entrapment is important. Mercuric chloride reduces the amount of glycogen which may be demonstrated.

Glycosaminoglycans (GAGS), proteoglycans and some mucins remain soluble after tissue has been fixed in formaldehyde or glutaraldehyde, where losses may reach 70% during processing. Although mucins can be demonstrated in routinely fixed surgical material, successful attempts have been made to improve their fixation. Various cationic dyes and cetylpyridiniums have been added to the fixative with increased retention of proteoglycans and GAGS, including alcian blue and cuprolinic blue (Scott 1980, van Kuppevelt et al 1984).

A more recently introduced technique to investigate mucosubstances involves the use of lectins, a group of substances derived largely from plants which bind to specific sugar residues on cell surfaces, in membrane development (Jamieson et al 1979) and their diagnostic usefulness is now being investigated in various areas (Söderström 1988). The effects of various fixatives on the subsequent binding of a number of lectins has been studied by Allison (1987). He found that formaldehyde-saline was an adequate fixative for the demonstration of lectin binding in routine paraffin sections. Staining can be enhanced by enzyme digestion of the tissue (see below).

Nucleic acids and nucleic proteins

Formaldehyde does not react with nucleic acids under normal fixation conditions (Hopwood 1975). Although some nucleic acids are lost during fixation, they are mostly trapped in the cells. The associated histones with the nucleic acids react with formaldehyde and this reaction with formaldehyde remains reversible for some time (McGhee & Hippez 1975).

The interest of the pathologist in nucleic acids is chiefly directed towards recognizing atypia and RNA-rich plasma cells. Formaldehyde is not the optimum fixative and usually Carnoy is the one recommended, especially if any quantitative work is envisaged. Other than haematoxylin and eosin, the commonly used histochemical techniques for demonstrating nucleic acids are methyl green pyronin and the Feulgen reaction. The latter can be used quantitatively. Flow cytometry is now available in some centres and it has been found possible to examine formaldehyde-fixed archival material using suitable preparative techniques (Quirke & Dyson 1986).

Lipids

The general pathologist's interest in lipids is limited. He notes the hole left by extracted lipids in steatosis of the liver, which, in many ways, denotes the attitude to this important class of substance.

A modest repertoire of methods for the histochemical demonstration of a variety of lipids and lipoproteins exists. These are outlined by Pearse (1984). In general the techniques for their demonstration call for frozen sections of unfixed or formaldehyde-fixed tissues. The methods for the fixation of lipids, making them insoluble, are few. The roles of osmium tetroxide and potassium dichromate are well known, although both of these alter the chemical reactivity. Formaldehyde has also been shown to react with various lipids. The reactions with those containing amine groups such as phosphatidyl ethanolamine, would be expected. Formaldehyde has been shown to react with ethylene double bonds to form 1,2 glycols (Jones 1973).

Enzymes

The ability of enzymes to withstand fixation varies enormously. The dehydrogenases are sensitive even to brief fixation in dilute aldehydes, whereas the peroxidase activity of haemoglobin can persist in red blood cells for several weeks (Torack 1974) in buffered formaldehyde. Biochemical and morphological studies, largely of aldehyde fixation of enzymes, have been pursued and these have been summarized by Hopwood (1972) and Pearse (1980).

Some form of fixation is desirable in general, although in the case of the dehydrogenases this may have to be limited to a short time with low fixative concentrations. Unbound enzymes are liable to be lost from the specimen. To overcome this difficulty, a variety of protective agents, such as polyvinyl-pyrrolidone, have been introduced into the incubation medium.

In general with aldehyde fixation the longer the fixation period and the higher the temperature at which it was carried out, the less enzyme activity is retained. However, washing fixed tissue in buffer restores enzyme activity considerably, while retaining good morphological preservation.

Phosphate-buffered 4% formaldehyde (with added sucrose) is a good general fixative for preservation of hydrolase activity (Holt & Hicks 1961) with subsequent storage in cold gum sucrose. Glutaraldehyde-paraformaldehyde mixtures (Karnovsky 1965) are useful for the ultrastructural demonstration of enzymes. If the commonly used methods of fixation do not work, there are other worth hile alternatives which could be tried, for example the possibilities of microwave fixation (Appendix 1) need to be explored more fully.

Immunocytochemistry

Fixation and processing of tissues for subsequent immunological techniques produces some problems of artefact and interpretation and no single method is applicable to all problems. Basically, there are the usual two horns of a dilemma: on the one hand, preservation of morphology and on the other, preservation of antigenicity without diffusion artefacts.

Immunofluorescence techniques

In our department, the routine practice for the investigation of patients' autoantibodies is to use fresh unfixed cryostat sections of animal (rat) tissue and an indirect immunofluorescent technique. Similarly, skin and kidney biopsies may be investigated as unfixed cryostat sections, using directly labelled fluorescent antibodies. On lymphoid tissue, a brief fixation with ethanol or acetone has two functions: to intensify the fluorescence and to reduce the background. Fixation can be successfully employed for immunofluorescent studies, but much depends on the structure of the antigen. Interestingly, osmium tetroxide has been shown to preserve the immunoreactivity of growth hormone and prolactin in pituitary (Baskin et al 1979). Leathem & Atkins (1980) have compared fixatives in their effects on morphological preservation of lymphoid tissue with the preservation of immunoglobulin antigenicity. Susa was found to be the best fixative for demonstrating specific staining with the minimum background staining. This was followed by Bouin and formol sublimate. Formalin fixation generally gave poor specificity and enhanced background staining.

Immunoperoxidase techniques

Immunoperoxidase techniques have established themselves in routine pathological investigations. They have a particular advantage over the fluorescent technique, in providing permanent preparations. For further views see Chapters 3 and 20, and Polak & Van Noorden (1986).

Many of the antigens investigated earlier were relatively stable and 'easy'. Smaller molecules are more labile and difficult to investigate. Such include the smaller peptides in the nervous system and the diffuse endocrine system in the gut and elsewhere but these too can now be successfully detected after formalin fixation (Polak & van Noorden 1986).

Xylene decreases immunoreactivity in frozen sections and in non-trypsinized material fixed in formaldehyde-calcium, formaldehyde-saline or Bouin, whereas chloroform or Inhibisol gives better results. For electron immunocytochemistry,

the problem of morphological and antigenic preservation are even more acute.

Tissue digestion

The effects of a number of proteases on paraffin sections have been found to be helpful in immunohistochemistry (Mepham et al 1979), the use of lectins and in the demonstration of mast cells. Trypsin was found to be the most useful. The ideal time of digestion depended on the fixative and the duration of fixation. The need for freshly prepared trypsin was stressed. This technique cannot be applied to all antigens, for example problems are encountered with complement (Huang et al 1976). In Dundee, we have used trypsin digestion with our routine immunohistology and to date we have had no problems, in line with other pathology departments. Purified proteases may also be used successfully.

The enhancing effect on lectin binding by pretreatment of the paraffin sections with proteolytic enzymes has been described by Leathem & Atkins (1983) and Jeffrey et al (1987). They investigated the effects of a number of proteases. With most lectins, the results were similar, no matter which enzyme was used. The pattern of staining in some tissues was markedly affected by the enzyme pretreatment used. It was concluded that controlled digestion with purified narrow spectrum enzymes were preferred and that tissue fixation may mask the carbohydrates which bind the lectins. Sialidase digestion may also be beneficial in the investigation of lectin binding.

Fixation has an important role in the differentiation of mucosal mast cells from tissue mast cells (Strobel et al 1981). Trypsinization may be helpful prior to staining (Wingren & Ennerback 1983).

Biogenic amines

Over the past 15 or so years, methods have been elaborated which use formaldehyde vapour both to fix tissue and produce a fluorescent product with biogenic amines. These have characteristic wavelengths for their fixation and emission spectra and are fully detailed by Falk & Owman (1965).

Greater sensitivity is reported by including magnesium ions (Lorén et al 1977). A simple rapid method for producing monoamine fluorescence has been described by de la Torre & Surgeon (1976), who used cryostat sections with glyoxylic acid-induced fluorescence. The complete process from obtaining the tissue to examining for fluorescence under the microscope takes under 20 minutes. The details are given in Appendix 2.

Polypeptide hormones

Tissue in which polypeptide-containing cells are to be demonstrated must be dealt with in a specific manner. Ideally, the specimen should be prepared for the following procedures, depending on the tissue.

1. Immunocytochemistry — formalin fixation (benzoquinone may be necessary on occasions).
2. Radioimmunoassay — frozen.
3. Electron microscopy — glutaraldehyde.
4. Fluorogenic amine content — formaldehyde vapour and glyoxylic acid.
5. Non-specific esterase or cholinesterase — cold formaldehyde calcium or frozen.
6. Lead haematoxylin, argyrophil — paraffin sections. Bouin masked metachromasia.
7. Recently, neuron-specific enolase and chromogranin has been found to be a marker for these cells (see Ch. 29).

Inorganic materials

The investigation and fixation of these substances has received scant attention. Some substances, such as iron, calcium and copper, are commonly looked for in routine formaldehyde-fixed material. Walker and his colleagues (1971) have shown that there was little correlation between histological grading of iron in biopsies and the concentration measured by atomic absorption spectroscopy. Part of the problem was iron loss into the fixative. In a similar way Renaud (1959) reported loss of calcium into aqueous formaldehyde but not into ethanol, and copper may be lost unless the biopsy is fixed in the presence of rubeanic acid. The presence of copper-binding protein using the prolonged orcein stain on formaldehyde-fixed material (Shikata et al 1974) is discussed in Chapter 15.

Relatively simple techniques have recently been described for the analysis of inorganic material in routine surgical material, even after H & E sections have been prepared. Crocker et al (1980) used microincineration of paraffin sections followed by scanning electron microscopy and X-ray energy spectroscopy. They have since shown that the microincineration stage may be omitted and the back-scattered electrons used for analysis (Crocker et al 1981).

Haematology

A number of histochemical techniques have been applied to the study of white and red cells. Those commonly used in this department are PAS, Sudan black B, acid and alkaline phosphatase and non-specific esterase. A variety of fixative methods have been proposed. Hayhoe & Quaglino (1988) suggest the use of formaldehyde vapour alone, but aqueous fixatives also work well (see also Chs 18 and 19).

Cytology

Cytological techniques have been subsequently applied for a number of years to gynaecological and other problems. The repertoire has been extended by aspiration cytology (Melcher & Linehan-Smith 1981 and Ch. 30). Various attempts have been made to quantify and automate the cell scanning. In the case of the cervix, this has been with the Feulgen reaction and naphthylamidase activity (Hussain & Millett 1979). Alternatively, cells may be collected on a Millipore filter for biochemical analysis.

Similar preparations of cells may be made from touch preparations or dabs from unfixed material such as lymph nodes and tumours where architectural relations may still remain. Smear preparations of brain biopsies have been used for many years and their use has been reviewed (Adams et al 1981).

Microwave fixation

For several years now, fixation or stabilization of tissues by microwaves has been used (Mayers 1970, Hopwood et al 1984). Early work using domestic microwave ovens gave acceptable results, but custom designed microwave processors are now available commercially with improved temperature control (Polaron/Bio Rad HZ500).

The size of block fixed by microwave irradiation depends on their depth of penetration. For the commonly used magnetron, this is 2–3 cm. For the same reasons microwave irradiation is not applicable for bulk processing of tissue. The tissue is placed in phosphate buffered saline or some other suitable buffer and irradiated (see Appendix 1). Postfixation can be used. The temperature to which the tissue is heated is important. We have found that for human material this is about 60°. At higher temperatures distortion occurs, at lower temperatures there is little or no fixation.

How fixation occurs with microwaves is uncertain. Using polyacrylamide gel electrophoresis for detection, no crosslinking is produced in purified chemicals (Hopwood et al 1988) but a little occurs in tissue homogenates (Hopwood et al 1984). This is probably related to mild melting of the protein molecules. The increased temperature will speed up diffusion processes. There is some evidence that microwaves produce loss of protein tertiary structure (Ortner et al 1981). Two specific benefits of microwave fixation are its speed — minutes as opposed to hours — and that a chemical fixative is not necessary.

What advantages does microwave irradiation have other than tissue fixation? This has been reviewed recently by Boon & Kok (1987). These include histoprocessing for paraffin and resins and tissue staining. Microwaves have also been found useful in reducing times of some neuropathological techniques (Bodian 20 hours to 90 minutes). Metallic methods in general are speeded up, including the Grimelius method (3 hours to 3 minutes). Microwave irradiation has also been found useful in immunostaining at the light and electron microscope levels, in preparation for electron microscopy and in decalcification. One drawback of the technique is the lysis of erythrocytes.

Plastic embedding

The use of plastic embedding materials has helped certain areas of pathological practice considerably, especially renal, lymph node, lung, needle and bone marrow biopsies. Early attempts to apply histochemical immunoperoxidase methods met with limited success and the main benefit is in morphological quality.

Hazards

Biological

Unfixed surgical specimens sent for frozen section pose a potential for infection to all those who handle the tissue, including the reception area. There are three agents which have caused most concern to pathologists in recent years, namely hepatitis B and HIV viruses and the tubercle bacillus.

The risk of acquiring hepatitis B infection has been shown to be highest to workers in haematology and chemical pathology laboratories (Grist 1980). Although most hospitals have some system for identifying specimens which are positive for HBsAg, workers remain exposed to two groups of material. These obviously include patients who are unscreened asymptomatic carriers. Known HBsAg-positive patients in whom the information is not passed on are a potential source for this hazard, clinical notes being kept by separate hospitals and departments.

Laver and his colleagues (1979) have shown the primary mode of transmission in the laboratory is for the hepatitis B virus to enter through trivial cuts and abrasions on the hands from contaminated material. Protection will be afforded by the use of waterproof dressings and gloves. Care should also be taken in the disposal of sharp objects. A further discussion of this topic is given in the Bulletin of the Royal College of Pathologists (1981).

Active tuberculosis remains a problem in unfixed surgical and necropsy specimens. The diagnosis may be unsuspected in up to one-third to one-half of the patients (Edlin 1978, Cameron & McCoogan 1981). Clearly, this raised the problem of contamination of cryostats and other

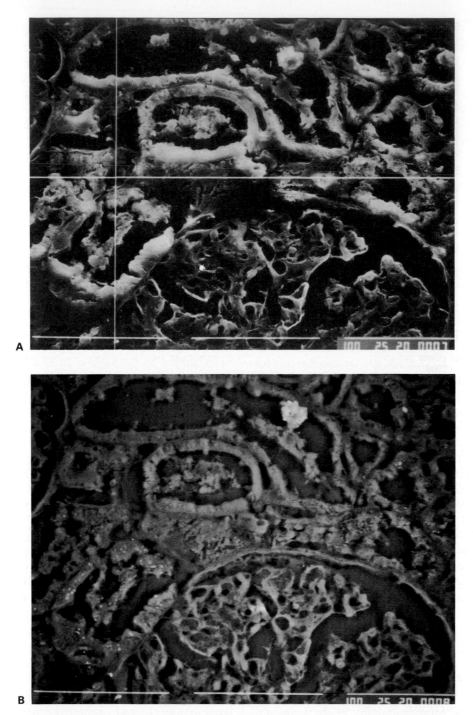

Fig. 1.1   A renal biopsy from a patient with rheumatoid arthritis treated with gold, who developed nephrotic syndrome and was found to have membranous glomerulonephritis.
A Paraffin wax section (deslipped H & E), lightly coated with carbon. ×350
B Same field, same magnification, back-scatter electron image. Note tiny bright granules in tubular epithelium, especially just to left of glomerulus. ×350

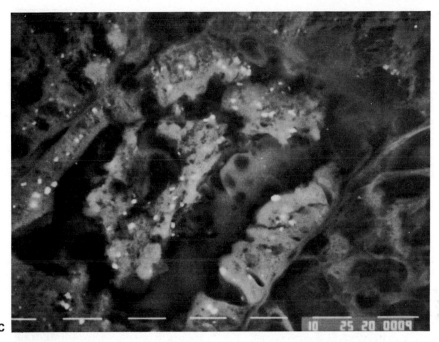

Fig. 1.1 C Higher-power picture of tubule with bright granules showing more clearly. On X-ray analysis, these granules proved to be gold. ×1000
(Photographs kindly supplied by D. A. Levison and P. R. Crocker, Department of Histopathology, St Bartholomew's Hospital)

instruments. Methods for this disinfection are detailed by Howie (1978) who also deals with the more general problem of the prevention of infection in laboratories.

The risks of the HIV viruses are best covered by following the procedures as for hepatitis B virus. In most laboratories, there is already a protocol to cover this problem. Note also that the slow virus of Creutzfeldt-Jacob disease is not inactivated by fomalin fixation, and the same may also apply to other slow viruses.

Chemical

The safety of various chemicals used in preparative techniques has been questioned. Problems may be considered in particular in relation to fixatives, buffers and embedding materials for electron microscopy.

Attention has been drawn to the toxic effects of formaldehyde vapour (Nature 1980). All aldehydes and their vapours should be regarded as toxic.

Attention should also be drawn to the buffer system often used in electron microscopy, namely cacodylate. This contains arsenic which may be absorbed through the skin, and the dust from the powder through the respiratory tract.

The epoxy resin embedding media, their hardener and catalysts are all toxic (especially Spurr's). In susceptible individuals, they cause dermatitis on the hands and the use of gloves for their handling is obviously necessary. The methacrylates, which are enjoying a new popularity for 1 μm sections, also produce toxic vapours.

Substances used in the staining reactions, histochemical and microscopic, also have their toxicities and carcinogenicities. Some of these, such as DAB and the dianisidines, are well known. The dangers of enzyme inhibitors should be self-evident and they should be handled with appropriate care.

Obviously, care must be taken in the storage of dangerous materials, that is, both the inflammable substances used preparatively and toxic chemicals. Attention should also be drawn to the hazard of storing inflammable materials in an ordinary domestic refrigerator with an internal thermostat which can spark.

Although some of the hazards of the more commonly used procedures have been mentioned here, this is obviously not the place for a complete and detailed description of chemical and biological hazards. These are dealt with more fully in Howie (1978), Bretherick (1980) and Weakley (1981).

Conclusion

It seems that we are still some distance from finding a universal fixative which will give both perfect or even acceptable morphological preservation and, at the same time, good biochemical or immunological activity. Until such a fixative is evolved, we, as pathologists, must inform our clinical colleagues of histochemical possibilities for diagnosis and the fixation procedures required. At the same time, in spite of all the ideal situations described for various techniques, it is often worth trying a method from tissue fixed under non-ideal conditions. It may well succeed (see Fig. 1.1).

REFERENCES

Adam J H, Graham D I, Doyle D 1981 Brain biopsy — the smear technique for neurosurgical biopsies. Chapman and Hall, London

Ainsworth S K 1977 An ultrastructural method for the use of polyvinylpyrrolidone and dextrans as electron opaque tracers. Journal of Histochemistry and Cytochemistry 25: 1254–1259

Allison RT 1987 The effects of various fixatives on subsequent lectin binding to tissue sections. Histochemical Journal 19: 65–74

Arborgh B, Bell P, Brunk U, Collins V P 1976 The osmotic effect of glutaraldehyde during fixation. Journal of Ultrastructural Research 56: 339–350.

Baskin D G, Erlandsen S L, Parsons J A 1979 Immunocytochemistry with osmium-fixed tissue. I. Light microscopic localization of growth hormone and prolactin with the unlabelled antibody enzyme method. Journal of Histochemistry and Cytochemistry 27: 867–872

Behnke O, Zelander T 1970 Preservation of intercellular substances by the cationic dye Alcian Blue in preparative procedures for electron microscopy. Journal of Ultrastructural Research 3l: 424–438

Boon ME, Kok LP 1987 Microwave cookbook of pathology. Coulomb Press Leyden, Leiden

Bretherick L (ed) 1980 Hazards in the chemical laboratory, 3rd edn. Chemical Society, London

Bulletin of the Royal College of Pathologists 1981 34: 3–7

Cameron H M, McGoogan E 1981 A prospective study of 1152 hospital autopsies. II. Analysis of inaccuracies in clinical diagnoses and their significance. Journal of Pathology 133: 285–300

Christie K N, Stoward P J 1974 A quantitative study of the fixation of acid phosphatase by formaldehyde and its relevance to histochemistry. Proceedings of Royal Society B 186: 137–164

Collins V P, Arborgh D, Brunk U 1977 A comparison of the effects of three widely-used glutaraldehyde fixatives on cellular volume and structure. Acta Pathologica et Microbiologica Scandinavica 85A: 157–168

Coupland R E, Hopwood D 1966 The mechanism of the differential staining reaction for adrenalin and nonadrenalin storing granules in tissues fixed in glutaraldehyde. Journal of Anatomy 100: 227–243

Crocker P R, Doyle D V, Levison D A 1980 A practical method for the identification of particulate and crystalline material in paraffin-embedded tissue specimens. Journal of Pathology 131: 165–173

Crocker P R, Toulson E, Levison D A 1981 The back-scattered electron image (BEI) for identification of particles in paraffin sections. Jeol News 19E: 10–14

De La Torre J C, Surgeon J W 1976 A methodological approach to rapid and sensitive monoamine histofluorescence using a modified glyoxylic acid technique: the SPG method. Histochemistry 49: 81–93

Edlin G P 1978 Active tuberculosis unrecognised until necropsy. Lancet 1: 650–652

Falk B, Owman C 1965 A detailed methodological description of fluorescence method for the cellular demonstration of biogenic amines. Acta Universitatis Lundensis II (7): 1

Grist N R 1980 Hepatitis in clinical laboratories 1977–78. Journal of Clinical Pathology 33: 471–473

Hardonk M J, Haarsma T J, Kijkhuis F W J, Poel M, Koudstaal J 1977 Influence of fixation and buffer treatment on the release of enzymes from the plasma membrane. Histochemistry 54: 57–66

Hayat M A 1973 Specimen preparation. In: Hayat M A (ed) Electron microscopy of enzymes. Van Nostrand Reinhold, London, vol 1

Hayhoe F G J, Quaglino D 1988 Haematological cytochemistry. Churchill Livingstone, Edinburgh

Holt S J, Hicks R M 1961 Journal of Biophysical and Biochemical Cytology 11: 31

Hopwood D 1969 Fixatives and fixation: a review. Histochemical Journal 1: 323–360

Hopwood D 1972 Theoretical and practical aspects of glutaraldehyde fixation. Histochemical Journal 4: 267–303

Hopwood D 1975 The reactions of glutaraldehyde with nucleic acids. Histochemical Journal 7: 267–276

Hopwood D 1985 Cell and tissue fixation. Histochemical Journal 17: 389–442

Hopwood D, Coghill G, Ramsay J, Milne G, Kerr M 1984 Microwave fixation: its potential for routine techniques, histochemistry, immunocytochemistry and electron microscopy. Histochemical Journal 16: 1171–1191

Hopwood D, Yeaman G, Milne G 1988 Differentiating the effects of microwave and heat on tissue proteins and their crosslinking by formaldehyde. Histochemical Journal 20: 341–347

Howie J 1978 Code of practice for the prevention of infection in clinical laboratories and post mortem rooms. HMSO, London

Huang S-N, Minassian H, More J D 1976 Application of immunofluorescent staining on paraffin sections improved by trypsin digestion. Laboratory Investigation 35: 383–390

Hudson I, Hopwood D 1986 Macrophages and mast cells in chronic cholecysytitis and 'normal' gallbladers. Journal of Clinical Pathology 39: 1082–1087

Hussain O A N, Millett J A 1979 The detection of malignancy in the cervix. In: Pattison J R, Bitensky L, Chayen J (eds) Quantitative cytochemistry and its applications. Academic Press, London

Jamieson J D, Hull B E, Galardy R E, Maylié-Pfenninger M F 1979 Acinar cells: relationship to secretagogue action in secretory mechanisms. Cambridge University Press, SEB Symposia 33

Jeffrey IJM, Mosley SM, Jones CJP, Stoddart RW 1987 Proteolysis and lectin histochemistry. Histochemical Journal 19: 269–275

Jones D 1973 Reactions of aldehydes with unsaturated fatty acids during histological fixation. In: P J Stoward (ed) Fixation in histochemistry. Chapman and Hall, London, p 1–45

Karnovsky M J 1965 A formaldehyde-glutaraldehyde fixative of high osmolarity for use in electron microscopy. Journal of Cell Biology 27: 137A

Karnovsky M J 1971 Use of ferrocyanide reduced osmium textroxide in electron microscopy. 11th Annual Meeting of the American Society of Cell Biology, p 146

Lake B D 1970 The histochemical evaluation of the glycogen storage diseases. A review of techniques and their limitations. Histochemical Journal 2: 441–450

Lake B D, Ellis R B 1976 What do you think you are quantifying? An appraisal of histochemical methods in the measurement of the activities of lysosomal enzymes. Histochemical Journal 8: 357–366

Laurila P, Virtanen I, Wartiovaara J, Stenman S 1978 Fluorescent antibodies and lectins stain intracellular structures in fixed cells treated with non-ionic detergent. Journal of Histochemistry and Cytochemistry 26: 251–257

Laver J I., Van Drunen N A, Washburn J W, Balfour H H 1979 Transmission of hepatitis B virus in clinical laboratory areas. Journal of Infectious Diseases 140: 513–576

Leatham A, Atkins N 1980 Fixation and immuno-histochemistry of lymphoid tissue. Journal of Clinical Pathology 33: 1010–1012

Leathem AJC, Atkins NJ 1983 Lectin binding to paraffin sections. In: Bullock GR, Petrusz P (eds) Techniques in immunocytochemistry 2. Academic Press, London, pp 39–70

Lorén I, Björklund A, Lindvall O 1977 Magnesium ions in catecholamine fluorescence histochemistry. Histochemistry 52: 223–239

McGhee J D, Hippez P H Von 1975 Formaldehyde as a probe of DNA structure. Biochemistry 14: 1281–1296

Mayers CP 1970 Histological fixation by microwave heating. Journal of Clinical Pathology 28: 273–275

Melcher D H, Linehan Smith R S 1981 Fine needle aspiration cytology. In: Anthony P P, MacSween R N W (eds) Recent advances in histopathology 11. Churchill Livingstone, Edinburgh

Mepham B L, Frater W, Mitchell B S 1979 The use of proteolytic enzymes to improve immunoglobulin staining by the PAP technique. Histochemical Journal 11: 345–357

Nature 1980 Academy recommends cut in formaldehyde exposure. 284: 587

Ortner MJ, Galvin MJ, Chignell CF, McCrcc DOI 1981 A circular dichroism study on human erythrocyte ghost proteins during exposure to 2450 MHZ microwave irradiation. Cell Biophysics 3: 325–347

Pearse A G E 1980 Histochemistry, theoretical and applied, 4th edn. Churchill Livingstone, Edinburgh, vol 1

Pearse A G E 1984 Histochemistry, theoretical and applied, 4th edn. Churchill Livingstone, Edinburgh, vol 2

Pentilla A, McDowell E M, Trump B F 1975 Effects of fixation and post-fixation treatments on volume of injured cells. Journal of Histochemistry and Cytochemistry 23: 251–270

Polak JM, Van Noorden S (eds) 1986 Immunocytochemistry. Modern methods and applications, 2nd edn. Wright, Bristol.

Quirke P, Dyson JED 1986 Flow cytometry. Methodology and applications in pathology. Journal of Pathology 149: 79–87

Renaud S 1959 Superiority of alcoholic over aqueous fixation in the histochemical dection of calcium. Stain Technology 34: 267–271

Scott JE 1980 Collagen-proteoglycan interactions. Localization of proteoglycans in tendon by electron microscopy. Biochemical Journal 187: 887–891

Shikata T, Uzawa T, Yoshiwara A, Akatsuka T, Yamazaki S 1974 Staining methods of Australian antigen in paraffin section. Detection of cytoplasmic inclusion bodies. Japanese Journal of Experimental Medicine 44: 25–36

Simionescu N, Simionescu M, Palade G E 1972 Permeability of intestinal capillaries: pathway followed by dextrans and glycogen. Journal of Cell Biology 58: 365–392

Slidders W, Hopwood D 1989 Buffered phenol formaldehyde (pH 7.0 and pH 7.5): improved fixation in an enclosed tissue processor. Medical Laboratory Sciences 46: 74–76

Smitherman M L, Lazarow A, Sorenson R L 1972 The effect of light microscopic fixatives on the retention of glycogen in protein matrices and the particulate state of native glycogen. Journal of Histochemistry and Cytochemistry 20: 463–471

Söderström K O 1988 Lectin binding to serous ovarian tumours. J Clin Path 41, 308–13

Strobel S, Miller HRP, Ferguson A (1981) Human intestinal mucosal mast cells: evaluation of fixation and staining techniques. Journal of Clinical Pathology 34: 851–858

Torack R M 1974 Peroxidase activity in autopsy material. Archives of Pathology 98: 233–236

van Kuppevelt THMSM, Domen JGW, Gremer FPM, Kuyper CMA 1984 Staining of proteoglycans in mouse lung alveoli II Characterization of the cuprolinic blue positive sites. Histochemical Journal 16: 671–686

Walker R J, Miller J P G, Dymock I W, Shilikin K B, Williams R 1971 Relationship of hepatic iron concentration to histochemical grading and to total chelatable body iron in conditions associated with iron overload. Gut 12: 1011–1014

Weakley B S 1981 A beginner's handbook in biological transmission electron microscopy, 2nd edn. Churchill Livingstone, Edinburgh

Wingren U, Enerback L 1983 Mucosal mast cells of the rat intestine: a re-evaluation of fixation and staining properties with special reference to protein blocking and solubility of the granular glycosaminoglycan. Histochemical Journal 15: 571–582

2. Histochemical methods for routine diagnostic histopathology

P. J. Stoward

INTRODUCTION

The methods available to the histopathologist for visualizing, and in some cases quantifying, the different kinds of substances and cells present in sections of tissue fall roughly into four groups of increasing complexity:

1. Simple routine histological methods, such as haematoxylin and eosin.

2. A more extended array of histological and quasihistochemical methods for staining particular cells more selectively than is possible with the older and simpler methods of the previous group.

3. Routine histochemical methods.

4. 'Special' histochemical and immunocytochemical methods for detecting specific pathological agents and features that are difficult, if not impossible, to visualize by standard methods, for example viruses and bacteria in tissue sections.

Most pathologists prefer haematoxylin and eosin, and perhaps a trichrome method in addition, for the routine staining of specimens of tissue presented for histological examination. Not only are the methods simple to perform but they stain many cells and tissue elements clearly. They have also stood the test of time and are adequate for recognizing the vast majority of common histological lesions. However, they do not, and cannot, reveal all diagnostically important features. Further, they give no or little information about the molecular changes occurring within cells during disease processes. Consequently, several histopathology laboratories now routinely employ a wider range of histological methods, such as those listed in Table 2.1.

Unfortunately, even these methods, numerous as they are, are inadequate for investigating some disorders of skeletal muscle, the neuroendocrine system, the gastrointestinal tract, liver, kidney and peripheral blood. For such so-called 'special' cases, a systematic histochemical analysis is appropriate. In this chapter, some reliable and well-tried histochemical methods for such an analysis are brought together in a logical order, together with comments on their underlying rationale.

In most histopathology laboratories, these methods are now increasingly being supplemented by an expanding battery of immunocytochemical procedures. One such procedure, in situ hybridization, is included in this chapter. Routine immunocytochemical procedures are reviewed in Chapter 3 of this book, and by Lee & DeLellis (1987).

SUBSTANCES IDENTIFIED BY ROUTINE HISTOCHEMICAL ANALYSIS

Most specialized cells in a section of tissue contain a preponderance of one particular macromolecular substance. The aim of a histochemical analysis is to visualize this substance relatively specifically so that the cell in which it is present can be recognized and distinguished from morphologically similar cells.

Five types of chemical substances are present in varying proportions in a mammalian cell:

1. Nucleic acids
2. Proteins and peptides, including enzymes and hormones
3. Mucosubstances

Table 2.1 Histological and quasihistochemical methods used routinely for the visualization of specific tissue components and histopathological features. The simplest and more informative ones are marked*

Histological feature visualized	Methods[1]	Histological feature visualized	Methods[1]
Blood cells (in smears and tissues)	Wright*, Romanowsky-Giemsa or Leishman[2]	Lipid deposits (including cholesterol)	*Oil red O, Sudan black or Sudan IV, osmium tetroxide (for unsaturated lipids), *Schultz's iron-sulphuric acid (for cholesterol), ferric haematoxylin (for phospholipids)
Brush borders of e.g. enterocytes and renal proximal tubules	Alkaline phosphatase		
		Melanogenesis	*DOPA oxidase
Calcification	Alkaline phosphatase, *Von Kossa	Micro-organisms	Gram, Ziehl-Neelsen, Gomori's methenamine silver, *PAS, Warthin-Starry, cresyl violet, Gimenes
Calcium/urate deposits	*Harris's haematoxylin, Von Kossa, Gomori's methenamine silver		
Central nervous system:		Mucus (especially in or on hypersecreting cells in respiratory and gastrointestinal tracts)	*Diastase-PAS, *alcian blue (pH 2.5 and 1.0) – PAS high iron diamine-alcian blue pH 2.5, PAS and variations[4]
a. Nerve cells and axons	Holmes's silver, Bielschowsky's silver, *Nissl's cresyl fast violet		
b. Glia	Mallory's phosphotungstic acid-haematoxylin (PTAH), Holzer's crystal violet, Cajal's gold-mercuric chloride	Muscle fibre typing	*Myofibrillar ATPase pH 9.5 after preincubation at pH 4.6 or 4.2, *NADH dehydrogenase,
c. Myelin	Solochrome cyanine, *Luxol fast blue Weigert-Pal haematoxylin (Kultschitzky's modification)	Phagocytic cells	*acid phosphatase
		Pigments: Melanin	Masson-Fontana ammoniacal silver, *DOPA oxidase
Cell proliferation	*Methyl green-pyronin (to reveal DNA and RNA), Feulgen-Schiff		
		Lipofuscin/ceroid	*Autofluorescence
Connective tissue:		Protein deposits, granules and inclusions (e.g. viruses):	
a. Macrophages	*Acid phosphatase	a. General	*Eosin (preceded by haematoxylin), *phloxine-tartrazine
b. Mast cells	*Metachromasia (towards e.g. azure A), Csaba's alcian blue-safranin, chloroacetate esterase		
c. Fibres (collagen, elastin, reticulin)	*Trichrome stains (e.g. Mallory, Masson), Weigert-French, Verhoeff or orcein (for elastin), silver impregnation (for reticulin)	b. Amyloid	Methyl violet, *Congo red, thioflavine T (preferably preceded by alcian blue)
		c. Copper associated protein (and hepatitis B antigen)	*Shikata's prolonged orcein[5]
Endocrine cells (especially in gastro-intestinal tract)	Diazonium salt coupling,[3] Masson-Fontana alkaline silver, Grimelius's silver, Gomori's aldehyde fuchsin, Solcia's lead haematoxylin	d. Fibrin	*MSB (Martius-scarlet-blue)[6]
		Resorption (e.g. in bone)	*Acid phosphatase
Glycogen accumulation	*Diastase (or saliva)-PAS	Vascular system	Alkaline phosphatase

Notes

[1] Except where indicated by a footnote, practical details of all these methods, original references and some background explanation may be found in Drury & Wallington (1980) or Bancroft & Stevens (1990). Lillie & Fullmer (1976) may be consulted for critical discussions of their relative specificity.

[2] Lillie & Fullmer (1976), pages 747–748. Wright's stain is faster than the other methods, and is suitable for automated staining of smears for routine haematology. The Romanowsky-Giemsa method is preferred for critical studies: see Wittekind and Kretschmer (1987) for optimal practical details. Chloroacetate esterase and alkaline phosphatase are also very useful for the routine identification of certain blood cells (see Table 2.7).

[3] Often referred to, incorrectly, as alkaline diazo methods

[4] Reid et al (1988)

[5] Shikata et al (1974)

[6] Lendrum et al (1962)

4. Lipids

5. Inorganic salts.

In addition, all cells have a unique sequence in their DNA and mRNA.

Various names are used in the histochemical literature to describe individual mucosubstances. Examples are mucin, mucoid, mucopolysaccharide, mucoprotein, sialoprotein, and sulphomucin. The term 'mucosubstance', although not officially recommended, has passed into common parlance for embracing all carbohydrate-containing macromolecular substances and is, therefore, used here. There are three kinds of mucosubstance:

1. Polysaccharides (composed entirely of carbohydrate)

2. Proteoglycans or glycosaminoglycans (consisting of long polysaccharide chains covalently attached to a relatively small protein core) and

3. Glycoproteins (proteins bearing numerous covalently-linked short oligosaccharide side chains).

Glycoproteins are divided further into neutral and acid glycoproteins, depending on whether or not they contain sialic acid or a sulphated sugar (or both) as a component. Nearly all glycosaminoglycans possess uronic acid and sulphate ester groups in various proportions, and thus are known as acid glycosaminoglycans (or acid mucopolysaccharides in older terminologies). Glycosaminoglycans bearing sulphate ester groups are commonly said to be sulphated.

Lipids are also classifiable into chemically well-defined substances (Bayliss High 1982), but in routine histopathology it is normally sufficient to divide them into two classes, acidic and unsaturated lipids as one class, and cholesterol and its esters as the other.

Each of the substances 1–4, all complex macromolecules, possesses chemical residues and end-groups often not present to any significant extent in the other three. Sometimes such groups, e.g. hydroxyl ($-OH$) or sulphate ester ($-OSO_3H$), are referred to as radicals in the histochemical and histological literature. This is wrong. A radical has an unpaired electron (for example, $\cdot OH$) and is highly injurious towards cell membranes. A chemical residue is a monomeric component such as an amino acid within a protein or a sugar within a complex saccharide.

Table 2.2 lists a selection of reliable histochemical techniques which may be used routinely for the detection of the principal residues and end-groups

Table 2.2 Principal chemical groups and residues identifiable in the substances commonly present in mammalian tissues, and recommended histochemical methods for visualizing them.

Substance	Identifiable groups and residues	Method
Nucleic acids	1. $-PO_3H$	Haematoxylin (as in H & E), methyl green-pyronin, acridine orange, Hoechst 33342
	2. Deoxyribose	Feulgen-Schiff
	3. Nucleotide sequences	In situ hybridization
Proteins	1. $-NH_2$	Eosin (as in H & E), phloxine-tartrazine
	2. $-SH$ $-SH$ & $-SS-$	Ferric ferricyanide Performic acid-alcian blue, or azure A, Shikata's prolonged orcein
	3. Tryptophan	DMAB-nitrite
Mucosubstances	1. *Vic*-glycols	PAS, diastase-PAS (for glycogen), chromic acid-methenamine silver
	2. $-SO_3H$	Azure A metachromasia, high iron diamine
	3. Uronic acid	Alcian blue (pH 2.5 and 1.0) $-PAS$
	Sialic acid	High iron diamine-alcian blue pH 2.5, high iron diamine-PAS, critical electrolyte concentration techniques (alcian blue + $MgCl_2$), hyaluronidase and neuraminidase digestion
	4. Specific saccharides	Lectins

continued overleaf

Table 2.2 Cont'd

Substance	Identifiable groups and residues	Method
Lipids	1. Long aliphatic chains	Oil red O (for all lipids)
		Nile blue sulphate
	2. Free fatty acids, olefin bonds, and phosphoproteins	
	3. Cholesterol & esters	Schultz's method, perchloric acid-naphthoquinone (PAN)
Metals	1. Iron	Perls' method
	2. Calcium	Von Kossa, alizarin, glyoxal-*bis* (2-hydroxyanil)
	3. Copper	Rubeanic acid, rhodanine
	4. Aluminium	Solochrome azurine

of the five types of substances referred to above. Most proteins are also antigenic and some have catalytic (enzymic) properties which can be exploited for their visualization as well.

GENERAL PRINCIPLES OF ROUTINE HISTOCHEMICAL METHODS

The histochemical methods listed in Table 2.2 are nearly all based on one of the four following principles:

1. Simple ionic interactions of either positively-charged basic dyes (B^+) or negatively-charged acid dyes (A^-) with groups of opposite charge in tissue macromolecules (Fig. 2.1).

2. Reactions of aldehydes with Schiff's reagent, or occasionally methenamine silver, to form a coloured product of unknown structure (Fig. 2.2).

3. Coupling of aromatic diazonium salts ($Ar.N^+_2$) with electron-rich centres in the aromatic residues (e.g. tyrosine) of cellular proteins and hormones (Fig. 2.3).

4. If the macromolecule being detected is an enzyme, conversion to an insoluble coloured

Fig. 2.1 Reactions of acid (A^-) and basic (B^+) dyes with, respectively, cationic and anionic macromolecular substances in tissue sections

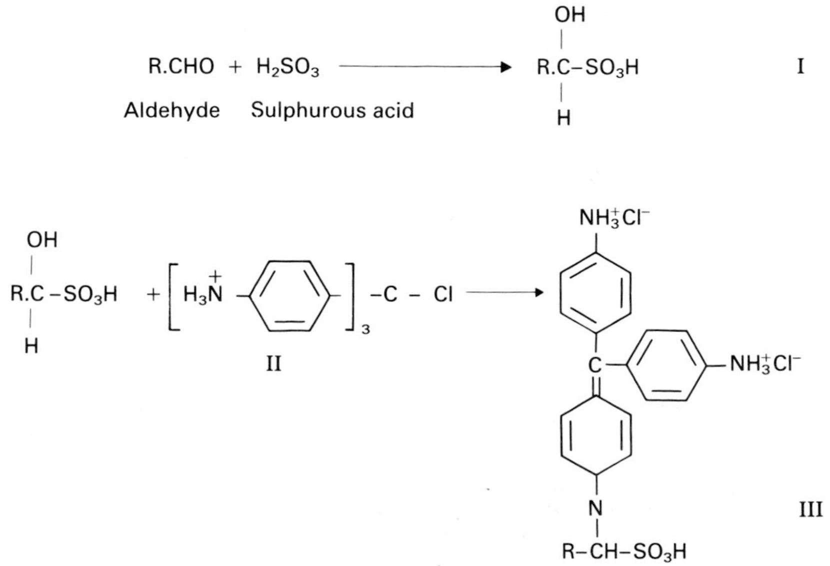

$$R.CHO + H_2SO_3 \longrightarrow R.\overset{\displaystyle OH}{\underset{\displaystyle H}{C}}-SO_3H \qquad I$$

Aldehyde   Sulphurous acid

Fig. 2.2 Possible course of reaction of tissue aldehydes (R.CHO) with Schiff's reagent. Aldehydes react first with the sulphurous acid present in the reagent to form a sulphonic acid intermediate (I) which then combines with II of pararosaniline hydrochloride, the principle dye constituent of Schiff's reagent to give the coloured derivative (III).

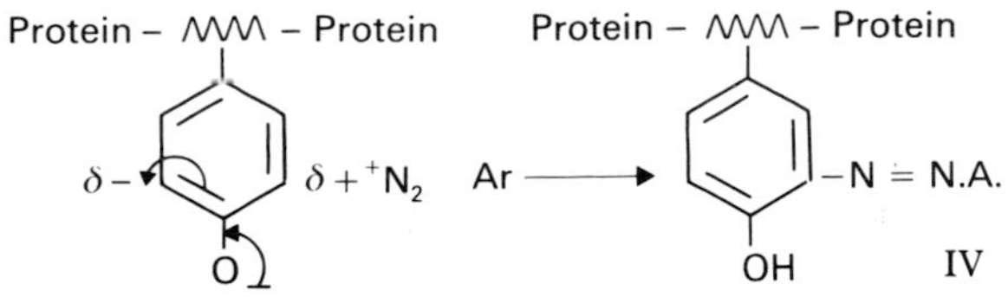

Fig. 2.3 Coupling of aromatic diazonium salts in alkaline conditions with the electron-rich centres of aromatic amino-acid residues of proteins to form coloured azo derivatives (IV)

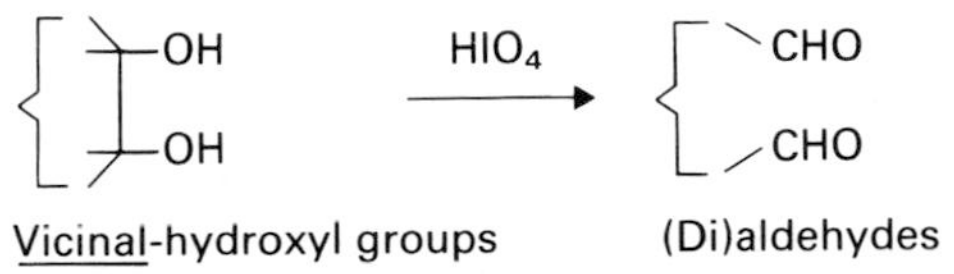

Vicinal-hydroxyl groups   (Di)aldehydes

Fig. 2.4 Selective cleavage of *vicinal*-hydroxyl groups (in the sugar residues of glycogen and glycoproteins) by periodic acid (HIO_4) to yield dialdehydes.

precipitate of the primary reaction products released by the catalytic activity of the enzyme acting on a suitable substrate.

When a particular chemical group or residue cannot take part in any of the first three of these reactions, it is first converted to a form in which it can. Glycogen, for example, cannot be visualized directly (except with mucicarmine) since its only chemically reactive groups, primary and *vicinal* hydroxyl groups, neither carry an electrical charge nor react with either Schiff's reagent or diazonium salts. However, the *vicinal* hydroxyl groups (also known as 1,2-glycols) can be made to react with Schiff's reagent by first oxidizing them to aldehydes with periodic acid (Fig. 2.4).

In some instances, reactions 1–3 can be rendered relatively specific for the substance one wishes to localize by controlling the reaction conditions (e.g. pH).

SUBSTANCES LOCALIZABLE

Nucleic acids

Both RNA and DNA can be localized in cells by the affinity of their negatively-charged phosphate ester groups for almost any basic dye, but particularly haematoxylin or methyl green and pyronin. Haematoxylin imparts a bluish-black colour to nuclei (containing DNA), but its staining of RNA is usually only apparent in cells whose cytoplasm is particularly rich in this nucleic acid (e.g. serous acinar cells). Haematoxylin is also taken up by other basophilic substances such as sulphated glycosaminoglycans. Therefore, it is not specific for nucleic acids exclusively.

Methyl green and pyronin, on the other hand, are more selective, staining DNA and RNA respectively. Because of the favourable molecular geometry of its molecule, methyl green intercalates between the stacked base-pairs of DNA and binds to its nucleotide phosphate groups. However, it is unable to fit easily onto single-stranded polynucleotides such as RNA and, therefore, stains this nucleic acid poorly. In contrast, pyronin, which has a flatter molecular shape, is taken up readily by RNA.

The specificity of the methyl green-pyronin technique (or any method involving basic dyes) is customarily confirmed by showing that in a control section which has been treated with ribonuclease before exposure to the dyes, the staining of the presumed cytoplasmic RNA is lost.

Fluorescent basic dyes are being increasingly used for revealing cellular DNA and RNA, particularly in automated searches for aberrant cells in, for example, cervical smears using fluorescence-activated cell sorting systems (Melamed & Darzynkiewicz 1981). The most widely-used dye for this purpose is acridine orange, which emits a green fluorescence when bound to undenatured DNA in nuclei, and a red fluorescence when taken up by cytoplasmic RNA. Other anionic constituents of cells, such as acid glycosaminoglycans, also bind this dye and therefore, it does not stain nucleic acids specifically. Fortunately, more selective fluorescent dyes are now available for DNA, such as ethidium bromide, propidium iodide and Hoechst 33342.

Nuclear DNA can also be visualized selectively, and quantified if desired, with the Feulgen-Schiff technique, in which sections of fixed tissue are initially treated with 5M-HCl at room temperature for about 10 min. This treatment is called Feulgen hydrolysis. The deoxyribose component of DNA is selectively hydrolysed by the acid to form an aldehyde, which is then converted to a magenta-coloured derivative by reaction with Schiff's reagent. DNA and RNA can be visualized in the same section by fixing tissue blocks or sections with Bouin's fixative solution (which also effects a Feulgen hydrolysis) and then staining successively with Schiff's reagent and Methylene Blue. Red DNA is readily distinguishable from the blue-coloured RNA.

In situ hybridization

All cells contain several unique nucleotide sequences in their DNA and mRNA. Such sequences can be visualized within individual cells at both the light and electron microscope levels using in situ hybridization techniques. This powerful technology allows the cytopathologist to trace cells expressing abnormal gene products. It is likely to be increasingly used in the next decade for detecting specific chromosomal aberrations, viruses, bacteria and other micro-organisms in diagnostic situations in which, hitherto, their identification has been either extremely difficult or time-consuming. Table 2.3 lists some key reviews and references of applications to date. Consult Penschow et al (1989) and Chapter 3 for practical details.

In situ hybridization techniques are based on the ability of exogenous DNA and RNA (including synthetic polynucleotides) to hybridize to their complementary sequences in the nucleic acids present in cells. Until the early 1980s, the nucleic acid probes were usually radioactively labelled because of their high sensitivity (reviewed by Coghlan et al 1985). The bound probe is then detected by autoradiography. This approach has several disadvantages. Long exposure times are required for the autoradiographs and the resolution is comparatively poor. Probes linked to non-radioactive 'reporter' molecules, such as peroxidase, biotin, or a heavy metal like mercury or gold, are now available. They give satisfactory results quickly and are easier to use. The bound labels are detected respectively with diaminobenzidine (DAB), streptavidin and silver amplification methods (see Cremers et al 1987, for references and practical details). Lewis et al (1987) have developed a very simple biotin-streptavidin-polyalkaline phosphatase method for routine use. Sections of paraffin-embedded tissue give satisfactory results provided precautions are taken to fix cellular RNA adequately (e.g. Loning et al 1986). If the developed probes are examined with a reflectance-contrast light microscope, the sensitivity and resolution approaches that obtainable with transmission electron microscopy (Cremers et al 1987).

Proteins

Amine groups

The simplest way of revealing basic proteins in sections of tissue is to stain them with solutions of an acid dye, such as eosin, at a pH below 6. As the pH is lowered, more protein terminal amino groups become protonated to form $-NH_3^+$ groups and consequently take up an increasing amount of dye. However, this may lead to so many tissue proteins becoming stained that cellular detail is lost. The staining can be made relatively more specific for densely packed proteins, such as viruses and Paneth cell granules, by extracting loosely bound dye with a suitable solvent (a procedure known as differentiation). The extraction can be controlled more reliably with phloxine-tartrazine than with eosin. The phloxine method is thus preferred for the routine diagnosis of protein inclusions.

Table 2.3 Diagnostic applications of in situ hybridization

Application	Key references and reviews
Diagnostic pathology (general reviews)	Syvanen (1986) Grody et al (1987) Hofler (1987) Wolfe (1988)
Endocrine disorders	Lloyd (1987)
Oncology	Stoner et al (1987)
Microbiology	Palva (1986) Zwadyk & Cooksey (1987)
Parasitology	Barker et al (1986) Rollinson et al (1986)
Virology (particularly papilloma- and cytomegaloviruses)	Loning et al (1986) Syrjanen et al (1986) Grody et al (1987)

For reasons that are not fully understood, a few acid dyes stain some proteins relatively more selectively than others. For example, orcein, when employed in Shikata's 'prolonged' method, seems to be taken up specifically by hepatitis B virus and copper-binding protein (in diseases of liver) and, therefore, this dye is useful for the diagnostic recognition of these cellular inclusions.

Thiol and disulphide groups

Most methods for demonstrating proteins rich in the sulphur-containing amino acid residues cystine, cysteine and methionine are based on reactions of their thiol groups ($-SH$). If such groups are initially absent (as in cystine residues), they are produced by treating sections with an alkaline solution of sodium thioglycollate, which chemically reduces the disulphide bonds to free thiol groups.

Both the revealed and endogenous thiol groups can be detected by reactions with either an organomercurial (such as mercury orange) or a maleimide, followed by coupling with a diazonium salt. The latter appears to be the most specific. Thiol groups can be visualized less specifically, but more easily, by their ability to reduce ferric ferricyanide to give ferrous ions which then react with unreduced ferricyanide to form an insoluble blue pigment (Turnbull's blue). Alternatively, thiol and disulphide-containing proteins can be revealed collectively by oxidizing them with a peracid, such as performic or peracetic acid, followed by staining of the sulphonic or sulphinic acids thus produced with a basic dye. This forms the basis of the performic acid-azure and -alcian blue methods listed in Table 2.2.

Tryptophan residues

Of all the other chemical end-groups and amino acid residues that it is possible to detect histochemically in proteins, only tryptophan and related indole residues are regularly investigated. They can be visualized specifically by reacting them with dimethyaminobenzaldehyde (DMAB) followed by nitrous acid to form a blue-coloured product.

Mucosubstances

Complex carbohydrate-containing substances are stored intracellularly in mammalian cells either as the polysaccharide glycogen or in combination with proteins as glycosoaminoglycans and glycoproteins. Carbohydrate residues (e.g. fucose, mannose and N-acetyglucosamine) are also essential components of most cellular membranes. Their detection in situ is based on reactions of their a) *vicinal* glycol groups, b) anionic groups (sulphate ester, uronic acid), c) anionic residues (e.g. sialic acid), or d) specific monosaccharide entities. In some cells, the relative amount and reactivity of these groups and individual sugars allow one to infer the predominant carbohydrate-containing substances present.

Vicinal glycol groups

Any substance containing these groups can be readily localized with the periodic acid-Schiff (PAS) reaction, which depends on the unique susceptibility of the groups to become oxidized to dialdehydes by periodic acid. The dialdehydes, like the aldehydes exposed in nuclear DNA after a Feulgen hydrolysis, are converted to strongly coloured products by treating them with Schiff's reagent, or alternatively, with methenamine silver (the PAMS technique).

The beauty of the PAS technique is its simplicity and reliability. It works equally well with glycogen and glycoproteins, but the presence of glycogen in a tissue can be distinguished by showing that the PAS reaction no longer occurs in sections treated beforehand with diastase (or saliva if a rapid confirmation is required). However, it is doubtful whether acid glycosaminoglycans are revealed with the PAS technique (as normally practised), even though they contain free *vicinal*-glycols. Fortunately, they are identified more easily with the basic dyes described below.

Chromic acid is also employed routinely in some laboratories for oxidising carbohydrate *vicinal*-glycols; it oxidizes the dialdehydes produced initially to carboxylic acid groups, which still react, however, with both Schiff's reagent and methenamine silver.

The chromic acid variant, especially when used with methenamine silver, is particularly useful in the diagnosis of renal diseases and fungal infections (Mowry 1981).

Anionic groups and residues

All acid glycosaminoglycans, in principle at least, can be detected by their affinity at an acid pH for the basic dye reagents azure A (or the virtually identical toluidine blue), alcian blue, high iron diamine and colloidal ferric iron.

At pH 3.5–4.0, azure A binds to most acid glycosaminoglycans and sialic acid-containing glycoproteins via their anionic groups. The bound dye is red (i.e. the dye exhibits metachromasia) in contrast to the orthochromatically-coloured blue dye taken up by other basophilic components of cells (e.g. nucleic acids). This enables acid mucosubstances to be identified rapidly in situ, but unfortunately the stained sections normally have to be mounted in water for examination. The usual dehydrating solvents destroy the metachromasia.

Alcian blue, or a comparable phthalocyanine dye, is perhaps the most useful basic dye reagent available at present for localising acid glycosaminoglycans and glycoproteins because by altering the staining conditions, it can be used to identify individual glycosaminoglycans selectively. This is usually accomplished by either adjusting the pH or the concentration of an inorganic salt ($MgCl_2$) in the dye solution. At pH 2.5, alcian blue stains both sulphated and carboxyl-containing acid glycosaminoglycans and glycoproteins, but at pH 1.0 only the more strongly ionized sulphated mucosubstances are stained.

Alcian blue is even more discriminating when it is dissolved in solutions of magnesium chloride of various concentrations (0.1–2.0M), and use is made of the 'critical electrolyte concentration' principle. The magnesium ions compete with the dye molecules for the anionic groups of the macromolecular carbohydrate complex and under the reaction conditions normally employed in histopathology laboratories (e.g. short staining times), this results in alcian blue 'staining' (i.e. being bound to) highly sulphated glycosaminoglycans selectively at high magnesium chloride concen-

trations ($\geqslant$ 1.0M), whereas the staining of hyaluronic acid- and sialic acid-containing complexes is extinguished at low magnesium chloride concentrations (e.g. 0.2M).

Alcian blue (at either pH 2.5 or 1.0) can be preceded by the high iron diamine reagent for demonstrating sulphated and carboxylated glycoproteins simultaneously in two or three different colours, or followed by the PAS technique for visualizing acid and neutral glycoproteins separately in the same preparation.

Sometimes not all the carbohydrate-containing substances present in a tissue react with any of these techniques because they are bound to, or are protected by, a protein masking their reactive endgroups. However, they usually react when treated briefly with mild alkali (e.g. potassium hydroxide dissolved in 75% ethanol), a procedure generally described as saponification.

The identity of some glycosaminoglycans, as inferred from any of the techniques outlined so far (particularly the critical electrolyte concentration method using alcian blue and magnesium chloride), can be confirmed by incubating sections of tissue before staining with a weakly buffered solution of a purified glycosidase, such as hyaluronidase, or one of the chondroitinases. This is particularly important for checking the identification of sulphated and hyaluronic acid-rich substances in situ. Sialic acid-rich glycoproteins can be confirmed by pre-treatment of tissue sections with neuraminidase. Prior saponification enhances the susceptibility of some sialomucins towards neuraminidase digestion, and also renders previously unreactive mucosubstances PAS-positive, thus enabling them to be differentiated further. This is useful in the diagnosis of those disorders of the gastrointestinal tract in which the distribution of different types of glycoprotein changes (Filipe 1989, Dawson 1981).

Specific saccharide residues and linkages

Increasingly in diagnostic histopathology, specific saccharides incorporated in membrane proteins are being detected with plant lectins for which they have a very strong, selective affinity. The lectins are usually conjugated with either horseradish peroxidase (which can be readily visualized with diaminobenzidine and hydrogen peroxide) or a fluorescent dye. Damjanov (1987) and Alroy et al (1988) have comprehensively reviewed their many applications in histopathology.

Lectins commonly employed in diagnostic histopathology are listed in Table 2.4, and their applications in Tables 2.5 and 2.6. In most diagnostic situations, it is worthwhile employing a

Table 2.4 Lectins used in pathological diagnoses (modified from Damjanov, 1987)

Common abbreviation	Source	Nominal saccharide specifity
Glucose/mannose group		
Con A	Jack bean (*Canavalia ensiformis*)	α Man $> \alpha$ Glc $\geqslant$ GlcNAc
LCA*	Lentil (*Lens culinaris*)	α Man $> \alpha$ Glc $>$ GlNAc
PSA	Pea (*Pisum sativum*)	α Man $> \alpha$ Glc $=$ GlcNAc
N-Acetylglucosamine group		
BSA II	*Bandeirea simplicifolia* seed	β-and α-GlcNAc
GSA-II	Griffonia seed (*Griffonia simplicifolia*)	α- and β-GlcNAc
DSA	Jimson weed (*Datura stramonium*)	GlcNAc ($\beta1$, 4-GlcNAc)$_{1-3}$ = Gal$\beta1$, 4-GlcNAc
PWM	Pokeweed (*Phytolacca americana*)	GlcNAc ($\beta1$, 4-GlcNAc)$_{1-5}$ = Gal$\beta1$, 4-GlcNAc$_{2-5}$
STA	Potato (*Solanum tuberosum*)	GlcNAc ($\beta1$, 4-GlcNAc)$_{1-4}$
UEA-II	Gorse seed (*Ulex europaeus II*)	L-Fuc$\alpha1$, 1-Gal$\beta1$, 4-GlcNAc $>$ GlcNAc ($\beta1$, 4-GlcNAc)$_{1-2}$
WGA	Wheat germ (*Triticum vulgaris*)	GlcNAc($\beta1$, 4-GlcNAc)$_{1-2}$ $\geqslant$ GlcNAc $>$ Neu5-Ac
N-Acetylgalactosamine/galactose group		
BPA	*Bauhinia purpurea* seed	α and β GalNAc $> \alpha$ and β Gal
BSA I-B$_4$	Bandeirea simplicifolia seed	α-D-Gal $> \alpha$-D-GalNAc
DBA	Horse gram (*Dolichos biflorus*)	α-D-Gal-NAc $\gg \alpha$-D-Gal
GSA-I	Griffonia seed (*Griffonia simplicifolia*)	α GalNAc $> \alpha$-D GalI-A$_4$
HPA	Edible snail (*Helix pomatia*)	GalNAc α 1, 3GalNAc $> \alpha$ GalNAc
LBA	Lima bean (*Phaseolus lunatus limensis*)	GalNAc α 1, 3[L-Fuc α 1, 2] Gal β GalNAc
MPA	Osage orange seed (*Maclura pomifera*)	αGalNAc $> \alpha$ Gal
PNA	Peanut (*Arachis hypogaea*)	Gal β 1, 3GalNAc $> \alpha$ and β Gal
RCA I	Castor bean (*Ricinus communis*)	β Gal $> \alpha$ Gal $\geqslant$ GalNAc
SBA	Soybean (*Glycine max*)	α and β GalNAc $> \alpha$ and β Gal
SJA	Japanese pagoda tree (*Sophora japonica*)	α and β GalNAc $> \alpha$ and β Gal
VVA	Hairy vetch (*Vicia villosa*)	GalNA α 1, 3Gal $= \alpha$ GalNAc α GalNAc
WFA	Wisteria seed (*Wisteria floribunda*)	GalNAc α 1, 6Gal $> \alpha$GalNAc $> \beta$GalNAc
L-Fucose group		
AAA	Orange peel fungus (*Aleuria aurantia*)	α L-Fuc
LTA	Asparagus pea (*Lotus tetragonolobus*)	α L-Fuc $>$ L-Fuc α 1, 2 Gal β 1, 4 GlcNAc $\geqslant$ L-Fuc α 1, 2 Gal $\beta-1$, 3GlcNAc
UEA-I	Gorse seed (*Ulex europaeus*)	α L-Fuc
Sialic acid group		
LFA	Slug (*Limax flavus*)	α Neu 5Ac $> \alpha$ Neu 5Gc
LPA	Horseshoe crab (*Limulus polyphemus*)	Neu 5Ac (or Gc) α 2, 6GalNAc $>$ Neu 5Ac

* Not to be confused with Leucocyte Common Antigen.
Abbreviations: Man — mannose, Glc — glucose, Ac — acetyl, GlcNAc — N-acetyl glucosamine, Gal — galactose, GalNAc — N-acetylgalactosamine, Neu — Sialic acid (neuraminic acid), Fuc — fucose, Gc — glucuronic acid.

Table 2.5 Examples of the use of lectins in human histopathology (see also Table 2.6)

Feature detected	Lectin	Reference
Blood cells		
Eosinophils	GSA-I, SBA	Lee et al (1987)
Megakaryocytes	WGA	Schick & Filmyer (1985)
Monocytes/macrophages	GSA-I	Holthofer et al (1984)
Blood group antigens	DBA, BSA-I	Judd (1980)
Degenerative diseases		
Alzheimer's	RCA-I, WGA etc	Szumanska et al (1987)
Fungi (in tissues)	RCA-I, SBA	Stoddart & Herbertson (1978)
	PWM, sucWGA*	Karayannopoulou et al (1988)
Histiocytes		
In malignant lymphomas	Con A,	Ree (1983)
	RCA	Ree et al (1983)
Benign in malignant histiocytosis	PNA, RCA	Ree & Kadin (1985)
In Langerhans cell histiocytosis	PNA	Ree & Kadin (1987)
Inflammatory diseases		
Crohn's	RCA-I, DBA, UEA-I	Jacobs & Huber (1985)
Ulcerative colitis	PNA	Cooper et al (1987)
Lymphocytes		
T-cell subsets	LCA, WGA	Boldt & Lyons (1979)
B-cell subsets	PNA	Rose et al (1981)
Myopathies and neuropathies	PNA, WGA, Con A	Bonilla et al (1980)
		Dunn et al (1982)
Storage diseases		
Fucosidosis	LTA, UEA-I	Virtanen et al (1980)
Hunter's syndrome	PNA, WGA	Faraggiana et al (1982)
I-cell disease	LTA, WGA	Aula & Virtanen (1981)
Salla disease	LPA	Virtanen et al (1980)
Various	Various	Alroy et al (1988)
Vascular endothelium and proliferation	UEA-I	Holthofer et al (1982),
		Suzuki et al (1986)

*Succinylated wheat germ agglutinin

battery of different lectins, including at least one specific for each of the principal naturally-occurring terminal saccharide groups (fucose, glucose/mannose, N-acetylglucosamine, N-acetygalactosamine/galactose and sialic acid). If the presence of N-acetylgalactosamine or galactose is suspected, tissue sections should be treated with neuraminidase first to remove sialic acid residues that may be blocking these groups. The exposed groups will then usually be detectable with, for example, peanut lectin. The absence of a particular saccharide may also be helpful in diagnosis (e.g. α-fucose in the vasculature of Wilm's tumour, Hennigar et al 1988). Lectins are particularly valuable for their potential in distinguishing neoplastic from non-neoplastic changes and malignant from benign tumours (Furmanski et al 1981). They also appear to enable the reoccurrence and invasiveness of some tumours, for example of the breast and the bladder, to be predicted (Cummings 1980, Leatham & Brooks 1987).

Lectins have one enormous advantage for the histopathologist compared to many of the other techniques at his disposal. The antigens they detect are very resistant to fixation and decalcification (Mukai et al 1986, Ordonez et al 1987). Thus lectins can be used on paraffin sections.

Endogenous lectin-like substances may also be present in the plasma membrane of mammalian cells, particularly in those prone to becoming neoplastic (see e.g. Gabius et al 1986a, b for reviews). Consequently tests for their presence are worth inclusion in the diagnostic histopathologist's armentarium.

Table 2.6 Applications of lectins for the diagnosis of neoplasias

Disease	Lectin	Reference
Adenocarcinomas	SBA, WGA, UEA-I GSA-I	Soderstrom (1987) Yen et al (1988)
Breast carcinoma	Con A HPA HPA, UEA-I PNA, WGA, Con A	Kahn & Baumel (1985) Leatham & Brooks (1987) Fenlon et al, (1988) Dansey et al (1988)
Fibrohistiocytic tumours	RCA-I	Ueda et al (1987)
Ganglioneuromas	PNA	Kahn et al (1988)
Gastro-intestinal tract	PNA PNA, GSA-II, UEA-I	Kahn & Baumel (1985) Rhodes et al (1986) Lee (1987)
Germ cell tumours	PNA, DBA	Teshima et al (1984) Malmi & Soderstrom (1988)
Leukaemias	PNA Various	Moller (1982) Gabius et al (1988)
Liver tumours	LCA	Sekine et al (1987)
Lung tumours	SucWGA, RCA-I	Kawai et al (1988)
Lymphomas	PNA RCA, WGA etc.	Ree & Hsu (1983) Ree et al (1983)
Melanomas	Con A, RCA-I	Kohchiyama et al (1987)
Nephroblastomas (Wilm's tumour)	UEA-I UEA-I etc.	Hennigar et al (1985) Yeger et al (1987)
Neuroblastomas	PNA	Kahn et al (1988)
Oral cavity tumours	JFA	Vijayan et al (1987)
Retinoblastomas	UEA-I	Bialasiewicz et al (1987)
Sarcomas	UEA-I	Leader et al (1986)
Skin tumours	UEA-I	Louis et al (1981)

Lipids

Many types of lipid can be identified histochemically in sections of unfixed fresh tissue (Bayliss High 1982), but in routine histopathology, it is sufficient to concentrate on detecting three broad classes.

All lipids

Oil red O stains most lipids, but not those in the solid state, and is thus the best reagent for their general screening. Two major classes of lipid can be distinguished if sections stained with this dye are viewed with polarized light: unstained (crystalline) lipids appear birefringent (anisotropic) whereas the stained liquid lipids are non-refringent (isotropic). The birefringent lipid is usually cholesterol or one of its solid esters.

Sudan black also stains most lipids and is commonly employed as an alternative to oil red O. The Sudan dyes also do not reveal free fatty acids, phosphoglycerides, certain protein-bound lipids and solid lipids. The first two named components are extracted by the dye solvent. These disadvantages can be largely avoided by treating tissues with bromine water first, which presumably brominates the olefinic bonds of unsaturated lipids, thus rendering them less soluble.

Acidic and unsaturated lipids

In water, the basic dye nile blue sulphate selectively binds to, or dissolves in, unsaturated hydrophobic lipids, free fatty acids and phospholipids, and is a good supplementary reagent to oil red O.

Cholesterol and its esters

Schultz's modification of the Liebermann Burchardt reaction is widely used in histopathology for revealing the presence of cholesterol and its esters. The chemistry of the reaction has not been clarified.

Because of the harshness of the reagent needed, gentler tests are performed in some laboratories. Of these, the perchloric acid-naphthoquinone (PAN) method seems to be the most popular.

Metals

The only metals, or rather their salts, that are commonly sought histologically are iron and calcium. Occasionally, there is a need to look for deposits of copper, aluminium or barium.

Ferric iron (usually in the form of haemosiderin deposits) is easily demonstrable with Perls' method, first described in 1867, in which sections are treated with a fresh solution of potassium ferrocyanide in dilute hydrochloric acid. The acid releases the ferric ions from their protein attachments, whence they immediately react with ferrocyanide ions to form insoluble Prussian blue.

Several tests exist for visualizing calcium salts. One of the oldest, and until recently perhaps the most frequently employed, is von Kossa's in which sections are placed in solutions of silver nitrate and exposed to light. Calcium phosphate deposits become coloured brown or red. Unfortunately, many other tissue components reduce silver salts, and thus the method is capricious and unspecific for calcium. As a result, there is a growing trend to apply more specific methods, particularly when the role of calcium in a disease process is being investigated. At present, the best seems to be alizarin and glyoxal bis-(2-hydroxyanil), which form coloured complexes with calcium.

Rubeanic acid is commonly employed for detecting copper (producing greenish-black complexes), and rhozodinate for lead and barium (to give intensely red deposits). Aluminium deposits are demonstrable with several dyes of the solochrome series, such as solochrome azurine.

Protein antigens

In principle, almost any protein or peptide can be localized specifically with immunocytochemical techniques if an antiserum against it is available. Such techniques are the only ones possible if standard histochemical techniques do not exist. Many peptide hormones and enzymic proteins have been localized by such means, and the list is growing. However, for the moment, the immunohistochemical localization of hormone and enzyme antigens should only be attempted for confirming certain rare syndromes, particularly those involving the diffuse neuroendocrine system (e.g. APUDomas), or in research investigations of the pathogenesis of a disease at the molecular level. On the other hand, the increasing availability of monoclonal antisera may result in immunohistochemical techniques replacing many existing histochemical and histological techniques in the near future.

Since immunocytochemical techniques and their applications are described in Chapter 3, they are not discussed further here.

Enzymes

Some enzymes exhibit a particularly high activity in certain specialized cells and organelles, and in these situations they are commonly either exploited for identifying the cells in sections of tissue or regarded as markers or indicators of their function, or if the activity is absent or substantially reduced, of cellular dysfunction. So far, about 30 such enzymes have been found valuable for the differential diagnosis of muscle diseases (Johnson

& Walton 1981), certain neoplasias, storage diseases, disorders of the gastrointestinal tract (Lojda 1981) and leukaemias (Catovsky et al 1981, Lojda 1981). These enzymes are listed in Table 2.7, together with the cell type and function they are assumed to signify. About 10 of the enzymes are worth identifying routinely.

Stoward & Pearse (1990) and Appendix 5 should be consulted for reliable histochemical methods for localizing the enzymes. Purists sometimes worry that these methods may not be specific. In my view, it is sufficient if the methods give results (i.e. staining patterns) which can be correlated empirically either with the presence (or absence) of particular types of cell or with a defined clinical stage of a disease.

One area of pathology where enzyme histochemical techniques could be usefully applied more often than at present is in difficult forensic cases, particularly for estimating the time at which wounds occur before death. The usefulness is based on the observation that certain enzymes, which normally have a low histochemical activity, appear in a regular chronological sequence in the peripheral wound zone of injured tissues such as a myocardial infarct (Raekallio 1970).

Monoamine oxidase, non-specific esterase and ATPase are the first to appear, becoming evident

1 hour after injury, whereas acid phosphatase and alkaline phosphatase do not become histochemically demonstrable until 4 hours after injury (Fig. 2.5). This enzyme 'clock' is a little slower in older subjects, and faster in younger ones (Raekallio & Makinen 1974).

LIMITATIONS OF HISTOCHEMISTRY IN ROUTINE HISTOPATHOLOGY

At best, the histochemical techniques suggested so far in this chapter will, if applied systematically, only tell the histopathologist what substances are present in a tissue and where they are situated. Unless certain criteria are met, they do not allow one to infer how much of the substance is present, or what the level of activity is if the substance is an enzyme. Even though one can estimate the intensity of 'staining' produced by a histochemical technique on a 1–5+ scale in different histological sites, it cannot be assumed that the intensity is necessarily related to the concentration of the reacting substance. Before one can do so, it must be shown that the histochemical technique is specific and quantitatively valid in the senses which have been defined rigorously elsewhere (Stoward 1980).

A consequence of this caveat is that one must be cautious in interpreting histochemical observations in detailed biochemical and physiological terms.

In order to investigate a pathological mechanism at the molecular level, a completely different approach is required. First, a hypothesis should be formulated in biochemical terms about the cellular changes taking place in the early stages of a disease. From this hypothesis, it usually becomes apparent what substances and enzymes it would be desirable to identify and perhaps quantify in situ. Unfortunately it often turns out that suitable cyto- and histochemical techniques do not exist for their localization. New methods must, therefore, be developed rather than subverting existing routine ones. It is also necessary to validate the devised methods before applying them to the investigation in hand.

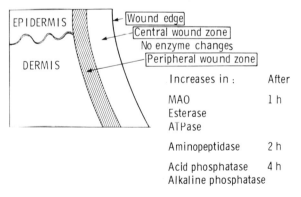

Fig. 2.5 Schematic representation of the time-dependent appearance of certain enzymes in the peripheral wound region of an injured tissue, e.g. cut skin (after Raekallio 1970).
MAO = monoamine oxidase

Table 2.7 Enzymes used as markers of cell type and function (or dysfunction) in diagnostic histopathology. Particularly useful ones are indicated by an asterisk (*)

Enzyme	Cell type or function indicated	Diagnostic applications [1]
*Acetylcholinesterase	Nerve fibre tracts	Hirschsprung's disease ($\uparrow$)
Acid esterase (E600-resistant)	T-lymphocytes (helper and suppressor subsets)	Wolman's disease (deficient)
Acid β-galactosidase	Degradation of mucosubstances within lysosomes	G_{M1}-gangliosidosis (deficient)
*Acid phosphatase	Active phagocytic cells (i.e. those containing numerous lysosomes, e.g. macrophages) Autolysis and necrosis Resorption (of bone)	Muscle fibre damage ($\uparrow$) Metastatic adenocarcinomas of prostate origin (phosphoryl choline as substrate) Identification of histiocytes, particularly osteoclasts
Alkaline phophatase	Epithelial cells with functioning brush borders Mature neutrophils of peripheral blood Osteoblasts, calcification	Renal tubule dysfunction ($\downarrow$) Chronic myeloid leukaemia ($\downarrow$) Myelofibrosis, leukaemoid reactions
Amine oxidase[2]	Sites of biogenic amine destruction	
Aminopeptidase M[3]	Activated macrophages Epithelial cells with functioning brush borders	Metastases originating from stomach, bile ducts, urinary bladder or kidney
ATPase (Mg^{2+}-activated)	Intra-epithelial lymphocytes Functioning bile canaliculi	Liver disease
*ATPase (myofibrillar)[4]	Fast contracting type II muscle fibres	Muscle fibre typing for myopathies and neuropathies Rhabdomyosarcoma
Catalase	Presence of peroxisomes	Zellweger syndrome ($\downarrow$), infantile Refsum's disease ($\downarrow$) and neonatal adrenoleucodystrophy ($\downarrow$)
Chloroacetate esterase	Neutrophils, mast cells, granulocytes	Differential diagnosis of leukaemias
Diaminopeptidyl peptidase IV (DPP IV)	T-lymphocyte subsets Capillary endothelial cells	Malignant lymphomas
*DOPA oxidase[5]	Melanocytes	Amelanotic melanoma
Elastase[6]	Neutrophils (but not eosinophils)	
Endopeptidase		Coeliac disease
Enterokinase		Malabsorption syndromes
Glucose-6-phosphatase	Hepatocytes, renal proximal tubular cells and enterocytes	Type 1A glycogenosis (deficient)
Glucose-6-phosphate dehydrogenase	Cells underoing rapid proliferation (i.e. comparatively rapid DNA and lipid synthesis)	
Glutamate dehydrogenase	Cells with damaged mitochondria	Recent tissue damage ($\uparrow$)
3-Hydroxybutyrate dehydrogenase	β-Oxidation of fatty acids (i.e. lipid metabolism)	
3α- and β-hydroxysteroid dehydrogenases	Steroid synthesis	

Table 2.7   Cont'd

Enzyme	Cell type or function indicated	Diagnostic applications[1]
*Lactase[7]	Absorption and digestion of disaccarides in brush borders, e.g. of enterocytes	Malabsorption syndrome (↓)
Maltase[7,8]	As lactase	Malabsorption syndrome (↓)
Malate dehydrogenase		Infarction (↓)
α-Mannosidase	Langerhans cells	Melanoblastomas, Langerhans cell histiocytosis
*NADH dehydrogenase[9]	Type 1 aerobic muscle fibres Aerobic metabolism	Muscle fibre typing for myopathies and neuropathies
NADPH dehydrogenase	Cytochrome electron transport Detoxification	
'Non-specific esterase'[10]	Cells containing lysosomes, monocytic markers and histocytes	Differential diagnosis of leukaemias Recent tissue injury
Peroxidase: eosinophil myelo* platelet	Eosinophils Intracellular cidosis, neutrophils, Platelets, megakaryocytes	Differential diagnosis of leukaemias
*Phosphorylase	Anaerobic metabolism	Ischaemia (↓) Type V glycogenosis (deficient)
Phosphofructokinase	Glycolysis	Type VII glycogenosis (deficient)
*Succinate dehydrogenase	Cells containing mitochondria Type I muscle fibres Aerobic metabolism	Ischaemia (↓) Liver damage (↓) Reye's syndrome (↓)
Sucrase[7]	As lactase	Malabsorption syndrome (↓)
Trehalase[7]	As lactase	Malabsorption syndrome (↓)

[1] The signs ↑ ↓ in parenthesis indicate that the activity is substantially increased or decreased respectively.

[2] Usually known in the histochemical literature as monoamine oxidase. Its EC recommended name is amine oxidase (flavin-containing).

[3] Normally referred to simply as aminopeptidase. However, several aminopeptidases, with different subcellular locations, are now known. This one, also called membrane aminopeptidase, is situated on brush borders, e.g. of enterocytes and renal proximal tubules and on the plasma membrane of activated macrophages.

[4] Myofibrillar calcium-activated ATPase demonstrable at pH 9.4 (Appendix 5). Usually demonstrated by *loss* of activity after preincubation of sections of unfixed muscle at pH 4.6 or 4.2 (see Appendix 5). However, Mg, Ca-activated actomyosin ATPase is probably a better enzyme for muscle typing. Mabuchi & Sréter's (1980) method for this enzyme enables the typing to be accomplished in one step without preincubation.

[5] EC recommended name is catechol oxidase.

[6] Normally referred to as elastase-like activity.

[7] Lojda (1981) recommends that in investigations of the malabsorption syndrome, the histochemical localization of lactase, sucrase and trehalase are all obligatory. Other brush border enzymes should also be investigated if possible (e.g. maltase, alkaline phosphatase).

[8] EC recommended name for maltase is α-D-glucosidase (3.2.1.20). Not the same as acid maltase (EC 3.2.1.3), a lysosomal enzyme.

[9] Normally called NADH-tetrazolium reductase in the histochemical literature. The name used here is the EC recommended one.

[10] There are many types of non-specific esterase. The enzyme referred to in this table is probably C-esterase because α-naphthyl acetate is the substrate usually employed for its histochemical localization.

REFERENCES

Alroy J, Ucci A A, Periera MEA 1988 Lectin histochemistry. An update. In:DeLellis R A (ed.) Advances in immunohistochemistry. Raven Press, New York, ch 5, p 93–131

Aula P, Virtanen I 1981 I-cell disease phenotype in vitro: Experiments with cell fusion and co-cultivation. Journal of Inherited Metabolic Disease 4: 153–154

Bancroft J D, Stevens A (eds) 1990 Theory and practice of histological techniques, 3rd edn. Churchill Livingstone, Edinburgh

Barker D C, Gibson L J, Kennedy W P, Nasser A A, Williams R H 1986 The potential of using recombinant DNA species-specific probes for the identification of tropical Leishmania. Parasitology 92 Supplement, S139–174

Bayliss High O 1982. Lipids. In: Bancroft J D, Stevens A (eds) Theory and practice of histological techniques, 2nd edn. Churchill Livingstone, Edinburgh, ch 12, p 217

Bialasiewicz A A, Salowsky A 1987 Binding patterns of lectins to human retinoblastoma cells: preliminary findings. Cancer Detection and Prevention 10: 63–70

Boldt D H, Lyons R D 1979 Fractionation of human lymphocytes with plant lectins. II. *Lens culinaris* lectin and wheat germ agglutinin identifying lymphocyte subclasses. Journal of Immunology 123: 808–816

Bonilla E, Schotland D L, Wakayama Y 1980 Application of lectin cytochemistry to the study of human neuromuscular disease. Muscle and Nerve 3: 28–35

Catovsky D, Crockard A D, Matutes E, O'Brien M 1981 Cytochemistry of leukaemic cells. In: Stoward P J, Polak J M (eds) Histochemistry: the widening horizons of its applications in the biomedical sciences. John Wiley, London, ch 6, p 67

Coghlan J P, Aldred P, Haralambidis J, Niall H D, Penschow J D, Tregear G W 1985 Hybridization histochemistry. Analytical Biochemistry 149: 1–28

Cooper H S, Farano P, Coapman R A, 1987 Peanut lectin binding in colons of patients with ulcerative colitis. Archives of Pathology and Laboratory Medicine 111: 270–275

Cremers A F M, Jansen in de Wal N, Wiegant J, Dirks R W, Weisbeek P, Van der Ploeg M, Landegent J E 1987 Non-radioactive in situ hybridization. A comparison of several immunocytochemical detection systems using reflection-contrast and electron microscopy. Histochemistry 86: 609–615

Cummings K B 1980 Carcinomas of the bladder: predictors. Cancer 45 (7 Supplement): 1849–1855.

Damjanov I 1987 Biology of disease. Lectin Cytochemistry and Histochemistry. Laboratory Investigation 57: 5–20

Dansey R, Murray J, Ninin D, Bezwoda W R 1988 Lectin binding in human breast cancer: clinical and pathologic correlations with fluorescein-conjugated peanut, wheatgerm and concanavlin A binding. Oncology 45: 300–302

Dawson I M P 1981 The value of histochemistry in the diagnosis and prognosis of gastrointestinal diseases. In: Stoward P J, Polak J M (eds) Histochemistry: the widening horizons of its applications in the biomedical sciences. John Wiley, London, ch 9, p 127

Drury R A B, Wallington E A 1980 Carleton's histological technique, 5th edn. Oxford University Press, Oxford

Dunn M J, Sewry C A, and Dubowitz V 1982 Cytochemical studies of lectin binding by diseased human muscle.

Journal of the Neurological Sciences 55: 147–159

Faraggiana T, Shen S, Childs C, Strauss L, Churg J 1982 Histochemical study of Hurler's disease by the use of peroxidase-labelled lectins. Histochemistry Journal 14: 655–664

Fenton S. Ellis I O, Bell J, Todd J H, Elston C W, Blarney R W 1988 *Helix pomatia* and *Ulex europaeus* lectin binding in human breast carcinoma. Journal of Pathology 152: 169–176

Filipe M I 1989 The histochemistry of intestinal mucins. Changes in disease In: Whitehead R (ed) Gastrointestinal and oesophageal pathology. Churchill Livingstone, Edinburgh, ch 7, p 65–89

Furmanski P, Kirkland W I, Gargala T, Rich M A 1981 Prognostic value of concanavalin A reactivity of primary human breast cancer cells. Cancer Research 41: 4087–4092

Gabius H J, Engelhardt R, Cramer F 1986a Endogenous tumor lectins: overview and perspectives. Anticancer Research 6: 573–578

Gabius H J, Engelhardt R, Cramer F 1986b Endogenous tumor lectins: a new class of tumor markers and targets for therapy? Medical Hypotheses 18: 47–50

Gabius H J, Vehmeyer K, Gabius S, Nagel G A 1988 Clinical application of various plant and endogeneous lectins to leukemia. Blut 56: 147–152

Grody W W, Cheng L, Lewin K J 1987 In situ viral DNA hybridization in diagnostic surgical pathology Human Pathology 18: 535–543

Hennigar R A, Sens D A, Othersen H B, Garvin A J 1988. Distribution of fucosubstance in kidney and related neoplasms. Absence of lectin-reactive alpha-fucose from the vasculature of bilateral Wilm's tumors. Archives of Pathology and Laboratory Medicine 112: 908–913

Hennigar R A, Sens D A, Spicer S S et al 1985 Lectin histochemistry of nephroblastoma (Wilm's tumour). Histochemical Journal 17: 1091–1110

Hofler H 1987 What's new in 'in situ hybridization'. Pathology, Research and Practice 182: 421–430

Holthofer M, Virtanen I, Kariniemi A L, Hormia M, Linder E, Miettinen A 1982 *Ulex europaeus* I lectin as a marker for vascular endothelium in human tissues. Laboratory Investigation 47: 60–66

Holthofer H, Virtanen I, Tornroth T, Miettinen A 1984 Lectins as markers for cells infiltrating human renal glomeruli. Virchows Archiv B. Cell Pathology 49: 119–126

Jacobs L R, Huber P W 1985 Regional distribution and alterations of lectin binding to colorectal mucin in mucosal biopsies from controls and subjects with inflammtory bowel disease. Journal of Clinical Investigation 75: 112–118

Johnson M A, Walton J N 1981 Histochemistry — its contribution to the study of normal and diseased muscle. In: Stoward P J, Polak J M (eds) Histochemistry: the widening horizons of its applications in the biomedical sciences. John Wiley, London, ch 11, p 183

Judd W J 1980 The role of lectins in blood group serology. CRC Critical Reviews of Clinical and Laboratory Science 12: 171–214

Kahn H J, Baumal R 1985 Differences in lectin binding in tissue sections of human and murine malignant tumours and their metastases. American Journal of Pathology 119: 420–429

Kahn H J, Baumal R, Thorner P S, Chan H 1988 Binding of peanut agglutinin to neuroblastomas and ganglioneuromas: a marker for differentiation of neuroblasts into ganglion cells. Pediatric Pathology 8: 83–93

Karayannopoulou G, Weiss J, Damjanov I 1988 Detection of fungi in tissue sections. Archives of Pathology and Laboratory Medicine 112: 746–748

Kawai T, Greenberg S D, Truong L D, Mattioli C A, Klima M 1988 Differences in lectin binding of malignant pleural mesothelioma and adenocarcinoma of the lung. American Journal of Pathology 130: 401–410

Kohchiyama A, Oka D, Ueki H 1987 Differing lectin-binding patterns of malignant melanoma and nervocellular and Spitz nevi. Archives of Dermatological Research 279: 226–231

Leader M, Collins M, Patel J, Henry K 1986 Staining for factor VIII related antigen and *Ulex europaeus* agglutinin I (UEA-I) in 230 tumors. An assessment of their specificity for angiosarcoma and Kaposi's sarcoma. Histopathology 10: 1153–1162

Leatham A J, Brooks S A 1987 Predictive value of lectin binding on breast-cancer recurrence and survival. Lancet i: 1054–1056

Lee A K, DeLelliss R A 1987 Immunohistochemical techniques and their applications to tissue diagnosis. In: Spicer S S (ed) Histochemistry in pathologic diagnosis. Marcel Decker, New York, ch 3, p 31

Lee M-C, Turcinov D, Damjanov I 1987 Lectins as markers for eosinophilic leukocytes. Histochemistry 86: 269–273

Lee Y S 1987 Lectin reactivity in human large bowel. Pathology 19: 397–401

Lendrum A C, Fraser D S, Slidders W, Henderson R 1962 Studies on the character and staining of fibrin. Journal of Clinical Pathology 15: 401–413

Lewis F A, Griffiths S, Dunnicliff R, Wells M, Dudding N, Bird C C 1987 Sensitive in situ hybridization technique using biotin-streptavidin-polyalkaline phosphatase complex. Journal of Clinical Pathology 40: 163–166

Lillie R D, Fullmer H M 1976 Histopathologic technique and practical histochemistry, 4th edn. McGraw-Hill, New York

Lloyd R V 1987 Use of molecular probes in the study of endocrine diseases. Human Pathology 18: 1199–1211

Lojda Z 1981 The applications of enzyme histochemistry in diagnostic pathology. In: Stoward P J, Polak J M (eds) Histochemistry: the widening horizons of its applications in the biomedical scienies. John Wiley, London, ch 12, p 205

Loning T, Milde K, Foss H D 1986 In situ hybridization for the detection of cytomegalovirus (CMV) infection. Application of biotinylated CMV DNA probes on paraffin-embedded specimens. Virchows Archiv A. Pathological Anatomy and Histopathology 409: 777–790

Louis C J, Wyllie R G, Chou S T, Sztynda T 1981 Lectin-binding affinities of human epidermal tumors and related conditions. American Journal of Clinical Pathology 75: 642–647

Mabuchi A, Streter F A 1980 Actomyosin ATPase. II. Fiber typing by histochemical ATPase reaction. Muscle Nerve 3: 233–239

Malmi R, Soderstrom K O 1988 Lectin binding to carcinoma in situ cells of the testis. A comparative study of CIS germ cells and seminoma cells. Virchows Archiv. A. Pathological Anatomy and Histopathology 413: 69–75

Melamed M R, Darzynkiewicz Z 1981 Acridine orange as a quantitative cytochemical probe for flow cytometry. In: Stoward P J, Polak J M (eds) Histochemistry: the widening horizons of its applications in the biomedical Sciences. John Wiley, London, ch 14, p 237

Moller P 1982 Peanut lectin: a useful tool for detecting Hodgkin cells in paraffin sections. Virchows Archiv A. Pathological Anatomy and Histopathology 396: 313–317

Mowry R W 1981 Contributions of practical carbohydrate histochemistry to the histopathological diagnosis of renal diseases, fungal infections, and some types of cancer. In: Stoward P J, Polak J M (eds) Histochemistry: the widening horizons of its applications in the biomedical sciences. John Wiley, London, ch 8, p 109

Mukai K, Yoshimura S, Anzai M 1986 Effects of decalcification on immunoperoxidase staining. American Journal of Surgical Pathology 10: 418–419

Ordonez N G, Brooks T, Thompson S, Batsakis J G 1987 Use of *Ulex europaeus* agglutinin I in the identification of lymphatic and blood vessel invasion in previously stained microscopic slides. American Journal of Clinical Pathology 11: 543–50

Palva A 1986 Microbial diagnosis by nucleic acid hybridization. Annals of Clinical Research 18: 327–336

Penschow J D, Haralambidis J, Darling P E et al 1989 Hybridisation histochemistry. In: Polak J M (ed) Regulatory peptides. Birkhauser, Basel, p 51–69

Raekallio J 1970 Enzyme histochemistry of wound healing. Progress in Histochemistry and Cytochemistry 1: no 2

Raekallio J, Makinen P L 1974 Effect of ageing on enzyme histochemical vital reactions. Zeitschrift für Rechtsmedizin 75: 105–111

Ree H J 1983 Lectin histochemistry of malignant tumors. II. Concanavalin A: A new histochemical marker for macrophage-histiocytes in follicular lymphoma. Cancer 51: 1639–1646

Ree H J, Hsu S-M 1983 Lectin histochemistry of malignant tumours. I. Peanut agglutinin (PNA) receptors in follicular lymphoma and follicular hyperplasia: an immunohistochemical study. Cancer 51: 1631–1638

Ree H J, Kadin M E 1985 Lectin distinction of benign from malignant histiocytes. Cancer 56: 2046–2050

Ree H J, Kadin M E 1987 The usefulness of peanut agglutinin for the detection of histiocytosis-X and interdigitating reticulum cells. Human Pathology 18: 309–310

Ree H J, Raine L, Crowley J P 1983 Lectin binding patterns in diffuse large cell lymphomas. Cancer 52: 2089–2099

Rhodes J M, Black R R, Savage A 1986 Glycoprotein abnormalities in colonic carcinomata, adenomata, and hyperplastic polyps shown by lectin peroxidase histochemistry. Journal of Clinical Pathology 39: 1331–1334

Rollinson D, Walker T K, Simpson A J 1986 The application of recombinant DNA technology to the problems of helminth identification. Parasitology 92 Supplement, S53–57

Rose M L, Habeshaw J A, Kennedy R, Sloane J, Wiltshaw E, Davies A J S 1981 Binding of peanut lectin to germinal-center cells: a marker for B-cell subsets of follicular lymphoma? British Journal of Cancer 44: 68–74

Schick P K, Filmyer W G 1985 Sialic acid in mature megakaryocytes: detection by wheat germ agglutinin. Blood 65: 1120–1126

Sekine C, Aoyagi Y, Suzaki Y, Ichida F 1987 The reactivity of alpha-1-antitrypsin with *Lens culinaris* agglutinin and its usefulness in the diagnosis of neoplastic diseases of the liver. British Journal of Cancer 56: 371–375

Shikata T, Uzawa T, Yoshiwara N, Akatsuka T, Yamazaki S 1974 Staining methods of Australian antigen in paraffin sections — detection of cytoplasmic inclusion bodies. Japanese Journal of Experimental Medicine 44: 25–36

Soderstrom K O 1987 Lectin binding to prostatic adenocarcinoma. Cancer 60: 1823–1831

Stoddart R W, Herbertson B M 1978 The use of fluorescein-labelled lectins in the detection and identification of fungi pathogenic for man: a preliminary study. Journal of Medical Microbiology 11: 315–324

Stoner G D, You M, Stouv J, Budd G C, Pansky B, Wang Y 1987 Detection of oncogene mRNA sequences in cultured cells by in situ hybridization. Annals of Clinical and Laboratory Science 17: 74–82

Stoward P J, 1980 Criteria for the validation of quantitative histochemical enzyme techniques. In: Trends in enzyme histochemistry and cytochemistry (Ciba Foundation Symposium 73). Excerpta Medica, Amsterdam, p 11

Stoward P J Pearse A G E (eds) 1990 Enzyme histochemistry. Histochemistry: theoretical and applied, vol. 3, 4th edn. Churchill Livingstone, Edinburgh

Suzuki Y, Hashimoto K, Crissman J, Kunzaki T, Nishiyame S 1986 The value of blood group-specific lectins and endothelial associated antibodies in the diagnosis of vascular proliferation. Journal of Cutaneous Pathology 13: 408–419

Syrjanen S M, Syrjanen K J, Lamberg M A 1986 Detection of human papillomavirus DNA in oral mucosa lesions using in situ DNA- hybridization applied on paraffin sections. Oral Pathology, Oral Medicine and Oral Pathology 62: 1660–1667

Syvanen A C 1986 Nucleic acid hybridization. Medical Biology 64: 313–324

Szumanska G, Vorbrodt A W, Mandybur II, Wisniewski H M 1987 Lectin histochemistry of plaques and tangles in Alzheimer's disease. Acta Neuropathologica 73: 1–11

Teshima S, Hirohashi S, Shimosato Y et al 1984 Histochemically demonstrable changes in cell surface carbohydrates in human germ cell tumors. Laboratory Investigation 50: 271–277

Uedo T, Aozasa K, Yamamura T, Tsujimoto M, Ono K, Matsumoto K 1987 Lectin histochemistry of malignant fibrohistiocytic tumors. American Journal of Surgical Pathology 11: 257–262

Vijayan K K, Remani P, Beevi V M et al 1987 Tissue binding patterns of lectins in premalignant and malignant lesions of the oral cavity. Journal of Experimental Pathology 3: 295–304

Virtanen I, Ekblom, P, Laurila P, Nordling S, Raivo K O, Aula P 1980 Characterization of storage material in cultured fibroblasts by specific lectin binding in lysosomal storage diseases. Pediatric Research 14: 1199–1203

Wittekind D H, Kretschmer V 1987 On the nature of Romanowsky-Giemsa staining and the Romanowsky-Giemsa effect II. A revised Romanowsky-Giemsa staining procedure. Histochemical Journal 19: 399–401

Wolfe H J 1988 DNA probes in diagnostic pathology. American Journal of Clinical Pathology 90: 340–344

Yeger H, Baumal R, Harason P, Phillips M J 1987 Lectin histochemistry of Wilm's tumor. Comparison with normal adult and fetal kidney. American Journal of Clinical Pathology 88: 278–285

Yen Y, Schmiemann C, Damjanov I 1988 Lectin histochemistry of adenocarcinomas. Archives of Pathology and Laboratory Medicine 112: 791–793

Zwadyk P, Cooksey R C 1987 Nucleic acid probes in clinical microbiology. CRC Critical Reviews of Clinical Laboratory Science 25: 71–103

3. Principles of immunostaining

S. Van Noorden

INTRODUCTION

Immunostaining provides a way of identifying substances in tissues using antigen-antibody reactions which can be made microscopically visible through the incorporation of a suitable label. The purpose of this chapter is to outline the techniques, including pitfalls and problems, and to recommend suitable procedures for immunostaining in a routine histopathology laboratory. The bias of the chapter will be towards light microscopical immunostaining, but the method is well adapted to use at the electron microscopical level and this will be touched on when it is applicable to pathological diagnosis. Techniques are detailed in Polak & Varndell (1984) and in Appendix 6.

CONDITIONS FOR IMMUNOSTAINING

The substance to be localized and the tissue itself must be preserved in such a way that the antigenic groups are available for reaction with the applied antibody. The antibodies must be of high affinity and react only with the substance being investigated. They must be fully labelled in order to achieve the maximum impact and the detection system for the label must be efficient.

Fixation

Light microscopy

Unfortunately, the histopathologist usually has to rely on formalin-fixed paraffin sections which are often far from ideal for immunocytochemistry. The antigenic sites of some substances are made immunologically inactive by routine formalin fixation, and for the cell surface markers of

lymphocytes, specially designed fixatives and careful processing should be employed if optimal staining is to be achieved (Holgate et al 1986, Pollard et al 1987, see also Ch. 19).

Fresh-frozen cryostat sections postfixed in methanol, acetone, or acetone mixed with chloroform are usually satisfactory for the preservation of some of these labile substances. Hall et al (1987) advocate 2 minutes in acetone followed by 8 minutes in cold periodate-lysine-paraformaldehyde (McLean & Nakane 1974). This allows immunostaining of lymphocyte surface antigens combined with improved tissue structure. Acetone-fixed cryostat sections are also useful in renal immunopathology, but routine paraffin sections are preferred (Ch. 17) for this purpose.

For other substances, such as neuropeptides, a mild cross-linking fixative such as formalin or p-benzoquinone (Pearse & Polak 1975, Bishop et al 1978) is essential. Without this they are soluble and leach from the tissue during the reaction. Yet others, such as the cytokeratin intermediate filaments of the cytoskeleton of epithelial cells, can be stained in unfixed preparations, but nevertheless survive fixation, though some epitopes may be destroyed.

That said, it is fortunate that, as a result of increasing demand, many diagnostically useful antibodies that do react with formalin-fixed paraffin-embedded material are now available commercially.

Electron microscopy

Routine fixation for electron microscopy, glutaraldehyde, or glutaraldehyde and formaldehyde mixtures, with further treatment in osmium tetroxide, may destroy or reduce immunoreactivity. In practice, omitting the osmication step and using a low concentration of glutaraldehyde and formaldehyde can often overcome some of the difficulties, although the ultrastructure of the tissue will be less than optimally preserved. Some antigens will remain immunoreactive after osmication.

Resin sections for light microscopy

Epoxy resin-embedded semithin (1 μm) sections can be immunostained in the same way as paraffin sections. The resin must first be removed, e.g. by alcoholic saturated sodium hydroxide (Lane & Europa 1966). Glycol methacrylate sections are often used for thin sectioning at light microscopical level and give excellent morphology. Unfortunately the polymerized methacrylate cannot be dissolved and antibodies do not reliably penetrate sections. However, hope for the future has been offered by a recent adaptation of the resin to yield a wider-meshed polymer that reportedly allows reliable immunostaining after careful fixation and processing (van Goor et al 1988).

Tissue preparation

This can be as damaging as fixation to antigens that do not survive solvent and heat treatment. In such cases, prefixed tissue can be frozen and cut in a cryostat or with a freezing microtome, or even cut unfrozen with a Vibratome or similar instrument to produce thick sections that can be used free-floating. Protease treatment and permeabilization may be necessary for such preparations (see Appendix 6).

Section preparation

Sections for immunocytochemistry should be picked up on slides coated with high molecular weight poly-L-lysine (Huang et al 1983), or other adhesive, particularly if protease treatment is to be used.

Paraffin sections

Paraffin sections should not be heated on a hot plate but dried for several hours, preferably over-night, at 37°C. They can be stored indefinitely without loss of antigen. Histologically stained sections from the archive can often be destained and then immunostained.

Cryostat sections

Cryostat sections should be air-dried at room temperature for several hours or overnight. They can then be stained, or stored at −20°C, wrapped in foil or 'cling film' in a sealed plastic bag containing silica gel or other desiccant. Before being opened, the entire bag is warmed up to room temperature to prevent condensation of atmospheric water onto the sections. The sections are then fixed according to the antigens to be immunostained and may then be air-dried before proceeding or placed directly in buffer. It is important that cryostat sections, once wet, are not allowed to dry throughout the immunostaining procedure, as this results in poor staining and disruption of tissue architecture.

Many antigens, or the parts of them that bind to certain monoclonal antibodies, are destroyed by paraffin or resin embedding but can be detected in cryostat sections, e.g. the antigen in proliferating cell nuclei stained by the Ki 67 antibody. Some T-lymphocyte antigens that were previously thought to be detectable only in cryostat sections can now be revealed in paraffin sections.

ANTIBODIES

The choice available is now enormous and can be baffling. Polyclonal (usually rabbit or goat) and monoclonal (mouse or rat) antibodies may be available to the same substance. When choosing an antibody it is wise to check with the manufacturer that it will react with the antigen in the fixation and processing system you intend to use and, if that information is not available, it is worth asking for a small sample to try out before committing yourself to large expenditure.

Monoclonal or polyclonal?

The advantage of monoclonal antibodies is that they are consistent and the same clone remains available; they are of a single, known, immunoglobulin subclass and have a defined specificity to one epitope of the antigen molecule. The disadvantage, compared with polyclonal antibodies, is that the particular epitope may not resist formalin fixation and paraffin embedding and that if the antibody is wrongly treated (e.g. frozen and thawed too often) the entire antibody may be destroyed. Polyclonal antisera, on the other hand, consist of a mixed population of antibodies which will be immunoreactive with a variety of epitopes on the antigen molecule, so the chances of at least some of these areas surviving fixation in a suitable state to react with the antibody are greater than for a monoclonal antibody. Polyclonal antibodies may also be more resistant to disaster, and even if some of the weaker members of the population are destroyed there may be enough left to react satisfactorily.

Dilution

Polyclonal antibodies are usually cheaper than monoclonal antibodies because they are less labour-intensive to produce and, in addition, they can usually be highly diluted if a sensitive detection method is used. Monoclonal antibodies in the form of tissue culture medium can usually be diluted 10–100 times and at least 1000 if prepared as ascites fluid. A useful immunoglobulin concentration to aim at is 10 μg/ml. Every antibody will have to be tested in the system in which it is to be used.

The purpose of dilution is not only for economy, but also, in the case of polyclonal antisera, to reduce to an insignificant minimum the concentration of unwanted antibodies (e.g. to the carrier protein used in immunization and to naturally encountered antigens) that might react with the tissue to be stained. In addition, the unlabelled antibody enzyme anti-enzyme methods (e.g. peroxidase antiperoxidase) will not work efficiently if the primary antibody is too concentrated with respect to the second antibody (see below).

Storage

Antibodies that are not destroyed by freezing can be snap-frozen undiluted, or at dilutions of 1:2–1:10, and stored at –20°C in aliquots suitable for subsequent dilution to a working concentration. Mixing the antibody 1:1 with glycerol will allow it to be stored at −20°C without freezing, which may be an advantage.

Antibodies that are destroyed by freezing (some monoclonal antibodies) can be stored at 4°C in tissue culture medium or a buffer solution containing some extra protein, such as 0.1% bovine serum albumin, and 0.1% sodium azide or merthiolate to prevent moulds and bacterial contamination. The added protein competes with the antibody for non-specific attachment sites on the walls of the storage vessel. Normal serum (from the species providing the second antibody of the method) at a concentration of 1% may be used instead of pure albumin.

It is sometimes convenient to store antibodies at their working dilution in buffer (phosphate- or Tris-buffered saline, pH 7.0–7.6) containing azide and albumin. The length of storage time will depend on the antibody and must be checked for each use. Some antibodies will remain fully active at high dilutions for several years, but this is not necessarily the general rule.

LABELS

Immunostaining reactions are made visible by the incorporation of a suitable label into the reaction. Fluorescent molecules, enzymes, biotin and colloidal gold are most commonly used. Primary antibodies may be labelled for the direct method (see below) but it is more usual to label the antibodies of the second stage. In the unlabelled antibody-enzyme methods the label is not chemically conjugated to the antibody but acts as an antigen, bound in a complex to the antibody.

Fluorescent labels

The most popular are fluorescein isothiocyanate, which fluoresces green in ultraviolet light (maximum emission at excitation wavelength of 490 nm) and tetra-methyl rhodamine isothiocyanate or Texas red, which fluoresce red in green light (maximum emission at excitation wavelength of 546 nm), Texas red giving a more stable fluorescence than rhodamine. Phycoerythrin gives a yellowish red fluorescence which can be viewed with the same filter combination as fluorescein.

Fluorescence methods have some advantages. They are rapid to perform because they require fewer steps than the enzyme methods, tissue fixation is less important because the immunoreaction is seen as brightly fluorescent against a dark background so cryostat sections are suitable. In addition, if double labelling is carried out using two antibodies with differently fluorescing labels, the different sites of reaction are seen separately and can be unequivocally distinguished by switching filters on the microscope. The disadvantages are that fluorescence fades, sometimes even during examination, it is difficult to see the background detail in the tissue and permanent preparations cannot be made because the fluorescence is destroyed by solvents and aqueous mountants must be used. Formalin-fixed sections often have high 'background' fluorescence which can sometimes be masked by a fluorescent counterstain such as Pontamine Sky Blue (Cowen et al 1985).

Enzyme labels

Horseradish peroxidase is the most widely used enzyme label. Alkaline phosphatase, glucose oxidase and β-D-galactosidase also have their particular applications. An enzyme label must produce a coloured end product in the final step of the immunostaining procedure and attention must be given to the correct pH, and substrate and chromogen concentrations for the incubation. Advantages over fluorescence methods are the production of permanent preparations that can be viewed with an ordinary light microscope and the possibility of counterstaining the tissue with histological methods to examine the antigenic site in context.

Blocking endogenous enzyme

Peroxidase. If there is active peroxidase in the tissue to be immunostained, it must be blocked before the peroxidase-labelled antibody is applied in order to avoid confusion as to whether the end-product is due to endogenous or antibody-associated enzyme. This is of particular importance if smears of fresh material or frozen sections are being used (e.g. peroxidase in myelocytes and macrophages), but even in tissue processed to paraffin the 'peroxidase' (catalase) in the red blood cells remains active and can produce an intrusive reaction. Blocking is usually carried out before application of the primary antibody, but can be done at a later stage as long as it is before the peroxidase-labelled reagent is used. For paraffin sections the blocking agent is usually hydrogen peroxide, which must be reasonably fresh. Frozen sections may be detached from slides by this, but less violent blocking is provided by substituting methanol, itself a partial inhibitor of peroxidase, for the water. Even so, a solution of this strength may injure the immunoreactivity of some antigens and a milder but effective method of blocking endogenous peroxidase is to use a combination of sodium azide and nascent hydrogen peroxide in very low concentration produced by the action of glucose oxidase on glucose (Andrew & Jasani 1987). I have found this a very useful method of blocking endogenous peroxidase in cryostat sections while not damaging their structure and preserving immunoreactivity of most antigens.

If endogenous peroxidase in paraffin sections is particularly strong and cannot be removed by the above means, stronger hydrogen peroxide can be used for a short time, followed by periodic acid (Heyderman 1979). These two steps are followed by treatment with borohydride solution which converts any aldehydes in the tissue to less 'sticky' alcohol groups. This treatment results in very clean preparations but may prevent a few antigens from reacting (e.g. leukocyte common antigen).

Alkaline phosphatase. Endogenous alkaline phosphatase is not a problem in paraffin sections since its activity except for the intestinal isoenzyme is destroyed by processing. In cultures, blood smears, frozen sections, etc., it can be inhibited at the enzyme development stage by adding levamisole to the incubating medium. Levamisole (1mM) inhibits all alkaline phosphatase isoenzymes except intestinal alkaline phosphatase, which is used to label immune reagents. Alkaline phosphatase methods are thus not suitable for use on frozen or paraffin sections of intestinal material. A short exposure to 20% acetic acid will inhibit the intestinal isoenzyme as well as the others, but is rather destructive to the tissue and some antigens (Ponder & Wilkinson 1981).

Glucose oxidase. This has the advantage of being a plant enzyme, absent from animal tissues, and thus no precautions are necessary.

β-D-galactosidase. The enzyme used for labelling is derived from bacteria and acts optimally at a different pH from the mammalian variety. This is not a suitable label for identifying bacterial antigens.

End-product

Most of these enzyme labels can be developed in different colours. Multiple staining can be carried out with combinations of antibodies labelled with different enzymes (see below).

Peroxidase. The most useful reaction product is that derived from the oxidative polymerization of 3, 3'-diaminobenzidine (DAB). The tetrahydrochloride is used in the reaction — the free base is not very soluble. The action of antibody-bound peroxidase on hydrogen peroxide supplies the oxidizing power, and the end-product deposited on the site of the antigen-antibody reaction is dark brown and insoluble, providing good contrast with the unreactive parts of the preparation. A light haematoxylin counterstain usually provides enough background detail and the preparations can be dehydrated and mounted in a permanent mountant. Although it had been suggested that DAB is carcinogenic, the compound has been taken off the list of carcinogens in the USA (Weisburger et al 1978). The possible hazard of frequent weighing of small amounts of the powder is removed by storing DAB as a concentrated solution in frozen aliquots for use in the development step (Pelliniemi et al 1980).

Alternative chromogens are 4-chloro-l-naphthol, which gives a blue-grey colour, and 3-amino-9-ethylcarbazole, also a potential carcinogen (Tubbs & Sheibani 1982), which gives a reddish-brown colour. Both these alternatives give end-products that are soluble in alcohol and the preparations must be mounted in aqueous mountants.

Alkaline phosphatase. This enzyme can be developed to give a bright blue colour with Naphthol AS-MX phosphate and Fast Blue BB (or red with Fast Red TR). These products are alcohol-soluble. Another bright red product resulting from incubation with Naphthol AS-TR phosphate and hexazotized New Fuchsin is somewhat less soluble, but mounting in an aqueous medium is probably wise. An insoluble blue-brown product of good contrast is produced from 5-bromo-4-chloro-3-indolyl phosphate and Nitro Blue tetrazolium (Nitro BT) (De Jong et al 1985).

Glucose oxidase. This enzyme with Nitro BT and glucose produces a dark blue deposit which can be permanently mounted (Suffin et al 1979).

β-D-galactosidase. Development with an indigogenic method gives a turquoise blue, insoluble end product (Bondi et al 1982).

Biotin

This non-enzymic label is finding increasing use in immunocytochemistry, in combination with avidin. Many molecules of biotin can be conjugated to the Fc portion of one immunoglobulin molecule and, since biotin combines very tightly with avidin, avidin labelled with fluorescent or enzyme labels can be used to reveal the site of antigen-antibody reaction. One avidin molecule can combine with four biotin molecules and a labelled avidin-biotin complex (ABC) can also be made (see below).

Colloidal metals

Colloidal particles of silver and gold, particularly the latter, have become well established labels for antibodies in immunocytochemical reactions. Colloidal gold markers were developed for electron microscopical immunocytochemistry, and can be produced in a range of sizes so that multiple immunoreactions can be carried out using antibodies labelled with gold particles of different diameter (De Mey 1986). Light microscopical immunostaining using colloidal gold labels has evolved as a spin-off from the original purpose.

Colloidal gold alone as a label produces a pink colour which is not usually strong but subsequent intensification with silver produces a dense black end-product, making this a particularly sensitive method (Holgate et al 1983a, b, Springall et al 1984). The gold label can be viewed (with or without silver intensification) with dark field microscopy, or with epipolarization combined with transmitted light illumination to allow combination of the back-scattered light from the gold label with ordinary staining on the same preparation (De Waele et al 1988).

METHODS

Little apparatus is needed. Sections on slides are placed flat on a rack in a humid chamber which can be as simple as a large Petri dish containing two wooden applicator sticks and some damp cottonwool. Antibody solutions are applied with Pasteur pipettes as drops to cover the sections.

Direct method

The single layer direct method using a labelled primary antibody is the simplest to use (Fig. 3.1). Fluorescent labels are the most usual — the method is probably not sensitive enough to give good contrast with enzyme labels. At electron microscope level a gold-labelled antibody can be used.

Applications

This method is particularly useful where a rapid result is required, provided that enough antigen is present to compensate for the low sensitivity. For example, fluorescein-labelled anti-human immunoglobulins and complement can be applied to cryostat sections (acetone-fixed) of kidney and skin biopsies. Possible non-specific binding sites (Fc receptors, hydrophobic and electrostatic binding sites) are blocked with a layer of normal rabbit serum before application of the primary (rabbit) antibody. Direct methods are also useful for double labelling.

Indirect method

The indirect method (Fig. 3.2) appears to be about 10 times more sensitive than the direct method. Non-specific binding sites in the tissue are blocked before application of the primary antibody with a layer of normal serum from the species providing the second, labelled antibody. The increase in sensitivity may be due to the fact that the initially bound primary antibody

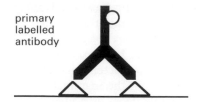

Fig. 3.1 Direct method

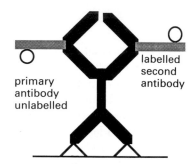

Fig. 3.2 Indirect method

molecules each have at least two binding sites for the conjugated second layer anti-immunoglobulin, but Sternberger (1986) suggests that the increase in sensitivity has more to do with the hyperimmunity of the second layer antisera.

As well as the increase in sensitivity this method has the advantage that any number of unlabelled primary antibodies raised in the same species can be detected with just one labelled antibody to the immunoglobulins of that species. In general, broad spectrum anti-immunoglobulins are used, reactive with IgG, IgA and IgM, so that the immunoglobulin type of the primary does not matter. In some cases, however, it is advantageous to use a more restricted second antibody, e.g. to the IgG subtype of a primary monoclonal antibody. This provides a more efficient (less wasteful) labelled antibody and could eliminate some background staining, such as might occur in detection of a human circulating autoantibody using the patient's serum on human tissue.

Three-layer techniques

Unlabelled antibody enzyme-antienzyme methods (Figs. 3.3 and 3.4)

A stable, cyclic peroxidase anti-peroxidase (PAP) complex was created (Sternberger et al 1970) by

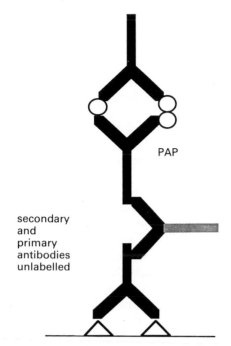

Fig. 3.3 Peroxidase anti-peroxidase (PAP) method

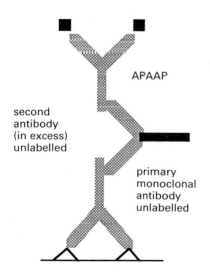

Fig. 3.4 Alkaline phosphatase anti-alkaline phosphatase (APAAP) method.

immunizing a rabbit with horseradish peroxidase and reacting the resulting antibody with its own enzyme antigen. The complex consists of two antiperoxidase immunoglobulin molecules bound by three peroxidase molecules. This complex is used as a third layer following a first layer of un-

labelled rabbit primary antibody and a second layer, also unlabelled, of antirabbit immunoglobulin. Since the PAP complex is rabbit immunoglobulin it acts as an antigen and combines with the unbound arms of the second layer antibody. The second layer antibody must always be applied in excess so that some binding sites are free to combine with the PAP complex. Nonspecific tissue binding sites are blocked at the beginning of the method with normal serum from the species providing the second antibody. As this will not be antirabbit immunoglobulin it will not react with the PAP complex of the third layer. The increase in sensitivity, again about 10 times more than the indirect method on a primary antibody dilution basis, is produced not only by the increased number of peroxidase molecules built up at the site of primary antibody reaction, but also because neither the enzyme nor any of the antibodies are chemically conjugated and therefore retain their full activity. In addition, the primary antibody can (and must) be highly diluted, thus reducing the possibility of unwanted reactions due to subpopulations of antibodies. The highlights of this method are thus an intense reaction combined with a low background.

Rabbit, goat and mouse PAP complexes are commercially available (the PAP complex must, of course, be of the same species as the primary antibody), as is mouse alkaline phosphatase anti-alkaline phosphatase (APAAP) and glucose oxidase anti-glucose oxidase (GAG) complexes from several species.

Avidin-Biotin techniques (Figs 3.5 and 3.6)

The primary antibody is unlabelled and again, very dilute. The second antibody is biotinylated with a large number of biotin molecules (conjugation with these very small molecules does not adversely affect its immunoreactivity) and the third reagent is either labelled avidin (or streptavidin), or a complex of avidin (streptavidin) with labelled biotin. One advantage of the method is that any species can be used to provide the biotinylated second layer and the third layer is 'universal' and not dependent on the species of the primary antibody. If the primary antibody is biotinylated a two-layer technique can be used.

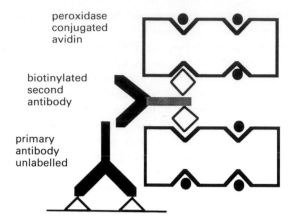

Fig. 3.5   Labelled avidin method.

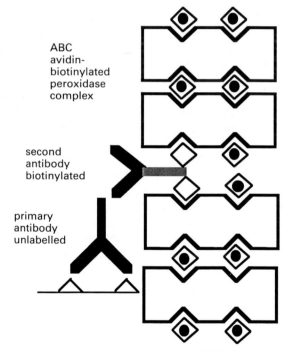

Fig. 3.6   Avidin-labelled biotin complex (ABC) method.

Some claim that the avidin-biotin complex (ABC) methods are even more sensitive than the enzyme antienzyme methods because of the very large amount of label that can be built up at the site of reaction. However, this may be true only when a very small amount of antigen is present, since the large size of the ABC complex will limit the number of complexes that can be present (Sternberger & Sternberger 1986).

Avidin can be conjugated with a fluorescent label, an enzyme or colloidal gold or, if a complex is required, bound to a biotinylated label. In preparation of the complex excess avidin in the mixture ensures that some biotin-binding sites will be free to combine with the biotin on the second antibody.

The disadvantages of avidin are that it is a glycoprotein and may bind to lectins in the tissue via its carbohydrate groups, its isoelectric point is 10, which may cause it to bind to charged sites in the tissue, and free biotin in the tissues may cause non-specific binding. The glycoprotein problem can be avoided by using the bacterial protein, streptavidin, instead of avidin. Streptavidin has an isoelectric point of 7, nearer that of tissue and thus less prone to attachment through charged binding sites. If streptavidin is not available, using a pH 9 buffer will help to prevent non-specific attachment (e.g. to mast cells) and will not affect the immunoreaction (Bussolati & Giuliotta 1983). Native biotin may be blocked before the reaction is carried out by incubation with unlabelled avidin followed by unlabelled biotin to fill any unoccupied sites on the avidin (Wood & Warnke 1981). These two additional steps are not always necessary.

Immunogold staining with silver intensification

As mentioned above, colloidal gold particles can be used to label antibodies (De Mey 1986, De Mey et al 1986). An immunogold reaction can be seen in the light microscope (deep pink to red) if the concentration of the immunogold reagent is high enough or if several layers of immunogold reagents are built up. Except in some cases of double immunostaining, where the pink gives a pleasing contrast (Gu et al 1981), this is not a very useful reaction because of its lack of intensity. The immunogold reaction can, however, be made extremely sensitive by use of silver intensification of the gold particles (Holgate et al 1983a and b).

The principle of the intensification is that in the presence of a reducing agent (hydroquinone) metallic gold acts as a catalyst to reduce silver ions from a solution of a silver salt (silver lactate or silver acetate) to metallic silver, which is deposited as a shell round the gold particles and subsequently itself acts as a catalyst for further reduction of the silver salt. An increasingly intense deposit is thus built up during the course of the reaction.

The advantages of the immunogold-silver staining (IGSS) method are several: the strong staining makes it excellent for low-power scanning of preparations and its high sensitivity means that the primary antibody can be very highly diluted, providing both economy and a clean immunostain; there is no need to block endogenous enzyme, the method is relatively quick to perform and no toxic agents are included. Paraffin or cryostat sections can be used and the method is also very suitable for staining smears or cells in suspension (De Waele et al 1988); the silver deposit is stable and will allow counterstaining by any histological/histochemical method. The disadvantages are that it requires some practice and dedication to produce reliable results without a background of silver deposition. As in all methods, the quality of the reagents is very important. Small sized (5 nm diameter) gold particles are recommended for the best penetration of the preparation. A standard method and a recent modification of the development step (Hacker et al 1988) are given in Appendix 6. The high concentrations of salt and albumin in the buffer are designed to reduce background staining and the optional inclusion of gum arabic in the silver solution slows the reaction for better control.

Applications

The IGSS method is useful in pathology where a particularly intense reaction is required for visualization of a small amount of antigen. In addition to the demonstration of cell surface antigens of lymphocytes, the identification of virus antigens in sections or cytological preparations can be facilitated by the method. Cytological and haematological preparations are well suited to the IGSS method for the identification of surface antigens because the absence of supporting tissue reduces the problems of background staining and because a large area of the surface of a whole cell preparation is available for silver deposition rather than just a section through the membrane. Blood cells can be stained in suspension before or after

fixation and deposited on a slide prior to silver intensification. The intense black of the IGSS reaction product is well seen against conventional May-Grünwald-Giemsa staining. In addition, if the preparation is viewed by epipolarized light, the reaction is even more intense. The gold/silver particles back-scatter the incident light and appear to shine brightly while simultaneous illumination with transmitted light can show the conventionally stained cell structure (De Waele et al 1986).

In paraffin sections a minimal quantity of peptide or neurotransmitter is preserved for immunostaining in fine nerve fibres. The IGSS method allows these fibres to be seen even at low power. Where an immunoreactive antigen is abundant in a cell section, the main advantage of the IGSS method is aesthetic — providing a very intense reaction and using a higher dilution of primary antibody than a three-layer technique. It can, however, provide enhanced immunostaining in cases where a peroxidase stain produces doubtful results due to poor preservation or low quantity of the antigen, and makes it easy to pick out a few positive cells in a mainly negative preparation.

Other ways of intensifying an immunostaining reaction

Repetition of layers

Immunostaining reactions can often be made to provide a stronger (more intense) reaction by repetition of the second and third layers of a three-layer PAP (or APAAP) technique (not possible with ABC because all binding sites for biotin on the third layer ABC are occupied) or in a two-layer, indirect method by applying PAP(APAAP) (if the second layer is concentrated enough) or peroxidase-(AP-) conjugated antisecond layer immunoglobulin as a third layer (Vacca 1982, Mason 1985).

Intensifying the end-product with heavy metals

The end-product of immunoperoxidase methods developed with DAB can be enhanced by including imidazole in the incubating medium (Straus 1982) or a heavy metal salt such as cobalt chloride or nickel sulphate (Hsu & Soban 1982, Shu et al 1988). Nickel sulphate is particularly effective in producing a dark blue-black colour.

Scopsi & Larsson (1986) compared available methods of developing and intensifying peroxidase immunoreactions, using a series of drops on paper containing antigen in graded quantities. They found the most sensitive methods (demonstrating smallest amounts of antigen) to be the addition of nickel sulphate to the DAB solution or a postsilvering of the DAB reaction product. They suggested that a methenamine silver solution (Rodriguez et al 1984) was very effective. Since this solution is part of a standard periodic acid/silver methenamine stain used routinely for revealing basement membranes in glomeruli in kidney biopsies it is readily available in histology laboratories. It is a one-step intensification and we have found it is an extremely simple and useful method of intensifying a weak DAB reaction (Figs. 3.7 and 3.8). It is used without the periodate oxidation step (see Appendix 6). The advantage of this postreaction blackening (as with other ones) is that no decision need be taken until the reaction has been completed, and a section can be left in distilled water indefinitely until the silver methenamine solution is prepared routinely. Coverslips can be removed from pre-

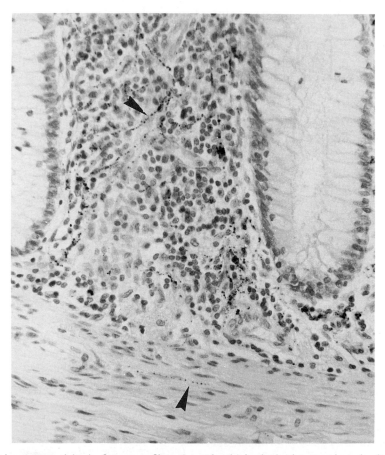

Fig. 3.7 Substance P immunoreactivity in fine nerve fibres (arrowheads) in the lamina propria and submucosa of human colon. The tissue was fixed in phosphate-buffered neutral formalin and embedded in paraffin. 3 μm section immunostained overnight with rabbit anti-substance P diluted 1:5000 followed by biotinylated swine anti-rabbit Ig (Dako, 1:500) for 30 min and peroxidase conjugated avidin (Dako, 1:800) for 30 min. Development of the peroxidase with DAB and H_2O_2 was followed by immersion in hexamine silver solution at 60°C for 10 min. Nuclei counterstained with haematoxylin. ×307

viously stained sections with a weak reaction and the reaction intensified (it is best to remove the nuclear counterstain first). We have shown (Peacock et al 1990) that lymphocyte surface membrane immunoglobulins can be as well stained in paraffin sections by this means as by the IGSS method, even if the tissue has been fixed in neutral-buffered formalin.

In silver intensification methods, reticulin fibres and other argentaffin structures may take up the silver. This problem can be prevented by treating the section with copper sulphate followed by hydrogen peroxide (Gallyas & Wolff 1986).

Alternatively the DAB reaction product can be intensified by postosmication followed by potassium ferrocyanide treatment (Lascano & Berria 1988). If any non-specific background labelling is present it will be enhanced in all these techniques.

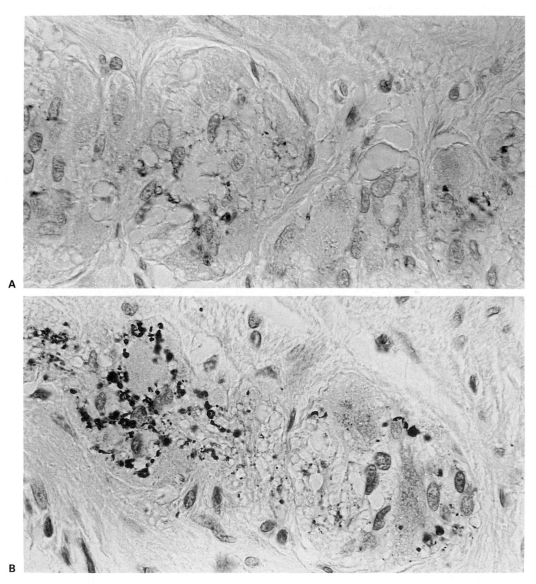

Fig. 3.8 Vasoactive intestinal peptide immunoreactivity in the myenteric plexus of human colon. Method as for Figure 3.7. Primary antibody diluted 1:20 000. **A** Without and **B** with postsilvering of the DAB reaction product. × 768

Protease digestion

An extremely useful method of improving immunostaining on formalin-fixed, paraffin-embedded material, even to the extent of showing an antigen where none was revealed by other means, is to treat the sections with a protease such as trypsin before immunostaining (Huang et al 1976). The enzyme breaks some of the protein cross-linkages introduced by formalin fixation, revealing antigenic sites that would otherwise be hidden. Whether or not protease pretreatment is necessary depends on the antigen, the fixation and the antibody. No absolute rules can be laid down — each system must be tested. However, intracellular immunoglobulins in neutral buffered formalin-fixed paraffin sections are usually more intensely stained after trypsin treatment and it may be essential for revealing those located on the cell surface. Trypsin is essential for showing factor VIII-related antigen (von Willebrand factor) in endothelial cells in paraffin sections. The intermediate filaments cytokeratin and desmin are usually better stained after trypsin digestion, vimentin is unaffected and the neurofilaments and glial fibrillary acidic protein are variable in their response. Trypsinization may destroy some of the antigen that survived the processing while revealing hidden epitopes, so it is usually advisable to stain non-trypsinized and trypsinized samples in parallel, and it is also necessary to establish optimal conditions for digestion. The concentration and length of time for optimal trypsinization may vary from specimen to specimen. Ideally a range of incubation times should be used on every occasion, but in practice this is too demanding and establishing an average optimum for every batch of enzyme is usually adequate. Over-trypsinization can result in section destruction.

Other proteases may be equally or more effective depending on what groups they act on. Protease type XXIV (Sigma) is superior to trypsin in revealing glomerular interstitial immunoglobulin in paraffin sections of renal biopsies. Neuraminidase is useful for revealing the Leu M 1 antigen in granulocytes and Reed-Sternberg cells.

Proteolytic enzyme digestion is never required for fresh frozen sections.

NON-SPECIFIC BACKGROUND STAINING

Staining of tissue that does not contain the antigen under investigation is known as background staining.

Background staining may be due to a) unwanted antibodies in the primary antiserum combining with unsuspected sites in the tissue, b) cross-reaction of second antibodies (anti-immunoglobulin) with native immunoglobulin in the tissue being stained, c) 'sticky' sites in the tissue attracting immunoglobulins, e.g. by hydrophobic/electrostatic reactions, d) necrotic tissue (which seems to attract antibodies non-specifically), e) Fc receptors attracting immunoglobulins (particularly in frozen sections or fresh cell smears or suspensions; Fc receptors are destroyed by paraffin processing), f) avidin-binding sites in the avidin-biotin methods, g) endogenous enzyme reactivity or h) crossreaction of a primary antibody with molecules related to the antigen.

Standard procedures for reducing background staining are incorporated into all immunostaining methods. Blocking of endogenous enzymes, biotin and aldehyde groups has already been dealt with. Preliminary blocking of non-specific attachment of immunoglobulins is carried out by treating the preparation with non-immune serum from the species providing the second antibody (or, in a direct method, the primary antibody). Protein from this solution will attach to the electrostatic/hydrophobic binding sites and (in cryostat sections) native immunoglobulin will bind to Fc receptors. These sites are thus occupied and as the serum is not rinsed off, merely drained from the preparation, the next layer (primary antibody) is prevented from attaching. If necessary (not usually) the blocking step can be repeated before application of the second antibody, which will find non-specific and Fc-receptor sites occupied by molecules of its own species with which it will not react. The enzyme antienzyme or ABC complex of a third layer will only react with second layer antibodies bound to the primary antibody. In paraffin sections, without Fc receptors, the blocking can be carried out with a strong protein solution such as egg albumin (not for biotin-avidin methods) or dried milk powder. This avoids the necessity of having different blocking sera available for the different species second layer antibodies.

Crossreactions of the second layer anti-immunoglobulin with tissue immunoglobulins can be prevented by preabsorption of the second layer antibody with immunoglobulins from the species to be stained. Many commercially available antibodies have been affinity purified to react only with one species' immunoglobulin, but in practice addition to the diluted second layer antibody of 5% of serum (decomplemented by heating to 56°C for 30 min) from the species to be stained is usually adequate. Crossreactive immunoglobulin binding can be tested by staining cryostat sections of spleen from the species to be stained with the second layer anti-immunoglobulin before and after absorption.

Crossreactivity of primary antibodies with related molecules is the most intractable type of 'non-specific' or rather, unwanted reaction. If possible, antibodies to unshared portions of the molecule should be used. Sometimes crossreactions are present but unsuspected, and the worker should be aware of this possibility.

Unwanted reactions from subpopulations of antibodies in the primary antiserum can often be reduced or eliminated by diluting the primary antibody. Here, a three-layer technique such as the PAP or ABC method has the advantage that the primary antibody must be very highly diluted.

CONTROLS

Histologists are used to the practice of carrying a positive control preparation through with every batch of staining and this is a wise precaution to take for immunostaining. Many steps are involved in one immunostaining procedure and each step must be correct for the reaction to be optimal. Without a positive control a negative result is meaningless. However, a negative test preparation in the presence of a positive control does not necessarily mean that the antigen sought is totally lacking. It may be that the quantity present is too small to be picked up by the method used, or that the epitope(s) recognized by the antibody have been destroyed or hidden by the processing of the tissue. The substance looked for may be so actively secreted that none is stored in the tissue, or it may be present in a precursor form not recognized by the antibody used. Thus a negative result need not always be taken as conclusive.

A negative control should also be carried out on each tissue being stained. Substituting for the primary antibody normal serum from the species providing the primary (polyclonal) antiserum, at the same dilution, or buffer or an inappropriate antibody in the case of a monoclonal antibody, allows the investigator to assess the level of background staining inherent in the method and particular to that specimen, and allow for it in looking at the test result. Of course, the negative control should be blank, but some tissues may have a high proportion of 'sticky' areas which might be thought to be positively stained, were it not for the similar staining pattern of the negative controls.

Doubts about the specificity of the primary antibody can be partially resolved by absorbing it with its specific antigen before use, which should remove the specific staining. Absorption with an unrelated substance and related but different substances should have no effect.

TIMING

Each laboratory will work out its own immunostaining schedules and these will vary depending on the urgency of the result. The time for each layer of the method to react depends partly on the dilution of the antibodies. It is possible to carry out the methods with 30 min or shorter incubation in the primary antibody, but the concentration will have to be relatively high; increasing the incubation time in the primary antibody to several hours or overnight allows at least a 10-fold increase in dilution which is useful (with polyclonal antibodies) in diluting out unwanted reactants, and in any case reduces the cost of the test. We have found it most convenient to incubate sections overnight with the primary antibodies. This means that immunostaining can be begun in the afternoon and finished by mid-morning the following day. Timing of each layer is not important (unless a controlled experiment is in progress). Antigen-antibody reactions are reversible and dilutions should be adjusted so that each reaction in the schedule reaches equilibrium in the time allowed, excess of applied antibody over antigen always being present. Prolonging the time in a standard dilution of the reagent will not then affect the result.

Kits versus individual antibodies

The advantage of the many commercially available kits is that the manufacturer has worked out suitable dilutions and timing for each step of the reaction, and the operator need only apply the right drop from the right bottle at the right time. However, this is probably an expensive and inflexible way of doing things, except for a laboratory which carries out immunostaining only infrequently. It is more satisfactory to have a battery of chosen primary antibodies — the shelf-life is often several years — and a few suitable second and third layer reagents, and find the dilutions and timing that suit your circumstances.

QUANTIFICATION

It is something of a misconception that strong immunostaining means an abundance of antigen and weak immunostaining a little. In a pathology laboratory, the time interval between excision of a tissue and placing it in fixative, the length and strength of fixation and the thickness of sections are too variable for this to be true. The most that can be said is that the preparation is positive or negative, although it may be that a larger area or greater number of cells is positive in one specimen for a particular antibody, compared with another treated simultaneously.

MULTIPLE LABELLING

Occasionally it is useful to identify two or more antigens in a single area and simple comparison of thin (2 μm) serial sections stained with different antibodies is probably the most reliable method. However, double staining is helpful in determining the immunoglobulin light chain clonality of a lymphoma. Visualization of kappa- and lambda-positive cells together gives a more immediate and less laborious way of deciding whether the tumour is of a single cell type or not.

Co-localization of regulatory peptides, the identification of mixed tumours, the presence of small pockets of tumour tissue in otherwise normal areas, or the need to examine the relationships between different types of cell are other reasons for using multiple staining techniques.

Several ways of carrying out multiple immunostaining have been devised. The double direct immunofluorescence method is reliable, with fluorescein and rhodamine using different filter combinations. Dual photographic exposure of a single microscopical field allows both reactions to be seen together as green and red, with areas of co-localization as yellow. Indirect methods can also be used with primary monoclonal antibodies of different IgG subclass and IgG class-specific second antibodies (Janossy et al 1986).

In the double immunoenzymatic method (Mason & Sammons 1978, Mason et al 1983) mouse and rabbit primary antibodies to different antigens are applied simultaneously and detected with a mixture of non-crossreacting, species-specific antibodies to mouse and rabbit immunoglobulins labelled with alkaline phosphatase and peroxidase. The enzymes are developed separately, e.g. alkaline phosphatase in blue (with Naphthol AS-MX phosphate and Fast Blue BB) and peroxidase in brown with H_2O_2 and DAB. The blue and brown reaction products can be seen separately, or as a purplish-grey colour if the two antigens are located in the same tissue structure. This is probably the most convenient, reliable and easily assessed of the double immunostaining methods although some subjectivity enters with the decision as to when the enzyme reactions should be stopped for a suitable balance of colours. The brown peroxidase reaction should not be allowed to go beyond a fairly light colour.

If two antigens are known to be present in different structures, the procedure is easier. A sequential immunoreaction can be carried out, regardless of species, provided that the first antigen is developed to give a very dark colour. This could be achieved with peroxidase developed with DAB and nickel sulphate, or with the postsilvered DAB or the IGSS method. The second antigen is then developed in any contrasting colour, and even if there is a build-up of crossreaction on the site of the first reaction, the intensity of the stain in the first site will be so great that the reaction product from the second reaction will not be noticeable and the contrast with the true colour of the second reaction will be easily seen (Sternberger & Joseph 1979, De Mey et al 1986).

HYBRIDIZATION HISTOCHEMISTRY

Immunocytochemical localization of any antigen can only demonstrate its presence at the moment of tissue fixation, and at best the relative quantity of the substance that may be there. It cannot indicate, except in conjunction with other methods such as electron microscopy, and then only by inference, whether the cell containing the antigen is in an active or storage phase or whether the antigen is made in the cell or merely taken up by it. A method that overcomes this problem, by identifying the nuclear gene (DNA) or cytoplasmic messenger RNA for the substance, is hybridization histochemistry. This technique, which has recently come to the fore after several years of modification, is still in its infancy as far as pathological diagnosis is concerned, but it is advancing rapidly and may even supersede immunocytochemistry as the methods become better established and the variety of available probes increases. A potential diagnostic use is the identification of viral DNAs which withstand paraffin embedding well (Burns et al 1987)

The method makes use of the power of a specific labelled (usually radiolabelled) complementary DNA or RNA probe to combine with the corresponding DNA or mRNA in a tissue section or other preparation. Thus the synthetic capability (DNA) of the cell is shown and the actual synthesis of the gene product is indicated by identification of mRNA, even when the cell is secreting its product so rapidly that none is available for demonstration by immunostaining. Comparative quantification of labelled mRNA/DNA can also be carried out by image analysis of the final autoradiographs, and gives an idea of the increase/decrease in production of the substance investigated. In combination with immunocytochemistry, the dynamic aspects of product synthesis and release can be assessed (Hoefler et al 1986, Hamid et al 1987).

Probes are labelled single strands of cDNA or antisense mRNA (cRNA). If the DNA of the genome is to be localized, the tissue must first be treated with RNase then denatured so that the two strands of nuclear DNA separate to allow the labelled cDNA to bind. For localization of RNA this denaturation is not necessary, since RNA is single-stranded and the native double-stranded DNA is not detected. Labels are incorporated into the nucleotides of the probes during synthesis. Radioactive labels are commonly used and are usually ^{32}P, ^{35}S or ^{3}H, revealed by autoradiography. ^{3}H provides high resolution but has a long autoradiographic exposure time (1–3 months); ^{35}S has reasonable resolution and a shorter exposure; ^{32}P is less well resolved but has a rapid exposure.

Biotinylated probes, revealed by labelled avidin, are less hazardous and provide a much shorter but less sensitive method than the radioactive labels but biotin's relatively large size produces problems with penetration. Enzymes such as peroxidase and alkaline phosphatase have also been used as labels by direct conjugation to DNA, and fluorescent labels have also been incorporated in probes with some success.

Conditions for successful in situ hybridization are similar to those for immunocytochemistry, but even more stringent. Thus the nucleic acids must be preserved (fixed) in a form suitable for hybridization, the tissue must be permeabilized so that the probes can penetrate easily and are sometimes treated with proteases to reveal target sites. Care must be taken not to contaminate the preparations with RNAase (e.g. from the skin — gloves must be worn). Radioactive probes, treated with the usual precautions, must be freshly made and used, since the radioactivity will decay. This precaution does not apply to biotinylated probes.

Careful controls must be included, for example, irrelevant and 'sense' RNA labelled in the same way as the antisense RNA in order to assess nonspecific background binding, including material known to contain and not to contain the nucleic acid sought, and digesting the nucleic acid from the preparation before applying the probe. As with immunocytochemistry, unsuspected crossreaction is difficult to detect and eliminate.

Fixation and tissue preparation methods cannot be standardized. As for immunocytochemistry, they seem to depend on the tissue, the nucleic acid species and the probe. Postfixed cryostat sections and frozen or Vibratome sections of tissue prefixed in paraformaldehyde are often suitable, but freeze-dried, paraformaldehyde vapour-fixed and conventional paraffin sections may also be used in some cases. The material should be fixed as soon as possible after removal to avoid autolytic degradation of nucleic acids.

The upsurge in genetic engineering means that specific probes can be readily manufactured so that in situ hybridization, at present mainly a research tool, will soon become part of the battery of routine histochemical techniques.

REFERENCES

Andrew S, Jasani B 1987 An improved method for the inhibition of endogenous peroxidase non-deleterious to lymphocyte surface markers. Application to immunoperoxidase studies on eosinophil-rich tissue preparations. Histochemical Journal 19: 426–430

Bishop A E, Polak J M, Bloom S R, Pearse A G E 1978 A new universal technique for the immunocytochemical localisation of peptidergic innervation. Journal of Endocrinology 77: 25P–26P

Bondi A, Chieregatti G, Eusebi V, Fulcheri E, Bussolati G 1982 The use of β-galactosidase as a tracer in immunohistochemistry. Histochemistry 76: 153–158

Burns J, Graham A K, Frank C, Fleming K A, Evans M F, McGee J O'D 1987 Detection of low copy human papilloma virus DNA and mRNA in routine paraffin sections of cervix by non-isotopic in situ hybridisation. Journal of Clinical Pathology 40: 858–864

Bussolati G, Gugliotta P 1983 Nonspecific staining of mast cells by avidin-biotin-peroxidase complexes (ABC). Journal of Histochemistry and Cytochemistry 31: 1419–1421

Cowen T, Haven A J, Burnstock B 1985 Pontamine sky blue, a counterstain for background autofluorescence in fluorescence and immunofluorescence histochemistry. Histochemistry 82: 205–208

De Jong A S H, van Kessel-van Vark M, Raap A K 1985 Sensitivity of various visualization methods for peroxidase and alkaline phosphatase activity in immunoenzyme histochemistry. Histochemical Journal 17: 1119–1130

De Mey J 1986 The preparation and use of gold probes. In: Polak J M, Van Noorden S (eds) Immunocytochemistry, modern methods and applications. Wright, Bristol, p 115–145

De Mey J, Hacker G W, De Waele M, Springall D R 1986 Gold probes in light microscopy. In: Polak J M, Van Noorden S (eds) Immunocytochemistry, modern methods and applications. Wright, Bristol, p 71–88

De Waele M, De Mey J, Renmans W, Labeur C, Reynaett P, Van Camp B 1986 An immunogold-silver staining method for the detection of cell surface antigens in light microscopy. Journal of Histochemistry and Cytochemistry 34: 935–939

De Waele M, Renmans W, Segers E, Jochmans K, Van Camp B 1988 Sensitive detection of immunogold-silver staining with darkfield and epi-polarization microscopy. Journal of Histochemistry and Cytochemistry 36: 679–683

Gallyas F, Wolff J R 1986 Metal-catalyzed oxidation renders silver intensification selective. Journal of Histochemistry and Cytochemistry 12: 1667–1672

Gu J, De Mey J, Moeremans M, Polak J M 1981 Sequential use of the PAP and immunogold staining methods for the light microscopical double staining of tissue antigens. Its application to the study of regulatory peptides in the gut. Regulatory Peptides 1: 365–374

Hacker G W, Grimelius L, Danscher G et al 1988 Silver acetate autometallography: an alternative enhancement technique for immunogold-silver staining (IGSS) and silver amplification of gold, silver, mercury and zinc in tissues. Journal of Histotechnology 11: 213–221

Hall P A , Stearn P M, Butler M G, Ardenne A J D 1987 Acetone/periodate-lysine-paraformaldehyde (PLP) fixation and improved morphology of cryostat sections for immunohistochemistry. Histopathology 11: 93–101

Hamid Q, Wharton J, Terenghi G et al 1987 Localization of atrial natriuretic peptide mRNA and immunoreactivity in the rat heart and human atrial appendage. Proceedings of the National Academy of Sciences USA 84: 6760–6764

Heyderman E 1979 Immunoperoxidase techniques in histopathology: applications, methods and controls. Journal of Clinical Pathology 32: 971–978

Hoefler H, Childers H, Montminy M R, Lechan R, Goodman R H, Wolfe H J 1986 In situ hybridization methods for the detection of somatostatin mRNA in tissue sections using antisense RNA probes. Histochemical Journal 18: 597–604

Holgate C, Jackson P, Cowen P, Bird C 1983a Immunogold-silver staining: new method of immunostaining with enhanced sensitivity. Journal of Histochemistry and Cytochemistry 31: 938–944

Holgate C, Jackson P, Lauder I, Cowen P N, Bird C C 1983b Surface membrane staining of immunoglobulins in paraffin sections of non-Hodgkin's lymphomas using an immunogold-silver technique. Journal of Clinical Pathology 36: 742–746

Holgate C S, Jackson P, Pollard K, Lunny D, Bird C C 1986 Effect of fixation on T and B lymphocyte surface membrane demonstration in paraffin processed tissue. Journal of Pathology 149: 293–300

Hsu S M, Soban E 1982 Color modification of diaminobenzidine (DAB) precipitation by metallic ions and its application to double immunohistochemistry. Journal of Histochemistry and Cytochemistry 30: 1079–1082

Huang S, Minassian H, More J D 1976 Application of immunofluorescent staining in paraffin sections improved by trypsin digestion. Laboratory Investigation 35: 383–391

Huang W M, Gibson S J, Facer P, Gu J, Polak J M 1983 Improved section adhesion for immunocytochemistry using high molecular weight polymers of L-lysine as a slide coating. Histochemistry 77: 275–279

Janossy G, Bofill M, Poulter L W 1986 Two-colour immunofluorescence: analysis of the lymphoid system with monoclonal antibodies. In: Polak J M, Van Noorden S (eds) Immunocytochemistry, modern methods and applications. Wright, Bristol, p 438–455

Lane B P, Europa D L 1965 Differential staining of ultrathin sections of epon-embedded tissues for light microscopy. Journal of Histochemistry and Cytochemistry 13: 579–582

Lascano E F, Berria M I 1988 PAP labeling enhancement by osmium tetroxide-potassium ferrocyanide treatment. Journal of Histochemistry and Cytochemistry 36: 679–699

Mason D Y 1985 Immunocytochemical labeling of monoclonal antibodies by the APAAP immunoalkaline phosphatase technique. In: Bullock G R, Petrusz P (eds) Techniques in immunocytochemistry, vol 3. Academic Press, London, p 25–42

Mason D Y, Sammons R E 1978 Alkaline phosphatase and peroxidase for double immunoenzymatic labelling of cellular constituents. Journal of Clinical Pathology 31: 454–462

Mason D Y, Abdulaziz Z, Falini B, Stein H 1983 Double immunoenzymatic labelling. In: Polak J M, Van Noorden S (eds) Immunocytochemistry, practical applications in pathology and biology. Wright, Bristol, p 113–128

McLean J W, Nakane P K 1974 Periodate-lysine-paraformaldehyde fixative, a new fixative for immuno-electron microscopy. Journal of Histochemistry and Cytochemistry 22: 1077–1083

Peacock C S, Thompson I W, Van Noorden S 1990 Silver enhancement of polymerised diaminobenzidine. Increased sensitivity for immunoperoxidase staining. (In preparation)

Pearse A G E, Polak J M 1975 Bifunctional reagents as vapour and liquid phase fixatives for immunohistochemistry. Histochemical Journal 7: 179–186

Pelliniemi L J, Dym M, Karnovsky M J 1980 Peroxidase histochemistry using diaminobenzidine tetrahydrochloride stored as a frozen solution. Journal of Histochemistry and Cytochemistry 28: 191–192

Polak J M, Varndell I M (eds) 1984 Immunolabelling for electron microscopy. Elsevier Science Publishers, Amsterdam

Pollard K, Lunny D, Holgate C S, Jackson P, Bird C C 1987 Fixation, processing and immunohistochemical reagent effects on preservation of T-lymphocyte surface membrane antigens in paraffin-embedded tissue. Journal of Histochemistry and Cytochemistry 35: 1329–1338

Ponder B A, Wilkinson M M 1981 Inhibition of endogenous tissue alkaline phosphatase with the use of alkaline phosphatase conjugates in immunohistochemistry. Journal of Histochemistry and Cytochemistry 29: 981–984

Rodriguez E M, Yulis R, Peruzzo B, Alvial G, Andrade R 1984 Standardization of various applications of methacrylate embedding and silver methenamine for light and electron microscopy immunocytochemistry. Histochemistry 81: 253–263

Scopsi L, Larsson L I 1986 Increased sensitivity in peroxidase immunocytochemistry. A comparative study of a number of peroxidase visualization methods employing a model system. Histochemistry 84: 221–230

Shu S, Gong J, Fan L 1988 The glucose oxidase-DAB-nickel method in peroxidase histochemistry of the nervous system. Neuroscience Letters 85: 169–171

Springall D R, Hacker G W, Grimelius L, Polak J M 1984 The potential of the immunogold-silver staining method for paraffin sections. Histochemistry 81: 603–608

Sternberger L A 1986 Immunocytochemistry, 3rd edn. Wiley, New York

Sternberger L A, Joseph F A 1979 The unlabeled antibody method. Contrasting color staining of paired pituitary hormones without antibody removal. Journal of Histochemistry and Cytochemistry 29: 1424–1429

Sternberger L A, Sternberger N H 1986 The unlabeled antibody method: comparison of peroxidase-antiperoxidase with avidin-biotin complex by a new method of quantification. Journal of Histochemistry and Cytochemistry 34: 599–605

Sternberger L A, Hardy Jr P H, Cuculis I J, Mayer H G 1970 The unlabeled antibody-enzyme method of immunohistochemistry. Preparation and properties of soluble antigen-antibody complex (horseradish peroxidase-antihorseradish peroxidase) and its use in identification of spirochetes. Journal of Histochemistry and Cytochemistry 18: 315–333

Straus W 1982 Imidazole increases the sensitivity of the cytochemical reaction for peroxidase with diaminobenzidine at a neutral pH. Journal of Histochemistry and Cytochemistry 30: 491–493

Suffin S C, Muck K B, Young J C, Lewin K, Porter D D 1979 Improvement of the glucose oxidase immunoenzyme technic. American Journal of Clinical Pathology 71: 492–496

Tubbs R R, Sheibani K 1982 Chromogens for immunohistochemistry. Archives of Pathology and Laboratory Medicine 106: 205

Vacca L L 1982 'Double bridge' techniques of immunocytochemistry. In: Bullock G R, Petrusz P (eds) Techniques in immunocytochemistry, vol 1. Academic Press, London, p 155–182

Van Goor H, Harms G, Gerrits P O, Kroese F G M, Poppema S, Grond J 1988 Immunohistochemical antigen demonstration in plastic-embedded lymphoid tissue. Journal of Histochemistry and Cytochemistry 36: 115–120

Welsburger E K, Russfield A B, Homburger F et al 1978 Testing of twenty-one environmental aromatic amines or derivatives for long-term toxicity or carcinogenicity. Journal of Environmental Pathology and Toxicology 2: 325–356

Wood G S, Warnke R 1981 Supression of endogenous avidin-binding activity in tissues and its relevance to biotin-avidin detection systems. Journal of Histochemistry and Cytochemistry 29: 1196–1204

4. General approach to tumour markers in diagnostic pathology

F. T. Bosman

INTRODUCTION

For those who entered the field of diagnostic pathology after the immunohistochemical revolution it would be hard to imagine what histopathology was like before. The possibility to detect specifically an almost unlimited array of antigens has drastically changed the practice of pathology. As a large proportion of the daily duty of a pathologist lies in the diagnosis of cancer, a very significant amount of energy has been invested into exploration of the potential applications of immunohistopathology in oncology. Originally, the idea was to look for tumour-specific antigens, based on the assumption that neoplastic transformation might lead to the emergence of detectable neoantigens on tumour cells. Carcinoembryonic antigen (CEA) remains one of the great successes of this approach, but also exemplifies an inherent shortcoming that none of the presently available tumour-associated antigens is truly tumour-specific. New enthusiasm was generated when the hybridoma technique for the production of monoclonal antibodies was developed. Expectations that this powerful technique would soon result in antibodies that would detect tumour-specific antigens were high. It is now clear, however, that also with this technique tumour specificity of the generated antibodies is far from optimal. A general consensus has emerged that tumour-specific antigens, if they exist at all, are exceedingly rare. This does not imply that immunohistochemical staining of antigens in tumours is a futile activity. Re-expression of fetal antigens such as CEA or alpha-fetoprotein (AFP) in neoplastic cells has appeared to be of significant diagnostic use and probably approximates best the original aims of the search for tumour antigens.

Normal cellular antigens can be used to assess the direction and the level of differentiation of neoplastic cells. Extracellular matrix antigens can be used to study invasive growth. These applications of immunohistochemistry in tumour diagnosis are quite different and, as a consequence, the term tumour marker has become rather poorly defined. In the context of this chapter, a tumour marker is regarded as any substance that can be detected by immunohistochemistry and the pattern of immunoreactivity can be used in the diagnosis or classification of neoplasia.

METHODOLOGICAL CONSIDERATIONS

The various immunohistochemical techniques, their principles and (dis)advantages are reviewed in Chapter 3 and will not be discussed here. It is important, however, to stress that the reliability of a diagnosis based upon immunohistochemical information rests primarily on the quality of the applied techniques. Therefore, it is important to re-emphasize some technical considerations, as they apply to the practice of diagnostic immunohistopathology. An easy reminder of the importance of proper technique is to go over the following four checkpoints for each immunohistochemical analysis.

1. The validity of the applied techniques

Reliable immunohistochemistry rests on two cornerstones: adequate tissue processing and appropriate immunohistochemical technique. Adequate tissue processing is absolutely essential for diagnostic histopathology. In a routine laboratory, where time and facilities do not allow specific tissue processing protocols for different antigens, standardization is a comparatively easy way to approximate this goal. Major variables with appreciable effects on antigen immunoreactivity are duration of fixation, the fixative used and the maximum temperature reached during embedding. These factors can be relatively easily controlled. The large majority of antigens of interest can be detected in formalin fixed and routinely processed paraffin sections, provided that the sections are adequately pretreated. This often requires the use of proteolytic enzymes (Curran & Gregory 1980). Conceivably each antigen might require a unique tissue processing protocol. As monoclonal antibodies react with only one epitope, and different epitopes on an antigen might respond variably to the tissue processing procedure, even each monoclonal antibody might require a unique protocol. It is therefore important for each new antigen and new monoclonal antibody to establish optimal conditions for immunohistochemistry. It is always advisable to quick-freeze an unfixed tissue specimen which can be used for cryosectioning and to make touch preparations from the surface of a freshly cut tumour.

A wide variety of immunohistochemical techniques is presently available (Van Noorden & Polak 1986, see also Ch. 3). It should be realized that the majority of the antigens useful for diagnostic purposes can be detected with relatively simple standard techniques. In practice it appears to be strongly advisable to use one technique, become familiar with its characteristics and fully exploit its potentials.

2. The use of adequate controls

An important general rule is that an immunohistopathological diagnosis cannot be made unless appropriate control procedures are performed. These will first of all concern elucidation of the patterns of reactivity of the applied primary antibody on a standard tissue panel under the tissue processing conditions of the laboratory. Furthermore for each case a positive control on a known positive tissue specimen, processed identically to the case material, should be included. Finally, negative controls on parallel sections should include checks for non-specific binding of the applied detection system (labelled second antibody, PAP complex, ABC complex etc.) by omission of the primary antibody, as well as for the specificity of the primary antibody, by applying preimmune serum or, preferably, immune serum preabsorbed with the antigen under investigation. The use of a panel of antibodies will test the integrity of the tissue and its processing. Omission of control procedures may lead to serious misinterpretations (Van Leeuwen 1986).

3. Knowledge of the distribution of the antigen

This aspect is of particular importance when monoclonal antibodies are used. The specificity claims of many antibodies in use for tissue diagnosis rest solely on immunoreactivity patterns determined in limited tissue panels. The extent of the test panel therefore determines the validity of the specificity claim. Many monoclonal antibodies originally claimed to be tumour-, organ- or cell-specific, upon exhaustive testing have shown a much wider reactivity pattern than initially claimed. An interesting example is the Leu-7 antibody, originally claimed to be specific for natural killer cells. The occurrence of the Leu-7 antigen on small cell lung cancer cells initially led to discussions concerning the histogenesis of this neoplasm, but it is now well established that the Leu-7 antigen is also a useful marker for neuroendocrine cells (Tischler et al 1986, see also Ch. 7). Another example is the occurrence of cytokeratin immunoreactivity, which is regarded as specific for epithelial cells, in developing muscle cells and myogenous tumours (Brown et al 1987). Similar events have occurred with conventional polyclonal antisera. Neuron-specific enolase, contrary to the original claims, appears not to be specific for neurons or neuroendocrine cells, but occurs also in a variety of other cell types (Pählmann et al 1986), and prostatic acid phosphatase does not only occur in the prostate, but also in B-cells in islets of Langerhans and some endocrine tumours, especially of the pancreas (Sobin et al 1986).

4. Validity of the working hypothesis

A central working hypothesis in tumour marker classification of neoplasms is that tumour cells mostly differentiate along the same lines as their normal counterparts and that markers of normal differentiation will also be useful as differentiation markers in neoplasia. It is quite clear, however, that tumours break rules. Many reports of unique cases in the histopathological literature illustrate the potential for aberrant differentiation of neoplastic cells. This notion calls for careful interpretation of tumour marker immunoreactivity patterns in the diagnosis and classification of neoplasia. Phenomena such as the occurrence of cytokeratins in smooth muscle tumours and some sarcomas (Salisbury & Isaacson 1985, Brown et al 1987), vimentin in carcinoma cells (McNutt et al 1985), chorionic gonadotrophin in a wide variety of carcinomas (Braunstein et al 1986) and neuro-endocrine cells in non-neuroendocrine tumours (Bosman 1984) illustrate that differentiation in neoplastic cells not infrequently follows unexpected directions. The immunophenotype of a neoplasm should therefore never be used as the sole argument in its classification.

Histopathological diagnosis of cancer implies careful evaluation of topographical, morphological and immunohistological information, and in that order.

DIAGNOSTIC USE OF TUMOUR MARKERS

Tumour marker immunohistochemistry can be used for a wide variety of diagnostic problems. These can be clustered into two main groups:

1. the classification of primary neoplasms and
2. the determination of the primary site in a metastasis of unknown origin.

1. Classification of primary neoplasms

In discussing the value of immunohistochemistry for tumour classification, it is important to emphasize that tumour classifications should be clinically relevant and they should be established by consensus. The clinical relevance of a diagnosis relies on the possibility of using a defined set of characteristics to predict the behaviour of the defined entity in individual patients. In practical terms this implies that each new characteristic, including a reactivity pattern for tumour markers,

should be carefully evaluated in clinicopathological studies in order to establish its clinical relevance. It would be essential, for example, to determine if a histologically undifferentiated tumour, which by virtue of the expression of a tumour marker is assigned to a histologically defined category, also behaves like more differentiated neoplasms in that category. In this area of tumour marker immunohistochemistry, the knowledge is far from complete and many more clinicopathological studies will have to be performed to establish the clinical significance of marker expression. This aspect is closely related to the necessity to establish tumour classifications by consensus. Consensus would also imply standardization in the use of reagents for immunohistochemical studies. If techniques and reagents are not standardized it will remain impossible to compare clinicopathological studies, which are necessary to develop new classifications with more clinical impact.

The following general problems in the histopathological classification of neoplasms might be solved by immunohistochemistry:

Is a neoplastic lesion invasive or not?

Many carcinomas are preceded by a phase of dysplasia of increasing severity, finally leading to carcinoma in situ, which may in turn progress to frankly invasive carcinoma. Examples are the cervical intraepithelial neoplasia sequence, adenomatous endometrium hyperplasia leading to endometrial carcinoma, laryngeal mucosal dysplasia followed by carcinoma in situ and squamous cell carcinoma, and papillary urothelial neoplasms of the bladder. In such cases it is important to be able to distinguish between invasive and non-(or pre-) invasive lesions. In recent years it has been demonstrated that immunohistochemical staining of the basement membrane can be quite useful in this situation (Bosman et al 1985). Basement membranes consist of a basic structure formed by a special type of collagen-type IV, which only occurs in the basement membrane. Furthermore all basement membranes contain laminin, a large glycoprotein which functions as an adhesive between type IV collagen and the adjacent epithelial cell (Martinez-Hernandez & Amenta 1983). Both basement membrane com-

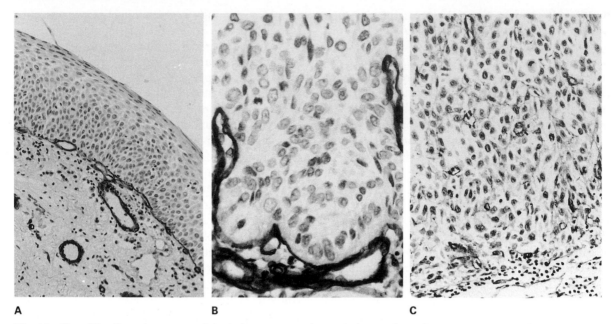

A   **B**   **C**

Fig. 4.1 Type IV collagen immunocreativity in basement membranes in laryngeal neoplasms.
A Hyperplasia. Note continuous basement membranes lining the hyperplastic epithelium. ×100
B Moderate dysplasia. The irregular cell nests are surrounded by a continuous basement membrane. ×400
C Squamous cell carcinoma. Basement membranes do not surround the tumour cell nodules. Stromal vascular structures do show a basement membrane. × 200

ponents can be stained specifically by immunohistochemistry (Havenith et al 1987) and, using this approach, invasive lesions have invariably shown basement membrane interruptions whereas non-invasive lesions mostly showed an intact basement membrane (Fig. 4.1). This approach has been used with success not only in cervical (Birembaut et al 1985), endometrial (Barsky et al 1983) and laryngeal (Visser et al 1986) neoplasms, but also for the distinction between tubular carcinoma and sclerosing adenoma of the breast (Willebrand et al 1987), or between adenomatous polyps and polypous carcinomas in the colon (Kellokumpu et al 1985). It is important, however, to stress that the apparent general rule that invasive lesions show basement membrane interruption or even complete absence of basement membranes is not always valid. Malignant melanomas, for example, almost invariably show very dense pericellular deposits of basement membrane material (Stenbäck & Wasenius 1986) and highly differentiated squamous cell carcinomas frequently show extensive basement membrane deposition. Conversely, in the presence of a dense inflam-

matory infiltrate an epithelial basement membrane may appear quite discontinuous without any signs of malignancy, due to leucocytes invading the epithelium (Visser et al 1986).

An additional aspect of interest with regard to basement membrane immunohistochemistry is the finding that the presence or absence of basement membrane antigens may be of prognostic significance. In colon carcinomas (Havenith et al 1988), as well as in bladder carcinomas (Daher et al 1987), it has been reported that a tendency of the tumour cells to deposit extensive basement membranes is a prognostically favourable sign.

Which type of tumour?

Usually the site of the tumour and the histological characteristics will lead to an initial assignment to one of the major categories: carcinoma, sarcoma, melanoma or lymphoma. In cases of doubt it is possible to distinguish between these categories through the use of cell lineage markers. The most frequently used markers are listed in Table 4.1. Carcinomas are characterized by cytokeratin inter-

Table 4.1 Markers for classification of undifferentiated neoplasms

Diagnosis	Subtype	Initial marker(s)	Additional markers(s)
Carcinoma		Cytokeratin, EMA, desmoplakin	
	Squamous cell carcinoma		Cytokeratin 5, 6
	Adenocarcinoma		Cytokeratin 8, 18, 19
Biphasic neoplasm	Synoviosarcoma	Cytokeratin, vimentin	
	Epithelioid sarcoma	Cytokeratin, vimentin	
	Mesothelioma	Cytokeratin, vimentin	
Lymphoma		Leucocyte common antigen (LCA)	
	B-cell type	Immunoglobulins	B-cell antigens
	T-cell type		T-cell antigens
Melanoma		S-100, NKI/C3	
Sarcoma basal lamina reactive		Vimentin Laminin, type IV collagen	
	Myosarcoma		Desmin, myosin, myoglobin
	Angiosarcoma		Factor VIII-related antigen
	Liposarcoma		
non-basal lamina reactive			
	Fibrosarcoma Malignant fibrous histiocytoma		

mediate filaments (Ramaekers et al 1983, Fig. 4.2), desmosomal plaque proteins (Moll et al 1986) and epithelial membrane antigen (Pinkus et al 1986). It is important to stress here that the presence of one or more of these antigens identifies the tumour as a carcinoma but the absence does not necessarily indicate that it is not a carcinoma! Lymphomas can be identified through the presence of the leucocyte common antigen (Lauder et al 1984). Immunohistochemically, sarcomas are diagnosed by exclusion, but do usually contain vimentin intermediate filaments (Altmannsberger et al 1986). A diagnosis of sarcoma will have to be further confirmed through identification of characteristic markers of the type of sarcoma. Melanomas have vimentin positive, cytokeratin negative intermediate filaments and can be positively identified by immunoreactivity for neuron specific enolase and S-100 protein (Löffel et al 1985).

Once a tumour has been identified as a carcinoma the question, 'which type of carcinoma' remains to be answered. As a first step the broad categories of squamous cell carcinoma, adenocarcinoma and neuroendocrine carcinoma can be distinguished through marker studies. For the distinction between adenocarcinomas and squamous cell carcinomas cytokeratin subtyping appears to be quite useful. Cytokeratin consists of a family of at least 19 polypeptides, which can be distinguished by two-dimensional gel electrophoresis (Moll et al 1982, see also Table 4.2). The basic cytokeratins 5 and 6 occur almost exclusively in squamous cells carcinomas whereas the (basic) cytokeratins 7 and 8 and the (acidic) cytokeratins 18 and 19 occur predominantly in adenocarcinomas, neuroendocrine carcinomas and transitional cell carcinoma (Moll & Franke 1986). Against many of these cytokeratin polypeptides specific monoclonal antibodies have been

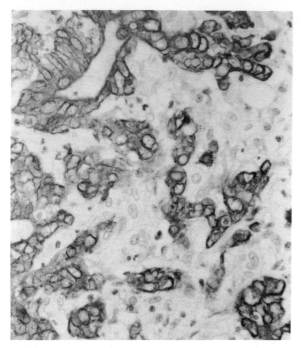

Fig. 4.2 Immunoreactivity for cytokeratin in a gastric adenocarcinoma using a broad-spectrum anticytokeratin antibody. ×100

generated, which allows their detection by immunohistochemistry. Based upon these principles gross subdivision of carcinomas can be achieved

(Moll 1988). A limitation of this approach, however, is that many of the monoclonal antibodies to the cytokeratins do not react with formalin-fixed, paraffin-embedded tissues.

Additional markers, which may be used for subclassification of carcinomas, are the oncofetal proteins carcinoembryonic antigen (CEA) and alpha-fetoprotein (AFP), secretory proteins such as secretory component (SC) and endocrine markers such as neuron-specific enolase and chromogranin (Heitz 1988, Fig. 4.3). The majority of the adenocarcinomas of the gastrointestinal tract, pancreas, the binary system, lung, breast, salivary glands, ovary and endometrium shows CEA immunoreactivity (Fig. 4.4). Some endocrine tumours, such as medullary carcinoma of the thyroid and carcinoid, small cell lung carcinoma and also some squamous cell carcinomas are CEA positive (DeLellis et al 1978, Wachter et al 1984). AFP was originally reported to occur exclusively in hepatocellular carcinomas and in endodermal sinus tumour elements in testicular and ovarian teratomas. Occasionally, however, adenocarcinomas of the lung, stomach, pancreas and colon also produce AFP (Kodama et al 1981). SC occurs in adenocarcinomas of glandular epithelia, including salivary gland tumours, gastrointestinal and pancreatic carcinomas and breast carcinomas (Rognum et al 1988). Neuron-

Table 4.2 The cytokeratins

Mol. wt (kDa)	Moll no. Basic	Acidic	Epithelial type		Antibody reactivity	
67	1					
	2			Squamous and ductal		
		9				AE3
60	3					
	4					
	5		Squamous			
		10				
	6					
		11				
55		12				
	7			Simple and ductal	AE1	
		13				
	8					
50		14, 15				
		16				CAM 5.2
		17				
45		18				
40		19				

The cytokeratins, showing the inter-relationships between molecular weight, the Moll catalogue number, presence in the different types of epithelium and their reactivity with some commonly used antibodies.

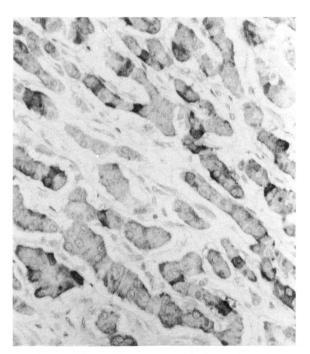

Fig. 4.3 Chromogranin immunoreactivity in a bronchial carcinoid. The basal subnuclear compartment of the cells is strongly stained. ×300

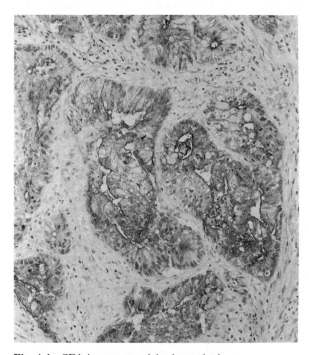

Fig. 4.4 CEA immunoreactivity in a colonic adenocarcinoma. Note the strong apical reactivity surrounding the gland lumina. ×200

specific enolase and chromogranin identify endocrine tumours (Heitz 1988) but also are found in some tumour cells, especially in adenocarcinomas of endodermal origin (Bosman 1983). Further classification of adenocarcinomas according to organ or cell specific markers would form the final step in the immunohistochemical analysis (see p. 57).

Once a neoplasm has been diagnosed as a lymphoma, based upon the expression of the leucocyte common antigen, further classification can be achieved in a combination of morphological and immunohistochemical parameters. Initial classification will be into the broad categories of Hodgkin's versus non-Hodgkin's lymphoma. The role of immunohistochemistry lies in the recognition of the exclusive B or T cell phenotype of non-Hodgkin's lymphoma, which can be achieved by application of monoclonal antibodies against B or T cell specific antigens. Originally this could only be performed on unfixed frozen sections. Recently, however, monoclonal antibodies have been reported which can detect these antigens in formalin fixed, paraffin embedded tissues (Dobson et al 1987). These antibodies have greatly expanded the possibilities for routine immunophenotyping of malignant lymphomas. It is important to mention, however, that their specificity for the B or T cell lineage is not absolute and therefore lymphoma classification requires careful assessment of a combination of morphological and immunohistochemical parameters (Warnke & Rouse 1985). Further subclassification of B and T cell lymphomas can be achieved by immunophenotyping according to a large set of differentiation antigens (see Ch. 19).

Further subclassification of a sarcoma can also be facilitated through immunohistochemical analysis. First of all, a category can be distinguished consisting of neoplasms which display a combination of carcinoma and sarcoma characteristics. This category includes mesothelioma, synovial sarcoma, chordoma and epitheloid sarcoma. In keeping with their mesenchymal origin, these tumours express not only vimentin intermediate filaments, but also cytokeratins and some even EMA (Coindre et al 1988). At electron-microscopical examination the epithelial characteristics are substantiated in the presence of desmosomes, formation of lumina and microvilli.

Further sarcoma classification can be performed using markers for specific sarcomas. Staining for the basement membrane antigens (type IV collagen and laminin) allows a distinction between sarcomas that do not (e.g. fibrosarcoma, malignant fibrous histiocytoma) and those that do deposit a basement membrane (e.g. leio- and rhabdomyosarcoma, liposarcoma, malignant nerve sheath tumours) (Miettinen et al 1983). Myosarcomas commonly express desmin intermediate filaments (Molenaar et al 1985) and myosin (De Jong et al 1984) but myoglobin only occurs in rhabdomyosarcomas (De Jong et al 1984). Hemangiosarcomas usually express factor VIII–related antigen (Little et al 1986). Malignant nerve sheath tumours express the S-100 antigen (Nakajima et al 1982). This protein, however, shows a rather broad distribution.

Which tumour grade?

It is well established that tumours classified by histological criteria into the same category (e.g. adenocarcinoma of the colon) may nevertheless be rather heterogeneous in regard of morphological characteristics, expression of differentiation antigens, response to therapy and prognosis. It has long been attempted to categorize neoplasms, according to the level of resemblance to the tissue in which they arose, and into different grades of differentiation (e.g. well, moderately well and poorly differentiated). Although grading appears to be a rather subjective activity, in carcinomas of the bladder, prostate, colorectum and ovary, tumour grading has appeared to be of prognostic significance. Attempts to improve the reproducibility of grading of differentiation have been made through immunohistochemical staining for differentiation antigens. It was found that for bladder carcinomas the expression of blood group antigens (Limas et al 1984) and for colorectal carcinomas expression of secretory component (Arends et al 1984) conveys prognostically relevant information. It is important, however, to realize that the tumour cells with a strong tendency to differentiate are probably not the cells with a high proliferation rate or of metastatic potential. This limits the significance of immunohistochemical studies for differentiation antigens.

A more promising approach is the assessment of the proportion of proliferating cells in a neoplasm. This is almost routinely performed nowadays through DNA-flow cytometry after retrieval of tumour cell nuclei from paraffin blocks (Schutte et al 1986). From the DNA histogram relatively simple algorithms allow the calculation of the fraction of cells in the S-phase. The discovery of the antigen defined by the Ki-67 monoclonal antibody, which is expressed only in cycling cells, has rapidly led to investigations concerning the predictive value of Ki-67 antigen expression for tumour behaviour. Indeed, some studies have substantiated the prognostic value of the Ki-67 antigen (Hall et al 1988). This area is, however, in rapid development. An interesting possibility for detailed cytokinetic analysis of tumour cells populations is in vitro or ex vivo labelling of DNA synthesizing cells with the synthetic nucleotide bromodeoxyuridine (BrdU), which can subsequently be detected by immunohistochemistry (Schutte et al 1987). In a strict sense BrdU cannot be regarded as a tumour marker. One of the most important characteristics of neoplastic cells, however, is the disregulation of proliferation and therefore it can be expected that the search for markers which specifically identify proliferating cells will lead to new biologically interesting and clinically useful antigens.

Does the neoplasm produce antigens which can be detected in body fluids and used for follow-up?

In most cases of cancer a reasonably detailed diagnosis is well anticipated before a final tissue diagnosis can be made on a resection specimen. In these cases the presence of circulating tumour associated antigens, which may be helpful in the postoperative monitoring of disease activity, will have been ascertained. Occasionally such information is not available. In such a case immunohistochemical detection of an antigen which may be released into the circulation, can be clinically important. This holds especially for CEA in gastrointestinal tract, pancreas, lung, breast and ovarian carcinomas, AFP and hCG in testicular and ovarian teratomas, AFP in hepatocellular carcinomas, prostate specific acid phosphatase in prostatic carcinomas and peptide hormones in neuroendocrine tumours.

2. Determining the primary site of a metastasis of unknown origin

This remains one of the most frustrating problems in diagnostic histopathology of cancer. Careful and stepwise immunohistochemical marker analysis can be of tremendous help towards solving this problem, although for squamous cell carcinoma and adenocarcinoma metastases often a final decision as to the primary site cannot be made.

If the metastasis is of an undifferentiated neoplasm, the first question will be 'which type of tumour?' The immunohistochemical approach to this problem has been discussed in relation to the classification of primary tumours. Epithelial markers, such as cytokeratin, epithelial membrane antigen or desmoplakin identify carcinomas; leucocyte common antigen identifies lymphomas; S-100 protein is compatible with melanoma and finally sarcomas are usually positive for vimentin, although this marker is not at all specific for sarcomas. For lymphomas, melanomas and sarcomas markers for the origin of the primary lesion do not exist. The diagnostic process for these categories will continue with attempts toward further subclassification. For neoplasms positive for carcinoma markers the possibility of a

mesothelioma, epitheloid sarcoma or synoviosarcoma will have to be considered. If this is excluded, which may involve additional (immuno)histochemical or ultrastructural studies, the lesion must be a carcinoma. The next step will then be to decide which type of carcinoma it is. Markers useful for the subclassification of carcinoma are listed in Table 4.3. As has been outlined earlier, cytokeratin polypeptide analysis is a useful approach to this problem. Squamous cell carcinomas express complex combinations of cytokeratin and peptides. As yet, immunohistochemical determination of the origin of a squamous cell carcinoma metastasis is impossible. Adenocarcinomas generally express cytokeratin 8, 18 and 19 (Moll 1988).

For further characterization of adenocarcinomas a range of organ specific markers is available. For the clinician the most important question will be whether or not an adenocarcinoma metastasis has arisen from a primary site for which a specific therapeutic regimen is available. This concerns adenocarcinomas from the breast, ovaries, thyroid and prostate. Breast carcinoma may be positive for oestrogen and/or progesterone receptor protein (Reiner et al 1986), which, however, can also occur in endometrial carcinoma. Not all breast carcinomas are steroid receptor positive, indicating that absence of immunoreactivity for these proteins does not exclude the breast as a primary

Table 4.3 Markers for subclassification of carcinoma

Tumour	Markers
Thyroid carcinoma	Thyroglobulin
Prostatic adenocarcinoma	Prostatic acid phosphatase, prostate-specific antigen
Hepatocellular carcinoma	Alpha-fetoprotein

Embryonal carcinomas
  with yolksac differentiation
  with trophoblastic differentiation
	Alpha-fetoprotein
	Chorionic gonadotrophin

Breast carcinoma	Oestrogen receptor protein
Ovarian serous cystadenocarcinoma	CA 125
Gastrointestinal carcinomas (and others)	Carcinoembryonic antigen (CEA)
Neuroendocrine carcinomas	Neuron-specific enolase, chromogranins, peptide hormones

site. For serous cystadenocarcinomas of the ovary CA-125 antigen is a fairly specific marker (Kawabat et al 1983). Mucinous cystadenocarcinomas of the ovary frequently contain neuroendocrine cells and mucins (Sporrong et al 1981), which also occur, however, in a wide variety of adenocarcinomas from other primary sites, most notably of endodermal origin. In cases of large cell undifferentiated carcinoma, an embryonal cell carcinoma may be considered which often is AFP positive. AFP also occurs in hepatocellular carcinoma and, less frequently, in digestive tract adenocarcinomas. For thyroid carcinomas of follicular cell origin, thyroglobulin is a highly specific marker. Finally, prostate-specific antigen (Purnell et al 1984) and prostatic acid phosphatase (Jöbsis et al 1978) are highly specific for prostatic adenocarcinoma (Fig. 4.5), although prostatic acid phosphatase also occurs in some endocrine tumours of the pancreas (Sobin et al 1986) and in adenomas of the middle ear.

In this chapter we have attempted to describe how, in general terms, tumour marker immunohistochemistry can be applied in the histopathological diagnosis of cancer. Undoubtedly, in this area immunohistochemistry has provided new solutions to old problems. It is also clear that the basis of tumour classification remains careful morphological analysis in the context of relevant clinical information, which as yet cannot be replaced by tumour marker

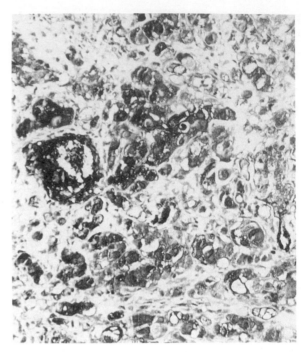

Fig. 4.5 Prostatic acid phosphatase immunoreactivity in a prostatic adenocarcinoma. × 100

immunohistochemistry. At best, tumour marker immunohistochemistry can offer reasonably reliable answers to specific questions raised through careful morphological analysis, which always comes first and for the time being continues to provide the basic frame of reference.

REFERENCES

Altmannsberger M, Dirk M, Osborn M, Weber K 1986 Immunohistochemistry of cytoskeletal filaments in the diagnosis of soft tissue tumors. Seminars in Diagnostic Pathology 3: 306

Arends J W, Wiggers T, Thijs C T, Verstijnen C, Swaen G J V, Bosman F T 1984 The value of secretory component (SC) immunoreactivity in diagnosis and prognosis of colorectal carcinomas. American Journal of Clinical Pathology 82: 267

Barsky H, Siegal G P, Jannotta F, Liotta L A 1983 Loss of basement membrane components by invasive tumors but not by their benign counterparts. Laboratory Investigation 49: 140

Birembaut P, Caron Y, Adnet J J, Foidart J M 1985 Usefulness of basement membrane markers in tumoral pathology. Journal of Pathology 145: 283

Bosman F T 1984 Neuroendocrine cells in non-neuroendocrine tumors. In: Falkmer S, Hakanson R, Sundler F (eds) Evolution and tumour-pathology of the neuroendocrine system. Elsevier Science Publication, Amsterdam p 519

Bosman F T, Havenith M, Cleutjens J P M 1985 Basement membranes in cancer. Ultrastructural Pathology 8: 291

Braunstein G D, Thompson R, Princler G L, McIntire K R 1986 Trophoblastic proteins as tumour markers in nonseminomatous germ cell tumours. Cancer 57: 1842

Brown C, Theaker J M, Banks P M, Gatter K C, Mason D Y 1987 Cytokeratin expression in smooth muscle and smooth muscle tumours. Histopathology 11: 477

Curran R C, Gregory J 1980 Effects of fixation and processing on immunohistochemical demonstration of immunoglobulin in paraffin sections of tonsil and bone marrow. Journal of Clinical Pathology 3: 1047

Daher N, Abourachi H, Bove N, Petit J, Burtin P 1987 Collagen IV staining pattern in bladder carcinomas — relationship to prognosis. British Journal of Cancer 55: 665

De Jong A S H, Van Vark M, Albus-Lutter C H, Van Raamsdonk W, Voute P A 1984 Myosin and myoglobin as tumor markers in the diagnosis of rhabdomyosarcoma. American Journal of Surgical Pathology 8: 521

DeLellis R A, Rule A H, Spiler I, Nathanson L, Tashijan A T, Wolfe H J 1978 Calcitonin and carcinoembryonic antigen as tumor markers in medullary thyroid carcinoma. American Journal of Clinical Pathology 70: 587

Dobson C M, Myskow M W, Krajewski A S, Carpenter F H, Horne C H W 1987 Immunohistochemical staining of non-Hodgkin's lymphoma in paraffin sections using the MB1 and MT1 monoclonal antibodies. Journal of Pathology 153: 303

Goldenberg D M, Sharkey R M, Primus F J 1978 Immunocytochemical detection of carcinoembryonic antigen (CEA) in conventional histopathologic specimens. Cancer 42: 1546

Hall P A, Richards M A, Gregory W W M, D'Ardenne A J, Lister T A, Stansfeld A G 1988 The prognostic value of Ki-67 immunostaining in non-Hodgkin's lymphoma. Journal of Pathology 154: 223

Havenith M G, Cleutjens J P M, Beek C, v.d. Linden E, De Goeij A F P M, Bosman F T 1987 Human specific anti-type IV collagen monoclonal antibodies characterization and application to immunohistochemistry. Histochemistry 87: 123

Havenith M G, Arends J W, Simon R E M, Volovics A, Wiggers T, Bosman F T 1988 Type IV collagen immunoreactivity in colorectal cancer: prognostic value of basement membrane deposition. Cancer 62: 2207

Heitz PhU 1988 Neuroendocrine tumor markers. In: Seifert G (ed) Morphological tumor markers. Current Topics in Pathology 77: 279

Jöbsis A C, De Vries G P, Anholt R R M, Sanders G T B 1978 Demonstration of the prostatic origin of metastases. An immunohistochemical method for formalin-fixed embedded tissue. Cancer 41: 1788

Kawabat S E, Bast R C, Welch W R, Knapp R C, Colvin R B 1983 Immunopathologic characterization of a monoclonal antibody that recognizes common surface antigens of human ovarian tumors of serous, endometroid, and clear cell types. Journal of Clinical Pathology 79: 98

Kellokumpu I, Ekblom P, Scheinin T M, Anderson L C 1985 Malignant transformation in human colorectal mucosa as monitored by distribution of laminin, a basement membrane glycoprotein. Acta Pathologica Microbiologica Immunologica Scandinavica Section A: 93: 285

Kodama T, Kameya T, Hirota T et al 1981 Production of alpha-fetoprotein, normal serum proteins and human chorionic gonadotrophin in stomach cancer. Cancer 48: 1647

Lauder I, Holland D, Mason D Y, Gowland G, Cunbliffe W J 1984 Identification of large cell undifferentiated tumors in lymph nodes using leucocyte common and keratin antibodies. Histopathology 8: 259

Limas C, Lange P 1982 A, B, H antigen detectability in normal and neoplastic urothelium. Cancer 49: 2476

Little D, Said J W, Siegel R J, Fealy M, Fishbein M C 1986 Endothelial cell markers in vascular specific antigens, 6-keto-PGF1 alpha, and Ulex europaeus I lectin. Journal of Pathology 149: 89

Löffel S C, Gillespie Y, Mirmiran S A et al 1985 Cellular immunolocalization of S100 protein within fixed tissue sections by monoclonal antibodies. Archives of Pathology and Laboratory Medicine 109: 117

Martinez Hernandez A, Amenta P S 1983 The basement membrane in pathology. Laboratory Investigation 48: 656

McNutt M A, Bolen J W, Gown A M, Hammar S P, Vogel A M 1985 Coexpression of intermediate filaments in human epithelial neoplasms. Ultrastructural Pathology 9: 31

Miettinen M 1984 Antibodies to epithelial membrane antigen and carcinoembryonic antigen in differential diagnosis. Archives of Pathology and Laboratory Medicine 18: 891

Miettinen M 1988 Immunoreactivity for cytokeratin and epithelial membrane antigen in leiomyosarcoma. Archives of Pathology and Laboratory Medicine 112: 637

Miettinen M, Foidart J M, Ekblom P 1983 Immunohistochemical demonstration of laminin, the major glycoprotein of basement membranes, as an aid in the diagnosis of soft tissue tumors. American Journal of Clinical Pathology 79: 306

Molenaar W M, Oosterhuis J W, Oosterhuis A M, Ramaekers F C S 1985 Mesenchymal and muscle-specific intermediate filaments (vimentin and desmin) in relation to differentiation in childhood rhabdomyosarcomas. Human Pathology 16: 838

Moll R 1988 Epithelial tumor markers: cytokeratin and tissue polypeptide antigen (TPA). In: Seifert G (ed) Morphological tumor markers. Current Topics in Pathology 77: 71

Moll R, Franke W W 1986 Cytochemical cell typing of metastatic tumors according to their cytoskeletal proteins. In: Lapis K, Liotta L A, Rabson A S (eds) Biochemistry and molecular genetics of cancer metastasis. Nijhoff, The Hague, p 101

Moll R, Franke W W, Schiller D L, Geieger B, Krepler R 1982 The catalog of human cytokeratins: patterns of expression in normal epithelia, tumors and cultured cells. Cell 31: 11

Moll R, Osborn M, Hartschuh W, Moll I, Mahrle G, Weber K 1986 Variability of expression and arrangement of cytokeratin and neurofilaments in cutaneous neuroendocrine carcinomas. Immunocytochemical and biochemical analysis of 12 cases. Ultrastructural Pathology 10: 473

Nakajima T, Watanabe S, Yuichi S, Kameye T, Hirota T, Shimosato Y 1982 An immunoperoxidase study of S-100 protein distributed in normal and neoplastic tissues. American Journal of Surgical Pathology 6: 715

Pählmann S, Escher T, Nilsson K 1986 Expression of gammasubunit of enolase: neuron-specific enolase, in human non-neuroendocrine tumors and derived cell lines. Laboratory Investigation 54: 554

Pinkus G, Etheridge C L, O'Connor E M 1986 Are keratin proteins a better tumor marker than epithelial membrane antigen? A comparative immunohistochemical study of various paraffin-embedded neoplasms using monoclonal and polyclonal antibodies. American Journal of Clinical Pathology 85: 269

Purnell D M, Heatfield B M, Trump B F 1984 Immunocytochemical evaluation of human prostatic carcinomas for carcinoembryonic antigen, nonspecific cross-reacting antigen, beta-chorionic gonadotrophin, and prostate-specific antigen. Cancer Research 44: 285

Ramaekers F C S, Puts J J G, Moesker O et al 1983 Antibodies to intermediate filament proteins in the immunohistochemical identification of human tumours: An overview. Histochemical Journal 15: 691

Reiner A, Spona J, Reiner G et al 1986 Estrogen receptor analysis of biopsies and fine-needle aspirates from human breast carcinoma: correlation of biochemical and immunohistochemical methods using monoclonal receptor antibodies. American Journal of Pathology 125: 443

Rognum T O, Thrane P S, Korsrud F R, Brandtzaeg P 1988 Epithelial tumor markers: Special markers of glandular differentiation. In: Seifert G (ed) Morphological tumor markers. Current Topics in Pathology 77: 133

Salisbury J R, Isaacson P G 1985 Synovial sarcomas. Journal of Pathology 147: 49

Schutte B, Reijnders M M J, Bosman F T, Blijham G H 1987 Studies with anti-bromodeoxyuridine antibodies. II. Simultaneous detection of DNA synthesis and antigen expression by immunocytochemistry. Journal of Histochemistry and Cytochemistry 35: 371

Schutte B, Reijnders M M J, van Assche C L M V J, Hupperets P S G J, Bosman F T, Blijham G H 1987 An improved method for immunocytochemical detection of BrdUrd labeled nuclei using flowcytometry. Cytometry 8: 372

Sobin L H, Hjermstad B M, Sesterhenn I A, Helwig E B 1986 Prostatic acid phosphatase activity in carcinoid tumors. Cancer 58: 136

Sporrong B, Alumets J, Clase L I et al 1981 Neurohormonal peptide immunoreactive cells in mucinous cystadenomas and cystadenocarcinomas of the ovary. Virchows Archives 392: 271

Stenbäck F, Wasenius V M 1986 Occurrence of basement membranes in pigment tumors of the skin, relation to cell type and clinical behaviour. Journal of Cutaneous Pathology 13: 175

Tischler A S, Mobtaker H, Mann K et al 1986 Anti-lymphocyte antibody Leu-7 (HNK-1) recognizes a constituent of neuroendocrine granule matrix. Journal of Histochemistry and Cytochemistry 34: 1213

Van Leeuwen F 1986 Pitfalls in immunocytochemistry. Histochemical Journal 16: 179

Van Noorden S, Polak J M 1983 Immunocytochemistry today. Techniques and practice. In: Polak J M, Van Noorden S (eds) Immunocytochemistry. Practical applications in pathology and biology. Wright, Bristol, p 11

Visser R, Van der Beek J M H, Havenith M G, Cleutjens J P M, Bosman F T 1986 Immunocytochemical detection of basement membrane antigens in the histopathological evaluation of laryngeal dysplasia and neoplasia. Histopathology 10: 171

Wachter R, Wittekind C, VonKleist S 1984 Localization of CEA, β-HCG, SP1, and keratin in the tissue of lung carcinomas. An immunohistochemical study. Virchows Archives 402: 415

Warnke R A, Rouse R V 1985 Limitations encountered in the applications of tissue section immunodiagnosis to the study of lymphomas and related disorders. Human Pathology 16: 236

Willebrand D, Bosman F T, De Goeij A F P M 1986 Patterns of basement membrane deposition in benign and malignant breast tumours. Histopathology 10: 1231

5. Non-immunohistochemical methods in tumour diagnosis

T. J. Palmer   M. Isabel Filipe

Most diagnoses in histopathology are made on haematoxylin and eosin stained sections. Three main features in particular are important for diagnosis:

1. Tissue architecture.
2. The degree of preservation of function.
3. The cytological details of the cells.

Usually the diagnosis is clear, but at times the pathologist is left with several conditions which will require further investigation. Special stains should be used to answer specific questions with regard to diagnosis. The special stains available can be divided into groups that correspond to the three features mentioned previously. In addition, stains may accentuate particular features or reveal details not seen initially.

STAINS OF TISSUE ARCHITECTURE

These are essentially the connective tissue stains and may be divided broadly into stains for collagen and those that show basement membrane.

Epithelial tissues, with the exception of the liver, contain basement membrane at the epithelio-mesenchymal junction. The distribution of basement membrane in mesodermally-derived tissues is restricted to vascular, muscular, neural and adipose tissues (see Table 5.1). The distribution of basement membrane may be helpful in tumour diagnosis since it may reflect the tissue of origin, epithelial tumours containing little or no basement membrane material, while tumours derived from mesoderm may have a prominent pericellular basement membrane (D'Ardenne 1989). Its presence will therefore distinguish mesodermally derived

Table 5.1 Distribution of basement membrane in normal tissues

Tissue	Distribution
Epithelia	Around groups of cells at interface between epithelium and supporting tissues
Connective tissue	
Nerves	Around individual Schwann cells
Muscles	Around individual muscle cells
Adipose tissue	Around individual fat cells
Blood vessels and lymphatics	Around pericytes; between endothelium and supporting tissues
Synovium	None
Fibroblasts	None
Lymphocytes	None
Histiocytes	None
Haemopoietic tissues	None

tumours from epithelial and fibrohistiocytic tumours and lymphomas (see also Ch. 7). However, malignant tumours may lose part or all of their ability to produce basement membranes.

There are only two commonly used non-immunohistochemical methods which stain basement membranes — the periodic acid Schiff reaction and the silver (recticulin) stains — neither of which are entirely specific.

The PAS reaction detects the glycoproteins of the basement membrane, type IV collagen and other carbohydrate residues. Periodic-reactive groups are more frequent in basement membrane, which therefore stains darker than the surrounding connective tissue.

The stromal response to tumours consists of inflammation and the formation of fibrous tissue,

61

and therefore collagen stains will emphasize the growth pattern of the tumour. Van Gieson's and reticulin stains are connective tissue stains specific for collagen types I, II and III, whereas the trichrome stains — Picro-Mallory, Martius-Scarlet-Blue and Masson's, for example — stain a variety of tissue structures. A reticulin stain will often emphasize the biphasic growth pattern of a synovial sarcoma, and similarly collagen stains may show more clearly the storiform appearance of a malignant fibrous histiocytoma. The trichrome stains show differentiation as well as growth pattern in connective tissue tumours, but require careful technique.

Basement membrane and connective tissue stains are not generally useful for determining the presence or absence of invasion, or for distinguishing between benign and malignant tumours, although the absence of basement membrane or an interrupted basement membrane are generally considered to be indicators of malignancy. However, the demonstration of a complete ring of reticulin around the tubules in microglandular adenosis of the breast enables differentiation from infiltrating carcinoma. Basement membrane material is not present in tubular carcinoma of the breast, whereas it is seen around the tubules in a radial scar.

Connective tissue stains may show the destruction of normal tissue architecture associated with malignancies. This is especially useful in the differentiation of lymphomas from benign lymphoid hyperplasias, where reticulin stains are an important adjunct to H and E stains. Vascular invasion is often more easily seen using a connective tissue stain. An elastic van Gieson stain is essential for detecting microangioinvasion in follicular thyroid tumours. Here, the presence of endothelium directly in contact with tumour is taken to indicate vascular invasion; the presence of even a thin strand of collagen shows that vascular invasion has not taken place. Vascular invasion may also be confirmed by demonstrating binding of *Ulex europaeus* 1 (UEA 1) lectin to endothelial cells, since this is a more sensitive method than using antibodies to factor VIII-related antigen (Ordonez & Batsakis 1984, Stephenson et al 1986).

STAINS FOR TISSUE DIFFERENTIATION

Once it has been decided whether a tumour is benign or malignant, epithelial or mesenchymal, the likely tissue of origin and its degree of differentiation must be ascertained. Both of these may be apparent on the haematoxylin and eosin stained sections, but sometimes further clarification is required.

Stains for epithelial differentiation

Squamous differentiation

This does not normally present a problem, the mainstays of diagnosis being growth pattern, the presence of intercellular bridges and keratin formation.

Adenocarcinomatous differentiation

A diagnosis of 'adenocarcinoma' is not usually sufficient in itself, although even after extensive investigation it may not be possible to say much more. Many glandular epithelia secrete mucins and the type present may alter during the transition of a cell from normal through dysplasia to malignancy. Mucins are divided into neutral mucins and acid mucins, which way be sialated or sulphated types, with highly sulphated mucins also distinguishable from weakly or non sulphated mucins. Commonly used methods are as follows:

*P*AS. Stains all neutral mucins, glycogen and the carbohydrate portion of glycoproteins and glycolipids. Staining for glycogen is abolished by predigestion with diastase.

Alcian blue. Alcian blue staining at varying pH, with or without pretreatment of the sections with hyaluronidase, or by the critical electrolyte concentration method, enables the various types of acid mucin to be distinguished (Table 5.2). This is mainly of relevance in the diagnosis of soft tissue tumours (Table 5.2), but can be helpful in the differentiation of carcinoma from malignant mesothelioma (see also Ch. 22).

Mucicarmine. A commonly used empirical but insensitive stain for epithelial mucin. A better al-

Table 5.2 Histochemical characterization of myxoid and chondromatous tumours

	Alcian blue pH 2.5	Alcian blue pH 2.5 + hyaluronidase	CEC method Alcian blue pH5.6 + MgCl$_2$
Normal			
Umbilical cord (Wharton's jelly)	+	−	<0.25M MgCl$_2$
Fetal cartilage	+	−	<0.55M MgCl$_2$
Adult cartilage	+	+	>0.55M MgCl$_2$
Tumours			
Myxoid tumours without chondromatous element			
Myxoid			
Myxoid lipoma			
Lipoblastoma			
Myxoid liposarcoma	+	−	<0.25M MgCl$_2$
Myxoid malignant fibrous histiocytoma			
Myxoid fibrosarcoma			
Myxoid nerve sheath tumours	+	±	<0.55M MgCl$_2$
Myxoid tumours with chondromatous element			
Benign chondromatous tumours	+	+	>0.55M MgCl$_2$
Chondrosarcoma			
well differentiated	+	+	>0.55M MgCl$_2$
poorly differentiated	+	±	<0.55M MgCl$_2$
extraskeletal myxoid	+	±	<0.55M MgCl$_2$
mesenchymal	+	±	<0.55M MgCl$_2$

CEC = Critical electrolyte concentration method.

ternative as a general mucin stain is the combined Alcian blue-PAS technique, which allows the important differentiation between acid and neutral mucin to be made in one procedure.

High iron diamine. Specific for highly sulphated mucin. Its presence in gastric mucosa showing intestinal metaplasia and the loss of HID staining in colonic mucosa are useful indicators of dysplasia. Highly sulphated mucins are also present in adenocarcinoma of the prostate but not in benign prostatic epithelium.

Mucins are not the only secretory products of adenocarcinomas. Acinic cell tumours of the salivary glands show exocrine differentiation and the secretory granules are PAS positive. The uncommon acinic cell carcinoma of the pancreas also contains secretory granules, which are PAS, phloxine-tartrazine and p-dimethylaminobenzaldehyde-nitrite (DMAB) positive.

All cells, normal or malignant, contain some glycogen, but its presence is particularly striking in certain malignant tumours. Glycogen is frequently found in 'clear' cell tumours, such as renal cell carcinoma, intercalated duct tumours of salivary glands and 'mesonephroid' ovarian carcinoma. Hepatocellular carcinoma contains glycogen together with bile pigment. Whilst bile pigment is visible on H and E staining as yellow–brown, van Gieson's stain will show bright green bile against a yellow background, making identification considerably easier.

Neuroendocrine tumours

An important group of tumours are those characterized by the presence on electron microscopy of dense core granules. Embryologically they are derived from neuroectoderm or endoderm. Neuroectodermal tumours include tumours of the adrenal medulla and paraganglia, Merkel cells, melanocytes, thyroid C cells, thymus, pituitary and pineal glands, whilst endodermal tumours encompass tumours arising from the gastrointestinal and urogenital tracts and their derivatives —lung, pancreas, prostate and ovary (see Ch. 27). Neuroendocrine tumours produce a variety of peptide hormones which can be characterized by immunohistochemistry. Light microscopy shows features variously called 'endocrine', 'carcinoid' or 'neurosecretory' and generally they stain positively by the Grimelius technique or one of its variants, and are therefore termed argyrophil. Some peptide-producing cells can reduce silver salts directly and are termed argentaffin. These cells are stained using the Masson-Hamperl or Masson-Fontana methods. Argentaffin cells may also be identified by their ability to couple with diazonium salts to produce insoluble diazo dyes. It is suggested that these cells contain large amounts of 5HT, and positivity is usually restricted to carcinoids of midgut origin. In addition, there are some histochemical techniques that may identify the peptides produced. Pancreatic islet cells can be identified on the basis of the pattern of reaction with a panel of four histochemical stains:

Aldehyde-fuchsin	B cells
Silver (Hellestrom technique)	D cells
Silver (Grimelius technique)	AD and PP cells
Lead haematoxylin (Solcia)	A and PP cells

The presence of melanin within tumour cells indicates a melanocytic tumour. Melanin is argentaffin, stains with the Masson-Fontana reaction, and may be differentiated from other argentaffin substances by bleaching.

The tumours of the adrenal medulla and extraadrenal paraganglia are conventionally divided into two groups on the basis of their chromaffin reaction. Fixation in a chromate-containing fixative and staining with Giemsa demonstrates green chromaffin granules. These granules contain noradrenaline and adrenaline together with their precursors. Chromaffin positivity is usually associated with tumours related to the sympathetic nervous system (e.g. phaeochromocytoma), whereas chromaffinnegative endocrine tumours are usually derived from the parasympathetic system (e.g. paraganglionoma), or from baroreceptors (e.g. chemodectoma).

Mesenchymal differentiation

Stains that reveal the differentiation of a tumour of mesodermal origin are limited. The demonstration of collagen or fat in a soft tissue tumour is of little diagnostic help. Glycogen is found in Ewing's tumour and primary muscle tumours but its presence is not diagnostic. It is useful in the differentiation of small round cell tumours of mesodermal origin from lymphomas which do not contain glycogen. The muscular origin of a tumour can be confirmed by demonstrating myofibrils using a trichrome or PTAH technique which will also reveal cross-striations in skeletal muscle. Endothelium of both lymphatics and blood vessels binds UEA 1 lectin and this can be used to determine the endothelial origin of a tumour. UEA 1 is more sensitive than factor VIII-related antigen, especially in staining the endothelium of capillary-sized vessels (Ordonez & Batsakis 1984, Stephenson et al 1986). Frequently, however, the morphology on H and E staining and the growth pattern of the tumour as demonstrated by reticulin stains are more helpful than elaborate histochemistry. Immunohistochemistry is of greater use than conventional histochemistry in the differential diagnosis of sarcomas, although mucin histochemistry may help to distinguish between the myxoid connective tissue tumours (Table 5.2, see also Ch. 21).

Lymphomas

The development of immunohistochemistry has largely displaced histochemistry from the diagnosis of lymphomas, although the PAS and Unna-Pappenheim (U-P) techniques are still useful. The PAS technique demonstrates intracellular immunoglobulin in lymphoplasmacytoid lymphomas and occasionally follicle centre cell lymphomas (Dutscher bodies), together with Russell bodies in lymphoplasmacytoid lymphoma and myeloma. The U-P highlights cells of immunoblastic and later stages of differentiation. Nevertheless, immunohistochemistry, interpreted in the light of the morphology and architecture, is considerably more helpful in the diagnosis of lymphoreticular neoplasms. Peanut agglutinin (PNA) has an affinity for cells of Langerhans cell histiocytosis while not detecting normal Langerhans cells or interdigitating reticulum cells. The pattern of cellular reactivity can be used to differentiate between the cells of Langerhans cell histiocytosis and macrophages (Hadju et al 1986, Ree & Kadin 1986, Jaffe 1987, see Chs. 11 and 20).

STAINS FOR CYTOLOGICAL FEATURES

There has recently been increased interest in the analysis of individual cells in histological sections or in tumour cell suspensions in an attempt to define more accurately the transition from normal to hyperplasia through dysplasia to malignancy. It is also hoped that information relating to diagnosis and prognosis will become available through cytological analysis. In parallel with the experimental evidence linking changes in DNA content of cells to malignant transformation, much effort has been put into examination of the tumour cells to find morphological correlates with the results of DNA analysis which show, for instance, oncogene activation, mutation and increased numbers of copies of certain genes. Abnormalities of DNA that lead to malignant transformation of a cell must also be reflected in transcription of DNA to RNA and its subsequent translation into proteins.

The techniques available for investigation of DNA content and expression include:

1. Cytophotometric analysis of DNA.
2. Flow cytometry of cell suspensions.
3. DNA probes for oncogenes.
4. Detection of gene rearrangements.
5. Transcription products of oncogenes.
6. Analysis of nucleolar structure and function.
7. Analysis of cell surface components.

The role of in situ hybridization with DNA probes in diagnostic pathology is at present limited by the availability of probes of known significance, and gene rearrangement studies cannot be performed on paraffin-processed tissue sections. However, it is becoming apparent that there is diagnostic and prognostic information to be gained from cytophotometric analysis of total DNA content of a cell. Two methods are currently used, first, cytophotometry of tissue sections stained by the Feulgen technique for DNA, and second, flow

cytometry using fluorochrome-stained DNA (Friedlander et al 1984, Quirke & Dyson 1986). Although a normal (diploid) DNA content does not exclude dysplasia or malignancy, the development of hypertetraploid or aneuploid cell lines is strongly predictive of malignant potential or change. The development of dysplasia and malignancy in ulcerative colitis and the adenoma-carcinoma sequence in the genesis of adenocarcinoma of the skin have been linked to abnormal DNA content (Cuvelier et al 1987, Hammarberg et al 1984, Goh & Jass 1986). Tumour DNA content has been linked to morphological indicators of prognosis, Cuvelier et al (1987) reporting aneuploid colonic carcinomas as having a poorer prognosis than those which were polypoid. Goh & Jass (1986) failed to show a correlation between ploidy and the degree of differentiation in large bowel tumours and a similar conflict of results has been noted for gastric carcinoma (Macartney et al 1986, Hattan et al 1984, Inokuchi et al 1983). It would appear that determination of DNA content is useful in identifying bladder tumours that have a higher risk of progression (Gustafson et al 1982a, b). In the future, however, total DNA content is likely to be more widely used as an adjunct to more conventional histochemical methods in tumour diagnosis.

Nucleolar function

The regions on the chromosomes associated with the ribosomol RNA genes — the nucleolar organizer regions (NORs) — can be identified in paraffin sections using a colloidal silver technique. This stains proteins associated with the chromosomal DNA at particular sites (Ag-NORs) (Plotan et al 1986). Such sites are present on five chromosomes in man, and differences in their staining characteristics, both size and number, between normal or hyperplastic tissues and their tumours might help in the diagnosis of borderline lesions and tumours (Crocker & Nar 1987, Egan et al 1987, Crocker & Skilbeck 1987, Egan & Crocker 1988, Smith & Crocker 1988). Significant differences in mean numbers of Ag-NORs per nucleus have been found between high and low grade lymphomas, benign and malignant melanocytic lesions and between Ewing's sarcoma, rhabdomyosarcomatous tumours and neuroblastomas in childhood. The differences described between various types of skin adnexal tumours are difficult to assess because of the small sample numbers. The results from the study of breast lesions do not show clear cut differences between benign and malignant lesions. It is apparent, however, that there are potential uses for this technique in tumour pathology. As stressed by Walker (1988), the results need to be confirmed by other workers. Further studies to define the distribution of Ag-NORs in tumour subgroups and in dysplasias (e.g. melanocytic dysplasia/'dysplastic naevi') would help consolidate the role of this technique, although an initial study on gastric mucosa and adenocarcinoma is disappointing (Suarez et al 1989).

The cell surface is important in intercellular communication and in interactions between cells and their connective tissue matrix. Changes in composition, such as increased negative charge and increased amounts of sialic acid, for example, have been noted in association with malignant transformation. In view of the likely role of cell surface components in determining invasive and metastatic potential, attempts have been made to analyse glycoproteins bound to the cell membrane and relate them to subsequent progress of the disease. Cell membrane glycoproteins consist of a protein core covered by oligosaccharides which bind lectins derived from a variety of plant and animal sources. As with antigen/antibody reactions, binding of lectins to their oligosaccharide targets is dependent upon spatial relationships and may be affected by fixation and subsequent treatment of the tissue (Bell & Skerrow 1984, Allison 1987, Alroy et al 1988).

In the assessment of metastatic potential in breast carcinoma the binding of helix pomatia agglutinin (HPA) has been found to correlate well with lymph node stage, time to first recurrence and survival time in premenopausal patients (Leathem et al 1984, Leathem et al 1985, Fenlon et al 1986, Leathem & Brooks 1987, Brooks & Leathem 1989). Binding of UEA1 also correlated with disease-free interval and survival (Fenlon et al 1986), whereas PNA showed no helpful correlations (van der Linden et al 1985, Walker et al 1985). Attempts have been made to show similar correlations between lectin binding and prognosis for tumours of the gastrointestinal tract, without conclusive results (see Ch. 12).

CONCLUSION

The non-immunohistochemical techniques described emphasize the use of 'special stains' as the next stage after H and E staining in the diagnosis of tumours, by demonstrating tumour structure and function. They are easy to perform and cheap and should not be forgotten in the approach to tumour diagnosis where reliance of a panel of antibodies is sometimes erroneously considered to be more important than morphology.

The introduction of techniques to measure DNA on formalin-fixed paraffin sections and the availability of a battery of lectins may, in the future, provide important information about malignant transformation and subsequent prognosis.

REFERENCES

Allison R T 1987 The effects of various fixatives on subsequent lectin binding to tissue sections. Histochemical Journal 19: 65–74

Alroy J, Ucci A A, Pereira M E A 1988 Lectin histochemistry. An update. In: DeLellis R A (ed) Advances in immunohistochemistry. Raven Press, New York, Ch. 5, 93–131

Atkin N B, Kay R 1979 Prognostic significance of modal DNA values and other factors in malignant tumours. British Journal of Cancer 40: 210–221

Bell C M, Skerrow C J 1984 Factors affecting the binding of lectins to normal human skin. British Journal of Dermatology 111: 517–26

Brooks S, Leathem A 1989 Histochemical demonstration of carbohydrate receptors in human tissues and their proposed role in secondary spread of breast cancer. Proceedings of the Royal Microscopical Society 24 (1): A53–54

Crocker J, Nar P 1987 Nucleolar organiser regions in lymphomas. Journal of Pathology 151: 111–118

Crocker J, Skilbeck N 1987 Nucleolar organiser region associated proteins in cutaneous melanotic lesions: a quantitative study. Journal of Clinical Pathology 40: 885–889

Cuvelier C A, Morson B C, Roels H J 1987 DNA content in cancer and dysplasia in chronic ulcerative colitis. Histopathology 11: 927–940

D'Ardenne A J 1989 Use of basement membrane markers in tumour diagnosis. Review. Journal of Clinical Pathology 42: 449–457

Egan M J, Crocker J 1988 Nucleolar organiser regions in cutaneous tumours. Journal of Pathology 154. 247–254

Egan M J, Raafat F, Croker J, Smith K 1987 Nucleolar organizer regions in small cell tumours of childhood. Journal of Pathology 153: 275–280

Fenlon S, Ellis I O, Elston C W, Bell J, Todd J, Blamey R W 1986 Investigation of Ulex europaeus I and Helix pomatia lectin binding in primary breast carcinoma. Journal of Pathology 149: 250A

Friedlander M L, Hedley D W, Taylor I W 1984 Clinical and biological significance of aneuploidy in human tumours. Journal of Clinical Pathology 37: 961–974

Goh H S, Jass J R 1986 DNA content and the adenoma-carcinoma sequence in the colorectum. Journal of Clinical Pathology 39: 387–392

Gonzalez-Campora R, Montero C, Martin-Lacave I, Galera H 1986 Demonstration of vascular endothelium in thyroid carcinomas using Ulex europaeus I agglutinin. Histopathology 10: 261–266

Gustafson H, Tribukait B, Espositi P L 1982a DNA pattern, histological grade and multiplicity related to recurrence in superficial bladder tumours. Scandinavian Journal of Urology and Nephrology 16: 135–139

Gustafson H, Tribukait B, Espositi P L 1982b The prognostic value of DNA analysis in primary carcinoma-in-situ of the urinary bladder. Scandinavian Journal of Urology and Nephrology 16: 141–146

Hadjin I, Zhang W, Gordon G B 1986 Peanut agglutinin binding as a histochemical tool for the diagnosis of eosinophilic granuloma. Archives of Pathology and Laboratory Medicine 110: 719–721

Hammarberg C, Slezak P, Tribukait B 1984 Early detection of malignancy in ulcerative colitis — a flow cytometric study. Cancer 53: 291–295

Leathem A J, Brooks S A 1987 Predictive value of lectin binding on breast cancer recurrence and survival. Lancet i: 1054–56

Leathem A, Dokal I, Atkins N 1984 Carbohydrate expression in breast cancer as an early indicator of metastatic potential. Journal of Pathology 142: 32A

Leathem A, Atkins N, Eisen T 1985 Breast cancer metastasis, survival and carbohydrate expression associated with lectin binding. Journal of Pathology 145: 73A

van der Linden J G, Baak J P A, Linderman J, Smeulders A W M, Meyer C J L M 1985 Carcinoembryonic antigen expression and peanut agglutinin binding in primary breast cancer and lymph node metastases: lack of correlation with clinical, histopathological, biochemical and morphometric features. Histopathology 9: 1051–1059

Macartney J C M, Camplejohn R S, Powell G 1986 DNA flow cytometry of histological material from human gastric cancer. Journal of Pathology 148: 273–277

Ordonez N G, Batsakis J G 1984 Comparison of Ulex europaeus I lectin and factor VIII-related antigen in vascular lesions. Archives of Pathology and Laboratory Medicine 108: 129–132

Ploton D, Menager M, Jeannesson P, Himber G, Pigeon F, Adnett J J 1986 Improvement in the staining and in the visualisation of the argyrophilic proteins of the nucleolar organiser regions at optical level. Histochemical Journal 18: 5–14

Quirke P, Dyson D 1986 Flow cytometry: methodology and applications in pathology. Journal of Pathology 149: 79–88

Quirke P, Dixon M F, Clayden A D et al 1987 Prognostic significance of DNA aneuploidy and cell proliferation in rectal adenocarcinomas. Journal of Pathology 151: 285–292

Ree H J, Kadin M E 1986 Peanut agglutinin: a useful marker for histiocytosis X and interdigitating reticulum cells. Cancer 57: 282–287

Suarez V, Newman J, Hiley C, Croker J, Collins M 1989 The value of NOR numbers in neoplastic and non-neoplastic epithelium of the stomach. Histopathology 14 (1): 61–66

Smith K, Crocker J 1988 Evaluation of nucleolar organiser region-associated proteins in breast malignancy. Histopathology 12: 113–125

Stephenson T J, Griffiths D W R, Mills P M 1986 Comparison of Ulex europaeus I lectin binding and factor VIII-related antigen as markers of vascular endothelium in follicular carcinoma of the thyroid. Histopathology 10: 251–260

Walker R A 1985 The use of lectins in histopathology. Histopathology 9: 1121–1124

Walker R A 1988 The histopathological evaluation of nucleolar organising regions. Histopathology 12: 221–223

Walker R A, Hawkins R A, Miller W R 1985 Lectin binding and steroid receptors in human breast carcinomas. Journal of Pathology 147: 103–106

6. Metabolic disorders — general view

B. D. Lake

COMMUNICATION

Metabolic disorders are generally rare and their investigation is usually carried out in centres specializing in such conditions. It may be possible to refer appropriately prepared tissue but unless there is an adequate degree of communication between the physician whose patient is under investigation, the surgeon who will take the biopsy and the pathologist who has to ensure that the biopsy is handled correctly, the appropriate preparative techniques cannot be anticipated. Figure 6.1 illustrates the triangle of communication.

Each should be aware of what is to be done, when it is to be done and what can be expected of the procedure. Without the necessary communication valuable tissue will be wasted, perhaps resulting in another operation with the attendant risks.

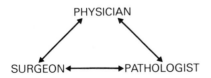

Fig. 6.1 The triangle of communication.

FREEZE SOME — FIX SOME

In the investigation of a biopsy from a patient with a suspected metabolic disorder, histology is probably the least important aspect. However, some portion of the tissue should be fixed in formalin for routine sections, to provide some morphological assessment in case the suspected diagnosis does not materialize. For example a needle biopsy of liver taken for the diagnosis of glycogen storage disease should be divided so that at least 50% of the tissue (weighing a minimum of 8 mg) is frozen for biochemical assay, 25% is taken for histochemistry and frozen for cryostat sections, and the remaining 25% is divided between electron microscopy and histology. Whatever the diagnosis might be, it is valuable to take a portion of the biopsy and freeze some for preparation of cryostat sections. This is true for all specimens, whether or not a metabolic disorder is considered.

Many pathologists feel that cryostat sections are often of inferior quality. However, provided that the tissue has been frozen properly, the knife in the cryostat is sharp, and the anti-roll plate correctly adjusted there is no reason why perfect sections cannot be obtained. With the almost unlimited selection of techniques available for demonstration of practically everything within the cell, cryostat sections offer the most potent means of microscopic diagnosis in the majority of metabolic disorders.

A sample of fresh tissue will allow not only most histochemical techniques but is also available for biochemical assay and for extraction of components for detection by thin-layer chromatography. Although fresh tissue is important for most thin-layer procedures, sphingomyelin, cerebrosides, sulphatides, cholesterol and its esters, and triglycerides may be adequately detected in a semiquantitative fashion in formalin-fixed tissue. By semiquantitative, I mean that the amount of the compound in question can be compared with that in normal tissue fixed and extracted in the same way provided that equivalent amounts of tissue are applied to the thin-layer plate (Lake & Goodwin 1976).

The variety of substances encountered in the metabolic disorders means that a variety of methods are necessary for their preservation and subsequent detection. Cystine may be relatively insoluble, but cystinotic tissues fixed in aqueous media, processed and sectioned by routine methods will have had nearly all the cystine extracted because of the large volumes of aqueous solutions washing over the few micrograms of cystine present in a section. Cryostat sections of snap-frozen tissue stained in an alcoholic dye solution, or sections from alcohol-fixed tissue floated out on alcohol are necessary for demonstration of cystine. There are many methods for mucosubstances applicable to routine sections but the mucopolysaccharide deposited in the several mucopolysaccharidoses is not protein-bound (i.e. not a proteoglycan) and is extremely water-soluble. Attempts to preserve the soluble mucopolysaccharides by fixing tissue blocks in Lindsay's fixative (picric-dioxan) are only partially successful and extremely wasteful in valuable tissue. Snap-frozen tissue offers the best preservation of mucopolysaccharides, but detection through metachromasia with toluidine blue in one guise or another, although sensitive, does not give good localization in tissue sections.

In the lipid storage diseases the deposited lipid survives formalin fixation and can be demonstrated in standard frozen sections. Cryostat sections of snap-frozen tissue can also be used. However the routine processing schedules of dehydration, clearing and impregnation with wax followed by sectioning, dewaxing etc. effectively removes most of the storage product. Thus the diagnosis of metachromatic leucodystrophy will be thwarted if routine sections of nerve are used. In some of the lipid storage diseases different cell types accumulate the substance in different forms, some of which resist standard procedures. The glial storage in Tay-Sachs disease is thus PAS-positive while the neuronal storage has been extracted to give a negative reaction (Fig. 6.2). The interstitial foamy cells in the kidney in Fabry's disease are PAS-positive in routine sections in contrast with the ballooned foamy glomerular epithelial cells which have had their stored hexosides extracted during processing. Smooth muscle deposits in Fabry's disease are retained in processing not only to show PAS and Sudan black positivity but also to exhibit

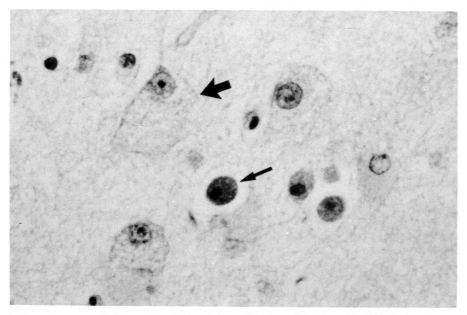

Fig. 6.2 Tay-Sachs disease. Routine section of brain stained by the PAS technique. Large ballooned neurons (large arrows) are unstained because the stored ganglioside has been extracted during processing. The glial cell storage (small arrows) resists extraction. ×600

birefringence. It is clear that wherever possible, sections of snap-frozen tissue are to be preferred, but in many instances formalin-fixed frozen sections can be used for diagnosis and in isolated cases routine paraffin wax-embedded tissue is adequate.

BIOPSY SITE

Metabolic disorders manifest their presence in many ways and it is not always necessary to biopsy the main target organ. In G_{M1}-gangliosidosis the brain stores the ganglioside within the neurons which are also accessible for diagnosis by suction rectal biopsy performed as an outpatient procedure. However a bone marrow aspirate will confirm the generalized nature of the disorder (if needed) or blood films can be examined (see Ch. 20) and the diagnosis made by finding vacuolated lymphocytes and an absence of β-galactosidase activity. For other disorders urine sediment, a skin biopsy, jejunal biopsy, conjunctival biopsy, or even hair might be the site of choice.

DETECTION OF ENZYME ACTIVITY

Many of the metabolic disorders are a result of a deficient lysosomal enzyme activity. In order to detect the activity of most lysosomal enzymes it is necessary to use tissue fixed in formal-calcium (at + 4°C for up to 18 h). The tissue is rinsed before freezing or can be washed and soaked in gum-sucrose prior to freezing. Free-floating frozen or cryostat sections can then be prepared and incubated in the appropriate medium with sections of a normal control tissue, so that a valid comparison may be made. To be able to diagnose a deficiency state there should be an adequate reaction in the control normal tissue and absence in the case under investigation. Some lysosomal enzyme activities are amenable to demonstration in sections of snap-frozen tissue but acid esterase activity in particular must be detected in sections of formal-calcium/gum sucrose-treated tissue.

Occasionally tissue will show an apparent increase in lysosomal enzyme activity. This increase may not be a quantitative increase but reflects a change in the state of the cell and indicates either greater catabolic activity or that the cell is affected by a lysosomal storage disease. In the latter instance the increase may be gross, even though the routine sections do not appear to show involvement. For example, an acid phosphatase reaction on cryostat sections is particularly helpful to show that the liver in Niemann-Pick type C and Farber's disease is involved in the disorder. Endothelial cell involvement in Hunter's disease (Fig. 6.3), smooth muscle cell involvement in Fabry's disease and the presence of storage cells in a bone marrow aspirate can also be easily detected by application of an acid phophatase reaction.

In the detection of enzyme activity some background knowledge of the biochemistry of the enzymes involved is advisable. Many enzymes are present in tissues in several isoenzyme forms and although, for example, sulphatase A may be absent in metachromatic leucodystrophy the presence of sulphatase B makes it impossible to use histochemical techniques for the diagnosis of sulphatase A deficiency. Similar considerations apply to α-mannosidase in mannosidosis and hexosaminidase in the G_{M2}-gangliosidoses for which adequate biochemical methods are otherwise available. In any test to detect a deficiency of some particular enzyme activity it is most important to know that it is possible to detect the deficient activity and a known deficient tissue should have been examined to validate the method. A positive control with adequate activity should also be included.

The specificity of the method should be examined, not only with a deficient case wherever possible, but to test that the reaction is showing what was intended. In studies of the reaction to demonstrate ornithine carbamylphosphate transferase (OCT) activity two complicating factors are apparent. Firstly carbamylphosphate spontaneously hydrolyses, releasing phosphate ions, and secondly carbamyl phosphate appears to act as a substrate for glucose-6-phosphatase. So attempts to demonstrate deficiency or even OCT activity by a lead-phosphate capture reaction are unlikely to be successsful in unfixed tissue. However, the method does appear valid in lightly fixed tissue and with modifications of the method, Wareham and her colleagues (1983) could detect deficiency in a strain of mice. In addition, it was also possible

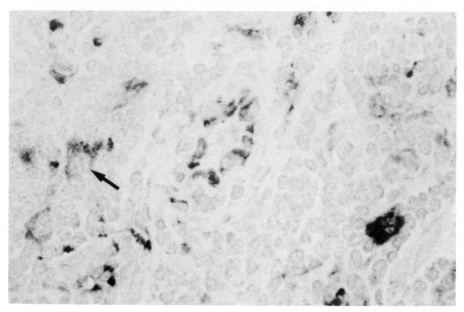

Fig. 6.3 Hunter's disease (mucopolysaccharidosis type 2). Cryostat section of adenoid stained to demonstrate acid phosphatase activity. Endothelial cells (centre), normally negative, show activity indicating their involvement in the disease. A large foamy storage cell is present (arrow). ×600

to show a mosaic of activity in the female carriers of this X-linked condition, with some hepatocytes active and others with deficient activity. This mosaic is the result of X chromosome inactivation predicted by the Lyon hypothesis.

FUNCTIONAL MARKERS

The metabolic disorders affect the whole range of cell organelles and it may be useful to use methods to detect particular organelles, either to show their presence or to indicate their activity. In Zellweger's cerebro-hepato-renal syndrome and in infantile Refsum's disease there is an absence of peroxisomes in liver and kidney. This absence can be demonstrated in tissue prepared for peroxidase activity. Although animal tissue gives a good reaction, human liver or kidney requires a rather more careful preparation to ensure that peroxisomes are reliably demonstrated (see Ch. 16). Defects in the urea cycle and other metabolic disorders in which an infection may precipitate raised ammonia levels, can be mistaken for Reye's syndrome. A liver biopsy taken for assay of one or other of the urea cycle enzymes can be used to exclude Reye's syndrome; the succinate dehydrogenase reaction is normal in the urea cycle defects and in disorders of fatty acid oxidation, but in Reye's syndrome no activity can be detected in an extremely fatty liver if the biopsy is taken within 48 hours of the onset of encephalopathy. The lipid droplets in Reye's syndrome are usually described as microglobular, in contrast with the macroglobular droplets found in more chronic conditions. The microglobular nature is probably an acute phenomenon which will become macroglobular with time. A variety of toxins can also damage mitochondrial function and lead to absent succinate dehydrogenase activity. In other situations where mitochondrial function is abnormal, and numbers of mitochondria are increased and their structure is altered, an increased succinate dehydrogenase activity may be apparent. For example in the mitochondrial cytopathies (Egger et al 1981) the accumulation of abnormally structured mitochondria in ragged-red muscle fibres is readily detected after short incubation times, before the normal fibres show any reaction product (Fig. 6.4).

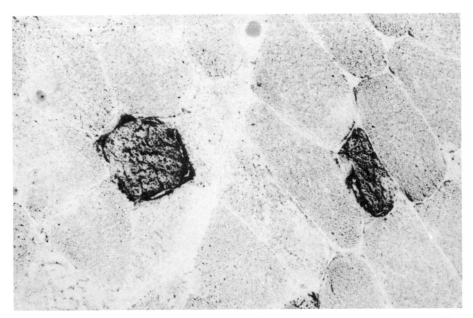

Fig. 6.4 Mitochondrial cytopathy. Cryostat section of a muscle biopsy stained to demonstrate succinate dehydrogenase activity. Two muscle fibres show excess activity indicating collections of abnormal mitochondria. These fibres correspond to the 'ragged-red' fibres seen in trichrome preparations. ×370

MULTIPLE STORAGE PRODUCTS

It used to be considered that in a defect of one particular enzyme only one storage product should accumulate, but it has become clear that many products are stored in most of the lysosomal enzyme deficiencies. There are two or possibly three main reasons for this phenomenon. The first is that the cell may find it difficult to accommodate a compound with a strongly polar end group or one with an awkward shape. To overcome the difficulty the cell may find it convenient to store the substance in association with other substances which may either neutralize the charge and/or help to pack the compound in an acceptable shape. Examples here are of the gangliosidoses where the gangliosides are stored as membranous cytoplasmic bodies (MCBs) with cholesterol and phospholipid. In Niemann-Pick disease (types A and B) the marked cholesterol content led some to believe at one time that the disorder should be considered as a cholesterol disorder as well as a sphingomyelin disorder. Subsequent studies have shown that deficient activity of sphingomyelinase is the cause

of the disorder and that the cholesterol is there to package the sphingomyelin. Since the substances involved in the majority of lysosomal storage disorders are lipids, and other lipids are soluble in the lipid complexes, it is not surprising that several lipid components may be isolated from the storage cell.

The second reason for multiple storage products is that most of the enzymes are not substrate-specific and only recognize a particular end group. Thus in β-galactosidase deficiency any compound with a terminal β-galactose link with will tend to accumulate. This will include G_{MI}-ganglioside and its asialo derivative G_{A1} in the brain, haematoside from red blood cell membranes in the liver and spleen and the hexasaccharide unit derived from the molecule shown in Figure 6.5. The deficiency of acid esterase in Wolman's disease leads to storage of cholesteryl esters and triglycerides (both are esters) and a variety of other minor esters (Fig. 6.6). Because of the variety of stored products no one special stain is the right one for the diagnosis but rather a range of special stains, coupled with careful assessment of the different cell types af-

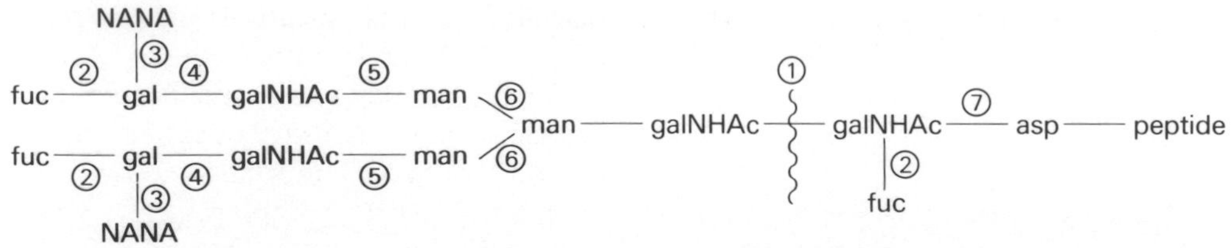

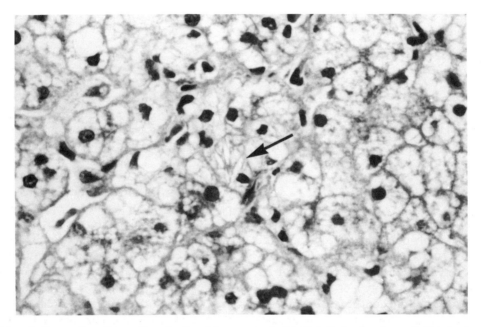

Fig. 6.5 A sialyloligosaccharide unit common to immunoglobulins and several plasma proteins, hormones etc. Cleavage by an endo-β-N-acetylglucosaminidase as shown at 1 leads to the 12-membered compound which is susceptible to hydrolysis by a number of glycosidases. 2 α-fucosidase, 3 sialidase, 4 β-galactosidase, 5 hexosaminidase B, 6 α-mannosidase, 7 aspartylglycosaminamidohydrolase. fuc = fucose, gal = galactose, galNHAc = N-acetylgalactosamine, man = mannose, asp = asparagine, NANA = N-acetylneuraminic acid.

Fig. 6.6 Wolman's disease. Rountine section of a liver biopsy stained with H & E showing foamy lipid-laden hepatocytes and storage cells. One of the storage cells (arrow) contains cholesteryl ester crystal shapes. ×600

fected and their morphologies, is required for accurate diagnosis. It should be pointed out at this stage that a set of staining properties coupled with a morphological appearance of, say neurons, does not necessarily mean that the diagnosis is the same as in another case with apparently similar characteristics. In Tay-Sachs disease (G_{M2}-gangliosidosis) the staining reactions in the brain are virtually indistinguishable from those in G_{M1}-gangliosidosis. Inspection of the structure of the two gangliosides will show the reasons for this. The two disorders can be distinguished by either thin-layer chromatography of solvent extracts of brain (Lake & Goodwin 1976) or by staining to demonstrate β-galactosidase activity which will be absent in G_{M1}-gangliosidosis.

Lectins may be used to detect a particular terminal saccharide moiety in the storage disorders and have been applied to cultured fibroblasts and cryostat and paraffin sections (Alroy et al 1988). However, the results are not always as predicted and the galactocerebroside in Krabbe's leucodystrophy is stained by succinylated wheat germ agglutinin which normally detects terminal N-acetyl-glucosamine.

FETAL PATHOLOGY AND THE PLACENTA

In the examination of fetal tissue for confirmation of storage disease predicted by enzyme assay of chorionic villus, cultured amniotic fluid cells, or by examination of amniotic fluid, it may be difficult to detect evidence of storage. This is particularly true in the mucopolysaccharidoses where, unless electron microscopy is undertaken, confirmation of the diagnosis by microscopy is rarely possible. Although in some disorders fetal tissue involvement may be relatively easy to detect (Lake 1977), with the advent of chorionic villus sampling, fetal tissue samples from termination of pregnancy are usually difficult to identify. However, in many instances (G_{MI}-gangliosidosis, I-cell disease, sialic acid storage disease, Niemann Pick disease type A and the mucopolysaccharidoses) the placenta affords plentiful evidence of disease (Fig. 6.7). Bone marrow from the femur of a fetus is also a helpful site to detect the presence of storage cells, and it is usually possible to prepare one or two marrow films from the site, even from a fetus of 11 weeks' gestation. The chorionic villus sample may also be used for morphological assessment to supplement the enzyme assay in some conditions (Lake et al 1989).

BIOCHEMICAL CONFIRMATION REQUIRED

In all metabolic disorders it is mandatory to have the diagnosis confirmed by appropriate enzyme assay. This can be performed on the frozen tissue from which the cryostat sections have been taken. Why reach a diagnosis by microscopy and biochemical assay? Tissue biopsies are often very small and allow only one or two assays. Histochemical assessment is valuable in establishing the identity of the tissue and in providing a working diagnosis which then only requires minimal biochemistry for confirmation.

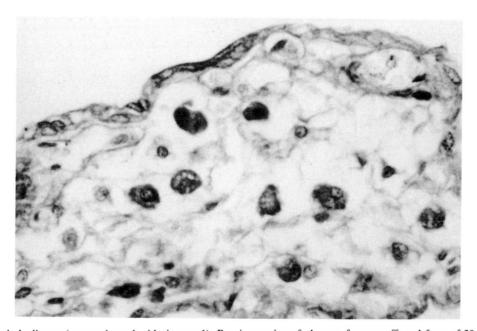

Fig. 6.7 Hurler's disease (mucopolysaccharidosis type 1). Routine section of placenta from an affected fetus of 20 weeks, stained with the colloidal iron technique. Large foamy storage cells are present in the stroma of a villus. ×650

REFERENCES

Alroy J, Ucci A A, Periera M E A 1988 Lectin histochemistry. An update. In: Delellis R A (ed) Advances in immunohistochemistry. Raven Press, New York, p 93–131

Egger J, Lake B D, Wilson J 1981 Mitochondrial cytopathy. A multisystem disorder with ragged-red fibres in muscle. Archives of Disease in Childhood 56: 741–752

Lake B D 1977 Histochemical and ultrastructural studies in the diagnosis of inborn errors of metabolism. Records of the Adelaide Children's Hospital 1: 337–345

Lake B D, Goodwin H J 1976 Lipids. In: Smith I, Seakins J W T (eds) Chromatographic and electrophoretic techniques, 4th edn. Heinemannn, London, vol 1, ch 14

Lake B D, Young E P, Nicolaides K 1989 Prenatal diagnosis of sialic acid storage disease in one of twins. Journal of Inherited Metabolic Disease 12: 152–156

Wareham K A, Howell S, Williams D, Williams E D 1983 Studies of X-chromosome inactivation with an improved histochemical technique for ornithine carbamoyltransferase. Histochemical Journal 15: 363–371

GENERAL BACKGROUND READING

Scriver C R, Beaudet A L, Sly W S, Yalle D (eds) 1989 The metabolic basis of inherited disease, 6th edn. McGraw Hill, New York

7. Immunohistochemistry in neuro-oncology

J. Lowe

The ability to locate specific cellular antigens in histological sections by immunohistochemistry has provided a valuable tool for evaluating tumours of the nervous system. At the most simplistic level of usage, immunohistochemistry can be used to replace many of the empirical staining methods in neurohistology and may be considered as a form of 'special stain'. A more important use of immunohistochemistry is in the establishment of the histogenesis of a tumour by evaluation of the expression of particular antigens.

Within the nervous system the main groups of tumours are those of astrocytic, meningothelial, oligodendroglial, neuronal, germ cell, lymphoid and epithelial origins. In all these instances, diagnosis can be assisted by the use of immunohistochemical staining for antigens characteristic of these cell types (Fig. 7.1). Less frequent tumours with specialized mesodermal elements such as skeletal muscle or endothelial origins are also encountered in the nervous system and their diagnosis is also made more certain by immunohistochemical methods. For detailed discussion of tumours of the nervous system the reader is referred to Russel & Rubinstein (1989).

The reasons for performing immunohistochemistry

There are four main situations where immunohistochemical staining of a nervous system tumour may be performed. In each instance there are different expectations of the technique.

The tumour is of uncertain origin with few distinguishing morphological features by light microscopy

This may be due to small biopsy size, an atypical area at the infiltating edge of a tumour, or a poorly differentiated lesion such as a small-cell anaplastic tumour. Here, immunohistochemistry is used with the expectation that a pattern of antigen expression will reveal the histogenesis of the neoplastic cells by the detection of antigens thought to be characteristic of certain cell types. In this instance the specificity of each antibody for a particular antigen must be known with confidence and each antigen must have some restricted expression to provide specificity for cell types.

The tumour is of known histological type but has multipotential lines of cellular differentiation which may be of prognostic importance

An example is in the medulloblastoma where astrocytic, neuronal and mesodermal lines of differentiation may be seen. In this instance immunohistochemistry is performed to look for expected patterns of cellular differentiation by the detection of antigens thought to be characteristic of certain cell types. Again it is necessary that the specificity of each antibody for a particular antigen is known with confidence, and each antigen has restricted expression to provide specificity for cell types.

The diagnosis has been narrowed down by conventional techniques and lies between possibly two or three lesions

Immunohistochemical demonstration of antigens restricted to certain tumour types can provide additional information on cellular phenotype which, taken into account with morphology, weights the diagnosis towards one type of histogenesis rather than another. In this instance it is not absolutely

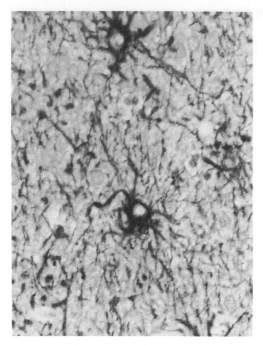

Fig. 7.1 Reactive astrocytes stained to show GFAP. Immunoperoxidase/haematoxylin. ×280

necessary for an antigen to be specific for a certain cell type, rather it is essential that its distribution in a large series of tumours is known, such that staining or lack of staining can be interpreted in terms of the probability of a particular tumour. One example of this is seen in the distinction between chondrosarcoma and chordoma where cytokeratin expression is seen in chordoma.

The tumour is of known histological type and immunohistochemical detection of a specific cellular antigen related to cellular proliferation is being performed to obtain information on the likely proliferative potential of a tumour

In this instance it is necessary to employ well characterized antibodies specific to proliferating cells to obtain a very specific end-point by counting the proportion of cells expressing the antigen. Here immunohistochemistry is providing information independent of tissue morphology, histogenesis, or differentiation, which can be of potential prognostic importance.

Requirements for diagnostic application of immunohistochemistry

There are three main requirements for the application of immunohistochemistry in diagnostic practice. The first is that the antibodies used are well characterized for a specific antigen. The second is that the pattern of staining of each antibody in a series of well characterized tumours has been performed. The third is that the interpretation of the immunohistochemistry is performed in the light of conventional morphology and clinical features of the tumour. Only if all three conditions are met can immunohistochemistry be used with any degree of certainty in a diagnostic setting.

It has been said that immunohistochemical stains should not be considered as specific markers but signposts to point to a direction of tumour differentiation (Rubinstein 1986). This is now even more true, as more thorough evaluation of antigen distributions in tumours of the nervous system has shown that antigens previously thought to be specific for a certain cell type are expressed by other cell types on rare occasions. It becomes essential that reliance is not placed on a single antigenic marker in assessing differentiation, but rather on a panel of antibodies known to be expressed preferentially by a certain cell type. In simple terms, if a tumour expresses four phenotypic markers of neuroendocrine differentiation, then there is more confidence of a neuroendocrine histogenesis than if it expresses only one. In addition, immunohistochemistry should not just be interpreted in terms of positive or negative staining for a particular tumour and interpretation should encompass the patterns of antigen expression in a tumour, both at a cellular and architectural level, which may be of diagnostic relevance.

The importance of critical interpretation of immunohistochemistry *in the light of all other information about a tumour* cannot be over-emphasized in view of the lack of any absolute antibody specificity.

There are problems in the use of immunohistochemical techniques in tumours of the nervous system as with other uses of this technique. At the outset, all interpretation must be performed in the understanding that neoplastic lesions can either lose or gain phenotypic markers. In practice this means that diagnostic evaluation is best performed by applying a panel of antibodies to each tumour to detect several types of differentiation antigens, for example typical of neuronal, glial and epithelial lineages. Most importantly, there are occasions when false positive reactions occur. These will usually become apparent by immunohistochemical evidence of differentiation being at variance with morphological data, for example apparent antineurofilament staining in an otherwise typical glial neoplasm. This may be due to genuine crossreaction, in which case other examples of similar crossreactivity should be sought. False negatives may be due to a lapse in technique, and it is important to use suitable controls (see Ch. 3). Certain mouse monoclonal antibodies are known to bind non-specifically to glial cells (Paasivuo & Saksela 1983, Perentes & Rubinstein 1986b). This can be investigated by using a monoclonal antibody of different specificity but of the same immunoglobulin class in control staining reactions, instead of the primary antiserum.

THE MAIN ANTIGENS USED IN DIAGNOSTIC IMMUNOHISTOCHEMISTRY OF THE NERVOUS SYSTEM

Neuron–specific enolase (NSE)

NSE is one form of a dimeric enzyme which has three possible subunits designated α, β, and γ. The γ-subunit is expressed at high levels in cells of neuroendocrine lineage and the $\gamma\gamma$-enolase isoenzyme was designated neuron-specific enolase for this reason.

While it is true that NSE is strongly expressed by many tumours and normal tissues of neuroendocrine origin (Royds et al 1982, Bishop et al 1982), the hope that NSE would be specific for cells of neuroendocrine lineage has not been realized. Astrocytomas, oligodendrogliomas, glioblastomas, meningiomas, ependymomas, choroid plexus papillomas, schwannomas and medulloblastomas may all exhibit focal areas of positivity on immunostaining with antibodies specific to the γ-subunit. In addition, carcinomas

of breast and lung which commonly metastasize to brain have been shown to exhibit NSE immunoreactivity (Ghobrial & Ross 1986, Cras et al 1988). NSE immunoreactivity with polyclonal antibodies cannot be used reliably as a specific marker of neuroendocrine lineage in CNS tumours; however, high levels of expression can be interpreted as part of a pattern of the neural phenotype. There has been a recent suggestion that monoclonal antibodies to NSE offer more specificity to cells of neuroendocrine lineage (Seshi et al 1988).

Protein gene product 9.5 (PGP 9.5)

PGP 9.5 is a soluble protein found in high concentrations in cells of neuroendocrine lineage (Doran et al 1983) and has been shown to be a ubiquitin C-terminal hydrolase (Wilkinson et al 1989). Monoclonal and polyclonal antibodies to this protein show staining in paraffin processed formalin fixed tissues in cells of neuroendocrine lineage but not normal astrocytes or oligodendrocytes. Normal ependymal cells exhibit focal positivity. Enteroendocrine cells of the small and large bowel show no immunoreactivity with polyclonal antibodies to PGP 9.5. Positive staining of normal tissues of non-neuroendocrine origin is seen in renal tubular epithelium, Leydig cells, spermatogonia and ovarian tissues (Wilson et al 1988).

PGP 9.5 is strongly expressed by many tumours of neuroendocrine lineage (Rode et al 1985). In tumours of the central nervous system an indirect immunoperoxidase technique using polyclonal antibody shows positive staining in neuroblastoma and medulloblastoma (Fig. 7.15), as well as strong diffuse cytoplasmic staining in astrocytomas, oligodendrogliomas, and glioblastomas (Fig. 7.11). In mixed tumours such as gangliogliomas, neural elements stain more intensely than the surrounding glial elements (Fig. 7.2). While this antibody is a sensitive marker for axons in tissues, it promises to be a useful marker of neuroendocrine differentiation in tumours only outside the central nervous system, provided morphological considerations are taken into account to exclude lesions of renal or ovarian origin.

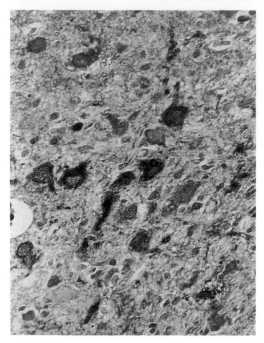

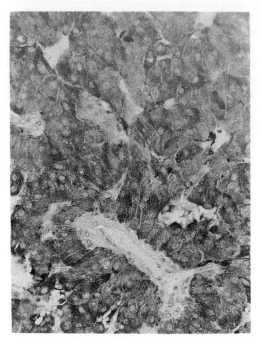

Fig. 7.2   Ganglioglioma stained to show PGP 9.5 localization. Neuronal elements in the tumour show dense reactivity, but surrounding glial elements also show immunoreactivity albeit at a lesser intensity. Immunoperoxidase/haematoxylin. ×280

Fig. 7.3   Immunohistochemical localization of synaptophysin in a paraganglioma arising in the filum terminale of the spinal cord. Cytoplasmic staining is granular reflecting the distribution of neurosecretory granules. Frozen section immunoperoxidase/haematoxylin. ×280

Synaptophysin

Synaptophysin is a membrane glycoprotein found in presynaptic vesicles (Weidenmann & Franke 1985) and can be demonstrated by immunohistochemical methods in many tumours of neuroendocrine origin (Gould et al 1986a). The intensity and extent of positive staining are dependent on the density of neurosecretory vesicles, hence tumours with small numbers of granules may not give obvious staining. Unlike chromogranin (see Ch. 29), this antigen is not restricted to granules of dense-core type.

Synaptophysin immunoreactivity is seen as a granular stain in the peripheral areas of neuronal cells, absent in the perikaryal region, and as a granular network corresponding to axonal and dendritic processes rich in synapses (Gould et al 1986a). Synaptophysin is a sensitive and specific marker for the presence of neuroendocrine granules in frozen section material (Fig. 7.3).

Careful tissue fixation in formalin also enables detection in paraffin processed tissues (Gould 1987).

Anti-Leu-7

Anti-Leu-7 recognizes antigens localized on normal lymphocytes with natural killer cell activity (Abo & Balch 1981), and also recognizes glycoproteins as part of myelin-associated glycoprotein (McGarry et al 1983), nervous system cell-adhesion molecules such as N-CAM (Kruse et al 1984), peripheral nerve associated glycolipid (Chou et al 1985), and a protein within neurosecretory granules (Tischler et al 1986). Immunohistochemical staining with anti-Leu-7 is seen in normal and neoplastic neuroendocrine cells (Caillaud et al 1984) and in prostatic epithelium (Rusthoven et al 1985).

In tumours of the nervous system, anti-Leu-7 stains a high proportion of schwannomas, most oligodendrogliomas, astrocytomas, glioblastomas and fewer ependymomas. Primitive neuroectodermal tumours of the CNS show small numbers of positive cells with the exception of medulloblastomas where numerous (30–40%) positive cells are seen (Perentes & Rubinstein 1986). Despite this broad range of expression there are several tumours which do not express anti-Leu-7 immunoreactivity, notably meningiomas (Motoi et al 1985, Perentes & Rubinstein 1985), fibrosarcomas and malignant fibrous histiocytomas (Perentes & Rubinstein 1985, Swanson et al 1987, Johnson et al 1988). Thus this antibody can be of use in distinguishing between neural and non-neural spindle cell tumours, particularly those arising in the meninges.

In the peripheral nervous system anti-Leu-7 is useful as part of a panel in the distinction of the histogenesis of peripheral nerve sheath tumours, showing positive staining in the majority of neurofibrosarcomas and neurilemmomas (Fig. 7.4), but absent staining in fibrosarcomas and the majority of leiomyosarcomas (Perentes & Rubinstein 1985, Johnson et al 1988, Swanson et

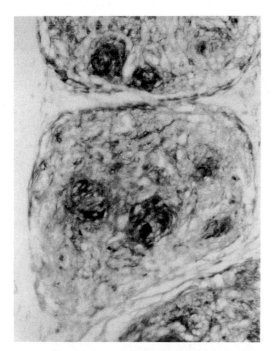

Fig. 7.4 Peripheral nerve sheath tumour showing positivity for Leu–7. Immunoperoxidase/haematoxylin. ×280

al 1987). The fact that positivity has been shown in a proportion of smooth muscle tumours means that caution needs to be exercised in interpreting Leu-7 positive staining of a spindle cell tumour as being indicative of nerve sheath origin (Swanson et al 1987).

Neurofilaments (NFP)

Neurofilaments are a class of intermediate filaments specific for the neural cytoskeleton. They are heteropolymers of three subunit proteins of molecular weights of around 68, 150, and 200 kD, which are chemically and antigenically distinct (Lee et al 1982). Neurofilament protein can be antigenically modified by post-translational phosphorylation (Carden et al 1985). This is particularly true of the 150 and 200 kD filament proteins, and antibodies raised to a phosphorylated epitope will not recognize non-phosphorylated epitopes (Sternberger & Sternberger 1983). Because of the variability in antigenicity of NFP and the differential expression of species with neuronal maturation, staining with a single monoclonal antibody to a specific NFP will not detect all types in all circumstances. By way of example, antibodies to 200kD NFP, frequently directed to phosphorylated epitopes, will demonstrate axonal staining but not perikaryal staining except in swollen neurons encountered in degenerative diseases (Dickson et al 1986). Many antibodies to NFP only work in fresh frozen material, particularly those to 68kD NFP; however, there are several available monoclonal antibodies which work reliably on paraffin sections.

Tumours of neural lineage which have poorly developed axonal processes (such as neuroblastoma) are seldom positive for the high molecular weight NFP. It has been reported that monospecific antibodies to 68kD NFP stain the majority of neural derived tumours including neuroblastoma cells not expressing 150 or 200kD types (Mukai et al 1986). Neurofilament protein expression has been shown in a wide range of neural derived tumours including paraganglioma, ganglioneuroma, ganglioglioma (Fig. 7.5), neuroblastoma, medulloblastoma, and phaeochromocytoma (Mukai 1986).

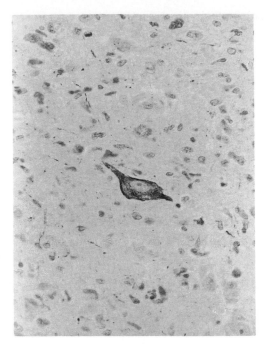

Fig. 7.5   Ganglioglioma stained to localize NFP 150. There is neuronal perikaryal staining and also fine staining of axons in the lesion. Glial elements are unstained. Immunoperoxidase/haematoxylin. ×280

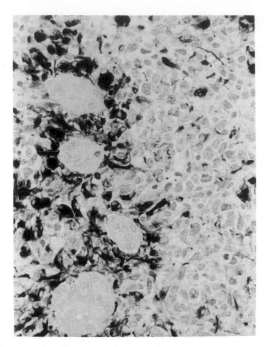

Fig. 7.6   Anaplastic astrocytoma of cerebral hemisphere showing GFAP expression by many cells, especially in a perivascular distribution in a manner reminiscent of so-called astroblastic differentiation. Many tumour cells do not express GFAP despite undoubted glial lineage. Immunoperoxidase/haematoxylin. ×280

Glial fibrillary acidic protein (GFAP)

GFAP is an intermediate filament protein which has a restricted distribution in normal tissues. It is especially expressed by developing and mature cells of astrocytic lineage (Fig. 7.1), but in addition is expressed by reactive ependymal cells. Unusual instances of extraneural immunoreactivity for GFAP have been reported in salivary gland (Nakazato et al 1982, Budka 1986), chondrocytes of epiglottic cartilage, cells of schwannomas (Gould et al 1986 b), metastatic renal adenocarcinoma, and meningioma (Budka 1986).

Antibodies to GFAP are commonly used to establish the glial nature of a neoplasm, being expressed in paraffin-processed formalin fixed tissues from astrocytomas of fibrillary (Fig. 7.6), protoplasmic, pilocytic (Fig. 7.7) and gemistocytic types (Bonnin & Rubinstein 1984). Not all cells within astrocytomas will express GFAP, particularly cells with few processes such as are seen in cellular areas of low grade cerebellar astrocytomas. Poorly differentiated glial neoplasms tend to have a smaller proportion of GFAP

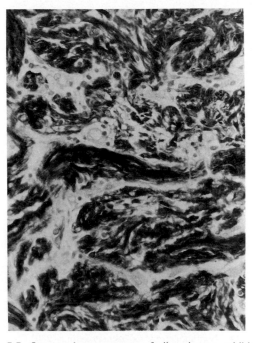

Fig. 7.7   Low grade astrocytoma of pilocytic type exhibiting dense GFAP immunoreactivity in most cells. Immunoperoxidase/haematoxylin. ×280

immunoreactive cells than well differentiated tumours. This is particularly true in the small infiltrating cells at the periphery of a tumour, which tend to have few cell processes, scanty cytoplasm and a low incidence of GFAP expression. Cells of glioblastoma and gliosarcoma (Fig. 7.10) may express GFAP, but, as in the case of poorly differentiated astrocytic tumours, the number of positive cells may be small. In most astroglial tumours neoplastic cells frequently express vimentin. In glioblastomas containing giant cells the large cells commonly strongly express vimentin without GFAP (Herpers et al 1986). Ependymomas express GFAP particularly in fibrillated perivascular pseudorosettes, but focal expression may also be seen in epithelial areas. The spinal densely fibrillated forms of ependymoma strongly express GFAP. Choroid plexus neoplasms may show focal expression of GFAP in addition to cytokeratin (Fig. 7.13)(Doglioni et al 1987). Oligodendrogliomas commonly show GFAP expression in small cells which resemble very small gemistocytic cells (Herpers & Budka 1984, Nakagawa et al 1986). GFAP expression may be demonstrated in primitive neuroectodermal neoplasms such as medulloblastoma (Fig. 7.16), where it has been interpreted as evidence for glial differentiation in a multipotential tumour (Palmer et al 1981, Herpers & Budka 1985, Burger et al 1987).

The detection of positive GFAP immunostaining has also allowed the delineation of unusual forms of glioma such as the pleomorphic xanthoastrocytoma (Kepes et al 1979) and lipidized glioblastoma (Kepes & Rubinstein 1981) previously interpreted as fibrohistiocytic in origin, and superficial spindle cell astrocytomas of infancy (Taratuto et al 1984), previously interpreted as low grade mesodermal tumours. Despite the delineation of these entities, there is no doubt that fibro-histiocytic tumours of the meninges exist and that reactive astrocytes may become entrapped. It is important that a tumour is not diagnosed as being of glial lineage on the result of GFAP staining alone in the absence of other morphological evidence.

Care in interpretation must be exercised as occasionally phagocytic cells may take up GFAP from degenerate brain tissue and thus come to express an inappropriate intermediate filament — a feature which should be appreciated from the cellular morphology. Another trap for the unwary is the presence of GFAP positive reactive astrocytes within non-glial tumours, in much the same way as fibroblasts form a stroma in tumours outside the nervous system. It is important to consider this possibility when interpreting GFAP positivity in any neural tumour, and ensure that the GFAP positive cells have the cytological features of the neoplastic cells forming the tumour (Paulus & Peiffer 1988). This is a particular problem in some cerebral lymphomas where massive overgrowth of a reactive astrocytic component may mask the atypical lymphoid cells (Figs 7.17–7.18).

Despite these potential pitfalls in interpretation, positive immunostaining for GFAP in a poorly differentiated tumour is a very strong indication of glial origin as non-glial specificities are either unusual or far removed from the central nervous system, and for this reason GFAP antisera are the most commonly used markers in immunocytochemistry of central nervous system tumours.

S100 protein

Named because of its solubility in 100% ammonium sulphate, S100 protein is a dimeric acid calcium binding protein composed of two proteins S100α and S100β (Isobe et al 1978). In normal neuronal tissues antibodies to S100 protein bind to glial cells and Schwann cells with little reactivity in neurons. Outside the nervous system S100 immunoreactivity is seen in Langerhans cells, nodal interdigitating reticulum cells, chondrocytes, and melanocytes.

S100 protein immunoreactivity is seen in melanocytic tumours, Schwann cell tumours, granular cell tumours, chondroid tumours, as well as some carcinomas of breast, bronchus and kidney.

In the nervous system the majority of astrocytomas, glioblastomas and ependymomas express S100 immunoreactivity (Van Eldik et al 1986). Meningiomas may also express S100 immunoreactivity (Nakamura et al 1983), but the majority do not show expression of this antigen.

Identification of S100 immunoreactivity with antibodies to either the S100α or S100β protein have remained useful in the differential diagnosis of tumours outside the central nervous system particularly nerve sheath tumours (Johnson et al 1988). They are also useful in the differentiation of malignant melanoma from carcinoma provided they are used as part of a panel of antibodies with anti-cytokeratin or epithelial membrane antigen.

Cytokeratins

Cytokeratins are one of the class of intermediate filaments characteristically found in cells of epithelial origin. In general, keratins of low molecular weight are expressed by simple epithelia, while keratins of high molecular weight are expressed in stratified epithelia (see Ch. 4).

Cytokeratin expression has been shown in craniopharyngiomas, meningiomas, choroid plexus tumours (Fig. 7.14), chordomas and focally in ependymomas (Fig. 7.12). Rare forms of malignant glioma with epithelial differentiation or metaplasia can also express cytokeratin (Mork et al 1988). Germ cell tumours may express cytokeratin, particularly embryonal carcinoma and yolk-sac tumour (see Ch. 24).

There are diagnostic pitfalls in the interpretation of cytokeratin expression. Some plasma cell tumours express cytokeratin positivity in the absence of leukocyte common antigen positivity (Wotherspoon et al 1989). A large proportion of well characterized tumours of smooth muscle origin and some rhabdomyosarcomas express cytokeratin as paranuclear 'dots' and occasionally demonstrate more widespread cellular immunoreactivity (Norton et al 1987, Coindre et al 1988). Importantly, certain monoclonal antibodies to cytokeratin including AE1 and AE3 have been shown to stain a high proportion of astrocytomas, underscoring the importance of using a panel of well characterized antibodies in a diagnostic setting (Cosgrove et al 1989, Ng and Lo 1989).

Epithelial membrane antigen (EMA)

Epithelial membrane antigen is the term used to describe high molecular weight glycoproteins expressed on the luminal surface of glandular epithelia and more weakly in other epithelia. Commercially available polyclonal and monoclonal antibodies to EMA have found great diagnostic use in pathology in determining the epithelial nature of poorly differentiated tumours as part of a panel of antibodies.

Within the nervous system EMA staining may be seen in chordomas, meningiomas, choroid plexus tumours, and focally in ependymomas using EMA E29 clone (Dako U.K.) (Cruz-Sanchez et al 1988a). Other antibodies well characterized for EMA may not, however, show immunostaining of ependymomas. Rare examples of epithelial metaplasia in malignant glial tumours have been found to express EMA immunoreactivity (Mork et al 1988). Oligodendrogliomas, most astrocytomas and schwannomas are EMA negative; however, cytoplasmic staining of astrocytomas has been reported (Perentes & Rubinstein 1987). In the differential diagnosis between chondrosarcoma and chordoma, chordomas are positive for EMA while chondrosarcomas are not (Coindre et al 1986). While most lymphomas do not stain for EMA, a proportion of plasmacytomas, T-cell lymphomas, and Reed-Sternberg cells of Hodgkins disease have been shown to stain for EMA (Delsol et al 1984).

Laminin

Laminin is a glycoprotein present only in basement membrane, and synthesized by the cells which are attached to the basement membrane such as endothelium, epithelium and schwann cells. Immunocytochemical studies on the distribution of laminin in the nervous system has shown vessel wall associated staining as well as staining of normal choroid plexus, arachnoid and pial-glial membrane. Staining between neuroepithelial cells is not seen (McComb & Bigner 1985).

Immunostaining for laminin is of diagnostic use in the differentiation between atypical papillary ependymomas and aggressive choroid plexus neoplasms. In papillary ependymomas, laminin staining is confined to vessel walls and not beneath the covering ependymal epithelium of papillae whereas in choroid plexus carcinoma laminin staining is associated with epithelium forming both surface of papillae and solid nests (Fig. 7.8) (Tarrant et al 1988).

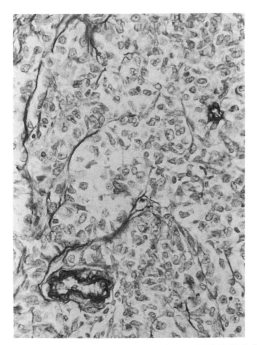

Fig. 7.8 Aggressive choroid plexus tumour stained to localize laminin. Immunoreactivity is seen around vessels, but is also present associated with epithelium as linear profiles in a way which is not seen in ependymoma. Immunoperoxidase/haematoxylin. ×280

Leucocyte common antigen (LCA)

LCA (CD45) is expressed on normal microglia (Lowe & Cox 1989) as well as on the majority of neoplasms derived from leucocytes, and may be identified reliably in paraffin sections. It is especially useful in distinguishing between lymphoid and non-lymphoid malignancies (Fig. 7.20) (Warnke et al 1983), and in the central nervous system is very useful in the identification of primary lymphoma of the CNS, particularly if neoplastic cells are hidden amongst reactive astrocytic cells at the edge of a tumour (Fig. 7.18).

This is also a very useful antibody to include in a panel for the investigation of poorly differentiated tumours, and aids in the differential diagnosis between small cell glioblastomas, anaplastic oligodendrogliomas and cerebral lymphoma in small biopsy fragments which lack diagnostic architectural features.

It is important to note that a small proportion of non-Hodgkins lymphomas, plasma cell tumours, and Reed-Sternberg cells fail to stain with antibodies to LCA. The use of antibodies to LCA and other markers for cells of lymphoid lineage is discussed fully in Chapter 19 on the lymphoid system. Primary cerebral lymphomas have trebled in frequency in the last 10 years and, in diagnostic pathology, detection of primary cerebral lymphoma will become more important as it has been predicted that over the next 5 years this tumour will become one of the most common neurological neoplasms as a result of the acquired immunodeficiency syndrome (Hochberg & Miller 1988).

Vimentin

Vimentin is one of the class of intermediate filaments and, although at first thought to be characteristic for cells of mesenchymal origin, it is also seen in many epithelial tissues (Azumi & Battifora 1987).

In normal tissues, neurons do not show immunoreactivity for vimentin (Azumi & Battifora 1987), while normal and reactive astrocytes coexpress GFAP and vimentin (Reifenberger et al 1987).

In astrocytic tumours, many of the cells express vimentin as well as GFAP (Herpers et al 1986, Schiffer et al 1986), and in glioblastomas where GFAP immunoreactivity is focal, vimentin is often the preferentially expressed intermediate filament (Fig. 7.9). Vimentin is particularly expressed in plump gemistocytic cells within astroglial tumours, is coexpressed with cytokeratin in choroid plexus neoplasms, and is also found in ependymomas (Doglioni et al 1987).

Vimentin is the major intermediate filament expressed by tumours of nerve sheath origin (Gould et al 1986b) and is also strongly expressed by meningiomas including those of haemangiopericytic type (Holden et al 1987). The virtually ubiquitous presence of vimentin in tumours of the nervous system means that it has no useful place in diagnosis by virtue of any limited expression. Immunostaining for vimentin is useful when used as a 'cytological stain' in tumours of astroglial origin as it highlights tissue architecture and cellular processes to provide additional morphological information.

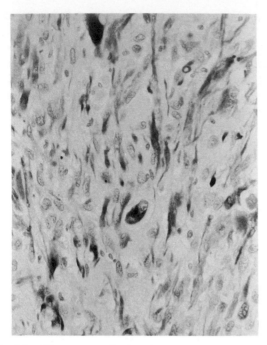

Fig. 7.9 Gliosarcoma stained to show localization of vimentin in atypical spindle cells. Immunoperoxidase/haematoxylin. ×280

S-Antigen

A 50kD protein found in photoreceptors of the retina and in pinealocytes, S-antigen has been seen in retinoblastoma (Mirshani et al 1986) as well as focally within pineoblastomas and pineocytomas (Perentes et al 1986). Cells in medulloblastoma may also express S-antigen immunoreactivity (Bonnin & Perentes 1988), hence this criterion alone may not be used to define pineoblastoma.

Immunohistochemical determination of cell proliferation

This has recently found use in pathological assessment of tumours of the nervous system.

Bromodeoxyuridine (BrdU), a thymidine analogue which is incorporated into cellular DNA of dividing cells, is administered to a patient prior to removal of a tumour, and cells which have divided between the time of administration and removal of the lesion become labelled with BrdU.

These cells can be detected in paraffin sections by monoclonal antibodies to BrdU and a labelling index can be derived for a tumour which reflects the proliferative potential of the tumour (Hoshino et al 1986).

Another technique uses the monoclonal antibody Ki67, which recognizes a protein expressed in cells in cycle (G_1, G_2, S, and M phases) (Gerdes et al 1984) and requires use of frozen sections. The proportion of positively stained cell nuclei is expressed as a labelling index. High grade gliomas have higher labelling indices than low grade tumours (Burger et al 1986), and for meningiomas this may be useful in identifying potentially recurrent lesions (Robson et al 1987).

Pituitary hormones

Pituitary tumours are common neoplasms involving the central nervous system. Their identification is discussed in Chapter 28.

IMMUNOHISTOCHEMICAL PROFILES OF THE COMMON TUMOURS

Astrocytomas

Astrocytomas are usually reliably diagnosed using conventional morphological techniques. Immunohistochemistry may be used to investigate mixed glial/neuronal tumours and may be necessary for diagnosis in small biopsy fragments of anaplastic astrocytic lesions. The immunohistochemical profile of astrocytomas is summarized in Table 7.1.

The vast majority of astrocytomas express GFAP both in the cytoplasm around the nucleus as well as in cell processes. In well differentiated astrocytomas, most tumour cells contain GFAP (Fig. 7.6); however, in anaplastic astrocytomas a proportion of tumour cells do not express this antigen (Fig. 7.7) and may express vimentin. Astrocytomas may also show focal immunoreactivity for S100 protein, PGP 9.5 and Leu-7; NSE is often present.

Astrocytomas do not express the markers of neuronal differentiation and so these markers can

Table 7.1 Immunohistochemical profiles of the common tumours of the nervous system

Tumour	Antigens												
	GFAP	Vimentin	S-100	PGP9.5	Leu-7	NSE	Cytokeratin	EMA	Chromogranin	Synaptophysin	NFP	LCA	Retinal S. antigen
Astrocytoma	++++	+++	+++	+	+	+	−	−	−	−	−	−	−
Glioblastoma	+++	+++	+++	++	++	+	±	±	−	−	−	−	
Oligodendroglioma	±	+	++		++	+	−	−	−	−	−	−	
Meningiomas	−	+++	+		−	+	+	+	−	−	−	−	
Ependymomas	+++	+	+		+		F	F	−	−	−	−	
Medulloblastoma	+		+	++	+	++				+	+	−	+
Neural derived elements	−			+		++			+	+	+	−	
Lymphoma	−	−	−	−	−	−	−	−	−	−	−	+	

Key: ± Very rare, ± rare, F focal.

be used to look for neuronal elements in glial tumours (Fig. 7.5). As noted earlier, certain monoclonal antibodies to cytokeratin stain some astrocytomas.

One trap for the unwary is in not recognizing infiltrating neoplastic lymphoid cells amongst pleomorphic, but reactive astrocytes in a biopsy from the edge of a cerebral lymphoma. Thus, in small biopsies, GFAP immunostaining should be combined with LCA staining to detect atypical cells as being lymphoid in origin (Fig. 7.18). It should be noted that reactive lymphoid cells are commonly present around blood vessels particularly in gemistocytic astrocytomas. Unusual entities such as spindle cell astrocytomas of infancy are distinguished from low grade mesodermal tumours by GFAP immunostaining (Taratuto et al 1984).

Glioblastomas

Glioblastomas have similar patterns of antigen expression to that seen in poorly differentiated astrocytomas as summarized in Table 7.1. The large giant cells of glioblastoma show variable expression for GFAP, while small anaplastic cells exhibiting a paucity of cell processes are commonly negative for GFAP.

Very rare forms of glioblastoma exhibit epithelial metaplasia and then these components can express cytokeratin or EMA. Positive GFAP staining identifies the glial nature of such tumours.

Lipidized glioblastoma (Kepes & Rubinstein 1981) and gliosarcoma are distinguished from

fibrohistiocytic tumours by virtue of positive GFAP immunostaining (Figs 7.9–7.11).

Oligodendrogliomas

Oligodendrogliomas are usually easy to recognize by conventional morphology; however, anaplastic oligodendrogliomas may be difficult to distinguish from other tumours such as anaplastic astrocytomas in small biopsy fragments. The

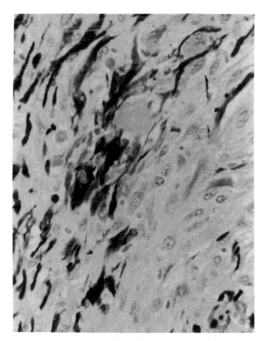

Fig. 7.10 Same tumour as in Figure 7.9 stained to show GFAP expression in a proportion of spindle cells. Immunoperoxidase/haematoxylin. ×280

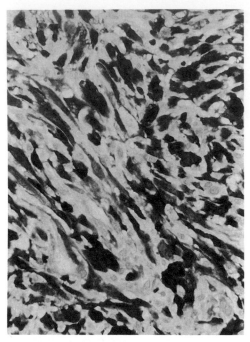

Fig. 7.11 Same tumour as Figure 7.9 stained to show PGP 9.5 immunoreactivity, seen as intense staining in the atypical spindle cells. Immunoperoxidase/haematoxylin. ×280

immunohistochemistry of oligodendrogliomas is summarized in Table 7.1.

In most pure oligodendrogliomas GFAP staining is seen only in reactive astrocytes within the lesion. There are, however, a significant number of GFAP positive oligodendrogliomas where oligodendrocytes take on a swollen eosinophilic appearance in conventional H&E preparations (Herpers & Budka 1984, Nakagawa et al 1986).

The differential diagnosis between lymphoma and oligodendroglioma in small samples is made by positive LCA staining in lymphoma. It is notable that immunohistochemical markers of benign oligodendrocytes such as carbonic anhydrase-C, myelin associated glycoprotein and myelin basic protein are unhelpful in the diagnosis of oligodendroglial tumours (Nakagawa et al 1987, Perentes & Rubinstein 1987).

Meningiomas

Meningiomas generally pose few diagnostic problems in identification, but in certain sites there may be confusion with other spindle cell tumours such as schwannomas. The immunohistochemical profile of meningiomas is summarized in Table 7.1.

In general these tumours exhibit markers of epithelial differentiation, particularly those of meningothelial or syncytial pattern. Tumours of fibroblastic pattern less frequently express cytokeratin than those of meningothelial pattern. Meningiomas may express EMA immunoreactivity, in contrast to schwannomas which do not (Perentes & Rubinstein 1987). In contrast to many other tumours of the nervous system, meningiomas do not stain for Leu-7 (Meis et al 1986, Perentes & Rubinstein 1987).

Ependymomas

Ependymomas may be composed of solid sheets of cells, or have acinar and papillary epithelial patterns. Some tumours, especially in the spinal cord are composed of spindle shaped cells which resemble low grade 'pilocytic' astrocytomas. The immunohistochemical profile of ependymomas is summarized in Table 7.1.

In general, most ependymomas strongly express GFAP, especially in cell processes forming pseudorosettes around blood vessels. Cytokeratin staining may be found focally in ependymoma confined to papillary areas (Mannoji & Becker 1988), particularly in myxopapillary ependymoma (Fig. 7.12) (Tarrant et al 1988). Any more than focal cytokeratin staining should raise doubts about the diagnosis of ependymoma, and if GFAP staining is evident in the same lesion (Fig. 7.13), the possibility of a tumour derived from choroid plexus should be considered.

The expression of cytokeratin in primitive tumours with ependymal differentiation (ependymoblastoma) is variable (Mannoji & Becker 1988, Cruz-Sanchez et al 1988 b).

In contrast to choroid plexus lesions, the epithelial components of ependymoma do not have a basal lamina demonstrated by immunohistochemical staining for laminin (Fig. 7.8) (Tarrant et al 1988).

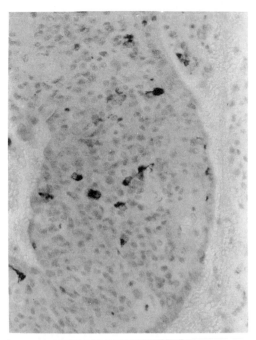

Fig. 7.12 Ependymoma showing cytokeratin expression by a few cells in a papillary area. Immunoperoxidase (CAM 5.2)/haematoxylin. ×280

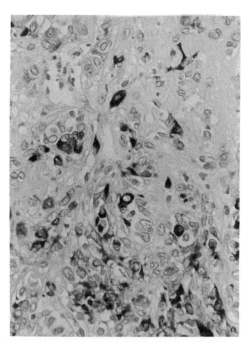

Fig. 7.13 Aggressive choroid plexus tumour stained to show GFAP expression. Immunoperoxidase/haematoxylin. ×280

Choroid plexus papilloma/carcinoma

These are mainly tumours of childhood which are derived from choroid plexus. In adult life, confusion with other papillary tumours metastatic to the central nervous system is a problem, although the finding of focal GFAP staining in the presence of cytokeratin expression is strongly suggestive of a choroid plexus origin (Figs 7.13–7.14).

In the differential diagnosis between atypical choroid plexus papillomas and papillary ependymomas, GFAP immunoreactivity is seen in both, but is usually very patchy and focal in choroid plexus lesions (Fig. 7.13). Cytokeratin immunoreactivity may be seen in both tumours, but is never more than focal and patchy in ependymomas (Fig. 7.12) (Mannoji & Becker 1988). Choroid plexus papilloma has a continuous basement membrane as shown by electron microscopy and immunoreactivity for laminin, in contrast to ependymona which does not (Fig. 7.8) (Tarrant et al 1988).

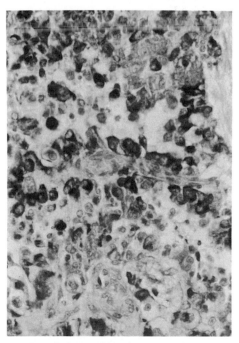

Fig. 7.14 Aggressive choroid plexus tumour stained to show cytokeratin expression.
Immunoperoxidase/haematoxylin. ×280

Medulloblastoma

The diagnosis of medulloblastoma is based on morphological features of a small celled primitive tumour arising in the region of the cerebellum. There is a possibility of confusion with other small cell tumours, particularly with lymphoma, but the lack of staining with LCA would be a strong factor against the diagnosis of lymphoma.

Medulloblastomas are primitive tumours of the nervous system and have the capacity to develop along several lines of differentiation, particularly glial and neuronal. There has been great interest in whether differentiation in medulloblastoma is related to prognosis or response to chemotherapy, and immunohistochemical evidence of neural (Fig. 7.15) or glial (Fig. 7.16) differentiation is currently under investigation in this respect (Burger et al 1987, Caputy et al 1987).

The immunoreactivity of primitive tumours arising from the pineal gland (pineoblastoma) is very similar to that of medulloblastoma.

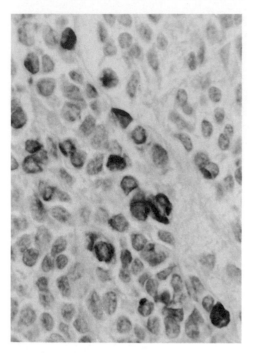

Fig. 7.16 Medulloblastoma stained to show GFAP expression. Staining is in cytologically neoplastic cells. Immunoperoxidase/haematoxylin. ×850

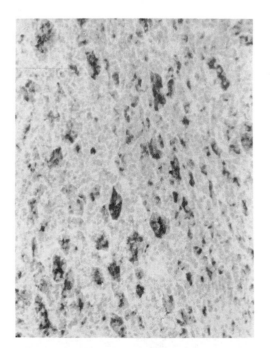

Fig. 7.15 Medulloblastoma stained to show PGP 9.5 expression. Small islands of positive cells correspond to primitive rosettes in which neurofilament protein expression can also be demonstrated. This seems to be a good stain to detect probable neural differentiation in this type of tumour. Immunoperoxidase/haematoxylin. ×175

Neural derived tumours

These include neuroblastoma, neurocytoma, paraganglioma and the neuronal component of ganglioglioma. The staining of neural derived tumours is summarized in Table 7.1.

Positive staining for neurofilament protein and antigens specifically related to neurosecretion are particularly useful in establishing the neural phenotype of a tumour (Fig. 7.5). Of particular note are the non-specificities of neuron-specific enolase and PGP 9.5 for the neural phenotype (see p. 79).

When differential diagnosis with other small cell malignant tumours is raised, the identification of antigens such as synaptophysin, chromogranin or neurofilament protein can greatly facilitate diagnosis. These antigens are also useful in diagnosis of mixed neuronal and glial tumours by picking out neuronal elements in gangliogliomas.

A particular problem in diagnosis is central neurocytoma, a form of low grade neural derived tumour closely resembling oligodendroglioma by light microscopy. Diagnosis of neurocytoma has been made by the ultrastructural demonstration of neuroendocrine granules (Nishio et al 1988). It is likely that many of these tumours are misdiagnosed as intraventricular oligodendrogliomas, and for this reason immunohistochemical demonstration of neural markers such as synaptophysin should be of great benefit in diagnosis.

Lymphoma

Lymphoma may arise as a primary tumour of the central nervous system as well as involving the central nervous system as part of spread of systemic lymphoma. These tumours are frequently deep seated lesions and tissue is obtained by needle biopsy thus providing a paucity of architectural features for diagnosis.

In many cases diagnosis is relatively straight forward and is part of the differential diagnosis of a malignant small cell tumours of the central nervous system, but in other cases lymphoma cells are masked by a significant reactive glial proliferation which can mimic an astrocytic tumour. The detection of LCA is an excellent and sensitive screening method for the identification of cerebral lymphoma (Figs. 7.17–7.18).

Primary lymphomas of the central nervous system are virtually all B-cell non-Hodgkins lymphomas and can be phenotypically classified by immunohistochemistry (see Ch. 19). The subject of primary central nervous system lymphoma is well reviewed by Hochberg & Miller (1988).

Chordomas

Chordomas are tumours derived from notochord which frequently involve the nervous system by virtue of their preferred sites of origin in the skull base and in the sacrum. The differential diagnosis includes chondrosarcoma and occasionally myxopapillary ependymoma. This can be achieved by the epithelial markers EMA and cytokeratin, which are expressed in the vast majority of chordomas but not chondrosarcoma (Abenoza & Sibley 1986, Meis & Giraldo 1988). The presence of GFAP in ependymoma, but not chordoma or chondrosarcoma, is also of value. The immunohis-

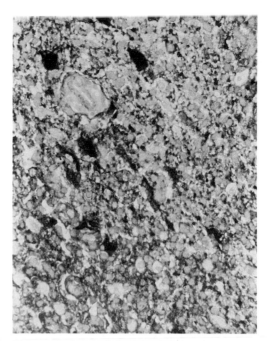

Fig. 7.17 Lymphoma of the central nervous system stained to show GFAP expression which highlights large astrocytic cells which can lead to the false impression that the lesion is glial in origin. Immunoperoxidase/haematoxylin. ×280.

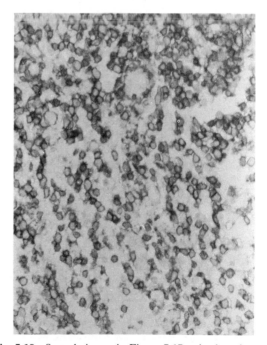

Fig. 7.18 Same lesion as in Figure 7.17 stained to show leucocyte common antigen which highlights atypical lymphoid cells by delicate membrane staining. Immunoperoxidase/haematoxylin. ×280

Table 7.2 Immunohistochemical profiles of chordomas, chondrosarcomas and ependymomas

	EMA	Cytokeratin	GFAP	S100
Chordoma	+	+	−	+
Chondrosarcoma	−	−	−	+
Ependymoma	focal	focal	+	+

tochemical profiles for useful antigens in this group of tumours is summarized in Table 7.2.

The entity of chondroid chordoma poses a further problem in diagnosis of tumours in this region and cases have been shown to have positive immunoreactivity for EMA and cytokeratin (Salisbury 1987). However, two of three chondroid chordomas failed to stain for these antigens in another study (Meis & Giraldo 1988).

Haemangioblastomas

These tumours of the central nervous system are composed of vascular endothelial cells, pericytic cells and stromal cells. They are common in the cerebellum, but may arise elsewhere and can be confused histologically with metastatic clear cell carcinoma, particularly renal carcinoma. Immunohistochemistry can help in this respect as stromal clear cells do not express EMA while epithelial cells of metastatic renal carcinoma do (Andrew & Gradwell 1986). The immunohistochemical profile of haemangioblastomas is complex because of the different cellular elements; GFAP is expressed by process-bearing stromal cells interpreted as reactive astrocytes, while NSE is seen in all stromal cells (Ironside et al 1988).

Germ cell tumours

Germ cell tumours of the central nervous system arise from the region of the pineal gland and occasionally from the region of the pituitary gland, and within the posterior fossa. These tumours are essentially the same as germ cell tumours arising in the ovary or testis and diagnosis may be facilitated by the detection of antigens expressed by specific elements of these tumours (see Ch. 24).

Placental-like alkaline phosphatase (PLAP) immunoreactivity is useful in small biopsies to detect germinoma cells amongst reactive lymphoid and macrophagic cells common to these lesions (Figs 7.19–7.20). EMA immunoreactivity has been seen in several germinomas of the pineal region (Perentes & Rubinstein 1987), hence caution should be exercised in attributing this to somatic epithelial elements in teratoma.

Tumours of nerve sheath origin

These tumours are derived from Schwann cells and perineurial cells. The common tumours derived from these cells include schwannomas, neurofibromas and neurofibrosarcomas.

S100 protein is a useful marker of spindle cell tumours, staining positively in most benign schwannomas but infrequently in malignant nerve sheath tumours (Johnson et al 1988).

Within the central nervous system schwannomas may be confused with meningiomas and here S100 immunoreactivity is not of help in differential diagnosis as some meningiomas may also express

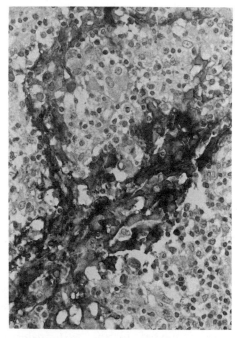

Fig. 7.19 Germinoma of pineal region stained to show localization of PLAP. Tumour cells stain positively in contrast to surrounding reactive macrophages and lymphoid cells. This is a useful method in small needle biopsy samples from this type of lesion. Immunoperoxidase/haematoxylin. ×280

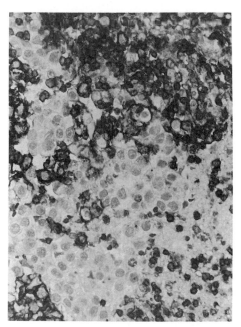

Fig. 7.20 Same lesion as in Figure 7.19 stained to show leucocyte common antigen. This also highlights the difference between neoplastic cells and reactive cells in the germinoma. Immunoperoxidase/haematoxylin. ×280

this marker. Anti-Leu-7 immunoreactivity is not seen in meningiomas (Motoi et al 1985, Perentes & Rubinstein 1985), but is seen in the majority of Schwann cell tumours. While meningiomas may express EMA, schwannomas do not (Perentes & Rubinstein 1987).

Positive staining for Leu-7 is seen in most schwannomas (Fig. 7.7) and some neurofibrosarcomas, but is absent in fibrosarcomas and the majority of leiomyosarcomas (Johnson et al 1988, Swanson et al 1987). However, Leu-7 positivity has been seen in some smooth muscle tumours and, hence, positive staining cannot be used as a specific marker of nerve sheath origin. Leiomyosarcoma has also been seen to express S100 protein (Swanson et al 1987). Cells of schwannoma have been seen to be GFAP positive (Gould et al 1986 b). EMA may be seen in perineurial cells surrounding nerve sheath tumours (Theaker & Fletcher 1989).

Metastatic tumours

Metastatic tumours are common in the central nervous system. Tumours of bronchus, breast, stomach, colon and melanomas are commonly encountered in the substance of the brain while lymphomas, and carcinomas of prostate, bronchus, breast, thyroid, or kidney are commonly encountered in the bone of the vertebral column compressing the spinal cord.

In poorly differentiated tumours, immunohistochemistry may be of value in establishing the likely primary site of a tumour. There are no completely rigid rules concerning antigen expression by tumours, rather there are relative probabilities of origin given expression or not for sets of antigens.

Melanoma is S100 and vimentin positive but EMA, LCA, and cytokeratin negative.

Carcinomas are commonly cytokeratin positive and express EMA. If a tumour does not express carcinoembryonic antigen (CEA) then it is unlikely to be of colonic origin and, in a similar manner, if a tumour does not express EMA then it is unlikely to be of breast origin. Prostatic origin for a metastatic carcinoma can be strongly suggested by expression of prostate-specific antigen (PSA) or prostate-specific acid phosphatase (PSAP) (see Ch. 25), although adenomas of the middle ear may also express PSA and PSAP (Filipe, unpublished observations).

The presence of lymphoma may be suggested by positive staining of tumour for LCA, although, as noted earlier, plasma cell tumours (myeloma) may not express LCA and may express EMA or cytokeratin. Carcinoma of the thyroid may be confirmed by detection of thyroglobulin in tumour (see Ch. 26).

SUMMARY

Immunohistochemistry has found wide use in neuropathology in the diagnosis of tumours involving the nervous system. Continued application of these techniques has shown that antigens thought to be specific to certain tumours may be expressed by other tumours on occasion. This means that pathologists using these techniques in a diagnostic setting should be aware of the reactivities of different antibodies, both as reported by other workers and in their own laboratories, as tested on classical well characterized tumours.

It is to be expected that further developments in the next few years will hone the mass of information on patterns of tumour immunoreactivity to a sharp set of well defined rules and truths of proven diagnostic use in neuro-oncology.

PANELS OF ANTIBODIES USEFUL IN DIAGNOSTIC DILEMMAS

Small cell tumours

LCA — lymphoma
Cytokeratin — carcinoma
Synaptophysin — neuroendocrine or primitive neuroectodermal tumours
Desmin — rhabdomyosarcoma or differentiating primitive neuroectodermal tumour
NFP — some neuroblastomas and primitive neuroectodermal tumours, e.g. medulloblastoma
GFAP — primitive neuroectodermal tumours e.g. medulloblastoma.

Papillary epithelial tumour in the CNS

GFAP — ependymoma (widespread), choroid plexus tumour (focal)
EMA — ependymoma, choroid plexus tumour, papillary meningioma, metastatic carcinoma
Cytokeratin — metastatic carcinoma, choroid plexus tumour, ependymoma
Laminin — subepithelial in choroid plexus tumours and metastatic carcinoma.

Spindle cell tumours

Leu-7 — not seen in meningioma, fibrosarcoma, or fibrous histiocytoma
S100 — present in schwannomas, meningiomas, gliosarcomas, rare smooth muscle tumours
GFAP — present in glial component of gliosarcoma, superficial astrocytomas in infancy
EMA — seen in meningiomas and some smooth muscle tumours

REFERENCES

Abenoza P, Sibley R K 1986 Chordoma: an immunohistological study. Human Pathology 17: 744–747
Abo T, Balch C M 1981 A differentiation antigen of human NK and K cells identified by a monoclonal antibody (HNK-1). Journal of Immunology 127: 1024–1029
Andrew S M, Gradwell E 1986 Immunoperoxidase labelled antibody staining in differential diagnosis of central nervous system haemangioblastomas and central nervous system metastases of renal carcinomas. Journal of Clinical Pathology 39(8): 917–919
Azumi N, Battifora H 1987 The distribution of vimentin and keratin in epithelial and non-epithelial neoplasms. American Journal of Clinical Pathology 88: 286–296
Bishop A E, Polak J M, Facer P, Ferri G L, Marangos P J, Pearse A G E 1982 Neurone specific enolase: a common marker for the endocrine cells and innervation of the gut and pancreas. Gastroenterology 83: 902–915
Bonnin J M, Rubinstein L J 1984 Immunohistochemistry of central nervous system tumours. Its contribution to neurosurgical diagnosis. Journal of Neurosurgery 60: 1121–1133
Bonnin J M, Perentes E 1988 Retinal–S antigen immunoreactivity in medulloblastomas. Acta Neuropathologica 76: 204–207
Budka H 1986 Non-glial specificities of the glial fibrillary acidic protein (GFAP). Acta Neuropathologica 72: 43–54
Burger P, Shibada T, Kleihues P 1986 The use of the monoclonal antibody Ki-67 in the identification of proliferating cells: applications to surgical neuropathology. American Journal of Surgical Pathology 10(9): 611–617
Burger P C, Grahmann F C, Bliestle A, Kleihues P 1987 Differentiation in the medulloblastoma: a histological and immunocytochemical study. Acta Neuropathologica 73: 115–123
Caillaud J, Benjelloun S, Bosq J, Braham K, Lipinski M 1984 HNK-1 defined antigen detected in paraffin embedded neuroectodermal tumours and those derived from cells of the amine precursor uptake and decarboxylation system. Cancer Research 44: 4432–4439
Caputy A J, McCullough D C, Manz H J, Patterson K, Hammock M K 1987 A review of the factors influencing prognosis of medulloblastoma: the importance of cell differentiation. Journal of Neurosurgery 66: 80–87
Carden M J, Schlaepfer W W, Lee V 1985 The structure, biochemical properties, and immunogenicity of neurofilament peripheral regions are determined by phosphorylation state. Journal of Biological Chemistry 260: 9805–9817
Chou K H, Ilyas A A, Evans J E, Quarles R H, Jungalwala F B 1985 Structure of a glycolipid reacting with monoclonal IgM in neuropathy and with HNK-1. Biochemical and Biophysical Research Communications 128: 383–388
Cocchia D, Michetti F, Donato R 1981 Immunochemical and immunocytochemical localisation of S100 antigen in normal human skin. Nature 294: 85–87
Coindre J, Rivel J, Trojani M, DeMascarel I, DeMascarel A 1986 Immunohistological study in chordomas. Journal of Pathology 150: 61–63
Coindre J, Mascarel A, Trojani M, Mascarel I, Pages A 1988 Immunohistochemical study of rhabdomyosarcoma. Unexpected staining with S100 protein and cytokeratin. Journal of Pathology 155: 127–132

Cosgrove M, Fitzgibbons P L, Sherrod A, Chandrasoma P T, Martin S E 1989 Intermediate filament expression in astrocytic neoplasms. American Journal of Surgical Pathology 13: 141–145

Cras P, Martin J J, Gheuens J 1988 γ-Enolase and glial fibrillary acidic protein in nervous system tumours. Acta Neuropathologica 75: 377–384

Cruz-Sanchez F F, Rossi M L, Esiri M M, Reading M 1988a Epithelial membrane antigen expression in ependymomas. Neuropathology and Applied Neurobiology 14: 197–207

Cruz-Sanchez F F, Haustein J, Rossi M L, Cervos-Navarro J, Hughes J T 1988b Ependymoblastoma: a histological, immunohistochemical and ultrastructural study of five cases. Histopathology 12: 17–27

Delsol G, Gatter K C, Stein H et al 1984 Human lymphoid cells express epithelial membrane antigen. Implications for diagnosis of human neoplasms. Lancet ii: 1124–1129

Dickson D W, Yen S-H, Suzuki K I, Davies P, Garcia J H, Hirano A 1986 Ballooned neurons in select neurodegenerative diseases contain phosphorylated neurofilament epitopes. Acta Neuropathologica 71: 216–223

Doglioni C, Dell'Orto P, Coggi G, Iuzzolino P, Bontempi L, Viale G 1987 Choroid plexus tumours. An immunocytochemical study with particular reference to co-expression of intermediate filament proteins. American Journal of Pathology 127: 519–529

Doran J F, Jackson P, Kynoch P A M, Thompson R J 1983 Isolation of PGP 9.5, a new human neurone-specific protein detected by high resolution two-dimensional electrophoresis. Journal of Neurochemistry 40: 1542–1547

Gerdes J, Lemke H, Baisch H, Wacker H-H, Schwab U, Stein H 1984 Cell cycle analysis of a cell proliferation-associated human nuclear antigen defined by the monoclonal antibody Ki-67. Journal of Immunology 133: 1710–1715

Ghobrial M, Ross E R 1986 Immunocytochemistry of neurone-specific enolase: a reevaluation. In: Zimmerman H M (ed) Progress in neuropathology, vol 6. Raven Press, New York, p 199–221

Gould V E 1987 Synaptophysin: a new and promising pan-neuroendocrine marker. Archives of Pathology and Laboratory Medicine 111: 791–794

Gould V E, Lee I, Wiedemann B, Moll R, Chejfec G, Franke W W 1986a Synaptophysin; a novel marker for neurons, certain neuro-endocrine cells, and their neoplasms. Human Pathology 17: 979–983

Gould V E, Moll R, Moll I, Lee I, Schwechheimer K, Franke W 1986b The intermediate complement of the spectrum of nerve sheath neoplasms. Laboratory Investigation 55: 463–474

Herpers M J H M, Budka H 1984 Glial fibrillary acidic protein (GFAP) in oligodendroglial tumours: gliofibrillary oligodendroglioma and transitional oligoastrocytoma as subtypes of oligodendroglioma. Acta Neuropathologica 64: 265–272

Herpers M J H M, Budka H 1985 Primitive neuroectodermal tumours including the medulloblastoma: glial differentiation signalled by immunoreactivity for GFAP is restricted to the pure desmoplastic medulloblastoma ('arachnoidal sarcoma of the cerebellum'). Clinical Neuropathology 4: 12–18

Herpers M J M H, Ramaekers F C S, Aldeweireldt J, Moesker O, Slooff J 1986 Co-expression of glial fibrillary acidic protein and vimentin intermediate filaments in human astrocytomas. Acta Neuropathologica 70: 333–339

Hochberg F H, Miller D C 1988 Primary central nervous system lymphoma. Journal of Neurosurgery 68: 835–853

Holden J, Dolman C L, Churg A 1987 Immunohistochemistry of meningiomas including the angioblastic type. Journal of Neuropathology and Experimental Neurology 46: 50–56

Hoshino T, Nagashima T, Murovic J A, Wilson C B, Davis R L 1986 Proliferative potential of human meningiomas of the brain: a cell kinetic study with bromodeoxyuridine. Cancer 58: 1466–1472

Ironside J W, Stephenson T J, Royds J A et al 1988 Stromal cells in cerebellar haemangioblastomas: an immunocytochemical study. Histopathology 12: 29–40

Isobe T, Tsugita A, Okuyama T 1978 The amino acid sequence and subunit structure of bovine brain S-100 protein (PAP1-b). Journal of Neurochemistry 30: 921–923

Johnson H D, Glick A D, Davis B W 1988 Immunohistochemical evaluation of Leu-7, Myelin basic protein, S100 protein, glial-fibrillary acidic protein and LN3 immunoreactivity in nerve sheath tumors and sarcomas. Archives of Pathology and Laboratory Medicine 112: 155–160

Kepes J J, Rubinstein L J 1981 Malignant gliomas with heavily lipidised (foamy) tumor cells. A report of three cases with immunoperoxidase study. Cancer 47: 2451–2459

Kepes J J, Rubinstein L J, Eng L F 1979 Pleomorphic xanthoastrocytoma: a distinctive meningocerebral glioma of young subjects with relatively favourable prognosis. A study of 12 cases. Cancer 44: 1839–1852

Kruse J, Mailhammer R, Wernecke H et al 1984 Neural cell adhesion molecules and myelin-associated glycoprotein share a common carbohydrate moiety recognised by monoclonal antibodies L2 and HNK-1. Nature 311: 153–155

Lee V, Wu H L, Schlaepfer W W 1982 Monoclonal antibodies recognise individual neurofilament triplet proteins. Proceedings of the National Academy of Sciences of the USA 79: 6089–6092

Lowe J, Cox G 1990 Neuropathological techniques. In: Bancroft J D, Stevens A (eds) Theory and practice of histological technique. Churchill Livingstone, Edinburgh

Mannoji H, Becker L E 1988 Ependymal and choroid plexus tumours: cytokeratin and GFAP expression. Cancer 61: 1377–1385

McComb R D, Bigner D D 1985 Immunolocalisation of laminin in neoplasms of the central and peripheral nervous system. Journal of Neuropathology and Experimental Neurology 44: 242–253

McGarry R C, Helfand S L, Quarles R H, Roder J C 1983 Recognition of myelin-associated glycoprotein by the monoclonal antibody HNK-1. Nature 306: 376–378

Meis J M, Giraldo A A 1988 Chordoma, an immunohistochemical study of 20 cases. Archives of Pathology and Laboratory Medicine 112: 553–556

Meis J M, Ordonez N G, Bruner J M 1986 Meningiomas: an immunohistochemical study of 50 cases. Archives of Pathology and Laboratory Medicine 110: 934–937

Mirshahi M, Boucheix C, Dhermy P, Haye C, Faure J-P 1986 Expression of the photoreceptor-specific S-antigen in human retinoblastoma. Cancer 57: 1497–1500

Mork S J, Rubinstein L J, Kepes J J et al 1988 Patterns of epithelial metaplasia in malignant gliomas. II. Squamous differentiation of epithelial-like formations in gliosarcomas and glioblastomas. Journal of Neuropathology & Experimental Neurology 47: 101–118

Motoi M, Yoshino T, Hayashi K, Nose Y, Horie Y, Ogawa K 1985 Immunohistochemical studies on human brain tumours using anti-Leu-7 monoclonal antibody in paraffin embedded specimens. Acta Neuropathologica 66: 75–77

Mukai M, Torikata C, Iri H et al 1986 Expression of neurofilament triplet proteins in human neural tumors. American Journal of Pathology 122: 28–35

Nakagawa Y, Perentes E, Rubinstein L J 1986 Immunohistochemical characterisation of oligodendrogliomas: an analysis of multiple markers. Acta Neuropathologica 72: 15–22

Nakagawa Y, Perentes E, Rubinstein L J 1987 Non-specificity of carbonic anhydrase C antibody as a marker in human neurooncology. Journal of Neuropathology and Experimental Neurology 46: 451–460

Nakajima T, Kameda T, Watanabe S, Hirota T, Sato Y, Shimosato Y 1982 An immunoperoxidase study of S100 protein distribution in normal and neoplastic tissues. American Journal of Surgical Pathology 6(8): 715–727

Nakamura T, Becker L E, Marks A 1983 Distribution and immunoreactivity of S100 protein in paediatric brain tumors. Journal of Neuropathology and Experimental Neurology 42: 136–145

Nakazato Y, Ishizeki J, Takahashi K, Yamaguchi H, Karnei T, Mori T 1982 Localisation of S100 protein and glial fibrillary acidic protein-related antigen in pleomorphic adenoma of the salivary glands. Laboratory Investigation 46(6): 621–626

Ng H K, Lo S T H 1989 Cytokeratin immunoreactivity in gliomas. Histopathology 14: 359–368

Nishio S, Tashima T, Takeshita I, Fukui M 1988 Intraventricular neurocytoma: clinicopathological features of six cases. Journal of Neurosurgery 68: 665–670

Norton A J, Thomas J A, Isaacson P G 1987 Cytokeratin-specific monoclonal antibodies are reactive with tumours of smooth muscle derivation: an immunocytochemical and biochemical study using antibodies to intermediate filament cytoskeletal proteins. Histopathology 11: 487–499

Paasivuo R, Saksela E 1983 Non-specific binding of mouse immunoglobulin by swollen bodied astrocytes: A potential cause of confusion in human brain immunohistochemistry. Acta Neuropathologica 59: 103–108

Palmer J Q, Kasselberg A G, Netsky M G 1981 Differentiation in medulloblastomas — studies including immunohistochemical localisation of glial fibrillary acidic protein. Journal of Neurosurgery 55: 161–169

Paulus W, Peiffer J 1988 Does the pleomorphic xanthoastrocytoma exist? Problems in the application of immunological techniques to the classification of brain tumours. Acta Neuropathologica 76(3): 245–253

Perentes E, Rubinstein L J 1985 Immunohistochemical recognition of human nerve sheath tumours by anti-Leu-7 (HNK-1) monoclonal antibody. Acta Neuropathologica 68: 319–324

Perentes E, Rubinstein L J 1986a Immunohistochemical recognition of human neuro-epithelial tumours by anti-Leu-7 (HNK-1) monoclonal antibody. Acta Neuropathologica 69: 227–233

Perentes E, Rubinstein L J 1986b Non-specific binding of mouse myeloma IgM immunoglobulins by human myelin

sheaths and astrocytes: A potential complication of nervous system immunoperoxidase histochemistry. Acta Neuropathologica 70: 284–288

Perentes E, Rubinstein L J 1987 Recent application of immunoperoxidase histochemistry in human neuro-oncology. Archives of Pathology and Laboratory Medicine 111: 796–812

Perentes E, Rubinstein L J, Herman M M, Donoso L A 1986 S-antigen immunoreactivity in human pineal glands and pineal parenchymal tumours. A monoclonal antibody study. Acta Neuropathologica 71: 224–227

Reifenberger G, Szymas J, Wechsler W 1987 Differential expression of glial- and neuronal-associated antigens in human tumors of the central and peripheral nervous system. Acta Neuropathologica 74: 105–123

Robson K, Lowe J, Thomas G, Eldridge P 1987 Ki-67 labelling index in meningiomas — histological correlates. Journal of Pathology 152(3): 183–184A

Rode J, Dhillon A P, Doran J F, Jackson P, Thompson R J 1985 PGP 9.5, a new marker for human neuroendocrine tumours. Histopathology 9: 147–158

Royds J A, Parsons M A, Taylor C B, Timperley W R 1982 Enolase isoenzyme distribution in the human brain and its tumours. Journal of Pathology 137: 37–49

Rubinstein L J 1986 Immunohistochemical signposts — not markers — in neural tumour differentiation. Neuropathology and Applied Neurobiology 12: 523–537

Russell D S, Rubinstein L J 1989 Pathology of tumours of the nervous system, 5th edn. Edward Arnold, London

Rusthoven J J, Robinson J B, Kolin A, Pinkerton P H 1985 The natural killer-cell-associated HNK-1 (Leu-7) antibody reacts with hypertrophic and malignant prostatic epithelium. Cancer 56: 289–293

Salisbury J R 1987 Demonstration of cytokeratins and an epithelial membrane antigen in chondroid chordoma. Journal of Pathology 153: 37–40

Seshi B, True L, Carter D, Rosai J 1988 Immunohistochemical characterisation of a set of monoclonal antibodies to human neuron-specific enolase. American Journal of Pathology 131: 258–269

Schiffer D, Giordana M T, Mauro A, Migheli A, Germano I, Giaccone G 1986 Immunohistochemical demonstration of vimentin in human cerebral tumours. Acta Neuropathologica 70: 209–219

Sternberger L A, Sternberger N H 1983 Monoclonal antibodies distinguish phosphorylated and non-phosphorylated forms of neurofilaments in situ. Proceedings of the National Academy of Sciences of the USA 80: 6126–6130

Swanson P E, Manivel J C, Wick M R 1987 Immunoreactivity for Leu-7 in neurofibrosarcoma and other spindle cell sarcomas of soft tissue. American Journal of Pathology 126: 546–560

Taratuto A L, Monges J, Lylyk P 1984 Superficial cerebral astrocytoma attached to dura — report of six cases in infants. Cancer 54: 2505–2512

Tarrant G S, Furness P N, Lowe J 1988 Basement membrane deposition and CAM 5.2 expression in choroid plexus carcinomas: a comparison with ependymomas. Journal of Pathology 155(4): 343A

Theaker J M, Fletcher C D M 1989 Epithelial membrane antigen expression by the perineurial cell. Further studies on peripheral nerve lesions Histopathology 14: 581–592

Tischler A S, Mobtaker H, Mann K et al 1986 Anti-lymphocyte antibody Leu-7 (HNK-1) recognises a constituent of neuro-endocrine granule matrix. Journal of Histochemistry and Cytochemistry 34. 1213–1216

Van Eldik L J, Jensen R A, Ehrenfried B A, Whetsell W O 1986 Immunohistochemical localisation of S100β in human nervous system tumors using monoclonal antibodies with specificity for the S100β polypeptide. Journal of Histochemistry and Cytochemistry 34: 977–982

Warnke R A, Gatter K C, Falini B et al 1983 Diagnosis of human lymphoma with monoclonal antileucocyte antibodies. New England Journal of Medicine 309: 1275–1281

Weidenmann B, Franke W W 1985 Identification and localisation of synaptophysin, and integral membrane glycoprotein of Mr 38 000 characteristic of pre-synaptic vesicles. Cell 45: 1017–1028

Wilkinson K D, Lee K, Deshpaude S et al 1989 The neuron specific protein PGP 9.5 is a ubiquitin carboxyl terminal hydrolase. Science 246: 670–673

Wilson P O G, Barber P C, Hamid Q A et al 1988 The immunolocalisation of protein gene product 9.5 using rabbit polyclonal and monoclonal antibodies. British Journal of Experimental Pathology 69: 91–104

Wotherspoon A C, Norton A J, Isaacson P G 1989 Immunoreactive cytokeratins in plasmacytomas. Histopathology 14: 141–150

8. Immunohistochemistry of peripheral nerve

S. Leibowitz

INTRODUCTION

The structure of peripheral nerve is relatively simple compared with that of the brain and the range of pathology is correspondingly limited. However, interpretation of the nerve biopsy is a difficult art and, until recently, was frequently uninformative. There were two main reasons for this. The first, which still applies, is that most biopsies are undertaken in patients with chronic disease and, as the nerve has only a limited range of response to injury, the endstage picture of one chronic disease may look very much like another. The second difficulty was the limited information to be derived from the conventional paraffin sections, even when supplemented by histochemistry. In recent years, however, the value of the biopsy has been greatly enhanced by the use of plastic embedded 1 μm sections; and electron microscopy, which allows very precise analysis of the morphological change on the subcellular as well as the cellular level; and the examination of teased fibres under the dissecting microscope. Finally, the introduction of immunohistological methods promises to add a new dimension to the analysis of pathological changes in peripheral nerve. The availability of specific monoclonal antibodies allows the identification of cell types on the basis of morphology to be supplemented by a wide range of highly discriminating cell markers. Not only can cells with the same morphology and different function, such as B and T lymphocytes, be distinguished, but different functional states in the same cell line. Even more striking is the ability to detect and localize chemical substances in tissues and cells. Immunocytochemical methods have already had an important impact on research and are gradually being applied to the routine biopsy. For the neuropathological descriptions of the pathology of peripheral nerves, see Dyck et al 1975 and Thomas et al (1984).

Nerve biopsy

Ideally the biopsy should be taken from a nerve only moderately involved in the disease, as changes in the more severely affected nerves may be so advanced as to make diagnosis difficult. In practice, however, the nerves that can be biopsied are limited by accessibility and motor nerve biopsies are difficult to justify. The most common nerves taken for biopsy are the cutaneous sensory branches of the sural nerve, or less commonly the radial nerve. These nerves have the additional advantage that their normal appearance is well known. In postmortem material, the choice is far wider and all the affected nerves may be sampled.

Unfixed nerve is very susceptible to damage and extreme care must be exercised by the surgeon in its removal. Up to 3–4 cm of nerve (less in a child) are removed and divided into three portions. The first is placed on a piece of card, with very slight stretching to hold it straight, and placed in 10% buffered formalin for paraffin sections. The second segment is orientated on a card in a globule of OCT compound and snap frozen in liquid nitrogen. The final portion, after fixation in glutaraldehyde, and postfixation in osmium tetroxide, provides plastic 1 μm sections, electron microscopy and teased fibres (for details see Jacobs & Love 1985). Plastic sections are preferred to paraffin sections for purely morphological observation since detail is much better, but the staining techniques are limited.

A complete examination of the nerve would, where appropriate, include the study of a) paraffin sections stained with H & E, a connective tissue stain, a myelin stain and a silver stain for axons, b) 1 μm plastic sections stained for morphology, c) electron microscopy, d) examination of teased fibres, e) relevant histochemical staining of frozen sections or paraffin embedded material, f) immunohistochemical studies on cryostat sections or paraffin sections.

Immunohistochemistry

Until recently most immunocytochemistry was done on cryostat sections, unfixed or treated with acetone or ethanol. In the case of nerve biopsies this method is still standard for antigens labile in paraffin processing. This may be quantitative, as in the case of immunoglobulin deposits, which can be detected in paraffin sections, albeit with some loss of sensitivity. Membrane antigens tend to be present in small amount compared with cytoplasmic components and are more difficult to detect. For this reason staining for immunoglobulin deposits and T cell subsets is normally performed on cryostat sections. For many purposes the simple direct immunostaining methods are adequate, but for increased sensitivity multiple layer methods, such as the peroxidase-immunoperoxidase (PAP) are used. The biotin-avidin system combines ex-

Table 8.1 Immunohistochemistry of peripheral nerve: list of useful reagents

Clone/specificity	Section type	Source	Specificity in peripheral nerve
PD7	CP	Dako	Common leucocyte marker (CD45)
UCHT1	C	Dako	Pan T
UCHL1	P	Dako	Pan T ($\pm$)
Anti-Leu-3a	C	Becton Dickinson	Ts (CD4)
Anti-Leu-2a	C	Becton Dickinson	Th (CD8)
HNK-1/Leu-7	CP	Becton Dickinson	NK cells. Myelin
MB2	P	Biotest	B lymphocytes
Tol5	C	Dako	B lymphocytes
L26	P	Dako	B lymphocytes
EBM11	C	Dako	Macrophages
MAC387	P	Dako	Macrophages
HAM 56	P	Enzobiochem	Macrophages
L243	C	Becton Dickinson	HLA class II
LN3	P	Biotest	HLA class II
Neurofilaments	CP	Sigma, Seralab.	Axons
S100 (polyclonal)	P	Dako	Schwann cells
GFAP (polyclonal)	CP	Dako	Some Schwann cells assoc. with small unmyelinated fibres
P1	C	Serotec	Myelin basic protein P1

C = cryostat sections, P = paraffin sections.

cellent localization with sensitivity and general convenience. However, immunofluorescence may be necessary when double staining is required, using fluorescein and rhodamine-labelled antibodies. For details of these techniques, see Chapter 3 and Appendix 6. Table 8.1 lists some antisera useful in the study of peripheral nerve pathology.

With one or two notable exceptions it is not possible to make a diagnosis on a biopsy by immunohistochemistry alone, but immunostaining may often be of assistance in the interpretation of pathological changes seen on light and electron microscopy. The immunocytochemistry can be considered under six headings: 1. immunoglobulin deposits, 2. inflammatory cells, 3. HLA expression, 4. Schwann cells, 5. axons, 6. myelin antigens.

IMMUNOGLOBULIN DEPOSITS

The cells of the endoneurium are separated from the plasma by the blood nerve barrier. The barrier is constituted by tight junctions between the en-

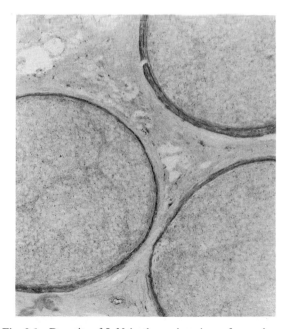

Fig. 8.1 Deposits of IgM in the perineurium of a sural nerve biopsy demonstrated by the direct immunoperoxidase method. This is a normal staining pattern and can be demonstrated in most biopsies and in postmortem nerve. ×110

dothelial cells of the small vessels and between the inner layers of the perineurial cells. Diffusion of plasma proteins into the endoneurial space is restricted and is inversely related to molecular size, with small molecules, such as albumin and IgG, entering more easily than the larger molecules of IgM (Van Lis & Jennekens 1977, Liebert et al 1985). The practical consequence of this is that the 'background' staining for IgG in nerve biopsies is higher than for IgM and immune deposits of IgG more difficult to detect. On the other hand, the large molecular weight substances tend to become trapped in the perineurium, which stains strongly for IgM in most normal as well as abnormal nerves (Fig. 8.1). The staining can be demonstrated with anti-μ and anti-light chain reagents in paraffin, as well as cryostat sections, and tends to be associated with the inner layer of perineurial cells. The staining is particularly pronounced in conditions in which there is marked subperineurial oedema, e.g. axonal degeneration and demyelination.

IgM paraproteinaemic neuropathy

The association of paraproteinaemia and neuropathy is not uncommon as 10% of all cases of idiopathic polyneuropathy have some form of gammopathy. In patients with macroglobulinaemia and neuropathy, 50% of the paraproteins have an immunological specificity for a carbohydrate determinant on myelin associated glycoprotein (MAG), a component of the myelin sheath. In biopsies from these patients IgM deposits can be demonstrated on the surface of surviving myelinated fibres (Mendell et al 1985). Nearly all belong to the category of benign monoclonal gammopathy, although occasional cases of Waldenstrom's macroglobulinaemia do occur.

The IgM deposits are encountered surprisingly often in routine biopsies, considering the comparative rarity of the disease. In part, this is a reflection of special interest in the disease by neurologists, but it is mainly because these cases fall into the category of chronic idiopathic polyneuropathy, which is usually an indication for biopsy. The patient develops a primary demyelinating neuropathy with some axonal loss and IgM can be demonstrated on the surviving myelin sheaths or on fibres undergoing remyelination (Fig. 8.2). In transverse sections staining takes the

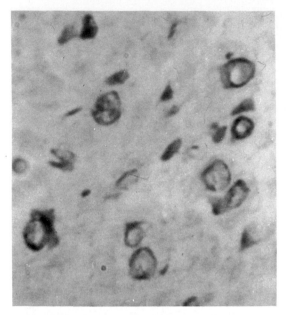

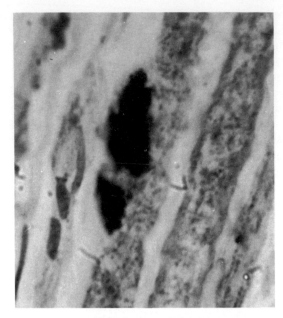

Fig. 8.2 IgM paraproteinaemic neuropathy. Deposits of IgM in a sural nerve biopsy demonstrated by the direct immunoperoxidase method. Ring staining of surviving myelin sheaths and/or remyelinating fibres. ×1100

Fig. 8.3 IgM paraproteinaemic neuropathy. Staining of myelin and Schwann cell pi granules in normal sciatic nerve by antiMAG IgM from a patient with benign monoclonal gammopathy and polyneuropathy. Indirect immunoperoxidase method. ×1100

form of tiny rings outlining the outer surface of the myelin. The deposits can be detected in acetone fixed cryostat sections using anti-human μ chain reagents or antisera or κ or λ light chains. The subclass of light chain will correspond with that of the circulating paraprotein. Similar deposits can be demonstrated in paraffin sections by the PAP method after pronase treatment, although there is some loss of sensitivity.

Until recently it was believed that complement components were not present. However, it has now been shown that immune reactive C3d is bound at the myelin surface together with IgM. C3d is a product of C3b degradation which remains attached to cell membranes as a result of complement activation (Hays et al 1988). C1q is also present and suggests activation by the classical pathway.

The serum of these patients reacts in very high titre with peripheral nerve antigens by complement fixation, and will stain myelin and pi granules in normal human nerve by the indirect immunoperoxidase method (Fig. 8.3). Confirmation of specificity for myelin associated glycoprotein is provided by an ELISA immunoassay and immunoblotting (Latov 1984, Gregson & Leibowitz 1985).

In virtually all the cases in which IgM can be demonstrated in the biopsy, the serum reacts with MAG. There are, however, exceptions, suggesting that at least some of these paraproteins are combining with other, as yet undetermined, myelin antigens (Nobile-Orazio et al 1987). The converse of this — anti-MAG activity in the serum with no deposits of IgM in the biopsy — is almost invariably due to absence of myelin.

Immunoglobulin in other conditions

In the acute Guillain-Barré syndrome, plasma proteins, including fibrin, may sometimes be demonstrated in the endoneurium, albeit with difficulty. They tend to be in the subperineurial space and perivascularly in nerve bundles showing cellular infiltration. Deposits of immunoglobulin and complement have been reported on the myelin sheaths (Luitjen & Faille-Kuyper 1972, Lisak et al 1975, Nyland et al 1981, Koski et al 1987), although our own experience with postmortem as well as biopsy material has not confirmed these findings. Recently, IgM, C3d and C1q have been described in some cases of chronic relapsing inflammatory neuropathy as well as in the acute disease (Hays et al 1983). Similar deposits of C3d, C1q and IgM associated with myelin are present in some cases of vasculitic neuropathy, and IgG, IgM and fibrin can be demonstrated in the walls of vessels affected by fibrinoid necrosis.

Recent work suggests that C3d may prove to be a sensitive indicator of complement activation in nerve. In acute and chronic inflammatory neuropathy and vasculitis neuropathy it is associated with C1q, suggesting complement activation by antibody binding. In metachromatic leucodystrophy C3d is deposited along the myelin sheath in the absence of C1q, C5 or immunoglobulin. In this case activation may be via the alternative pathway due to the action of sphingolipids or to the binding of C1q by sulphatide (Hays et al 1988).

INFLAMMATORY CELLS

The infiltration of peripheral nerve by inflammatory cells may be observed in acute and chronic inflammatory disorders, such as the Guillain-Barré syndrome, chronic inflammatory polyneuropathy and leprosy and in various forms of vasculitis. The cells are mainly lymphocytes and macrophages and can be identified using monoclonal antibodies to leucocyte common antigens. The most valuable of these, PD7, can be used in paraffin sections as well as frozen material.

In the Guillain-Barré syndrome accumulations of inflammatory cells may be seen in the endoneurium (Fig. 8.4) either diffusely scattered or in a perivascular or subperineurial distribution. They are also present in the epineurium around the small vessels. This is the usual picture of the Guillain-Barré syndrome as described by Arnason and others (Arnason 1984), although in many cases the cellular infiltration is much less prominent.

Endoneurial and epineurial inflammatory cells may also be found in the chronic or recurrent inflammatory polyneuropathies and is, indeed, one of the criteria for the diagnosis. Immunohistochemistry in these cases permits the rapid

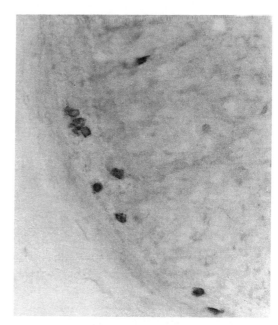

Fig. 8.4 Acute inflammatory demyelinating polyneuropathy (Guillain-Barré syndrome). Inflammatory cells in the endoneurium detected with a monoclonal leucocyte common antigen antibody (PD7). The cells in this field are mainly subperineurial. Indirect immunoperoxidase. ×440

scanning of sections for inflammatory cells which may be sparse and difficult to identify in conventionally stained sections.

A similar problem may arise in the diagnosis of vasculitis, a common cause of mononeuritis multiplex. In these cases ischaemic changes affecting the nerve fibres may be seen in the endoneurium, but the diagnosis depends upon evidence of acute or chronic vasculitis in the epineurial vessels. An acute necrotizing vasculitis, with fibrin in the vessel wall, presents no problem, apart from the vexed matter of sampling error in a focal lesion occurring randomly along a length of nerve. In the less florid lesions, however, the diagnosis rests on evidence of scarring, thrombosis or haemorrhage, and the presence of inflammatory cells in or around the vessel wall.

Lymphocytes

A large proportion of the cells in the inflammatory neuropathies and vasculitides are T-lymphocytes. The number staining with B cell markers is surprisingly low. Both CD4 and CD8 lymphocytes are present in the endoneurium and epineurium in the Guillain-Barré syndrome and in smaller numbers in the chronic inflammatory demyelinating neuropathies. In the demyelinating neuropathies associated with AIDS and the AIDS-related complex, CD8 cells predominate (De la Monte et al 1988). With a few notable exceptions, such as UCHL1, the Pan-T and the T subset markers can only be used on frozen sections. UCHL1 stains T cells and is a valuable adjunct to the use of the leucocyte common marker, PD7 in paraffin sections.

Macrophages

A variety of monoclonal markers for macrophages is now available which allows satisfactory staining of these cells in paraffin as well as cryostat sections. However, many of the reagents react with a limited range of cells (Hofman et al 1984), being restricted to macrophages in particular stages of their development. Antimuramidase antibody, for example, will detect monocytes of the blood and macrophages recently derived from the circulation, but will not react with the resident tissue histiocytes or epithelioid cells. Several recent markers detect a sufficiently wide range of macrophages found in pathological nerve. These are a) epineurial interstitial and perivascular mononuclear cells, b) inflammatory macrophages in the endoneurium, c) vacuolated intraneural tube and other macrophages associated with nerve fibre degeneration or demyelination, d) flattened macrophages of the perineurium and e) cells of the macrophage series associated with inflammatory granulomata.

The monoclonal antibody EBM 11 stains the widest range of cells (Fig. 8.5), but its use is limited to cryostat sections. Antibodies MAC 387 and HAM 56 can be used on paraffin sections, with HAM 56 having a wider range. These reagents make it possible to delineate the infiltrating macrophage population in the inflammatory neuropathies and vasculitides as well as to identify some of the macrophage type cells involved in the phagocytosis of myelin debris resulting from primary demyelination or secondarily from axonal degeneration (Fig. 8.6).

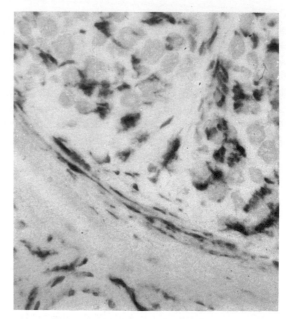

Fig. 8.5 Acute inflammatory demyelinating neuropathy (Guillain-Barré syndrome). Macrophages in the endoneurium, perineurium and epineurium are demonstrated with monoclonal antibody EBM11. Indirect immunoperoxidase. ×440

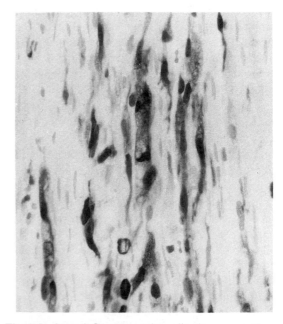

Fig. 8.6 Acute inflammatory demyelinating neuropathy. Endoneurial macrophages in a longitudinal section stained with MAC 387. Sural nerve biopsy. Indirect immunoperoxidase. ×440

HLA-DR

In normal nerve HLA class II antigens are expressed on vascular endothelium and occasional mononuclear and perineurial cells, and can readily be demonstrated using the monoclonal antibody L243 in cryostat sections. In Guillain-Barré syndrome, chronic inflammatory demyelinating polyneuropathy, systemic lupus erythematosus and AIDS-related peripheral neuropathy there is an increase in the intensity of staining with L243 on vascular endothelium and perineurium, and of the expression of HLA-DR on cells in the endoneurium (Fig. 8.7) and on many of the inflammatory cells in the epineurium. The endoneurial staining is due to expression on some Schwann cells as well as on activated macrophages (Pollard et al 1986, Pollard et al 1987, de la Monte et al 1988). Information on HLA-DR expression has, until recently, been limited to inflammatory/ immune disorders, such as the Guillain-Barré syndrome, and has provided no clear picture of neuropathies in general. Our own experience confirms that of Mancardi et al (1988) that this emphasis is misleading and that HLA-DR expression on the Schwann cell is not restricted to inflammatory or immune lesions, but may occur in many forms of peripheral nerve disease, including hereditary degenerative disorders and toxic and metabolic neuropathies.

SCHWANN CELLS

The most widely used marker for Schwann cells is the acidic protein S100. Antisera to this protein detect normal and abnormal Schwann cells associated with myelinated, as well as unmyelinated fibres. The flattened laminated leaves of the 'onion bulbs' of hypertrophic neuropathy (Fig. 8.8), and some malignant tumours of peripheral nerve such as neurolemmoma, neurofibroma and nerve sheath tumours express S100 protein which can be demonstrated in paraffin sections.

A Schwann cell marker of a different kind is the glial fibrillary acidic protein (GFAP). This is the

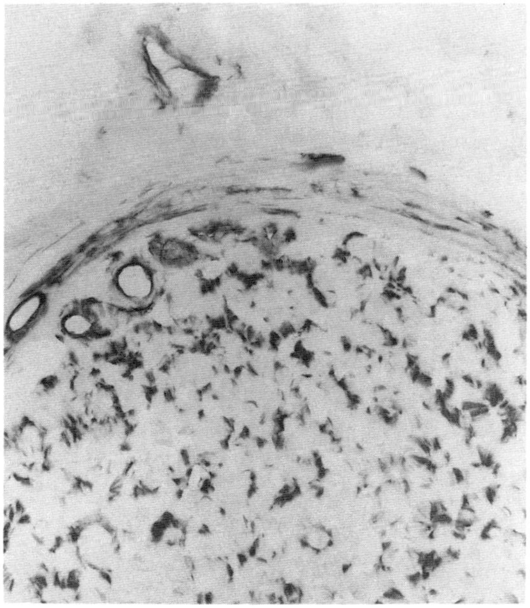

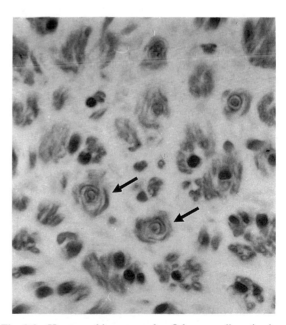

Fig. 8.7 Acute inflammatory demyelinating polyneuropathy. HLA-DR expression on cells in the endoneurium. There is staining of Schwann cells, macrophages and some flattened cells in the perineurium. Staining of the endothelium of endoneurial and epineurial vessels is present in normal nerves. Indirect immunoperoxidase. ×220

Fig. 8.8 Hypertrophic neuropathy. Schwann cells stained by antibody to S100 protein. Transverse section of sural nerve from patient with chronic demyelinating polyneuropathy and onion bulb formation (arrow heads). These are spherical laminated structures formed by concentric layering of flattened attenuated Schwann cells around a central myelinated or unmyelinated axon. Indirect immunoperoxidase. ×440

intermediate filament protein characteristic of astrocytes in the CNS, which can be detected by polyclonal antisera in paraffin sections. In normal peripheral nerve there is very little staining, GFAP being restricted to a small proportion of Schwann cells associated with small diameter unmyelinated fibres. In pathological states where active nerve sprouting or nerve regeneration is taking place there may be an increase in GFAP positive Schwann cells. In transverse sections this shows as tiny clusters of positively stained fibres.

AXONS

The characteristic intermediate filaments of the nerve cell are the neurofilaments , which are made up of three crossreacting protein subunits with molecular weights of 68 kD, 145 kD and 200 kD. A number of highly specific monoclonal antibodies are available. In peripheral nerve these monoclonal antibodies stain axons very effectively in cryostat sections as well as paraffin sections. The results are excellent and the method is simple and reliable, and may well replace the capricious silver impregnation techniques routinely used in histology (Gambetti et al 1981). Apart from demonstrating loss or degeneration of axons these monoclonal antibodies will also detect the clusters of small regenerating fibres which may be seen in many different pathological states.

MYELIN ANTIGENS

The antigens of the myelin sheath are protein, glycoprotein or glycolipid. In peripheral nerve the major components are proteolipid protein Po, the basic proteins P1 and P2, MAG and the glycolipids, galactocerebroside and sulphatide. Po is the most important structural protein of myelin. P1 is a small molecular weight basic protein, which is the antigen involved in experimental allergic encephalomyelitis in animals and is present in the CNS as well as in peripheral nerve. P2 is a similar protein found only in the PNS. Po, P1 and P2 are constituents of compact myelin, whereas MAG is restricted to Schwann cell membranes in the periaxonal and paranodal regions and in Schmidt-Lanterman clefts. In inflammatory demyelinating neuropathy the earliest lesions show abnormalities of Po, P1 and P2 staining mainly in the paranodal myelin. Changes in MAG immunostaining occur only in regions where myelin staining is already abnormal. This suggests that the primary target is myelin and not the Schwann cell (Schober et al 1981). Changes in the myelin sheath can be detected earlier by immunohistochemical methods than by conventional staining techniques.

Using specific antisera, P1 and P2 may be detected in cryostat sections fixed with alcohol or formalin alcohol. Myelin-associated protein can be demonstrated using anti-MAG sera from paraproteinaemic neuropathy patients or by using the antileucocyte monoclonal antibody Leu-7 (HNK-1). The latter defines a subset of T cells containing the NK cell and reacts with a carbohydrate determinant on MAG . This same determinant is also present on the myelin glycolipid, N-acetyl galactosyl glucuronyl globoside.

Specific antisera to galactocerebroside and sulphatide can also be used to stain myelin in alcohol fixed cryostat sections, but only with difficulty and in low titre.

REFERENCES

Dyck P J, Thomas P K, Lambert E H (eds) 1975 Peripheral neuropathy, vols I & II. Saunders, Philadelphia

Gambetti P, Autillio-Gambetti L, Papasozomenos S C 1981 Bodians silver method stains neurofilament polypeptides. Science 213: 1521–1522

Gregson N A, Leibowitz S 1985 IgM paraproteinaemia, polyneuropathy and myelin-associated glycoprotein (MAG). Neuropathology and Applied Neurobiology 11: 329–347

Hays A P, Lee S S L, Latov N 1988 Immune reactive C3d on the surface of myelin sheaths in neuropathy. Journal of Neuroimmunology 18: 231–244

Hofman F M, Lopez D, Husmann L et al 1984 Heterogeneity of macrophage populations in human lymphoid tissue and peripheral blood. Cellular Immunology 88: 61–74

Jacobs J M, Love S 1985 Qualitative and quantitative morphology of human sural nerve at different ages. Brain 108: 897–924

Latov N 1984 Immunological abnormalities associated with chronic peripheral neuropathies, plasma cell dyscrasia and neuropathy. In: Behan P, Spreafico F (eds) Neuroimmunology. Raven Press, New York, p 261–273

Liebert U G, Seitz R J, Weber I, Wechsler D W 1985 Immunocytochemical studies of serum proteins and immunoglobulins in human sural nerve biopsies. Acta Neuropathologica 68: 39–47

Lisak R P, Zweiman B, Norman M 1975 Anti-myelin antibodies in neurologic disease — immunofluorescence demonstration. Archives of Neurology 32: 163–167

Luitjen J A F M, Faille-Kuyper E B H 1972 The occurrence of IgM and complement factors along myelin sheaths of peripheral nerves. An immunohistochemical study of the Guillain-Barré syndrome. Journal of the Neurological Sciences 15: 219–224

McCombe P A, McLeod J G, Pollard J D, Guo Y-P, Ingall T J 1987 Peripheral sensorimotor and autonomic neuropathy associated with systemic lupus erythematosus. Brain 110: 533–549

Manicardi G L, Cadoni A, Zicca A et al 1988 HLA-DR Schwann cell reactivity in peripheral neuropathies of different origins. Neurology 38: 848–851

Mendell J R, Sahenk Z, Whitaker J N et al 1985 Polyneuropathy and IgM monoclonal gammopathy: studies on the pathogenetic role of anti-myelin-associated glycoprotein antibody. Annals of Neurology 17: 243–254

Monte de la S M et al 1988 Peripheral neuropathy in acquired immunodeficiency syndrome. Annals of Neurology 23: 485–492

Nakajima T, Watanabe S, Sato Y et al 1982 An immunoperoxidase study of S100 protein in normal and neoplastic tissues. American Journal of Surgical Pathology 6: 715–726

Nobile-Orazio E, Marmiroli P, Baldini L et al 1987 Peripheral neuropathy in macroglobulinaemia: incidence and antigen-specificity of M proteins. Neurology 37: 1506–1504

Nyland H, Matre R, Mork S 1981 Immunological characterisation of sural nerve biopsies from patients with Guillain-Barré syndrome. Annals of Neurology 9 (supplement): 80–86

Ota K, Ire H, Takahashi K 1987 T cell subsets and Ia Positive cells in the sciatic nerve during the course of experimental allergic neuritis. Journal of Neuroimmunology 13: 283–341

Polak J M, Van Noorden S (eds) 1986 Immunocytochemistry. Modern methods and applications, 2nd edn. Wright, Bristol

Pollard J D, Baverstock J, McLeod J G 1987 Class II antigen expression and inflammatory cells in the Guillain-Barré syndrome. Annals of Neurology 21: 337–341

Pollard J D, McCombe P A, Baverstock J, Gatenby P A, McLeod J G 1986 Class II antigen expression and T lymphocyte subsets in chronic inflammatory demyelinating neuropathy. Journal of Neuroimmunology 13: 123–134

Rhodes R H 1986 Diagnostic immunostaining of the nervous system. In: Taylor C R (ed) Immunomicroscopy: A diagnostic tool for the surgical pathologist. Saunders, Philadelphia, p 333–362

Schober R, Itoyama Y, Sternberger N H et al 1981 Immunocytochemical study of Po glycoprotein, P1 and P2 basic proteins and myelin associated protein (MAG) in lesions of idiopathic polyneuritis. Neuropathology and Applied Neurobiology 7: 421–434

Stefansson K, Wollman R, Jerkovic M 1982 S-100 protein in soft tissue tumours derived from schwann cells and melanocytes. American Journal of Pathology 106: 261–268

Taylor C R (ed) 1986 Immunomicroscopy: A diagnostic tool for the surgical pathologist. Saunders, Philadelphia

Thomas P K, Landon D N, King R H M 1984 Diseases of the peripheral nerves. In: Adams J H, Corsellis JAN, Duchen L W (eds) Greenfields neuropathology, 4th edn. Edward Arnold, London, p 807–920

Van Lis J M J, Jennekens F G I 1977 Plasma proteins in human peripheral nerve. Journal of the Neurological Sciences 34: 329–341

9. Metabolic disorders of the central and peripheral nervous system

B. D. Lake

A group of diseases that were once thought to be degenerative in origin are now recognised as inherited disorders caused by the deficient activity of certain enzymes (usually lysosomal) involved with lipid and mucopolysaccharide metabolism. Impaired enzyme activity at a particular point in any metabolic chain results in the accumulation of metabolites at the defective stage in the pathway. The impairment of enzyme activity can be due to a variety of factors, which includes defective enzyme protein, absent enzyme protein, or absent cofactor. The defects variously involve cells of the MP (mononuclear phagocyte) series, perivascular or endothelial cells, as well as those of the nervous system, which are the primary concern of this chapter.

The most important group of metabolic disorders that affect the nervous system are the sphingolipidoses. Sphingolipids include sphingomyelins, cerebrosides, sulphatides and gangliosides, all of which are important constituents of the normal cell (Fig. 9.1). Structurally they have in common a ceramide moiety derived from the unsaturated aminoalcohol sphingosine, in which one of the amino group hydrogen atoms is substituted by a long-chain fatty acid (see Fig. 9.2). Gangliosides, composed of ceramide, hexose molecules, sialic acid and hexosamine, are the lipids responsible for neuronal accumulations in Tay-Sachs disease and the other gangliosidoses. Gangliosides are additionally implicated in neuronal changes secondary to systemic disturb-

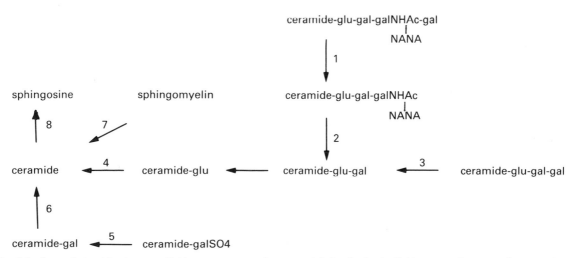

Fig. 9.1 Interrelationships between lipid components and enzyme deficiencies in the lipid storage disease. 1. β-galactosidase; G_{M1}-gangliosidosis; 2. β-hexosaminidase A; G_{M2}-gangliosidosis (Tay-Sachs); 3. α-galactosidase; Fabry; 4. β-glucocerebrosidase; Gaucher; 5. arylsulphatase A; metachromatic leucodystrophy; 6. galactocerebrosidase; Krabbe leucodystrophy; 7. sphingomyelinase; Niemann-Pick; 8. ceramidase; Farber.
gal = galactose; glu = glucose; galNHAc = N-acetylgalactosamine; NANA = N-acetylneuraminic acid (sialic acid)

ance of mucopolysaccharide metabolism in some types of mucopolysaccharidosis.

Other lipids that accumulate abnormally in the nervous system include cholesterol esters in Wolman's disease and phytanic acid in Refsum's disease. A further important group of disorders of lipid metabolism is Batten's disease in which lipofuscin-like pigments are deposited in neurons and elsewhere. The major categories of the metabolic disorders involving the nervous system are presented in Table 9.1 to indicate the enzyme defect and metabolites implicated in each case, their sites of deposition, together with selected histochemical methods for their identification in the relevant tissues or body fluids indicated.

The histochemical techniques referred to in Table 9.1 include the highly specific method for demonstrating the activity of the lysosomal enzyme β-galactosidase which has only been feasible in recent years since the appropriate substrate has become commercially available. Among the histochemical methods listed are some relatively non-specific staining techniques such as toluidine blue and the periodic acid-Schiff reaction, which, with judicious use of blockades, extractions and other controls, can become selective for sulphatide glycolipids and sialic acid. On the other hand the dichromate-acid haematein method is specific for choline-containing lipids — lecithins and sphingomyelins — and the reaction can be confined to sphingomyelins if sections are first treated with sodium hydroxide to remove alkali-labile lecithins. The colour of the reaction may vary in relation to the length and degree of unsaturation of the fatty acid side chain, and can range from blue to brown. The ferric haematoxylin method of Elleder and Lojda (1973) is more reliable. The perchloric acid-naphthoquinone (PAN) method for cholesterol (Adams 1961) utilizes a chemical reaction that is sensitive and highly specific for cholesterol and related sterols. Similarly the calcium-lipase method for triglycerides is selective so long as the lipase preparation is pure and uncontaminated by esterases and provided that control sections are in-

Table 9.1 Metabolic disorders of the nervous system

Disorder	Defect	Substances stored	Sites of storage	Tissue for diagnosis by microscopy	Staining methods and other tests
G_{M1}-gangliosidosis (two types)	β-Galactosidase	G_{M1}-ganglioside, oligosaccharides, ceramide tetrahexoside	Neurons, liver, spleen, kidney	Blood, bone marrow, rectal biopsy, liver	PAS, thionin, β-galactosidase, TLC, EM
G_{M2}-gangliosidosis (several types: Tay-Sachs, Sandhoff, etc.)	Hexosaminidase	G_{M2}-ganglioside, ceramide trihexoside	Neurons	Rectal biopsy	PAS, thionin, TLC, EM
Batten's disease (ceroid lipofuscinosis) several types	Unknown	Lipofuscin-like substances, dolichyl phosphate, oligosaccharides, lipid-binding protein of ATP synthase	Neurons, muscle, liver, kidney, pancreas, blood cells, sweat glands	Rectal biopsy, blood	PAS, Sudan black, luxol fast blue, acid phosphatase, autofluorescence, EM
Gaucher's disease (three types)	Glucocerebrosidase (β-glucosidase)	Glucocerebroside	Spleen, liver, brain stem nuclei (infantile form)	Bone marrow, liver	PAS, acid phosphatase, TLC, EM
Krabbe's leucodystrophy	Galactocerebrosidase	Galacto-cerebroside	White matter of brain in globoid cells	Brain, peripheral nerve	PAS, acid phosphatase, nerve teasing, TLC, EM
Metachromatic leucodystrophy (sulphatidosis)	Aryl sulphatase A	Sulphatides	White matter of brain, peripheral nerve, kidney, gall bladder	Urine, peripheral nerve (skin biopsy)	Cresyl fast violet, toluidine blue, acriflavine-DMAB, TLC

Table 9.1 Cont'd

Disorder	Defect	Substances stored	Sites of storage	Tissue for diagnosis by microscopy	Staining methods and other tests
Niemann-Pick disease type A (infantile)	Sphingomyelinase	Sphingomyelin, cholesterol	Neurons, white matter, spleen, liver, smooth muscle, endothelia	Blood, bone marrow, liver, rectal biopsy	Sudan black, ferric haematoxylin, acid phosphatase, TLC, EM
Niemann-Pick disease type B (adult non-neurological)	Sphingomyelinase	Sphingomyelin, cholesterol	Spleen, liver, smooth muscle	Bone marrow, liver, rectal biopsy (to exclude neuronal involvement)	Sudan black, ferric haematoxylin, acid phosphatase, EM, TLC
Niemann-Pick type C	Defective cholesterol esterification	Water-soluble acidic oligosaccharides of unknown composition, phospholipids	Neurons, liver, spleen	Bone marrow, rectal biopsy, liver	Cell. PAS, cell. dig. PAS, acid phosphatase, thionin, TLC, EM
Fabry's disease	α-Galactosidase	Ceramide trihexoside	Kidney, smooth muscle, spleen	Bone marrow, kidney, skin, urine	Sudan black, polarized light, EM, TLC of urine
Pompe's disease (GSD 2)	Acid α-glucosidase (acid maltase)	Glycogen	In lysosomes everywhere, brain stem nuclei, liver, heart spleen, muscle, endothelia	Blood, muscle, heart	Cell. PAS, acid phosphatase, EM
Mucopolysaccharidosis (several types)	See Chapter 18	Acid muco-polysaccharides (heparan sulphate, chondroitin sulphate, keratan sulphate) several gangliosides	Neurons (not mps), liver, spleen, heart	Urine, blood, liver, cultured fibroblasts	Toluidine blue, PAS, colloidal iron, EM, urinary GAG assay
Mannosidosis	α-Mannosidase	Oligosaccharides containing mannose	Neurons, liver, spleen, endothelia	Blood, bone marrow	Cell. PAS, EM, TLC
Farber's disease	Ceramidase	Ceramides, G_{M3}-ganglioside	Neurons, liver, lymph nodes	Lymph nodes, liver, rectal biopsy	Sudan black, polarized light, PAS, thionin, TLC, EM
Adreno-leucodystrophy	Lignoceroyl CoA ligase	VLCFA	Adrenal, white matter	Conjunctiva	EM, HPLC, oil red O & extractions
Zellweger's cerebrohepatorenal syndrome	Multiple peroxisomal enzymes	VLCFA	Liver	Liver	Catalase (absent peroxisomes), EM, HPLC
Infantile Refsum's disease	Multiple peroxisomal enzymes	VLCFA	Liver	Liver	Catalase (absent peroxisomes), EM, HPLC
Canavan's disease (spongy degeneration)	Aspartoacylase			Brain	N-acetylaspartic acid in urine

Key: TLC = thin layer chromatography, EM = electron microscopy, HPLC = high performance liquid chromatography, GAG = glycosaminoglycan (mucopolysaccharide), VLCFA = very long chain fatty acids, cell = celloidinized, dig = digested.

cluded to distinguish possible crossreaction of free fatty acids. These are just examples of the methods that are recommended for the demonstration of tissue lipids microscopically. Details of their methodology are given in Appendix 4.

Before performing the forementioned histochemical procedures it may be useful to confirm that lipid is actually present in the tissue to be examined. Sudan black B in propylene glycol is suitable in most instances but for more critical work the bromine-Sudan black B method (Appendix 4) is more effective.

Bromine not only reduces the solubility of phospholipids and free fatty acids in the alcoholic staining solution but also converts crystalline-free cholesterol to its liquid bromo-derivatives, which are intensely sudanophilic.

By no means all histochemical methods can be relied upon to identify lipids in the strictly chemical sense but when histochemical observations are taken in conjunction with clinical and morphological evidence, a characteristic pattern will emerge that is diagnostic of a particular disease. In most instances biochemical enzyme assays are available and necessary for the precise diagnosis in these cases, since prenatal diagnosis is possible by enzyme assay of chorionic villus samples, or cultured amniotic cells. For this reason enzyme assay is mandatory for the diagnosis. Histochemical study can give relatively precise diagnoses but cannot distinguish between the various types of G_{M2}-gangliosidoses. Its use lies in producing a working diagnosis and in saving unnecessary assays. Consequently histochemistry remains a useful diagnostic tool in this field with the one distinct advantage over the biochemical approach — namely the ability to localize metabolites at the cellular level.

Whenever possible snap-frozen tissue, formalin-fixed tissue and wax-embedded tissue should be prepared for examination. Where a choice must be made — due to the small size of the specimen — snap-frozen tissue will provide most information. Thin-layer chromatography of solvent extracts prepared from fresh frozen tissue will effect definite identification of all lipids (Lake & Goodwin 1976) and may be used in some instances on formalin-fixed tissue.

Because the substances involved in this group of conditions are soluble in lipid solvents, routine sections are useful only for morphology. The staining reactions used for identification of the various substances must be performed on sections of snap-frozen or formalin-fixed tissues.

For general background reading, Hers & van Hoof (1973), Scriver et al (1989), Glew et al (1985) and Lake (1984) provide clinical, biochemical and pathological descriptions of most of the disorders mentioned in this chapter.

Gangliosidoses

The ballooned neurons present in the CNS and PNS contain either ganglioside G_{M1} (generalized gangliosidosis) or ganglioside G_{M2} (Tay-Sachs disease and variants). These two gangliosides differ from each other only by a galactose residue and it will be appreciated from Figure 9.2 that the two compounds will have very similar staining characteristics and that it will be virtually impossible to differentiate one from the other. Both are readily extracted during processing and formalin-fixed frozen sections, or sections of snap-frozen tissue are necessary to demonstrate neuronal ganglioside accumulation. The ganglioside in glial cells is firmly bound and resists extraction. The high

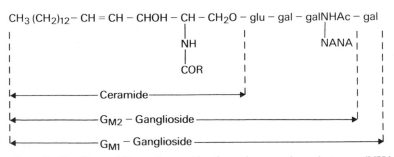

Fig. 9.2 The structure of gangliosides G_{M1} and G_{M2} and ceramide. glu = glucose; gal = galactose; galNHAc = N-acetylgalactosamine; NANA = N-acetylneuraminic acid (sialic acid)

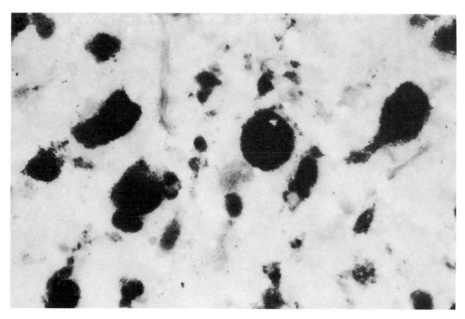

Fig. 9.3 Tay-Sachs disease; cryostat section of brain showing ballooned neurons and glial cells filled with PAS-positive ganglioside. The sections have been lightly counterstained with Carrazzi's haematoxylin. ×350

hexose content of gangliosides gives rise to a strong PAS reaction unaffected by diastase digestion (Fig. 9.3). The gangliosides are readily extracted from cryostat sections by chloroform: methanol 2:1 v/v, in contrast with the neuronal storage substance of Batten's disease, which is unaffected by this extraction and retains its PAS positivity. The presence of sialic acid is shown by an immediate rose metachromasia with Feyrter's thionin 'enclosure' method. This can be confirmed by the Roberts method in which dilute sodium metaperiodate reacts specifically with sialic acid releasing aldehyde groups detectable by the Schiff reagent (Appendix 3). Glycogen and other oligosaccharides remain unreactive under these conditions. Gangliosides, being particularly polar molecules, are packaged in lysosomes with cholesterol and phospholipid giving rise to the characteristic membranous cytoplasmic body (MCB) seen under the electron microscope. Stains to demonstrate cholesterol are positive and the luxol fast blue-neutral red method also indicates storage of some component (? protein) both in frozen and processed tissue. Neurons in the CNS or elsewhere (rectum) have the same appearance and staining reactions in all types of gangliosidosis. In G_{M1}-gangliosidosis neuronal, epithelial, en-dothelial and most mononuclear phagocytic cells show deficient β-galactosidase activity (Fig. 9.4). Reactive astrocytes, however, retain some β-galactosidase activity which may be of a different isoenzyme. In the lamina propria of the gut many PAS-positive histiocytes are present (frozen or fixed tissue) which show metachromasia (frozen tissue only) with Haust and Landing's toluidine blue method. These histiocytes are not present in G_{M2}-gangliosidosis. Methods for showing hexosaminidase activity do not differentiate between isoenzymes A and B and cannot be used for diagnosis of Tay-Sachs disease. However in Sandhoff's disease no hexosaminidase activity can be detected in brain, rectum, fibroblasts and white blood cells and histochemical methods for hexosaminidase activity reliably show the deficiency. A positive control is of course necessary (Fig. 9.5).

Differentiation of the individual gangliosides can only be achieved by thin-layer chromatography of solvent extracts of fresh (frozen) tissue. Formalin fixation alters the gangliosides and thus fixed tissue is less reliable for chromatography, although the gross excesses of G_{M1}- and G_{M2}-gangliosides are usually sufficient to mask the effect of slow formalin-induced changes.

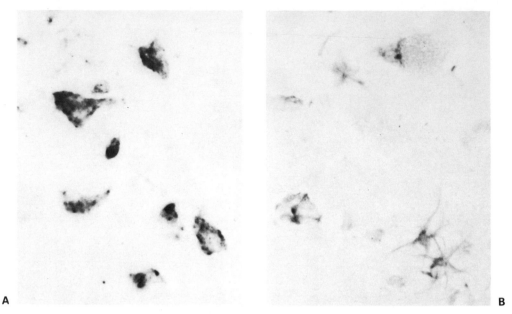

A

B

Fig. 9.4 Cryostat sections of brain stained to show β-galactosidase activity in: **A** normal cortex, **B** cortex from a patient with G_{M1}-gangliosidosis.
Normal neurons show strong activity. The ballooned neurons of G_{M1}-gangliosidosis are negative, and only a weak staining is present in reactive astrocytes. The sections have been lightly counterstained with neutral red. ×570

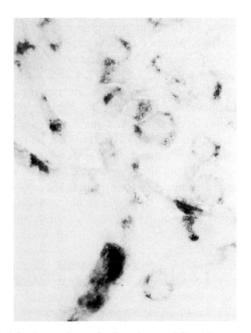

Fig. 9.5 A monolayer of cultured normal fibroblasts stained to show total hexosaminidase activity. No activity can be found in Sandhoff's disease. ×300

Prenatal diagnosis by enzyme assay of chorionic villus samples is possible and fetal tissue from termination of affected pregnancies will show an excess of the ganglioside by TLC, but routine sections and cryostat sections of snap-frozen tissue show no evidence of neuronal storage. Electron microscopy reveals formation of MCBs, particularly in the brain stem, even at 12 weeks' gestation. Sections of placenta show marked vacuolation of the syncytio-trophoblast, fibroblasts, endothelial cells and macrophages in G_{M1}-gangliosidosis, with little of note in G_{M2}-gangliosidosis.

Metachromatic leucodystrophy

Macrophages throughout the demyelinated white matter of the brain contain deposits of galactocerebroside sulphate (sulphatide) which in frozen sections imparts metachromasia to a wide variety of dyes. Those most commonly used are toluidine blue and cresyl fast violet although thionin may also be used. Wherever sulphatides occur (brain, peripheral nerve, gall bladder, kidney), the metachromasia is yellow-brown with cresyl fast violet (Hirsch Pfeiffer method). However, with the more permanent toluidine blue method (Bodian & Lake 1963) the colour varies from red in the brain, yellow red and purple in the macrophages and Schwann cells of peripheral nerves, to yellow and brown in the gall bladder and kidney tubules (Fig. 9.6). Neurons of the cortex and gastrointestinal tract do not show storage of sulphatide, but those of the basal ganglia and dentate nucleus in particular are distended with sulphatide deposition. Sulphatides are also demonstrable with the acriflavine-DMAB method, and have an alcian blue CEC greater than 0.5 mol l magnesium chloride. They are not retained in routinely processed tissue.

The histochemical demonstration of the sulphatase A deficiency is not a reliable procedure because of the presence of sulphatase B which has similar characteristics.

In mucosulphatidosis (multiple sulphatase deficiency), in which there is also urinary mucopolysaccharide excretion, neurons show marked evidence of gangliosidic storage identical to that in the gangliosidoses, in addition to sul-phatide accumulation identical to that in metachromatic leucodystrophy.

Krabbe's leucodystrophy

The collections of globoid cells in the demyelinated white matter are PAS-positive in frozen or routine sections (Fig. 9.7). In frozen sections they show no metachromasia, are only weakly sudanophilic, but show strong acid phosphatase activity. No cholesterol or esters are present in the globoid cells. No globoid cells are found in the peripheral nervous system although segmental demyelination of peripheral nerves is evident. Few macrophages are ever seen in peripheral nerves but the acid phosphatase as reaction in Schwann cells is strong, as in all forms of segmental demyelination (Fig. 9.8). This is evident in even the small nerves in skin and muscle and may be used as a diagnostic pointer where the enzyme assay is not available.

Although the deficient enzyme — galacto-cerebrosidase — is a β-galactosidase, it is substrate-specific and the histochemical substrates are not suitable to demonstrate its activity, or lack of it.

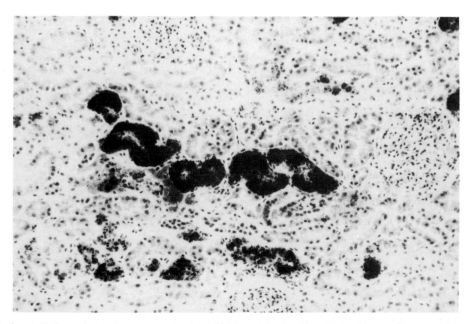

Fig. 9.6 Metachromatic leucodystrophy; cryostat section of kidney stained with toluidine blue to show metachromatic deposits of sulphatide in distal tubules. ×150

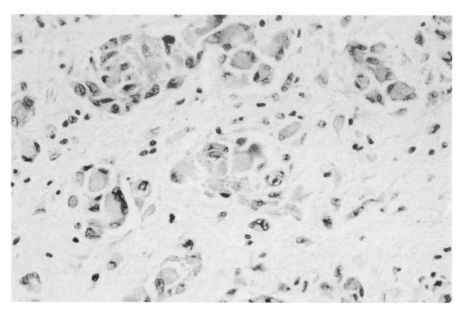

Fig. 9.7   Krabbe's leucodystrophy. Routine section of brain showing weakly PAS-positive perivascular collections of globoid cells in the demyelinated white matter. The section has been lightly counterstained with Carrazzi's haematoxylin. ×370

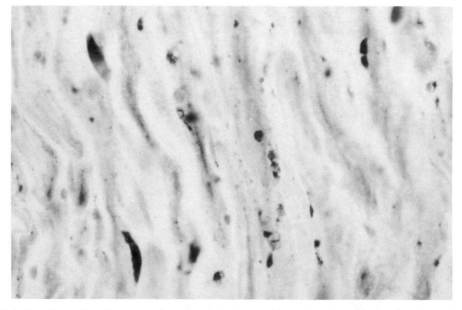

Fig. 9.8   Krabbe's leucodystrophy. Cryostat section of peripheral nerve biopsy showing acid phosphatase activity in a Schwann cell. The section has been lightly counterstained with Carrazzi's haematoxylin. ×880

Alexander's leucodystrophy

In the younger cases of Alexander's leuco-dystrophy the strongly eosinophilic Rosenthal fibres, which are present in abundance in the sub-pial regions, perivascularly in the demyelinated white matter and scattered throughout the cortex, are strongly stained with luxol fast blue (Fig. 9.9). Rosenthal fibres are fibrillary deposits in fibrous astrocytes and are positive with antibodies to glial

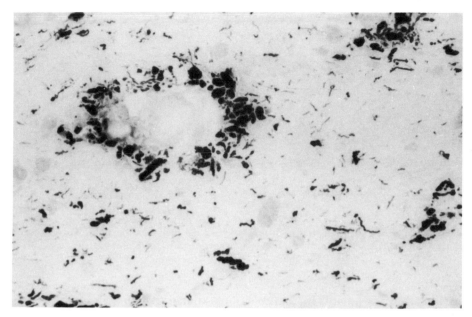

Fig. 9.9 Alexander's leycodystrophy; cryostat section of a brain biopsy stained with luxol fast blue-neutral red showing the numerous perivascular and scattered Rosenthal fibres in the demyelinated white matter. ×350

fibrillary acidic protein (GFAP). The older cases show fewer Rosenthal fibres which stain weakly with luxol fast blue but have the staining characteristics of fibrin. No metachromasia can be shown and neutral fat is only rarely present in macrophages.

Spongy degeneration (Canavan, von Bogaert and Bertrand)

The white matter appears spongy with multiple vacuoles which are also present perivascularly in the cortex. The vacuoles are empty and contain no demonstrable substance. Glycogen deposition may be prominent in perivascular astrocytes, particularly in the cortex. Abnormally structured mitochondria are found in the astrocytes and oligodendroglia (Adachi et al 1973). The recently discovered defect in aspartoacylase activity is not amenable to histochemical methods (Matalon et al 1988).

Batten's disease

This group of disorders has a bewildering variety of names associated with it which include neur-

onal ceroidlipofuscinosis, Jansky, Bielschowsky, Sjögren, Spielmeyer, etc. (see Brett 1990). They are best classified on the basis of clinical presentation and age, and fall into four main groups: infantile, late infantile, juvenile and adult. They share the common feature of neuronal deposition of a substance (substances) which has the staining characteristics of ceroid or lipofuscin but which is neither of these entities.

This material is resistant to processing, and in routine sections of brain stains with Sudan black, PAS, luxol fast blue and shows autofluorescence. The morphological changes in postmortem brain of the infantile form are characteristic with marked atrophy (walnut brain) and almost total loss of neurons and marked astrocytic storage. However, earlier in the course of the infantile form and in the late-infantile and juvenile forms there is little in the staining reactions of the neuronal storage material to differentiate one form from another. Batten's disease may be differentiated from the gangliosidoses by selective extraction techniques on cryostat sections of brain followed by staining with the PAS reaction. The gangliosides of the gangliosidoses are readily extracted by chloroform: methanol (2:1 v/v), a procedure which has little or

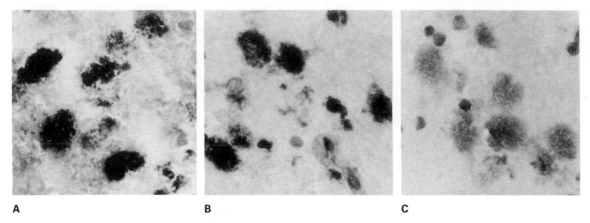

A   **B**   **C**

Fig. 9.10 Juvenile Batten's disease. Cryostat sections of brain cortex stained with PAS and Carrazzi's haematoxylin after: **A** no extraction **B** extraction with chloroform:methanol 2:1 v/v, **C** extraction with chloroform:methanol:water 1:1:0.3 v/v.

no effect on the PAS positivity in Batten's disease. In contrast, the PAS positivity in neurons and glial cells in Batten's disease (and of neuronal wear and tear pigment of the elderly) is abolished by prior extraction with the solvent chloroform: methanol: water (1:1: 0.3 v/v) (Fig. 9.10). This corresponds to the presence of dolichyl phosphate-linked oligosaccharides (Lake et al 1988) known to be in high concentration in Batten's disease (Hall & Patrick 1988). Formalin fixed tissue is unsuitable for these techniques. The autofluorescent part of the storage substance may correspond to the lipid binding protein of mitochondrial ATP synthase which is a major component of isolated storage bodies in the sheep model (Palmer et al 1989) and in the late infantile and juvenile forms of human Batten's disease. (Palmer, Lake and Hall, in preparation).

Electron microscopy shows ultrastructural characteristics which serve to differentiate the various forms. Granular osmiophilic deposits (GROD) are present in infantile Batten's disease; curvilinear bodies accumulate in late infantile Batten's disease; and predominantly fingerprint bodies are found in the juvenile form (Lake 1989). This is true for brain, skin, lymphocytes and rectum but other sites (skeletal muscle for example), although accumulating storage substance, may show different ultrastructure.

The substance is stored in neurons everywhere and diagnosis of the disorder and its type can be made by examination of cryostat sections of snap-

Table 9.2 Rectal biopsy: neuronal staining patterns in Batten's disease

	Infantile type	Late infantile type	Juvenile type
PAS	++	±	++
Sudan black	++	+	++
Luxol fast blue	—	+	++
Ultrastructural	Granular osmiophilic deposits	Curvilinear bodies	Finger-print bodies
Autofluorescence (excitation 360 nm* barrier 410 nm)	++	+	++

Similar depositis are found in smooth muscle cells and endothelial cells.
Histiocytes, staining strongly for acid phosphatase activity, are present among the smooth muscle cells in the juvenile type.
*Best results are obtained with the Zeiss UG5 filter and dark ground illumination.

frozen biopsies of rectum. In this site the neuronal staining reactions vary from one type to another, and these together with evidence of smooth muscle cell storage and the presence or absence of histiocytes among smooth muscle cells make the differential diagnosis of the different types possible (Figs 9.11, 9.12 and 9.13). Table 9.2 illustrates the patterns of staining.

The disease is systemic and deposits of the ceroid- lipofuscin like material may be found in liver, pancreatic acinar cells, skeletal muscle, kidney tubules, sweat glands and lymphocytes. The staining reactions vary slightly but the most reliable indicator in all sites, except lymphocytes, is autofluorescence of the stored material best seen with dark ground illumination.

Niemann-Pick disease

Niemann-Pick disease can be divided into two main groups. Group I is characterized by deficient sphingomyelinase activity and storage of sphingomyelin in liver and spleen (types A and B) and in neurons (type A only). In group II there is no deficiency of sphingomyelinase and sphingomyelin

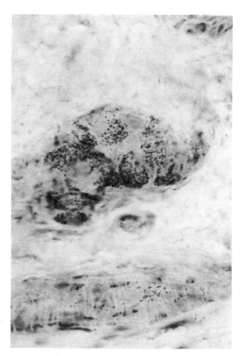

Fig. 9.11   Juvenile Batten's disease; cryostat section of a rectal biopsy stained with Sudan black (and carmalum) showing granular sudanophilic deposits in submucosal neurons in smooth muscle cells of an arteriole. × 425

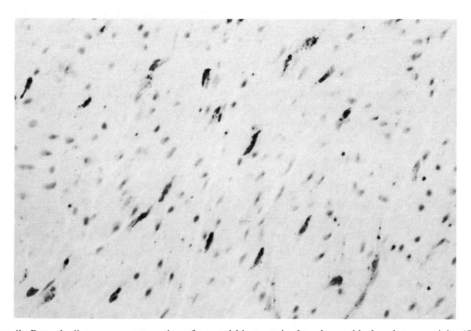

Fig. 9.12   Juvenile Batten's disease; cryostat section of a rectal biopsy stained to show acid phosphatase activity (Gomori). Many small strongly-staining histiocytes are present among the smooth muscle cells of the circular muscle cells of the circular muscle layer. These histiocytes are not present in the late infantile or infantile forms of Batten's disease. The section has been lightly counterstained. ×425

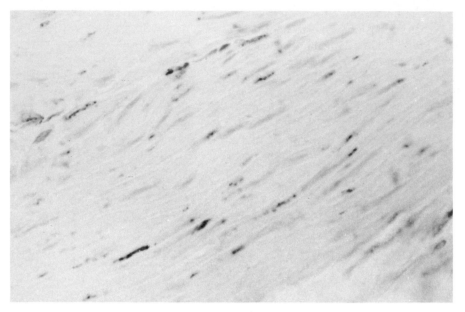

Fig. 9.13 Late infantile Batten's disease; cryostat section of a rectal biopsy stained with Sudan black (and carmalum) showing paranuclear sudanophilic deposits in smooth muscle cells of the circular muscle layer. ×600

accumulation is not seen in brain or liver, although it is usually present in the spleen. The patients in this group have Niemann-Pick disease type C and have a defect in cholesterol esterification (Vanier et al 1988).

Niemann-Pick disease types A and B

There is neuronal storage only in type A, type B being entirely visceral. The general appearance of the neurons is that of the classic ballooned storage cell usually associated with Tay-Sachs disease. Sphingomyelin is stored together with cholesterol and a variable amount of gangliosides, within the neurons and within the numerous histiocytes in the demyelinated white matter. These substances are also found within histiocytes in the spleen, lymph nodes, thymus, lamina propria of the gut and within smooth muscle cells and endothelia.

The presence of gangliosides gives rise to red metachromasia with Feyrter's thionin and a positive PAS reaction. The Schultz and PAN methods for cholesterol are also positive. Sphingomyelin is stained blue with the ferric haematoxylin reaction, and resists alkaline hydrolysis. The colour of the acid haematein reaction may vary from site to site

and, in some organs, the reaction may be yellow-brown and this may reflect differing degrees of unsaturation of the fatty acid side chains. Sphingomyelin is stained weakly by Sudan black, which imparts a red birefringence in polarized light.

All the affected cells stain for acid phosphatase activity, the reaction product outlining the numerous storage vacuoles in each cell.

Niemann-Pick disease type C

Although the bone marrow cells in this condition have some similarities to those of Niemann-Pick disease types A and B, and to other disorders in which foamy cells occur, Niemann-Pick disease type C represents a separate entity. Part of the confused classification of this disorder, placing it with Niemann-Pick disease, arose because Crocker and Farber included some cases which are now clearly within this group in their monograph on Niemann-Pick disease (Crocker & Farber 1958).

Within the brain there is morphological evidence of neuronal storage which may be as marked as in classical Tay-Sachs, or may be minimal. Routine sections show no evidence of storage substance, which appears to be a water-soluble oligosaccharide with acidic residues. Formalin-fixed frozen sections have also lost most of the substance, which can best be demonstrated in protected cryostat sections. In celloidinised sections stained with the PAS reaction the stored material is seen within neurons and in axonal swellings. The cells display strong acid phosphatase activity and exhibit metachromasia with Feyrter's thionin. Luxol fast blue and Sudan black stains are negative and there is no autofluorescence except for the occasional lipofuscin wear-and-tear pigment granules. There is no evidence of demyelination. Axonal spheroids may be common in the brain stem nuclei. Neurons throughout the gastro-intestinal tract show storage and a diagnosis can be made on a suction rectal biopsy (Fig. 9.14). Serial sections are stained by several methods. The neurons show strong acid phosphatase activity, are positive with the protected PAS method (cell PAS) and show metachromasia with Feyrter's thionin method. No staining is observed with Sudan black, luxol fast blue or in sections fixed in formol-calcium prior to staining with PAS. Smooth muscle cells are not involved, and endothelial cells are only positive for acid phosphatase activity in the cases presenting with neurological symptoms in the infantile to late-infantile age group. Foamy histiocytes are present in the submucosa in all groups.

It is vital to establish neuronal involvement because most of the patients present with only visceral symptoms which precede the onset of neurological deterioration by several years. Conjunctival biopsies have been used in diagnosis by study of the ultrastructure of the various cell populations present, but it should be noted that in heterozygote carriers of the disease, not only are characteristic foamy storage cells present in bone marrow aspirates (see Ch. 20), but also that the ultrastructural changes of Niemann-Pick disease type C can be found in skin (Ceuterick et al 1986) and conjunctival biopsies. Thus skin, conjunctiva and bone marrow biopsies should be interpreted with caution and within the clinical setting, and should not be used as the sole diagnostic test in the investigation of younger clinically unaffected siblings. It is my firm belief that in this condition particularly, evidence of neuronal storage — or at least clear signs of neurological deterioration — is essential before the diagnosis can be established.

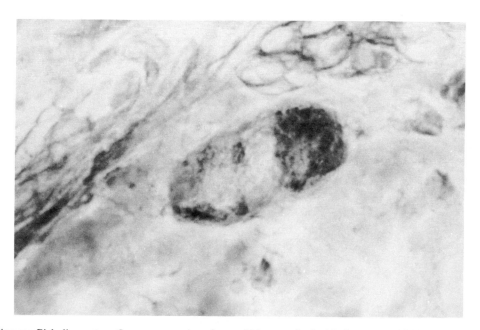

Fig. 9.14 Niemann-Pick disease type C; cryostat section of a rectal biopsy stained with the protected PAS method showing storage in submucosal neurons. This substance is extracted by aqueous media. ×1000

Gaucher's disease

There is no neuronal storage demonstrable by light microscopy in neurons of cerebral cortex or of the gastrointestinal tract. Neurons of basal ganglia show some PAS positivity indicative of storage of glucocerebroside only in the infantile form. Occasional perivascular Gaucher cells may be found, and these have the same staining characteristics as those found in the bone marrow, and are readily detected by an acid phosphatase reaction (Fig. 9.15).

Pompe's disease (GSD2)

The accumulation of intralysosomal glycogen, best shown in cryostat sections by the celloidin-PAS method, is not in evidence in the neurons of the cerebral cortex, but may be seen in the basal ganglia, brain stem and spinal cord (Figs. 9.16 and 9.17). Neurons of the gastrointestinal tract show storage throughout. In the cerebral cortex storage of glycogen can be demonstrated in astrocytes. Elsewhere there is gross accumulation of glycogen in the liver, monocyte-phagocyte series of cells of

the liver and spleen and in smooth muscle cells everywhere. Endothelial cells also exhibit glycogen storage. Skeletal muscle and heart muscle cells store glycogen in such vast quantities that it may be difficult to recognise the tissue. The glycogen deposition is associated with acid phosphatase activity and there may also be a substance which has the characteristics of an acid mucosubstance demonstrable with alcian blue or toluidine blue. This may be an unusual and so far unrecognized mucopolysaccharide or a glycogen-phosphate compound. The diagnosis is made from blood films in which glycogen is present in vacuolated lymphocytes (Ch. 20).

In adult GSD2 (van der Walt et al 1987), there is no evidence of storage in the brain, heart, spinal cord or liver in spite of very low levels of acid maltase activity. Skeletal muscles show variable glycogen storage, with the respiratory muscles severely affected, while muscles of the leg and arm are only minimally involved. Endothelial cells show no storage, but the smooth muscle of arterioles and bowel wall contain excess glycogen. The 'acid mucosubstance' present in the infantile form is not found in the adult type.

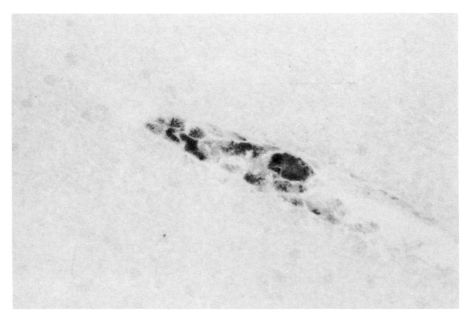

Fig. 9.15   Infantile Gaucher's disease; cryostat section of brain showing a small collection of perivascular Gaucher cells in the white matter. Stained for acid phosphatase activity (Gomori); counterstained with Carrazzi's haematoxylin. ×350

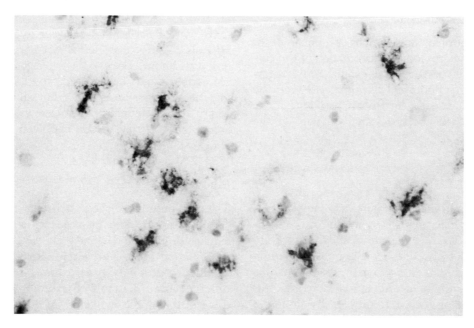

Fig. 9.16 Pompe's disease; cryostat section of brain showing glycogen accumulation in astrocytes in the cortex. No neuronal deposition is found in the cortex. Stained by the protected PAS method with Carrazzi's haematoxylin counterstain. ×450

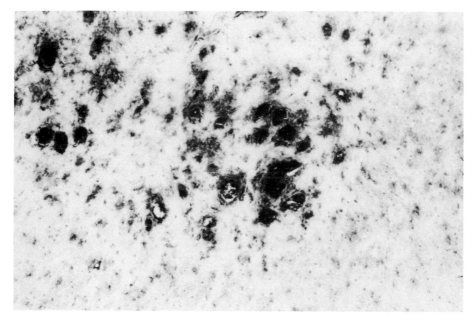

Fig. 9.17 Pompe's disease; cryostat section of spinal cord showing massive glycogen deposition in anterior horn cells. Stained by the protected PAS method. ×90

Aspartylglucosaminuria

Although neuronal storage may be apparent to routine light microscopical examination of the cortex and nuclei of the basal ganglia, only a few granules of a lipofuscin-like material can be demonstrated in the neuronal perikarya. The vacuolar contents do not stain with any of the usual methods.

Fucosidosis

Widespread neuronal ballooning with storage of a soluble oligosaccharide is evident. Fucose-containing compounds are widespread and deposits are found also in histiocytes in spleen, liver, lymph nodes and in the lung where granuloma-like nodules are prominent. All epithelia is affected. With adequate protection to prevent dissolution of the oligosaccharides the contents of the vacuoles should be PAS-positive. Since the condition is caused by the deficiency of a lysosomal enzyme, α-fucosidase, an acid phosphatase reaction should indicate those cells affected. The histochemical

method for the demonstration of α-fucosidase activity, with 1-naphthyl-α-fucoside as substrate and HPR as coupler, may be helpful but has not been tested in known cases.

Lectins have been used to detect storage material in fucosidosis and other storage diseases where other methods fail to show any accumulation. For a review of this approach, see Alroy et al (1988).

Mucolipidoses (ML)

The mucolipidoses show neuronal storage of a ceroid-lipofuscin-like material which can appear similar to that in Batten's disease. The clinical presentations with corneal clouding (ML4), Hurler-like features (ML2, ML3) are, however, quite different from Batten's disease. In ML2 (I-cell disease) not much evidence of storage is seen in brain or liver, but the heart shows deposition of a PAS-positive substance and strong acid phosphatase activity is present in the myofibres (Fig. 9.18). These changes in the heart are less florid than those of Pompe's disease, where the

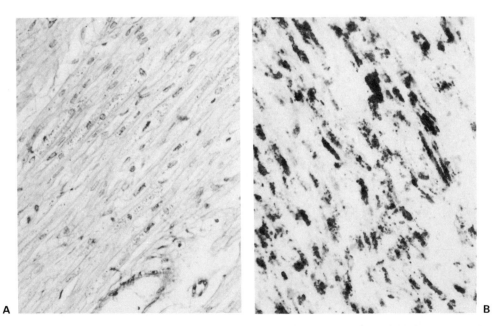

Fig. 9.18   I-cell disease (mucolipidosis 2); heart: **A** routine section stained with PAS, showing lipofuscin-like granules in vacuolated muscle cells; **B** cryostat section stained for acid phosphatase activity (Gomori) showing strong activity in muscle cells which would normally show very little activity. ×290

myofibres are markedly distended with glycogen deposition, appearing as though the whole heart were a rhabdomyoma.

In ML1 type I (sialidosis, cherry-red spot myoclonus syndrome) there is neuronal storage of a gangliosidic substance giving an appearance and staining reactions identical to those of Tay-Sachs disease. The gross neuronal storage may be found in young patients (by suction rectal biopsy) many years before the onset of dementia.

Adrenoleucodystrophy

This X-linked condition presents in the juvenile age range as a leucodystrophy with adrenal insufficiency (Powers & Schaumberg 1974, Schaumberg et al 1975). The cerebral cortex is normal, but the white matter shows changes of demyelination with sparing of the subcortical fibres. Foci of perivascular inflammatory cells are common. Numerous lipid-laden macrophages may be found throughout the demyelinated white matter. Adrenal atrophy is found in all cases and the cells in the zona reticularis and inner fasciculata are ballooned and contain cytoplasmic striations. The zona glomerulosa is not apparently involved. The striated inclusions, which are also present in the macrophages of the brain, peripheral nerve, testis and many other sites, are birefringent in frozen sections and represent very long chain fatty acid esters of cholesterol (Johnson et al 1976). These esters do not stain with oil red O and are not extracted with acetone, ethanol or methanol. They are soluble in n-hexane, xylene and chloroform. To demonstrate their presence three serial sections are required, stained and extracted

as shown in Table 9.3. A neonatal, autosomal recessive form has also been described in which a micronodular cirrhosis of the liver is found (Jaffe et al 1982). Both the X-linked and neonatal forms of adrenoleucodystrophy belong to the increasing list of disorders of the peroxisome (Schutgens et al 1986). Infantile Refsum's disease and the cerebrohepatorenal syndrome of Zellweger are also peroxisomal disorders and the histochemical aspects of their pathology are discussed in Chapter 16.

Mucopolysaccharidosis

In those types of mucopolysaccharidosis which have dementia, mental retardation and other signs of central nervous system involvement, the neurons of the CNS and PNS show marked storage resembling Tay-Sachs disease. The neuronal storage is of a mixture of gangliosides and no mucopolysaccharide can be detected. The neurons, in frozen or cryostat sections, stain positively with PAS, Sudan black and luxol fast blue. Deposition of acid mucopolysaccharide is found in perivascular regions, particularly in the white matter, and can only be demonstrated in cryostat sections of snap-frozen tissue. The method of choice is the toluidine blue method of Haust & Landing (1961). Macroscopic evidence may be found in brain slices where the perivascular accumulation of mucopolysaccharide has been extracted leaving holes or pits in the white matter (Lake 1984).

The lamina propria of the gut contains numerous macrophages filled with soluble mucopolysaccharide which is best detected with

Table 9.3 Demonstration of esters in adrenoleucodystrophy

Section no	Treatment	Result
1	Stain with oil red O	Neutral fat stained Long-chain esters birefringent
2	Extract with acetone Stain with oil red O	Neutral fat extracted Long-chain esters birefringent
3	Extract with acetone Extract with n-hexane Stain with oil red O	Neutral fat extracted Long-chain esters extracted No birefringence

the Haust & Landing method in cryostat sections. The neurons of the myenteric and submucosal plexuses contain a mixture of gangliosides and stain as the neurons in the brain.

Mannosidosis

Very few reports of the neuropathology of mannosidosis have appeared. Those that have, record the widespread neuronal ballooning, which in routine sections appears very similar to Tay-Sachs disease. In addition there is glial and endothelial storage. The mannosides deposited are very water-soluble and none remain in routine sections. Protected sections of snap-frozen tissue stained with PAS should show the stored material which will have no other staining properties. Lectins may also be used to detect mannose residues (Alroy et al 1988).

Neuroaxonal dystrophy

This disorder, delineated by Cowen and Olmstead in 1963, is readily diagnosed on postmortem histology when the axonal spheroids are prominent in the brain stem. In a biopsy of frontal cortex the spheroids (aggregations of microtubules and microfilaments) are not usually recognized by light microscopy and the diagnosis has to be made by electron microscopy. The axonal spheroids are present in peripheral nerves and it has been suggested that ultrastructural examination of the nerves in conjuctival biopsy will make the diagnosis. Although some cases have been diagnosed by this means the consistency of the procedure has not yet been proved (Raemakers et al 1987). El-leder and Jirasec (1983) have shown that the spheroids exhibit strong non-specific esterase activity in a brain biopsy and I have been able to confirm this in three further cases. Short incubation (10 minutes) of cryostat sections shows the spheroids present particularly in the molecular layer of the cortex and also scattered throughout the cortex. Neurons show much weaker activity. The typical clinical and neuropathological features

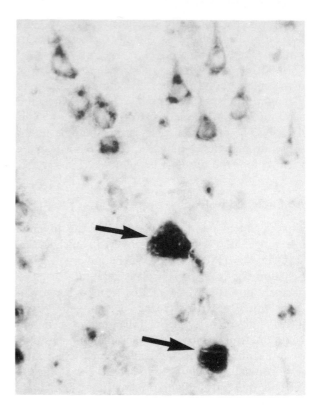

Fig. 9.19 Neuroaxonal dystrophy; cryostat section of a brain biopsy stained for non-specific esterase activity, showing strongly staining axonal spheroids (arrows) with their characteristic paracrystalline clefts. Neurons show much less activity. ×900

of neuroaxonal dystrophy were found in two brothers with a deficiency of α-N-acetylgalac-tosaminidase, a defect not found in eight other patients with neuroaxonal dystrophy (Schindler et al 1989). This enzyme activity should be detectable histochemically, although interference by lysosomal α-galactosidase might make the results uninterpretable.

Axonal spheroids are also found in Niemann-Pick disease type C, and it is becoming apparent that this is another microscopic appearance which does not justify a separate disease label. A similar situation occurs in Leigh's disease, where the same pathology is found in lactic acidosis resulting from different enzymatic deficiencies.

REFERENCES

Adachi M, Schneck L, Cara J, Volk B W 1973 Spongy degeneration of the central nervous system (van Bogaert and Bertrand type; Canavan's disease). A review. Human Pathology 4: 331–347

Adams C W M 1961 A perchloric acid naphthoquinone method for the histochemical localization of cholesterol. Nature (London) 192: 331

Alroy J, Ucci A A, Periera M E A 1988 Lectin histochemistry; an update. In: DeLellis R A (ed) Advances in immunohistochemistry. Raven Press, New York, p 93–131

Brett E M (ed) 1990 Paediatric neurology, 2nd edn. Churchill Livingstone, Edinburgh

Bodian M, Lake B D 1963 The rectal approach to neuropathology. British Journal of Surgery 50: 702–714

Ceuterick C, Martin J J, Fouland M 1986 Niemann-Pick disease type C. Skin biopsies in parents. Neuropediatrics 17: 111–112

Cowen D, Olmstead E V 1963 Infantile neuroaxonal dystrophy. Journal of Neuropathology and Experimental Neurology 22: 175–236

Crocker A C, Farber S 1958 Niemann-Pick disease; a review of 18 patients. Medicine 37: 1–95

Elleder M, Lojda Z 1973 Studies in lipid histochemistry. XI New rapid simple and selective method for the demonstration of phospholipids. Histochemie 36: 149–166

Elleder M, Jirasek A 1983 New enzymatic findings in infantile neuroaxonal dystrophy. Acta Neuropathologica 60: 153–155

Glew R H, Basu A, Prence E M, Remeley A T 1985 Lysosomal storage diseases. Laboratory Investigation 53: 250–269

Hall N A, Patrick A D 1988 Accumulation of dolichol-linked oligosaccharides in ceroid-lipofuscinosis (Batten disease). American Journal of Medical Genetics Supplement 5: 221–232

Jaffe R, Crumrine P, Hashida Y, Moser H 1982 Neonatal adrenoleukodystrophy. Clinical, pathologic and biochemical delineation of a syndrome affecting both males and females. American Journal of Pathology 108: 100–111

Lake B D 1984 Lysosomal enzyme deficiencies. In: Adams J H, Corsellis J A N, Duchen L W (eds) Greenfields neuropathology, 4th edn. Edward Arnold, London, ch 12

Lake B D 1989 Metabolic disorders. General considerations. In: Berry C L (ed) Paediatric pathology, 2nd edn. Springer, Berlin, ch. 14

Lake B D, Hall N A, Patrick A D 1988 Dolichyl pyrophosphate oligosaccharides are increased in Batten's disease. Proceedings of the Royal Microscopical Society 23: 45

Matalon R, Michals K, Sebesta D, Deanching M, Gashkoff P, Casanova J 1988 Aspartoacylase deficiency and N-acetylaspartic aciduria in patients with Canavan disease. American Journal of Medical Genetics 29: 463–471

Palmer D N, Martinus R D, Cooper S M, Midwinter G G, Reid J C, Jolly R D 1989 Ovine ceroid-lipofuscinosis. The major lipopigment protein is an N-terminal portion of the lipid binding subunit of mitochondrial ATP synthase. Journal of Biological Chemistry 264: 5736–5740

Ramaekers V Th, Lake B D, Harding B et al 1987 Diagnostic difficulties in infantile neuroaxonal dystrophy. A clinico-pathological study of eight cases. Neuropediatrics 18: 170–175

Schindler D, Bishop D F, Wolfe D E et al 1989 Neuroaxonal dystrophy due to lysosomal α-N-acetylgalactosaminidase deficiency. New England Journal of Medicine 320: 1735–1740

Schutgens R B H, Heymans H S A, Wanders R J A, Bosch H v D, Tager J M 1986 Peroxisomal disorders. A newly recognised group of genetic diseases. European Journal of Paediatrics 144: 430–440

Scriver C R, Beaudet A L, Sly W S, Valle D (eds) 1989 The metabolic basis of inherited disease, 6th edn. McGraw-Hill, New York

Vanier M T, Wenger D A, Comly M E, Rousson R, Brady R O, Pentchev P G 1988 Niemann-Pick disease group C; clinical variability and diagnosis based on defective cholesterol esterification. A collaborative study on 70 patients. Clinical Genetics 33: 331–348

van der Walt J D, Swash M, Leake J, Cox E L 1987 The pattern of involvement of adult-onset acid maltase deficiency at autopsy. Muscle and Nerve 10: 272–281

10. Skeletal muscle

M. A. Johnson

INTRODUCTION

Muscle pathology has become one of those areas of pathology where histochemical techniques have proved to be not only useful but absolutely indispensable in diagnostic practice. One of the main reasons for the particular value of a histochemical approach to the investigation of muscle disease lies in the nature of skeletal muscle itself. Muscle is not a uniform tissue; its constituent fibres are differentiated into fibre types whose contractile properties and other metabolic characteristics show striking functional differences. Many of these differences can be demonstrated histochemically and this ability to discriminate between the main muscle fibre types in tissue sections can be exploited in the investigation of muscle disorders.

Many neuromuscular diseases selectively affect one or other of the major fibre types and the detection of such selective processes is obviously dependent upon the use of the appropriate histochemical techniques. The normal spatial arrangement of the various fibre types in skeletal muscle is random, but in many neurogenic disorders this random distribution may be superseded by grouping of fibres of uniform type due to remodelling of motor units. Not infrequently this sort of change is the only indication of an underlying neurogenic abnormality and histochemical fibre typing procedures are clearly essential for its detection.

A substantial proportion of congenital myopathies, whose differentiating features are indistinguishable on routine histological examination, owe their initial recognition to the use of histochemical techniques and the number of specific metabolic disorders known to involve skeletal muscle is constantly increasing due to the identification of additional enzyme defects, many of which can be demonstrated histochemically.

The purpose of this chapter is to give a brief account of ways in which histochemical techniques can be used to increase the diagnostic potential of muscle biopsy in the investigation of a wide range of neuromuscular disorders. For detailed descriptions of the pathology of individual muscle diseases, the reader is directed to the various comprehensive monographs on the subject (Dubowitz 1985, Carpenter & Karpati 1984, Mastaglia & Walton 1982). Within the framework of the present chapter, only sufficient histopathological detail has been included as will allow the histochemical aspects of the disorders to be set in context. It should be emphasized that, particularly in muscle pathology, adequate clinical details and the results of investigations are necessary before the histopathological and histochemical features of a biopsy can properly be evaluated.

TISSUE PREPARATION

The preparative techniques used in processing muscle tissue for subsequent histochemical examination are a good deal less time-consuming than those needed for routine histology. Almost all the histochemical techniques used in diagnostic muscle pathology can be carried out on cryostat sections of fresh frozen tissue, though in some instances tissue sections will need to be postfixed subsequently.

Muscle biopsy techniques

Two types of biopsy procedure are in current use. The first is generally referred to as an 'open' biopsy since the muscle is removed through a comparatively large incision. In the 'needle' biopsy procedure only a very small incision is required and the muscle is sampled using a Bergstrom needle (or one of its modifications) or by conchotome (Dietrichson et al 1987). Correct orientation of the muscle prior to freezing is of paramount importance. If the biopsy has been obtained by means of an 'open' biopsy this generally poses no problem as the longitudinal direction of the muscle fibres should be clearly identifiable. If the specimen has been removed using sutures, any compressed portions from the ends of the biopsy should be removed. The use of muscle biopsy clamps, though invaluable for achieving good fixation in histological and electron microscopical preparations, is not necessary for tissue destined to be frozen and merely results in wasted tissue due to compression.

'Open' biopsies should be trimmed to give blocks not exceeding 8 mm^3 so that uniformly good freezing can be achieved, and the height of the blocks should not exceed their width otherwise the frozen specimens will tend to be unstable. Most diagnostic work entails assessment of fibre size and for this, good transversely oriented sections are obligatory; longitudinally orientated blocks of tissue are not often required and need not be provided routinely. The tissue obtained from 'needle' biopsies rarely exceeds 3 mm^3 and is frequently fragmented; correct orientation can best be achieved using a dissecting microscope. The fragments are not often large enough to make satisfactory blocks on their own, but a convenient way to overcome this problem is to sandwich the muscle, correctly orientated, between two blocks of gelatin before freezing.

Tissue freezing and sectioning

Skeletal muscle is not so susceptible to tissue autolysis as some other tissues, so immediate freezing is not strictly necessary. However, no more than 30 minutes should elapse between excision of the biopsy and freezing, and the tissue should be kept cool and not allowed to dehydrate.

Muscle blocks should be placed firmly on filter paper strips; composite blocks of gelatin plus sandwiched needle biopsy fragments should be treated in the same way. Freezing is optimally carried out using isopentane cooled to about −150°C in liquid nitrogen. Liquid nitrogen alone, though adequate for freezing most other tissues, is not recommended for freezing muscle which is particularly prone to ice-crystal artefact. Care should be taken to ensure that the temperature of the freezing bath is sufficiently low. This is most easily achieved if a small container of coolant is first completely frozen in liquid nitrogen and then used as it is thawing. Tissue blocks should be rapidly immersed in the coolant and agitated gently for about 15 seconds.

Sectioning of frozen muscle is best carried out at a temperature of −20 to −23°C. A section thickness of 10 μm is suitable for most of the commonly used diagnostic techniques.

HISTOCHEMICAL METHODS

The choice of methods to be used in diagnostic histochemical screening of muscle biopsy tissue is governed by the need to fulfil several objectives. One of the most important of these is to provide a basic 'fibre typing' system which will allow the main metabolic fibre types to be differentiated with reliability in both normal and pathological tissue.

Muscle fibre typing

The method which has proved to be of greatest value in this respect is the myofibrillar ATPase technique. Levels of ATPase activity reflect the basic differences in contractile properties between slow-twitch (type 1) and fast-twitch (type 2) fibres. The low ATPase activity in type 1 fibres is accompanied by high levels of mitochondrial oxidative enzyme activity, a feature which confers some degree of fatigue-resistance on this fibre type. Type 2 fibres, on the other hand, have comparatively low levels of mitochondrial oxidative enzymes, which renders them more prone to fatigue. The low mitochondrial oxidative enzyme content of type 2 fibres is offset by high myophosphorylase activity so that anaerobic glycolysis represents an important mechanism of energy production in these fibres (Dubowitz & Pearse

Table 10.1 Histochemical fibre types in human muscle

	Type 1	Type 2A	Type 2B	Type 2C★
ATPase pH 9.5				
after pH 10.2 preincubation	+	++	+++	+(+)
after pH 4.6 preincubation	+++	+	++	++(+)
after pH 4.3 preincubation	+++	−	−	+(+)
SDH/NADH-dehydrogenase	+++	++	+	++(+)
Myophosphorylase	+	++	+++	+(+)

★Normal adult muscle should contain <3% 2C fibres

1960). The enzyme profiles of type 1 and type 2 fibres are summarized in Table 10.1. From this table it can be seen that type 2 fibres can be classified into two main subtypes on the basis of their ATPase characteristics, glycolytic activity and mitochondrial oxidative enzyme content. Type 2A fibres have lower levels of myophosphorylase than type 2B, but higher mitochondrial oxidative enzyme activity. The distinction between types 2A and 2B can be made using the ATPase technique following preincubation at pH 10.2 or at pH 4.6 (see Appendix 5 for details of methods). The latter method is generally known as ATPase 'reversal' since after acid preincubation the ATPase activity of type 1 fibres is higher than that of type 2 fibres (Brooke & Kaiser 1970). The ATPase characteristics of normal muscle fibre types and their correlation with mitochondrial enzyme activity are shown in Figure 10.1.

Marker enzyme techniques

Another aim of a diagnostic protocol is to monitor the state of various muscle components and organelles through the use of 'marker enzyme' techniques. Thus succinate dehydrogenase (SDH) represents a suitable 'marker enzyme' for studying abnormalities in the distribution and activity of mitochondria (Pearse 1972). NADH-dehydrogenase is also commonly used as an indicator but it should be borne in mind that, since the sarcoplasmic reticulum also contains some NADH-dehydrogenase activity, the distribution of enzyme activity is not exclusively mitochondrial.

Myofibrillar ATPase, whose usefulness in muscle fibre typing has already been discussed, also serves as a marker enzyme for the detection of abnormalities in the myofibrils, e.g. areas of focal myofibril disorganization resulting in loss of ATPase activity.

The distribution of muscle lysosomes can be monitored using techniques for the demonstration of acid phosphatase activity. Simultaneous coupling azo dye methods (Burstone 1958) are to be preferred since it has been demonstrated that non-specific binding of lead salts to myofibrils and other muscle components may occur. In normal human muscle lysosomes are comparatively rare, especially in children's muscle. In young adults a few acid phosphatase-containing granules can be found adjacent to muscle nuclei or immediately under the plasma membrane and the deposition of lipofuscin-like pigments inside lysosomes in later life is accompanied by a gradual increase in the number and size of acid phosphatase-positive granules. The intensity of phagocytic cell reactions in diseased muscle can also be gauged effectively using lysosomal marker techniques. As an alternative to the acid phosphatase technique, phagocytic activity may be demonstrated using methods for the localization of esterase activity. Moreover, if α-naphthyl acetate is used as substrate it is possible to visualize not only histiocytes and phagocytic activity, but also the sites of endplate cholinesterase.

Detection of metabolic disorders

An important objective in diagnostic muscle pathology is to be able to detect evidence of metabolic disorders, either through the identification of specific enzyme defects or by detection of abnormal storage of tissue components. Classic examples of the latter are the use of the PAS reaction to screen biopsies for abnormal amounts of stored glycogen in cases of glycogenosis and Sudan black screening to detect evidence of lipid storage. An increasing number of metabolic defects may be precisely diagnosed because there is a suitable cytochemical method for demonstration of the catalytic activity of the affected enzyme. This list

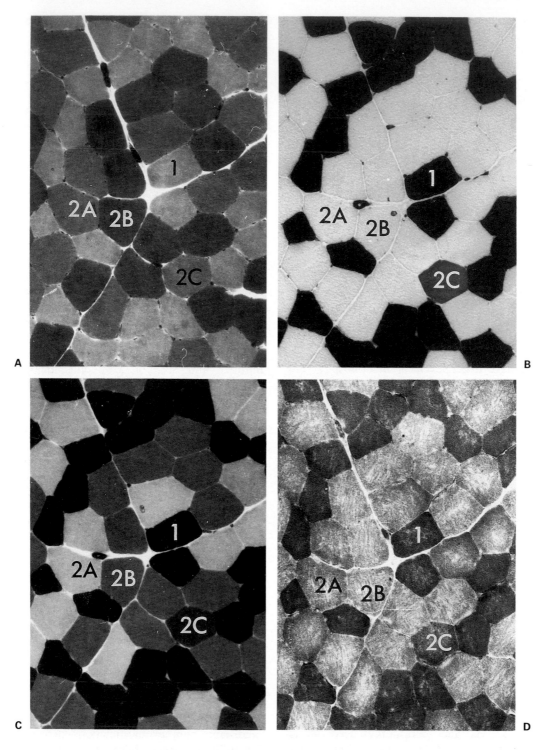

Fig. 10.1 Normal muscle fibre types showing correlation of ATPase characteristics and oxidative enzyme activity. **A** ATPase after alkaline (pH 10.2) preincubation, **B** ATPase after pH 4.3 preincubation, **C** ATPase after pH 4.6 preincubation, D NADH-dehydrogenase. All × 150

includes defects of myophosphorylase, phosphofructokinase, cytochrome oxidase and myoadenylate deaminase.

Diagnostic protocol for muscle biopsies

A routine 'battery' of histochemical techniques should include most, if not all, of those listed below:

1. ATPase — including acid preincubation to allow detailed fibre typing.
2. SDH or NADH-dehydrogenase — for detection of abnormalities of mitochondrial distribution and/or activity.
3. PAS — to screen for abnormal amounts of stored glycogen or other polysaccharides.
4. Myophosphorylase — to indicate glycolytic activity.
5. Sudan black — to screen for excess accumulation of neutral lipid and/or phospholipid.
6. Acid phosphatase and/or esterase — for assessment of lysosomal changes in muscle and extent of phagocytic cell infiltration.

These techniques should preferably be carried out on serial sections since it is not uncommon to need to refer from one histochemical preparation to another in order to assess the significance of observed changes, e.g. to determine whether mitochondrial abnormalities (SDH preparations) are confined to type 1 fibres (ATPase preparations). However, all the above techniques can be completed by one person within 3 hours so the demands on laboratory time are not excessive. Sections of the frozen tissue should also be stained using haematoxylin and eosin and/or haematoxylin-Van Gieson and the modified Gomori trichrome method to allow the direct correlation of histochemical and histological features.

Immunocytochemical techniques

The status of immunocytochemical techniques in neuromuscular pathology is at present one of importance in research rather than in routine diagnosis, though there are notable exceptions which will be discussed later. Much current immunocytochemical work centres round the discrimination of isoforms of the major contractile proteins, particularly myosin (Fig. 10.2), actin and troponin.

Correlation of immunocytochemical reactivity with conventional fibre typing reactions is important in its own right, but the potential uses of immunolabelling extend far beyond the ability to provide alternative fibre typing methods. Myopathies involving the absence of specific proteins, or alternatively the production and storage of such proteins to excess, are the particular domain of immunocytochemical investigation and are often inaccessible by other means.

Morphometric techniques

In diagnostic muscle pathology, it is necessary to be able to define objectively the limits of normality of fibre diameter, numerical fibre type proportions and spatial fibre type distribution. It is all too easy, on subjective assessment, to mistake, for example, a population of normal diameter and hypertrophied fibres for one of atrophied and normal diameter fibres.

The method of fibre diameter measurement most commonly used is that of Song et al (1963) who described the technique of determination of 'orthogonal diameters'. This gives a reliable estimate of fibre size, which is independent of any obliqueness in the plane of sectioning. In general terms the fibres of normal adult male subjects are approximately 40–80 μm and those of normal adult females are within the range 30–70 μm. The mean diameter of the muscle fibres of infants at 1 year is approximately 16 μm and increases by about 2 μm per year for the first 5 years and thereafter by 3–4 μm per year until puberty, when adult fibre diameters are attained (Brooke & Engel 1969a). The normal ranges are generally defined as mean ± 2 SD and it is useful to remember that a roughly two-fold variation is characteristic of normal adult muscle fibre populations while for infants and children fibre sizes are relatively uniform. Calculation of atrophy factors (AF) and hypertrophy factors (HF) as described by Brooke and Engel (1969b) represents a convenient method of assessment of abnormalities of fibre size distribution.

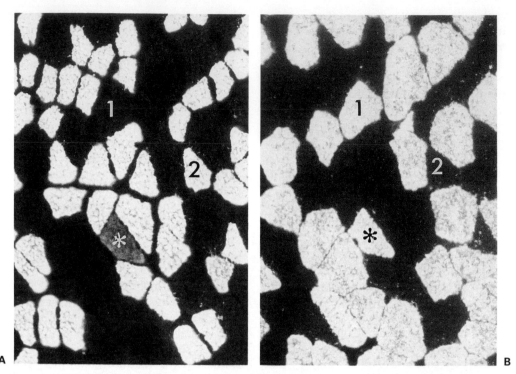

Fig. 10.2 Normal muscle immunolabelled for slow and fast myosins. **A** Antihuman fast myosin (mouse monoclonal antibody) visualized by rhodamine conjugated rabbit antimouse immunoglobulins; type 2 fibres are labelled. **B** Antihuman slow myosin (mouse monoclonal antibody) visualized as in **A**; type 1 fibres are labelled. Occasional fibres may show positive labelling with both antisera (asterisk). Fluorescence microscopy with epiillumination. ×150 (Courtesy of Dr Marion Ecob-Prince and Dr Mark Hill, Muscular Dystrophy Laboratories, Newcastle-upon-Tyne)

The numerical proportions of the main histochemical fibre types vary from muscle to muscle and it is important to recognize this fact when examining biopsy material from different sites. Normal ranges of percentages of type 1 fibres for the most commonly biopsied muscles are as follows: quadriceps 28–48%, deltoid 43–63%, biceps 34–51%, gastrocnemius 37–51% and tibialis anterior 62–84% (Johnson et al 1973). It will be noticed that the proportion of type 1 fibres is normally high in predominantly postural muscles such as tibialis anterior and low in more dynamic muscles such as biceps and quadriceps.

The spatial distribution of histochemical fibre types is random in normal muscle, but departures from this are common in neurogenic disorders (see below) and objective methods of assessing this abnormality should be applied. Commonly used methods are 1) the 'enclosed fibre' technique (Johnson et al 1973) and 2) the codispersion index

(Lester et al 1983), both of which provide an estimate of the non-randomness of fibre type distributions.

DENERVATING DISORDERS

Disorders of skeletal muscle in which denervation is the predominant pathological feature have numerous factors in common even though the site of the lesion may be the anterior horn cells (spinal muscular atrophies) or the peripheral nerve axons or Schwann cells (peripheral neuropathies). It is therefore worth considering some of these common features before examining individual disorders in more detail.

General histochemical features

The main fibre types in normal human muscle (types 1, 2A and 2B) are randomly arranged, reflecting the random distribution of fibres in normal motor units. The fibres in each motor unit are histochemically uniform, but because motor unit territories overlap extensively, the result is a mosaic of the various histochemical fibre types. If only a few motor units are affected by denervation, the distribution of the atrophied fibres will be scattered. There is generally no selective effect of denervating processes on either type 1 or type 2 motor units, so the atrophied fibres will consist of both fibre types. If a large proportion of motor units is affected by denervation, groups of atrophied fibres will be encountered; again these will consist of both type 1 and type 2 fibres. If, however, these denervated fibres become reinnervated by branches of a single axon the result will be a group of fibres of uniform histochemical fibre type. This phenomenon of 'uniform fibre type grouping' is an invaluable indication of reinnervation in denervating disorders (Brooke & Engel 1966). Not infrequently muscle biopsies in well-compensated (i.e. reinnervated) denervating disorders may show no signs of atrophied fibres

and diagnosis rests on the histochemical detection of fibre type grouping (Fig. 10.3). If subsequent cycles of denervation occur affecting reinnervated fibres, these fibres will be 'grouped' in the physical sense since the territory of a reinnervated motor unit is compact.

However, they will also show histochemical fibre type grouping, a feature which serves to distinguish them from groups of atrophied fibres which have resulted from a single cycle of denervation only.

During the process of reinnervation fibres will change from one fibre type to another if the reinnervating axons are of a different motor type from the original innervating axons. Fibres in the process of conversion from type 1 into type 2 or vice versa obviously must pass through a transitional stage. This stage is easy to detect in ATPase preparations after preincubation at pH 4.3 (see Table 10.1) and such fibres are generally referred to as '2C' fibres. Whereas the ATPase of normal type 2A or 2B fibres is totally inhibited after pH 4.3 preincubation, '2C' fibres show ac-

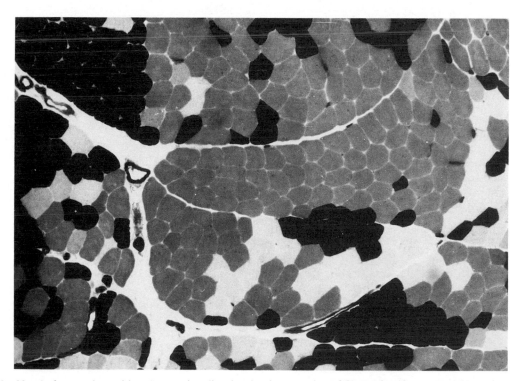

Fig. 10.3 Muscle from patient with a denervating disorder showing grouping of fibres of uniform type due to reinnervation. ATPase after pH 4.6 preincubation. ×150

tivity intermediate between that of type 1 and type 2 fibres. A low percentage (< 3%) of '2C' fibres may be found in normal human muscle, but in the context of denervating disorders, an increased number of '2C' fibres is generally indicative of ongoing fibre type conversion as a result of reinnervation. The '2C' ATPase characteristics are also found in various other classes of muscle fibre, e.g. in regenerating fibres, since they are indicative of incomplete or transitional states of histochemical fibre type differentiation.

Mitochondrial oxidative enzyme preparations frequently reveal another distinctive abnormality in muscle from many denervating disorders. Affected fibres do not show the usual homogeneous distribution of mitochondrial enzyme activity. Instead, three concentric zones of low activity (inner zone), high activity (intermediate zone) and normal activity (outer zone) are found — these fibres are known as 'target' fibres (Engel 1961). This type of abnormality is almost exclusively confined to denervating disorders (Fig. 10.4) and may be associated with reorganization of muscle organelles

as a consequence of changes in fibre type. Two-zoned 'targetoid' fibres, in which the intermediate high activity zone is absent, are found in a wider range of pathological situations.

Spinal muscular atrophies

The most severe form of spinal muscular atrophy (SMA) has its onset in early infancy (Werdnig-Hoffman disease). The juvenile form, with onset in later childhood, has a much more benign clinical course (Kugelberg-Welander SMA). There is also an intermediate form of SMA which shares the early onset of Werdnig-Hoffmann disease but is characterized by more gradual progression. In all forms an autosomal recessive mode of inheritance is usual.

Werdnig-Hoffman disease

This form of SMA involves widespread severe muscle fibre atrophy with no evidence of reinnervation. Surviving motor units show hypertrophy of

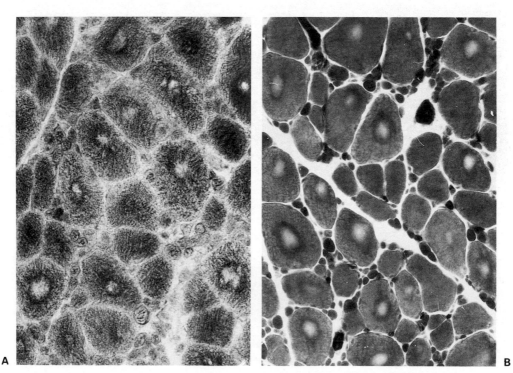

Fig. 10.4 Muscle from patient with the 'intermediate' form of spinal muscular atrophy showing 'target' fibres: **A** NADH dehydrogenase, **B** ATPase pH 9.5. ×385

fibres which are almost invariably type 1 (Fig. 10.5). Since there is no evidence of selective sparing of type 1 motor units, the most likely explanation for this is that surviving units, irrespective of their original fibre type, perform a basically tonic function in the severely weakened muscles and their histochemical profile is adapted accordingly.

Intermediate SMA

Whereas the Werdnig-Hoffmann form of SMA generally proves fatal before the age of 2 years, the more benign 'intermediate' form is associated with survival sometimes into the second decade. The more gradual progression of muscle weakness in the disease is associated with ability of some surviving motor units to enlarge their territories by reinnervating previously denervated fibres. Histochemical evidence of this is provided by the frequent occurrence of uniform fibre type grouping, generally involving type 1 fibres, though less frequently type 2 reinnervated motor units may be

found. 'Target' fibres are not uncommon whereas in Werdnig-Hoffmann disease they are never seen.

Kugelberg-Welander SMA

This form of SMA may have a very gradual clinical course indeed, extending over several decades. Especially in the early stages of the disorder, reinnervation may effectively compensate for the loss of motor units. The more effective the process of reinnervation, the less likely are atrophied fibres to be found in the muscle biopsy sample. Under these circumstances the application of histochemical techniques is essential in order to be able to detect the presence of fibre type grouping in samples of muscle which may appear histologically normal.

Peripheral neuropathies

Although denervating changes can be produced in skeletal muscle as a result of peripheral neuropathies of both axonal and

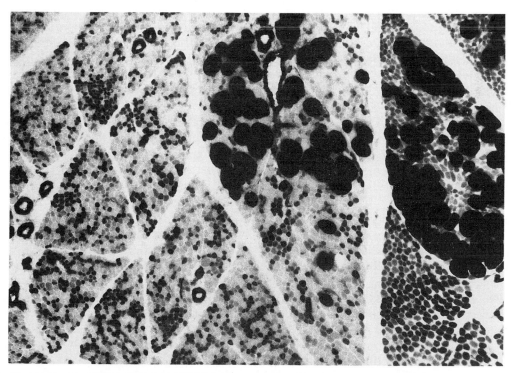

Fig. 10.5 Muscle from patient with Werdnig-Hoffman disease showing widespread atrophy affecting all fibre types; fibres of surviving motor units are hypertrophied and are all type 1. ATPase after pH 4.6 preincubation. ×150

demyelinating types, the extent of the denervation produced is generally more restricted in distribution than that associated with central denervating disorders. In denervation of central origin it is not uncommon to see numerous contiguous muscle fascicles consisting totally of atrophied fibres; this finding would not be encountered in any of the peripheral neuropathies. Nevertheless the extent of fibre type grouping may be considerable since multiple cycles of denervation and reinnervation may result from periodic remissions and exacerbations of the disease process. Angulated atrophied fibres frequently show high levels of NADH-dehydrogenase irrespective of fibre type due to the selective loss of myofibrillar protein and the relative preservation of mitochondrial contents.

MUSCULAR DYSTROPHIES

The muscular dystrophies pose many diagnostic problems for the pathologist, not least because they are diseases of unknown aetiology and the present classification undoubtedly includes disorders within this group which differ widely as regards pathogenesis. Recent progress in defining the molecular defect in the X-linked dystrophies enables previous diagnostic practice based on 'pattern recognition' to be replaced by the identification of an absent specific protein, using immunolabelling techniques. This illustrates the fulfilment of an important aim in diagnostic pathology: to replace, wherever possible, non-specific or empirical techniques with methods based on the precise identification of underlying defects.

Although the dystrophies have numerous histological features in common, histochemical examination tends rather to reveal the disparate nature of the individual disorders. All the dystrophies are progressive and eventually result in severe loss of muscle bulk. As in chronic denervating disorders, this is commonly associated with increased amounts of interstitial fibrous connective tissue and adipose tissue. However, evidence of frank denervation is lacking; fibre type grouping is not seen and 'target' fibres are consistently absent. Individual forms of muscular dystrophy show

different patterns of fibre type involvement which are useful in differentiation and diagnosis.

Duchenne muscular dystrophy (severe X-linked MD)

Duchenne muscular dystrophy (DMD) is a severe degenerative myopathy which presents in early childhood and generally proves fatal before the end of the second decade of life. The genetic abnormality is found on the short arm of the X chromosome (Xp2l locus); up to one-third of cases of DMD are due to new mutations, but in the remainder the 'carrier' state in females can be detected.

The random variation in fibre size seen in Duchenne MD affects both fibre types but in many biopsies the proportion of type 1 fibres is much greater than normal. Biopsies from preclinical cases tend to show more normal fibre type proportions so this feature appears to be progressive and is suggestive either of selective loss of type 2 fibres or alternatively of functional conversion of fibre types. There is also a progressive loss of the usual clear distinction between type 1 and type 2 fibres both in ATPase preparations and in mitochondrial oxidative enzyme preparations. This is partly due to the presence of a high proportion of '2C' fibres which have histochemical characteristics intermediate between those of type 1 and 2 (Fig. 10.6). Some of the '2C' fibres are clearly identifiable as regenerating fibres whose high cytoplasmic RNA content can be demonstrated using methyl green-pyronin or similar RNA methods. Other '2C' fibres are of more normal diameter and may represent fibres in the process of conversion from type 2 into type 1. There is also a severe deficit in numbers of type 2B fibres in particular; instead of roughly equal proportions of types 2A and 2B, the type 2B fibres may comprise only about 10% of the total number of type 2 fibres.

Multiple foci of active necrosis and phagocytosis are features of the early clinical stages of the disorder and are strikingly demonstrated using marker techniques for lysosomal activity such as acid phosphatase or esterase.

Numerous rounded intensely-staining fibres can be identified in histological preparations of

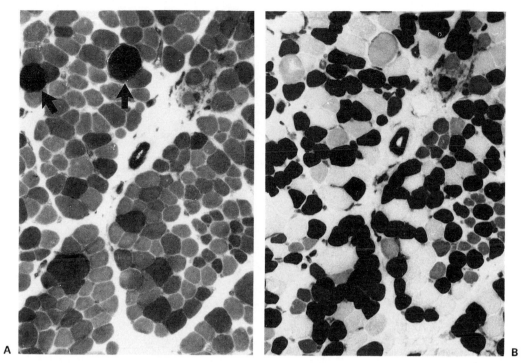

Fig. 10.6 Muscle from patient with Duchenne muscular dystrophy showing rounded, overcontracted fibres (arrows) and numerous '2C' fibres. **A** ATPase pH 9.5, **B** ATPase after pH 4.3 preincubation. ×120

Duchenne muscle. These fibres give an abnormally high ATPase reaction (Fig. 10.6) and this appears to be due to overcontraction of segments of the affected fibres. A likely cause of this is the influx of calcium due to localized membrane damage; foci of calcium accumulation can sometimes be detected in these fibres by means of the von Kossa, alizarin or fluorescent Morin methods.

Recent work in molecular genetics has resulted in the identification of the normal gene product, a 400 kDa protein referred to as dystrophin, the absence of which gives rise to DMD (Hoffman et al 1987). Dystrophin can be localized by conventional immunolabelling techniques (indirect immunofluorescence, fluorochrome- or peroxidase-linked avidin-biotin methods) and appears to be associated with the cell surface in normal individuals (Zubrzycka-Gaarn et al 1988). Primary antisera currently employed have been raised either to the products of cloned fragments of the gene or to proteins synthesized in accordance with known genetic sequences. In DMD total absence of specific immunolabelling is seen, affecting all muscle fibres (Fig. 10.7). Using dystrophin antisera, Arahata et al (1989) showed a mosaic of positive and negatively stained muscle fibres in biopsies from symptomatic female carriers of DMD. This is another example of the morphological demonstration of Lyonization and may be helpful in genetic counselling. The function of dystrophin in the maintenance of muscle homeostasis is not known but it is possible that its role is particularly crucial in type 2B fibres which tend to become selectively depleted at an early stage in the disorder.

Becker muscular dystrophy (benign X-linked MD)

Becker muscular dystrophy (BMD) is allelic with DMD but presents later in childhood and is associated with longer survival. One feature associated with the more protracted time-course of the disorder is the greater degree of compensatory hypertrophy which is observed, affecting mainly

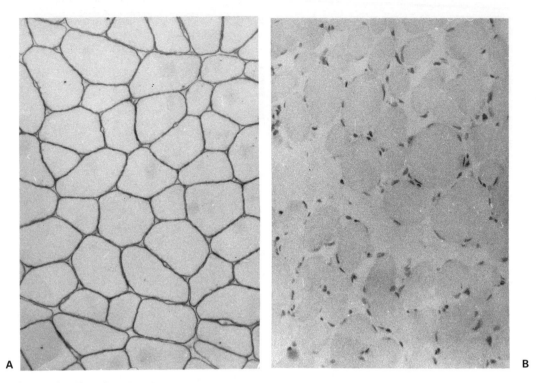

Fig. 10.7 Immunolabelling of the protein dystrophin in normal and dystrophic muscle. **A** Normal human muscle labelled with mouse monoclonal anti-dystrophin followed by peroxidase-labelled rabbit anti-mouse immunoglobulins. **B** Duchenne muscular dystrophy: note absence of immunolabelling. A nuclear counterstain has been used so that muscle fibre outlines are detectable. ×385 (Anti-dystrophin sera were made available by courtesy of Dr Louise Nicholson, Muscular Dystrophy Laboratories, Newcastle-upon-Tyne).

type 2 fibres. Although a type 1 fibre predominance is seen as in Duchenne MD, this is not accompanied by a type 2B fibre deficit. Moreover, the distinction between the histochemical characteristics of the various fibre types is well maintained.

Overcontracted fibres giving abnormally high ATPase reactions are not common and are certainly not present to the same extent as in Duchenne dystrophy. However, two other varieties of abnormal fibre are prevalent. The first is detectable in mitochondrial enzyme reactions (SDH and NADH-dehydrogenase) and consists of marked irregularities in mitochondrial distribution. This type of abnormality is usually referred to by the whimsical but descriptive term, 'motheaten fibres' (Brooke & Engel 1966).

The other type of abnormal fibre is visible on examination of ATPase preparations, but is even more striking in methods which demonstrate constituents of aqueous sarcoplasm, e.g. PAS. Displacement of peripheral myofibrils results in the formation of striated rings in affected fibres, hence the term 'ring-fibres', (see Fig. 10.8A). It is emphasized that 'ring-fibres' and 'moth-eaten fibres' are not confined to Becker dystrophy but are common in many chronic dystrophic processes.

Investigations of the protein dystrophin in the muscle of BMD patients, using SDS-polyacrylamide gel electrophoresis and immunoblotting, have shown that its abundance is generally decreased to 40–80% of normal and that the protein may be abnormal in size. Immunolabelling of tissue sections often shows an incomplete reaction round the fibre periphery, compatible with a partial loss of the protein (Bonilla et al 1988). Anti-dystrophin labelling of 'disease controls' including cases of myotonic dystrophy and facio-

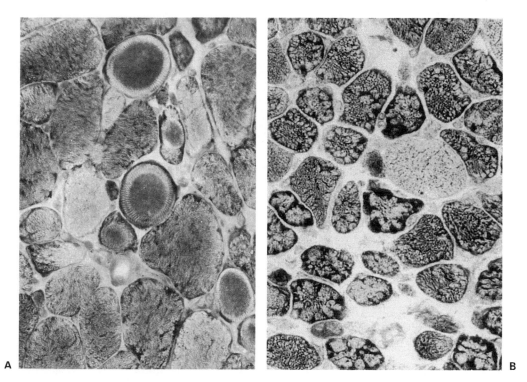

Fig. 10.8 A Muscle from patient with Becker muscular dystrophy showing 'ring fibres' containing striated annulets. Periodic acid-Schiff. **B** Muscle from patient with facioscapulohumeral syndrome showing abnormal 'lobulated' fibres. NADH-dehydrogenase. ×385

scapulohumeral muscular dystrophy has so far shown totally normal immunolabelling results.

Late onset muscular dystrophies

Many patients suffering from late-onset muscular dystrophies present with limb-girdle or facioscapulohumeral patterns of weakness. The clinical and histopathological diagnosis of these patients is complex, since within both limb girdle and facioscapulohumeral syndromes are found instances of atypical spinal muscular atrophies as well as adult-onset dystrophies. Any significant neurogenic component should be detectable using techniques for the analysis of the spatial distribution of fibre types, as outlined above. In most patients, however, the predominant histopathological findings are myopathic and are particularly characteristic of chronically damaged muscle reacting under stress. 'Ring fibres' are common and consist of the displacement of peripheral myofibrils, which coil corkscrew-fashion round the rest of the muscle fibre (see Fig. 10.8A). A lobulated or floccular distribution of mitochondria is often found in late-onset dystrophies, especially the facioscapulohumeral (FSH) and scapuloperoneal varieties (Fig. 10.8B). Discrete foci of inflammatory cells including lymphocytes are rare in disorders other than true inflammatory myopathies, but are sometimes encountered in FSH dystrophy. Their significance is unclear, but it has been speculated that they may be involved in secondary sensitization to muscle proteins liberated due to the primary myopathic lesion.

Limb-girdle syndrome

Weakness of pelvic and femoral musculature is the main feature of a syndrome which has frequently been labelled 'limb-girdle muscular dystrophy'. More recent opinion however, based on assessment of pathological as well as clinical data, tends to regard most cases as variants of either Becker dystrophy or juvenile SMA. It is not surprising therefore to find that the histological features in a proportion of cases are indistinguishable from Becker MD. Similarly, in other cases a bimodal size distribution typical of a neurogenic atrophy is discernible though 'myopathic' features including necrosis and phagocytosis may complicate interpretation of biopsy findings.

MYOTONIC DISORDERS

The main disorders in which myotonia is the predominant clinical feature are myotonic dystrophy (Steinert's disease) and myotonia congenita. These conditions are very dissimilar both in their clinical features as well as histologically and histochemically so that their differentiation presents no real problems (Engel & Brooke 1966).

Myotonic dystrophy

The considerable disparity in fibre size found in this disorder is due to atrophy of type 1 fibres accompanied by type 2 fibre hypertrophy affecting both 2A and 2B subtypes. Selective atrophy of type 1 fibres is much less common in neuromuscular pathology than type 2 atrophy, which is encountered whenever phasic activity is restricted, e.g. in simple disuse. Other characteristic features in myotonic distrophy include the presence of chains of internal muscle nuclei and the presence of numerous ring fibres. These are frequently associated with 'sarcoplasmic masses', which are zones of afibrillar sarcoplasm outside the rings of misorientated myofibrils. Because these areas contain abnormally dense concentrations of mitochondria and aqueous sarcoplasm they are particularly conspicuous in SDH and PAS preparations; since they contain no myofibrils they appear negative in the ATPase technique. Some authors report on acid phosphatase-positive deposits within muscle fibres, particularly in the infantile form.

Myotonia congenita

In contrast to the findings in myotonic dystrophy, muscle biopsies from patients with myotonia congenita are often histologically normal apart from showing generalized muscle hypertrophy. On histochemical examination, however, a total absence of type 2B fibres is a frequent finding, although the total percentage of type 2 fibres remains within normal limits. These findings are common to both autosomal dominant and recessive forms of myotonia congenita and also occur in several other situations including congenital myotonic dystrophy.

CONGENITAL MYOPATHIES

The group of disorders included under this heading are relatively non-progressive diseases as compared with the muscular dystrophies or central denervating disorders, but are present from birth. The various forms are often difficult or impossible to distinguish clinically, both from each other and from juvenile SMA. Moreover, routine histological examination is generally insufficient to enable their differentiating features to be recognized. Histochemical techniques have proved indispensable for this purpose and have been responsible for the initial recognition of more than one variety of congenital myopathy.

Central core disease

In many cases of this disorder, the muscle biopsy could readily be dismissed as showing no abnormalities in routine histological preparations and the range of fibre size is frequently within normal limits. However, in SHD or NADH-dehydrogenase preparations, numerous corelike areas devoid of enzyme activity are seen to be present. Sometimes there is only one such area per fibre, or alternatively there may by two, three or even more per fibre (see Fig. 10.9). Cores are usually present in a high proportion of fibres ($> 30\%$) and extend along the fibres for distances exceeding 1 mm. Ultrastructural examination shows these areas contain virtually no mitochondria.

It is common for all the fibres in cases of central core disease to be uniformly type 1, but less frequently cases are found in which there is merely a type 1 fibre predominance, or even where the fibre type proportions are within normal limits. Sometimes older members of a family show only

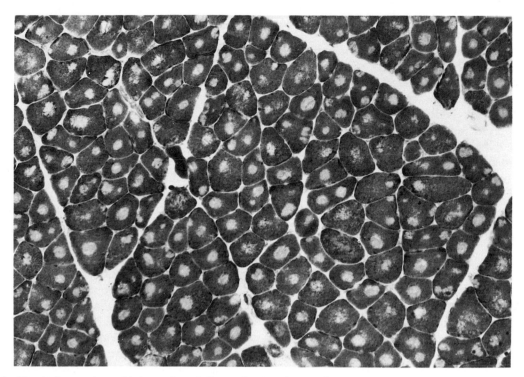

Fig. 10.9 Muscle from patient with 'central core' disease showing absence of enzyme activity from the cores. NADH-dehydrogenase. ×150

type 1 fibres on muscle biopsy, whereas younger relatives retain some type 2 fibres. This finding and the fact that raised numbers of '2C' fibres are seen in the younger patients is suggestive of fibre type conversion during the course of the disease. Cores are normally confined to type 1 fibres or to '2C' fibres. Frequently the cores are not detectable in ATPase preparations. This type of core is described as 'structured' because the myofibrils in it are either normal or only minimally disrupted. In 'unstructured' cores, the myofibrillar architecture is severely abnormal, resulting in low ATPase activity in the area of the cores (Neville & Brooke 1971).

Multicore disease (Minicore disease)

Multiple small core-like structures are found in another variety of congenital myopathy which has been duly named 'multicore' or 'minicore' disease. Its clinical features differ from those of central core disease in that it tends to be somewhat more severe with a certain degree of muscle wasting.

Multicores resemble small unstructured cores extending for less than 100 μm and are thus visible as areas of low activity in ATPase preparations. The disruption of myofibrils is sometimes so severe as to be visible in H & E stained sections. The margins of the cores are normally much less distinct in mitochondrial enzyme preparations than they are in central core disease. This correlates with ultrastructural findings that mitochondria, though reduced in number, are not totally absent from the cores. A type 1 fibre predominance is an almost invariable finding in this disorder, but multicores are found in both fibre types.

Centronuclear myopathies

This clinically heterogeneous group of disorders is characterized by the presence of numerous central nuclei in a high proportion of muscle fibres (Bethlem et al 1970). Initially, the name 'myotubular myopathy' was given to one variant of this disorder in which the nuclei were situated in a central zone devoid of myofibrils. These fibres have undoubted morphological similarities with myotubes, but, whereas true myotubes are invariably '2C' in histochemical profile since they are incompletely differentiated, the vast majority of these fibres have the histochemical characteristics of mature type 1 fibres. The term 'centronuclear myopathy' is, therefore, preferable. The central afibrillar zones are prominent in SDH, NADH-dehydrogenase, myophosphorylase and PAS preparations since they contain high concentrations of mitochondria and aqueous sarcoplasm. These zones are devoid of ATPase activity.

Another similar disorder is typified by the presence of very small type 1 fibres (see Fig. 10.10) in addition to the incidence of central nucleation. It is difficult to know whether these fibres represent the effects of retarded growth (hypotrophy) or whether some loss in diameter has taken place (atrophy), because few cases have been investigated by means of serial biopsy. This type of centronuclear myopathy appears to be more rapidly progressive than the variant without fibre size disparity, but continued careful documentation of these disorders is necessary so that the variants can be more surely identified and prognosis more confidently given.

Nemaline myopathy

This type of congenital myopathy takes its name from the rod-like protein-containing structures which are found in the disorder often in association with abnormal Z-line material. Although the rods can be seen in trichrome-stained sections, they are not usually apparent in H & E preparations. Severely affected fibres are frequently atrophied as compared with fibres which contain few rods. In the majority of cases type 1 fibres are selectively affected, but this is not an invariable finding and cases have been reported in which both fibre types contain nemaline rods or even where there was selective involvement of type 2 fibres. Aggregates of nemaline rods show up as negative areas in ATPase sections.

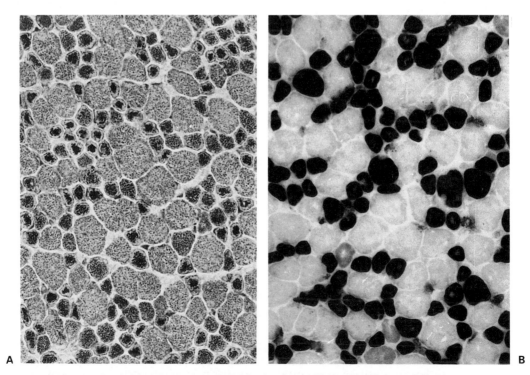

Fig. 10.10 Muscle from patient with centronuclear myopathy showing small myotube-like type I fibres. **A** NADH-dehydrogenase, **B** ATPase after pH 4.3 preincubation. ×308

The precise nature of the nemaline rods is still a matter of conjecture, though the fact that methods for tryptophan give negative results and those for arginine and tyrosine are positive has been interpreted as evidence that they resembled Z-line material in composition. Immunocytochemical studies have suggested that a major component may be α-actinin (Sugita et al 1973). Although similar structures may occasionally be found in other neuromuscular disorders, there is little doubt that nemaline myopathy represents a distinct clinicopathological entity. Its mode of inheritance appears to be autosomal recessive in the severe infantile form and autosomal dominant in the milder form (Martinez & Lake 1987). In families showing the latter type of genetic transmission the age of onset may vary widely.

Congenital fibre type disproportion

This class of muscle disorder is characterized by disparity in fibre *size* of the main histochemical muscle fibres. Unlike the other categories of congenital myopathy included in this section, CFTD is, however, associated with no specific muscle fibre abnormality other than this disproportion in fibre size (Brooke 1973). Moreover, it is not certain whether the condition represents the effects of aberrant maturation of the motor unit as a whole, rather than a purely myopathic defect. All reported cases share a common clinical picture of congenital hypotonia which is static, slowly ameliorating or in some cases fatal. The first recognized cases were typified by the relative smallness of type 1 fibres, with type 2B fibres having the largest diameter. Subsequently, however, other variants with identical clinical histories have been found, in which type 1 or type 2A fibres are significantly larger than the other fibre types. A 'significant' degree of difference, in the context of CFTD, is defined as at least 12% between the largest mean diameters of largest and smallest fibre types.

Recent evidence suggests that CFTD may be a histological phenomenon and may not necessarily be related to a specific clinical entity. However, it is particularly important that this type of disorder should effectively be distinguished from atypical cases of infantile SMA. Whereas SMA does not show the selectivity for individual fibre types which is characteristic of CFTD, biopsies from infants less than 3 months old with SMA may nevertheless fail to show the pattern of severe atrophy affecting all fibre types which is diagnostic of the Werdnig-Hoffmann form of SMA. The differential histological diagnosis should, however, be straightforward in biopsies taken after the age of 3 months.

METABOLIC MYOPATHIES

Skeletal muscle is involved in a wide range of metabolic disturbances; some of these affect muscle exclusively, while in others additional tissues are involved. Specific enzyme defects in carbohydrate or lipid metabolism and mitochondrial function have been identified; in many cases histochemical methods exist for the demonstration of the enzymes involved and hence the defects can be precisely detected. In many other instances the results of histochemical screening can at least indicate the general area in which further biochemical investigation should be concentrated. An important advantage of histochemical screening is that it is remarkably economical as regards use of tissue, since multiple metabolic processes can be investigated on very small tissue samples.

Glycogenoses

Skeletal muscle is affected in a majority of the established forms of glycogenosis. Some of these are characterized by the late onset of neuromuscular symptoms and are compatible with a normal lifespan, though the capacity for strenuous exercise may be severely curtailed. In others, however, the onset is in the immediate postnatal period and the prognosis much less favourable (Rowland et al 1971). The main histopathological feature of the glycogenoses is vacuolation of muscle fibres, the severity of which is roughly proportional to the extent of glycogen storage. If individual vacuoles are very large, the stored glycogen cannot be retained in cryostat sections except by protecting them by means of prior celloidinization. Muscle is not primarily affected in glycogenoses types 1 and 4.

Type 2 glycogenosis — acid maltase (α-1,4-glucosidase) deficiency

This deficiency exists in a severe infantile form (Pompe's disease) in which cardiac muscle is also affected, and more benign late-onset and adult forms affecting skeletal muscle only. The infantile form is characterized by severe neonatal hypotonia and, on histological examination, vacuolation of the muscle is frequently so severe that only a small proportion of the sarcoplasm of each fibre remains, giving the tissue a lace-like appearance. Celloidinized PAS-stained sections show that the vacuolation is due to massive glycogen storage. Both fibre types seem equally affected. There is also some minor storage of a substance with the characteristics of an acid mucosubstance.

In the adult form (Fig. 10.11), the extent of glycogen storage and consequent vacuolation is much less severe. In each form storage is within distended lysosome-like organelles which have high acid phosphatase activity associated with their limiting membranes. In the adult form es-pecially, lipofuscin and other dense phospholipid-containing debris may be found in the vacuoles, giving a positive PAS reaction after diastase diges-tion. Glycogen storage tends to be more prevalent in type 1 fibres; a possible explanation for this may be that glycogen levels in type 2 fibres are decreased to a greater extent by the action of myophosphorylase which is unaffected in this dis-order. Glycogen deposition in the adult form may be minimal (Trend et al 1985). Biochemical con-firmation of the suspected enzyme defect is necessary, since there is no histochemical method for its demonstration. Glycogen deposition in lym-phocytes (see Ch. 20) is a constant feature of all forms and is a useful confirmatory test for acid maltase deficiency.

Type 3 glycogenosis — debranching enzyme (amylo-1, 6-glucosidase) deficiency

This disorder presents in infancy with hepato-megaly and hypotonia. The former may subside

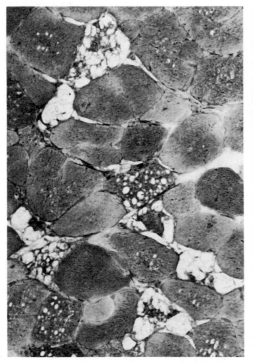 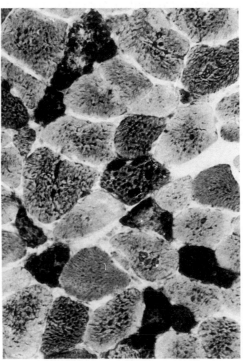

Fig. 10.11 Muscle from patient with adult form of type 2 glycogenosis (acid maltase deficiency); vacuolation of fibres is due to glycogen storage. **A** PAS without celloidinization. **B** PAS after celloidinization; note retention of glycogen in vacuoles. ×200

but residual muscle flaccidity frequently persists. The stored glycogen is highly abnormal in that it contains short external chains and an increased number of branch points (limit dextrin) and is more soluble than normal glycogen. Isolated cases may show type 1 fibre predominance, and multi-core-like structures have also been reported in this disorder (Pellissier et al 1979). Again, the specific enzyme defect is not demonstrable histochemically and biochemical confirmation of the suspected diagnosis is necessary.

Type 5 glycogenosis — myophosphorylase deficiency

This type of glycogenosis (McArdle's disease) generally becomes apparent in the teens or twenties because of weakness and pain on exertion. The enzyme deficiency is confined to skeletal muscle phosphorylase and on histochemical examination its activity is completely absent. However, it can be seen that smooth muscle phosphorylase activity in arterioles is unaffected (Fig. 10.12). If

regenerating myofibres are present, these are found to contain apparently normal phosphorylase activity. The explanation for this is that a 'fetal' isoenzyme is present in myotubes prior to their maturation and that this isoenzyme is unaffected. Because of the occurrence of necrosis and regeneration, '2C' fibres may be common in McArdle's disease. Vacuolation is frequently inapparent unless carefully searched for, and is most usually subsarcolemmal.

Type 7 glycogenosis — phosphofructokinase deficiency

The clinical symptoms of this disorder are similar to those of McArdle's disease, with cramps and exercise tolerance being associated with the inability to utilize muscle glycogen fully as an energy source. The glycogen content is variable and may appear to be within normal limits. An amylopectin-like glycogen (insoluble and indigestible) may be present in some fibres. The enzyme defect can

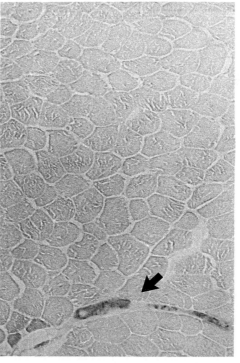

A   **B**

Fig. 10.12   Demonstration of the activity of myophosphorylase in: **A** Normal muscle, **B** muscle from patient with type 5 glycogenosis (McArdle's disease); note positive reaction only in smooth muscle of arteriole (arrow). ×150

be demonstrated histochemically (Bonilla & Schotland 1970), since PFK activity can be visualized using a multistep tetrazolium reduction method (see Appendix 5). The enzyme steps involved are PFK itself, aldolase, NAD-linked phosphoglyceraldehyde dehydrogenase and finally NADH-dehydrogenase. It is obviously necessary to demonstrate that the enzyme pathway from aldolase onwards is functioning normally before concluding that a negative result is due to PFK deficiency. Thus histochemical assays of PFK and aldolase activity should be carried out in tandem, a negative result in the former and positive in the latter indicating a PFK deficiency.

Lipid storage disorders

Although there are numerous conditions in which accumulation of intracellular lipid in skeletal muscle is a predominant feature, the underlying enzyme abnormalities have been identified in comparatively few instances. Utilization of substrates derived from fatty acids in oxidative phosphorylation is more prevalent in type 1 fibres than in type 2. Consequently, it is not surprising that most forms of lipid storage myopathy preferentially affect type 1 fibres, though this is not invariably the case. Screening of muscle biopsies for the presence of excess lipid is most conveniently carried out using Sudan black staining with propylene glycol as solvent since this procedure will allow the demonstration of both neutral and phospholipids. Differentiation of these classes of lipid can then be carried out using oil red O (Fig. 10.13) and acid haematein methods (Bourgeois & Hubbard 1965), either separately or in conjunction in the same tissue sections. On purely histological examination of cryostat sections, neutral lipid droplets are often visible as tiny round vacuoles evenly dispersed throughout the affected fibres.

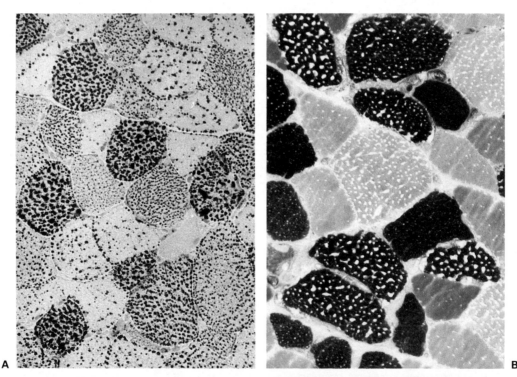

Fig. 10.13 Lipid storage myopathy due to short chain acyl-CoA dehydrogenase deficiency. **A** Excess lipid deposition shown using the Oil Red O method. **B** Lipid storage is particularly severe in some but not all type I fibres. ATPase after pH 4.6 preincubation. ×320

Carnitine deficiencies

Low levels of muscle carnitine (γ-trimethylamino-β-hydroxybutyrate) are commonly encountered in lipid storage myopathies (Engel & Angelini 1973), but it is likely that this deficiency is secondary to defects of mitochondrial β-oxidation. Specific defects of both short-chain and medium-chain acyl-CoA dehydrogenases have been identified (Turnbull et al 1984) in disorders previously labelled as 'carnitine deficiencies' presenting with a wide range of clinical features. There are currently no appropriate enzyme histochemical methods for the demonstration of these β-oxidation defects, but the availability of specific antisera to some acyl-CoA dehydrogenases enables immunocytochemical screening procedures to be undertaken. However, immunolabelling methods can never totally compensate for the lack of a cytochemical assay of catalytic activity. An enzyme deficiency *may* be due to the absence of the enzyme protein but, alternatively, loss of catalytic activity may be associated with very minor alterations of protein configuration which may not be amenable to investigation by immunolabelling.

Other lipid storage myopathies have been shown to be associated with glutaric aciduria and deficiencies of the electron transport flavoprotein (ETF) responsible for the subsequent passage of electrons from various acyl-CoA dehydrogenases (Turnbull et al 1988). In some instances treatment of affected patients with high doses of riboflavin has proved beneficial and has resulted in significant decrease in muscle lipid storage.

Carnitine palmityltransferase (CPT) deficiencies

Two components of CPT have been identified and together participate in the transport of fatty acids across mitochondrial membranes prior to oxidation. CPT I is situated on one side of the inner mitochondrial membrane and catalyses the conversion of acyl CoA into acylcarnitine; CPT II is situated on the other side of the membrane and catalyses the reverse reaction. Deficiencies in skeletal muscle may involve either or both enzymes (Scholte et al 1979) and fibroblasts and leucocytes are frequently affected. Although the deficiency in muscle causes severe cramps associated with exercise, lipid storage is minimal and may be inapparent except during attacks. The morphology of the muscle fibres is otherwise normal.

Batten's disease (ceroid lipofuscinosis)

There are several forms of Batten's disease (see Ch. 9), in which deposition of ceroid- or lipofuscin-like pigments takes place not only in neurons, but also in skeletal muscle and other tissues (Carpenter et al 1972). Although deterioration in motor ability is purely secondary to severe diffuse brain damage, muscle biopsy is sometimes undertaken as a diagnostic procedure in preference to biopsy of rectum or appendix. Vast quantities of abnormal pigment-containing material is stored within lysosomes in muscle and is detectable by its autofluorescence, the PAS technique and the Schmörl reaction. Whereas very few lysosome-like organelles are present in the muscle of normal children and the acid phosphatase activity is low, in these disorders the grossly enlarged lysosomes contain high levels of acid phosphatase. Vitamin E deficiency also leads to increased lipofuscin, but this deposition can be differentiated from that in Batten's disease by its orange-yellow autofluorescence (in contrast to the yellow autofluorescence of Batten's disease) and by the clinical features in which a peripheral neuropathy is often present.

Refsum's syndrome

Moderate storage of neutral lipid may be seen in muscle biopsies from cases of this disorder, which is due to a peroxisomal defect in the metabolism of phytanic acid (tetramethylhexadecanoic acid), a common constituent of green vegetables and dairy products. Lipid storage in muscle is limited to type 1 fibres and there is frequently evidence of uniform fibre type grouping resulting from the accompanying peripheral neuropathy.

Respiratory chain abnormalities

Defects of the enzymes of the respiratory chain are responsible for many of the disorders commonly labelled 'mitochondrial myopathies'. However, since it is now apparent that many lipid storage myopathies are due to defects of mitochondrial β-oxidation, the term 'mitochondrial myopathy' has become somewhat ambiguous. Nevertheless, from the histopathological standpoint, respiratory chain disorders are much more likely to give rise to gross morphological mitochondrial abnormalities. It was on this basis that a general 'diagnosis' of 'mitochondrial myopathy' was given for at least a decade before the underlying abnormalities began to be defined.

A common morphological finding is the 'ragged-red' fibre, so-called because of the characteristic appearance of peripheral mitochondrial clustering in the modified Gomori trichrome staining method. Such abnormalities are confirmed by demonstration of enzyme 'markers' for mitochondria, such as SDH. Respiratory chain defects also frequently give rise to bizarre ultrastructural abnormalities of mitochondria, including paracrystalline inclusions, disruption of cristae and gross distortions of normal size and shape. Un-fortunately, none of these features appear to correlate with specific enzyme defects and therefore cannot be exploited as diagnostic aids. It is also possible to find quite severe impairment of respiratory chain activity in the absence of any morphological mitochondrial abnormalities, so the lack of 'ragged-red' fibres, etc. on routine screening cannot be taken as a guarantee of mitochondrial integrity.

Cytochrome oxidase deficiencies

Defects of cytochrome oxidase (complex IV) activity are relatively straightforward to detect cytochemically, since there is a reliable method (Seligman et al 1963) for the demonstration of the enzyme. Total absence of activity is found in the fatal infantile form of cytochrome oxidase deficiency, which is characterized by severe hypotonia and lactic acidosis developing in the first few weeks of life (Van Biervliet et al 1977). Muscle biopsy shows minimal histological changes though scattered fibres show dense accumulation of reac-

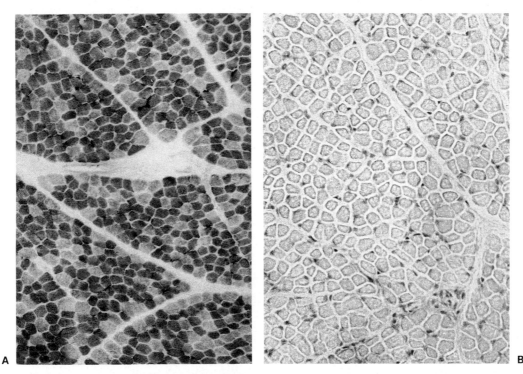

Fig. 10.14 Mitochondrial myopathy due to total cytochrome oxidase deficiency. **A** Control muscle showing normal cytochrome oxidase activity. **B** Complete absence of activity in muscle of affected infant. ×385

tion product in SDH preparations. Cytochrome oxidase activity is uniformly absent from all fibres (see Fig. 10.14).

The benign infantile cytochrome oxidase deficiency (Di Mauro et al 1983) is also characterized by total absence of the enzyme in skeletal muscle from birth, but a gradual increase in activity occurs linked to clinical improvement and resolution of hypotonia and lactic acidaemia. Differentiation from the fatal infantile form can be accomplished by serial needle biopsies and cytochemical enzyme assays which provide the most appropriate method of monitoring the progress of this disorder in affected infants.

A significant proportion of children with Leigh's disease (subacute necrotizing encephalomyelopathy) show decreased or absent cytochrome oxidase activity in skeletal muscle; biochemical assays on mitochondrial fractions and microphotometric cytochemical assays both show activity is frequently < 25% of that seen in normal muscle.

Partial deficiencies of cytochrome oxidase are also encountered in adult patients with oculocraniosomatic syndromes, such as the Kearns-Sayre syndrome and chronic progressive external ophthalmoplegia (CPEO). However, muscle fibres vary greatly in their expression of the defect (Johnson et al 1983). Some fibres contain no detectable enzyme activity, whereas in others near-normal activity is recorded using microphotometric assay techniques. Many enzyme-deficient fibres show peripheral aggregations of abnormal mitochondria (Fig. 10.15), but patients with up to 30% cytochrome oxidase-negative fibres have been found to display no overt morphological abnormalities.

Demonstration of cytochrome oxidase activity in skeletal muscle should be carried out using the Seligman method (see Appendix 5). Use of the Burstone technique or its variants is not recommended since these techniques cannot be made totally independent of endogenous cytochrome c. Since the amount of endogenous cytochrome c

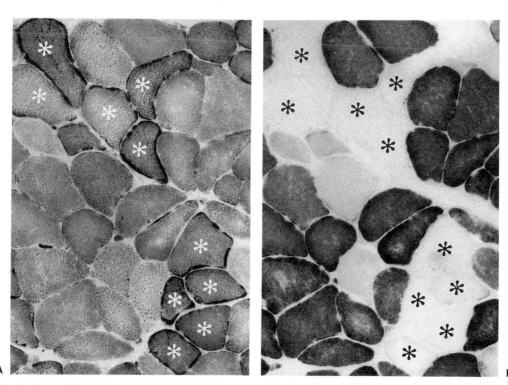

Fig. 10.15 Partial cytochrome oxidase deficiency in patient with chronic progressive external ophthalmoplegia. **A** Some fibres (asterisks) show peripheral aggregations of mitochondria, SDH. **B** Serial section showing absence of cytochrome oxidase activity in these and other fibres. ×150

varies, not only in pathological muscle but also in normal muscle, estimation of cytochrome oxidase activity cannot be accurate under these conditions.

It has been shown that defects at more than one site in the respiratory chain may exist in the same patients. An association of complex I (NADH-ubiquinone-reductase) and complex IV (cytochrome oxidase) defects is common, and abnormalities of complex V (mitochondrial ATPase) have been shown to coexist with complex IV deficiency. The common factor here may be that some of the protein components of all these individual respiratory complexes are coded by the mitochondrial genome. Deletions of a significant proportion of this genome have recently been demonstrated by Holt et al (1988) in the muscle mitochondria of some patients with proven complex I deficiencies. Cytochemical identification of complex I deficiency (Morgan-Hughes et al 1979) is at present hampered by the lack of a specific assay for NADH-ubiquinone reductase activity, but immunolabelling using anti-complex I sera is capable of demonstrating a severe decrease in immunoreactive protein in the muscle of affected patients.

Defects of purine metabolism

Myoadenylate deaminase is important in muscle metabolism as a key enzyme in the regulation of the balance of ATP and ADP in muscle fibres and is also important since muscle has no other anaplerotic reaction. Whereas moderate reductions in the level of this enzyme may be seen in various neuromuscular diseases, including muscular dystrophies and denervating disorders, some patients appear to have a total myoadenylate deaminase deficiency (Fishbein et al 1978). This condition may give rise to cramping pains and weakness on exertion, but muscle biopsy reveals no histological abnormalities. The enzyme defect may be demonstrated histochemically using a method which involves reduction of a tetrazolium salt by ammonia in the presence of dithiothreitol (see Appendix 5).

Muscle cramps and pain on exercise are also encountered in another defect of purine metabolism, namely xanthine oxidase deficiency. However, levels of xanthine oxidase are very low even in normal muscle and the defect is more reliably demonstrated in biopsies of jejunum. Occasionally crystalline inclusions, which probably contain xanthine and hypoxanthine, can be found in the intermyofibrillar spaces in muscle fibres.

ELECTROLYTE ABNORMALITIES

Studies of serum potassium levels in cases of periodic paralysis have shown that there are hypokalaemic, hyperkalaemic and normokalaemic forms of this disorder. Hyperkalaemic periodic paralysis is closely allied, though not precisely identical, to paramyotonia congenita. The typical pathological findings in periodic paralysis consist of vacuolation of muscle fibres, which may be widespread during attacks but at other times muscle biopsy may be diagnostically non-contributory. The vacuoles contain mainly water and electrolytes; glycogen is present only in small amounts due to leakage from the surrounding sarcoplasm. The vacuoles are therefore only weakly PAS-positive and are not likely to be confused with vacuolation due to any of the glycogenoses (see p. 145). The hyperkalaemic and normokalaemic forms of periodic paralysis are more likely to develop a permanent myopathy typified by gross fibre atrophy with some compensatory hypertrophy, type I fibre predominance and a high incidence of internal nucleation.

Both hypo- and hyperkalaemic forms are prone to striking cellular abnormalities referred to as 'tubular aggregates' (Engel et al 1970). These are most conspicuous in NADH-dehydrogenase preparations, where they appear as areas of extremely dense reaction product (Fig. 10.16). They are also PAS-positive and may be subsarcolemmal or central in position and may occupy up to one-third of the crossectional area of the fibre. The absence of SDH activity from the aggregates allows them to be distinguished with certainty from large clusters of mitochondria. Ultrastructural examination shows them to be composed of double-walled tubules about 52 nm in diameter packed in hexagonal lattice array. Their high phospholipid content is demonstrable using Sudan black or Baker's acid haematein methods. Their origin from the sarcoplasmic reticulum has been shown by electron microscopy. The aggregates are confined almost exclusively to type 2B fibres.

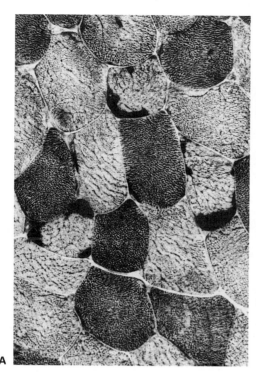
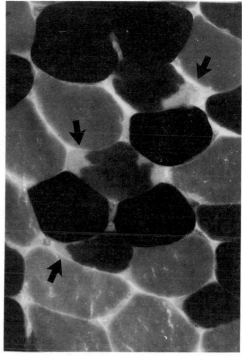

A

B

Fig. 10.16 Muscle from patient with hyperkalaemic periodic paralysis showing 'tubular aggregates' in type 2B fibres (arrows). **A** NADH-dehydrogenase. **B** ATPase after pH 4.6 peincubation. ×385

ENDOCRINE MYOPATHIES

This group of neuromuscular disorders is the source of numerous diagnostic problems because the effects of hormonal abnormalities on muscle may be exerted directly, via the central or peripheral nervous system or via the muscle vasculature. On occasions the effects of a combination of more than one of these mechanisms may be encountered in the same patient. Histochemical techniques have proved invaluable in this context, in differentiating between selective fibre type atrophies and those due to neurogenic mechanisms, and in detecting hormonal effects on particular enzymes.

Steroid myopathies

Muscle weakness and wasting are well-documented side-effects of steroid therapy and are also frequently associated with the endogenous glucocorticoid excess found in Cushing's syndrome. Selective type 2 fibre atrophy is the most common manifestation of both conditions, though in severe Cushing's syndrome atrophy of type 1 fibres may also occur (Pleasure et al 1970). The atrophied type 2 fibres generally show reduced levels of myophosphorylase, a feature which helps to distinguish the type 2 atrophy in this situation from that due to simple disuse. In both Cushing's syndrome and iatrogenic steroid myopathies, moderate neutral lipid accumulation may also occur, affecting type I fibres almost exclusively.

Neuromuscular abnormalities in diabetes

The direct effects of insulin deficiency on muscle tend, not surprisingly, to parallel those of glucocorticoid excess. Again, type 2 atrophy is the predominant pathological feature and the myophosphorylase activity of these fibres may be severely reduced. The distribution mitochondria is often irregular and some degree of lipid storage is also a common feature. These 'direct' effects of diabetes on the muscle fibres may, however, be complicated by the presence of a neuropathy, the underlying cause of which is damage to the vasculature of the peripheral nerves. Under these circumstances small group atrophy involving both type 1 and type 2 fibres is seen, but uniform fibre type grouping or large group atrophy are not normally encountered.

Dysthyroid myopathies

Abnormal thyroid hormone levels are responsible for a variety of pathological changes in skeletal muscle. Neuropathies of the compression type may be encountered in severe hypothyroidism (myxoedema) and mixed motor and sensory neuropathies have also been documented in hyperthyroidism. In these cases the pathological findings are indistinguishable from other peripheral neuropathies. However, in the absence of any electrophysiological evidence of neuropathy, it is often found that fibre type proportions are abnor-

mal in dysthyroid patients. The number of type 1 fibres is generally raised in hypothyroid subjects, whereas hyperthyroid patients often show increased numbers of type 2 fibres. The mechanism of fibre type conversation appears to be a neurally mediated effect of iodothyronine levels on ATPase activity in muscle.

INFLAMMATORY MYOPATHIES

It might be thought that the inflammatory myopathies would represent one area of muscle pathology in which histological techniques alone should suffice for the purposes of diagnosis. However, the pathogenesis of the various forms of myositis varies greatly and histochemical techniques are particularly useful in analysing the patterns of muscle damage and inflammatory cell infiltration (Fig. 10.17) seen in each. The dermatomyositis of childhood and adolescence is a fairly well-defined clinicopathological entity, in

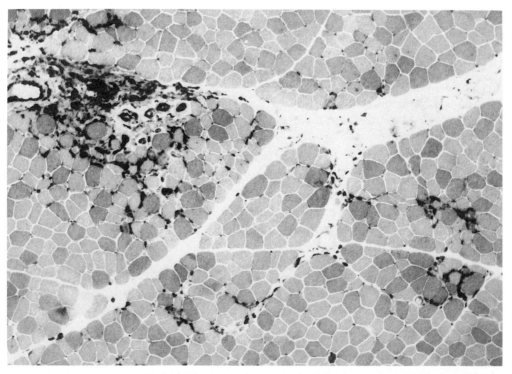

Fig. 10.17 Muscle from patient with inflammatory myopathy showing distribution of infiltrate in perivascular and endomysial sites. Non-specific esterase method. ×150

the pathogenesis of which vascular damage plays an important role (Carpenter et al 1976). In adult dermatomyositis a more variable pattern of muscle destruction is apparent with only some cases showing evidence of vascular damage. This group has a high incidence of associated malignant disease and considerable overlap with the clinical and immunological features of connective tissue disease. In adult polymyositis, uncomplicated by skin changes or other evidence of multisystem disorders, the muscle pathology does not seem to be associated with vascular lesions but the underlying mechanism is at present not known.

Juvenile dermatomyositis

This disorder is characterized by recurrent capillary damage affecting skin, skeletal muscle and the alimentary tract. If muscle biopsies are taken during an early phase of capillary necrosis the vascular impairment can be seen to cause loss of ATPase and mitochondrial oxidative activity from the centre of affected fibres. The distribution of these fibres is often perifascicular and the detection of this type of change may enable the diagnosis to be made even in the absence of inflammatory cell infiltrates which may prove elusive. A subsequent stage of muscle damage often involves necrosis and regeneration of these fibres. The regenerating fibres are readily distinguishable from normal fibres on account of their small diameter and high sarcoplasmic RNA content. This feature is clearly demonstrated in methyl green-pyronin preparations; the prominent nucleoli of regenerating fibre nuclei also show a very high RNA content. During maturation these fibres appear as '2C' fibres in ATPase preparations; differentiation into mature type 1 or type 2 fibres occurs after reinnervation, but their central nucleation and small diameter tend to persist. If restitution of capillary damage is sufficiently rapid it is probable that total muscle fibre necrosis does not occur, but such fibres may show moderate to severe atrophy. Perifascicular atrophy in a quiescent stage of the disorder may be the only indication of previous muscle damage. The precise mechanism of vascular damage in dermatomyositis is not known, but a high incidence of positive immunostaining for IgG, IgM and C_3

in the vessels of the juvenile form and some adult cases is suggestive of complex-mediated cell damage (Whitaker & Engel 1972).

Adult dermatomyositis

In dermatomyositis with onset in adult life, the incidence of capillary necrosis varies greatly from case to case. In certain circumstances it is widespread enough to cause muscle infarction affecting whole fascicles. Loss of myophosphorylase and glycogen are the earliest indicators of irreversible muscle necrosis; subsequently ATPase and mitochondrial enzyme activity is lost and the fibres degenerate totally. The postacute stage of this type of change is marked by the prevalence of regenerating myogenic elements, demonstrating high RNA content. Regenerating capillaries are conspicuous because of their high ATPase activity.

Large foci of inflammatory cells may be found in dermatomyositis complicating connective tissue disorders such as rheumatoid arthritis or systemic lupus erythematosus. Almost invariably there are signs of muscle damage adjacent to these foci which, if not detectable histologically, are seen clearly enough in histochemical preparations because of the enzyme abnormalities involved. This is an important factor in the differentiation of myositis occurring in association with connective tissue disorders from polymyopathies where inflammatory foci, e.g. rheumatoid nodules, are not associated with any muscle destruction.

Polymyositis

Recent studies of inflammatory cell populations in polymyositis and dermatomyositis (Arahata & Engel 1984) have provided fresh insight into pathogenetic mechanisms in these disorders. Using serial sections and a panel of monoclonal immunocytochemical markers for B and T lymphocyte populations, these investigations analyzed the predominant features of the cellular infiltration. It was found that B lymphocytes were most abundant in perivascular sites in polymyositis patients and least abundant in endomysial sites, whereas the converse applied to the population of T lymphocytes. It has been suggested that sensitisation of T^{8+} (cytotoxic) cells to a muscle-associated antigen was a likely major cause of muscle damage, but that perivascular B cells and T^{4+} (helper) cells may participate in ancillary humoral mechanisms of muscle fibre necrosis. The findings in polymyositis were substantially different from those in dermatomyositis where the observed high percentage of B cells and high T^{4+}/T^{8+} cell ratio was in favour of predominating humoral mechanisms.

Because all inflammatory muscle disorders are non-uniform in distribution the choice of muscle biopsy procedure is particularly important. If needle biopsy is chosen in preference to an open biopsy it must be recognized that the chances of making a positive diagnosis, especially in an early stage of the disease, are correspondingly decreased. Multiple needle biopsies have sometimes been used to monitor the response of the muscle to steroid therapy or other forms of treatment. Obviously, multiple open biopsies are not a practical proposition, but the results of 'follow-up' needle biopsies must be interpreted with caution: an abnormal histopathological picture is sig-

nificant, but a 'normal' biopsy may be due to sampling error. It is important to make maximal use of other diagnostic procedures in this situation. For example, a normal serum creatine kinase level in conjunction with a 'normal' follow-up biopsy gives some reassurance that the latter is not due to unrepresentative sampling.

Virological studies are particularly relevant in current studies of the aetiology of inflammatory muscle disorders. Coxsackie B virus-specific RNA has been found in the muscle of a series of patients with dermatomyositis or polymyositis, using blot hybridization techniques (Bowles et al 1987). Mumps virus antigen has been demonstrated immunocytochemically in the muscle of patients with inclusion body myositis, a type of inflammatory myopathy characterized by lack of response to steroid therapy (Chou 1986). It is likely that a combination of immunocytochemical techniques, to locate viral proteins and in situ hybridization methods for the detection of specific nucleic acid sequences may be of value in attempting to clarify the role of previous viral infection in this complex group of disorders.

REFERENCES

Arahata K, Engel A G 1984 Monoclonal antibody analysis of mononuclear cells in myopathies. I Quantitation of subsets according to diagnosis and sites of accumulation and demonstration and counts of muscle fibers invaded by T cells. Annals of Neurology 16: 193–208

Arahata K, Ishihara T, Kamakura K et al 1989 Mosaic expression of dystrophin in symptomatic carriers of Duchenne's muscular dystrophy. New England Journal of Medicine 320: 138–142

Bethlem J, Van Wijngaarden G K, Mumenthaler M, Meijer A E F H 1970 Centronuclear myopathy with type 1 fiber atrophy and 'myotubes'. Archives of Neurology 23: 70–73

Bonilla E, Schotland D L 1970 Histochemical diagnosis of muscle phosphofructokinase deficiency. Archives of Neurology 22: 8–12

Bonilla E, Samitt C E, Miranda A F et al 1988 Duchenne muscular dystrophy: deficiency of dystrophin at the muscle cell surface. Cell 54: 447–452

Bourgeois C, Hubbard B 1965 A method for the simultaneous demonstration of choline-containing phospholipids and neutral lipids in tissue sections. Journal of Histochemistry and Cytochemistry 13: 561–578

Bowles N E, Sewry C A, Dubowitz V, Archard L C 1987 Dermatomyositis, polymyositis and Coxsackie B virus infection. Lancet i: 1004–1007

Brooke M H 1973 Congenital fiber type disproportion. In: Kakulas B A (ed) Clinical studies in myology. Excerpta Medica, Amsterdam, p 147–159

Brooke M H, Engel W K 1966 The histologic diagnosis of neuromuscular diseases: a review of 79 biopsies. Archives of Physical Medicine and Rehabilitation 47: 99–121

Brooke M H, Engel W K 1969a The histographic analysis of human muscle biopsies with regard to fiber types, IV Children's biopsies. Neurology 19: 591–605

Brooke M H, Engel W K 1969b The histographic analysis of human muscle biopsies with regard to fiber types, II Diseases of the upper and lower motor neurons. Neurology 19: 378–393

Brooke M H, Kaiser K K 1970 Three 'myosin adenosine triphosphatase' systems. The nature of their pH lability and sulfhydryl dependence. Journal of Histochemistry and Cytochemistry 18: 670–672

Burstone M S 1958 Histochemical demonstration of acid phosphatases with naphthol AS phosphates. Journal of the National Cancer Institute 21: 523–531

Carpenter S, Karpati G 1984 Pathology of skeletal muscle. Churchill Livingstone, New York

Carpenter S, Karpati G, Andermann F 1972 Specific involvement of muscle, nerve and skin in late infantile and juvenile amaurotic idiocy. Neurology (Minneapolis) 22: 170–186

Carpenter S, Karpati G, Rothman S, Watters G 1976 The childhood type of dermatomyositis. Neurology (Minneapolis) 26: 952–962

Chou S M 1986 Mumps antigen in inclusion-body myositis. Muscle and Nerve 9 (Suppl): 36

Dietrichson P, Coakley J, Smith P E M, Griffiths R D, Helliwell T R, Edwards R T H 1987 Conchotome and needle percutaneous biopsy of skeletal muscle. Journal of Neurology, Neurosurgery and Psychiatry 50: 1461–1467

Di Mauro S et al 1980 Fatal infantile mitochondrial myopathy and renal dysfunction due to cytochrome c oxidase deficiency. Neurology (Minneapolis) 30: 795–804

Dubowitz V 1985 Muscle biopsy: a modern approach, 2nd edn. Baillière-Tindall, London

Dubowitz V, Pearse A G E 1960 Reciprocal relationship of phosphorylase and oxidative enzymes in skeletal muscle. Nature (London) 185: 707–702

Engel A G, Angelini C 1973 Carnitine deficiency of human skeletal muscle with associated lipid storage myopathy. Science 173: 889–902

Engel W K 1961 Muscle target fibres, a newly recognised sign of denervation. Nature (London) 191: 389–390

Engel W K, Brooke M H 1966 Histochemistry of the myotonic disorders. In: Kuhn E (ed) Progressive Muskeldystrophie, Myotonie, Myasthenie. Springer-Verlag, Stuttgart, p 203–222

Engel W K, Bishop D W, Cunningham G C 1970 Tubular aggregates in type II muscle fibers: ultrastructural and histochemical correlation. Journal of Ultrastructural Research 31: 507–525

Fishbein W N, Armbrustmacher V W, Griffin J L 1978 Myoadenylate deaminase deficiency: a new disease of muscle. Science 200: 545–548

Hoffman E P, Brown B H, Kunkel L M 1987 Dystrophin: the protein product of the Duchenne muscular dystrophy gene. Cell 51: 919–928

Holt I J, Harding A E, Morgan-Hughes J A 1988 Deletions of muscle mitochondrial DNA in patients with mitochondrial myopathies. Nature 331: 717–719

Johnson M A, Polgar J, Weightman D, Appleton D 1973 Data on the distribution of fibre types in thirty-six human muscles. An autopsy study. Journal of the Neurological Sciences 18: 111–129

Johnson M A, Turnbull D M, Dick D J, Sherratt H S A 1983 A partial deficiency of cytochrome c oxidase in chronic progressive external ophthalmoplegia. Journal of the Neurological Sciences 60: 31–53

Lester J M, Silber D I, Cohen M H et al 1983 The codispersion index for the measurement of fiber type distribution patterns. Muscle and Nerve 6: 581–587

Martinez B A, Lake B D 1987 Childhood nemaline myopathy. A review of clinical presentation in relation to prognosis. Developmental Medicine and Child Neurology 29: 815–820

Mastaglia F L, Walton Sir J (eds) 1982 Skeletal muscle pathology. Churchill Livingstone, Edinburgh

Morgan-Hughes J A, Darveniza P, Landon D N, Land J M, Clark J B 1979 A mitochondrial myopathy with a deficiency of respiratory chain NADH-CoQ reductase activity. Journal of the Neurological Sciences 43: 27–46

Neville H E, Brooke M W 1971 Central core fibers: structured and unstructured. In: Kakulas B A (ed) Basic research in myology. Exerpta Medica, Amsterdam, p 97–511

Pearse A G E 1972 Histochemistry, theoretical and applied, 3rd edn. Churchill Livingstone, Edinburgh

Pellisier J F, DeBarsy T, Faugere M C, Rebuffel P 1979 Type III glycogenosis with multicore structures. Muscle and Nerve 2: 124–132

Pleasure D E, Walsh G O, Engel W K 1970 Atrophy of skeletal muscle in patients with Cushing's syndrome. Archives of Neurology 22: 118–125

Rowland L P, Di Mauro S, Bank W J 1971 Glycogen storage disease of muscle. Problems in biochemical genetics. Birth Defects Original Articles Series 7: 43–51

Scholte H R, Jennekens F C I, Bouvy J J B 1979 Carnitine palmityltransferase II deficiency with normal carnitine palmityltransferase I in skeletal muscle and leucocytes. Journal of the Neurological Sciences 40: 39–51

Seligman A M, Karnovsky M J, Wasserkrug H L, Hanker J S 1968 Nondroplet ultrastructural demonstration of cytochrome oxidase activity with a polymerising osmiophilic reagent, diaminobenzidine (DAB). Journal of Cell Biology 38: 1–14

Song S K, Shimada N, Anderson P J 1963 Orthogonal diameters in the analysis of muscle fibre size and form. Nature (Lond) 200: 1220–1221

Sugita H, Masaki T, Ebashi S, Pearson C M 1973 Protein composition of rods in nemaline myopathy. In: Kakulas B A (ed) Basic research in myology. Excerpta Medica, Amsterdam, p 298–302

Trend P St J, Wiles C M, Spencer G T, Morgan-Hughes J A, Lake B D, Patrick A D 1985 Acid maltase deficiency in adults. Diagnosis and management in five cases. Brain 108: 845–860

Turnbull D M, Bartlett K, Stevens D L et al 1984 Short-chain acyl-CoA dehydrogenase deficiency associated with a lipid storage myopathy and secondary carnitine deficiency. New England Journal of Medicine 311: 1232–1236

Turnbull D M, Bartlett K, Eyre J A et al 1988 Lipid storage myopathy due to glutaric aciduria type II. Treatment of a potentially fatal myopathy. Developmental Medicine and Child Neurology 30: 667–672

Van Biervliet J P G M, Bruinris L, Ketting D et al 1977 Hereditary mitochondrial myopathy with lactic acidaemia, a deToni-Fanconi-Debré syndrome and a defective respiratory chain in voluntary striated muscles. Pediatric Research 11: 1008–1092

Whitaker J N, Engel W K 1972 Vascular deposits of immunoglobulin and complement in idiopathic inflammatory myopathy. New England Journal of Medicine 286: 333–338

Zubrzycka-Gaarn E E, Bulman D E, Karpati G et al 1988 The Duchenne muscular dystrophy gene product is localized in the sarcolemma of human skeletal muscle. Nature 333: 466–469

11. Immunohistochemistry and the skin

A. C. Chu   B. Bhogal   J. McLelland

INTRODUCTION

The concept of the skin as an inert barrier to infection and an organ of temperature regulation has given way to the modern idea of the skin as an important immunological organ. Mounting evidence points to the skin as being a T-cell organ where T-lymphocytes preferentially migrate and may undergo limited degrees of maturation and education (Chu et al 1987a). This change in concept has been brought about by recent advances in immunological understanding and technology.

Our understanding of cutaneous disease has similarly advanced over the last decade and the individual disciplines of histopathology, immunohistochemistry and electron microscopy have given us enormous insight into the pathogenesis of various dermatoses.

In this chapter we will describe the use of immunohistochemistry in the diagnosis of the autoimmune bullous dermatoses and how immunological techniques have enabled us to unravel the pathogenesis of these diseases. We will also discuss the use of monoclonal antibodies in the recognition of cells in the skin and how this has helped in the diagnosis of cellular infiltrates in the skin and increased our knowledge of the biology of various cells present in the skin.

Specimen collection and processing

Some of the techniques used in immunohistochemistry demand the use of fresh frozen tissue. This is particularly true of investigation of bullous dermatoses and cellular infiltrates in the skin. Some markers can be used on paraffin sections, but these tend to be the exception rather than the norm. The site of skin biopsy is important in many diseases as is the timing of the biopsy, particularly in the cutaneous vasculitides where immunoreactants are no longer identifiable in involved vessels after 12 hours.

In the investigation of bullous dermatoses the general rule is to take a biopsy through a fresh bullous lesion to include both normal surrounding skin and the margin of the blister. Most of the bullous dermatoses which are of autoimmune origin will show immunoreactants in normal or perilesional skin, but very often in lesional skin the cellular infiltrate will remove immunoreactants and give rise to a false negative result. Biopsies for immunohistochemistry should be snap frozen either in liquid nitrogen or after embedding in OCT compound in n-hexane cooled in a CO_2 bath. If tissue is to be stored for any length of time it should be embedded in OCT embedding medium rather than just kept wrapped in aluminium foil as this will eventually lead to freeze artifacts and difficulty with processing. In the investigation of cellular infiltrates the most indurated involved area of skin should be biopsied and the tissue frozen in the same way.

Immunohistochemical techniques

Direct immunofluorescence

This technique is generally used for detection of immunoreactants in lesional skin. Five sections are taken from each skin biopsy and stained with a fluorescein conjugated antiserum against IgM, IgG, IgA, C3 and fibrin. Sections are incubated for 30 minutes at 37°C in a moist chamber. They are then washed three times in phosphate buffer and mounted in buffered glycerol, before examination under a fluorescence microscope.

Indirect immunofluorescence

In this technique the patient's serum is examined for the presence of circulating antibodies against structures within the skin. A large number of different substrates have been used for this technique including guinea pig lip or oesophagus, normal human skin and monkey oesophagus. The optimal tissues are human skin and monkey oesophagus.

In certain conditions, however, i.e. cicatricial pemphigoid, the autoantibodies may have a tissue specificity and will react with human mucosa but to no other tissues. In this technique two cover slips are generally stained for each serum. The tissue is first incubated for 30 minutes at 37°C in a moist chamber with a 1/10 or 1/80 dilution of the serum. After three washes the sections are incubated with a fluorescein conjugated antihuman immunoglobulin, which is normally IgG but in the case of linear IgA disease would be an IgA, at 37°C for a further 30 minutes. After a final three washes the sections are mounted in buffered glycerol and examined under a fluorescence microscope.

Indirect C3 method

This method is useful in a) determining whether circulating antibodies are capable of fixing complement, which will give some idea as to the pathogenesis of the disease, and b) to amplify staining of a conventional indirect immunofluorescence technique, in which the circulating autoantibodies are of low titre but will fix complement and the detection of complement fixed by the antibodies will amplify the staining and allow detection. In this technique normal human skin is used as a substrate. Two sections are incubated with either neat serum or 1/4 dilution of the serum for 30 minutes at 37°C in a moist chamber, washed and then incubated with a source of complement which is usually normal human serum diluted 1/5 in complement-diluting buffer for 30 minutes. After a further wash the sections are stained with fluorescein conjugated anti-C3 for a further 30 minutes before final washing, mounting in

buffered glycerol and examination under a fluorescence microscope.

Alkaline phosphatase anti-alkaline phosphatase technique

The APAAP technique (Cordell et al 1984) is especially useful for monoclonal antibody studies in skin.

While immunoperoxidase methods give permanent records of staining, and are more sensitive than immunofluorescence, they do present problems with interpretation, particularly where pigmentary incontinence or a heavily pigmented skin is involved. The APAAP technique gives bright red staining and can be counterstained with haematoxylin. The technique is very easy and if mistakes are made during the process the tissue can still be salvaged by going back a step and repeating the procedures (see Appendix 6).

The addition of levamisole to the incubating medium inhibits the alkaline phosphatases of skin, leaving that of the conjugate unaffected.

BULLOUS DERMATOSES

The bullous dermatoses are a very heterogeneous group of diseases which are linked by the clinical manifestation of bullae or blisters on the skin surface. In many of these diseases the clinical appearance of the patient will suggest the diagnosis to the dermatologist. However, many bullous diseases look clinically identical and the clinician relies heavily on histopathology and, in particular, immunohistochemistry to establish the diagnosis. Immunohistochemistry is particularly valuable in making or confirming a diagnosis of the autoimmune bullous diseases, but may also be of value in the diagnosis of the some of the nonautoimmune bullous dermatoses.

Pemphigus vulgaris

Pemphigus vulgaris is a disease of the fourth and fifth decades and is characterized by the presence of flaccid bullae which easily rupture and give rise to enlarging erosions on any part of the body. Mucous membranes are often affected and may be the first site of involvement. Before the advent of steroid therapy pemphigus vulgaris carried a very high mortality and even now the mortality rate is about 10%. Diagnosis and rapid treatment are therefore of enormous importance and immunohistochemistry has proven invaluable in this.

Histopathology

The hallmark of pemphigus is the loss of adhesion of the keratinocytes in the epidermis which then round up and lie freely within the blister fluid, a process known as acantholysis.

Acantholysis is not, however, pathognomonic for pemphigus vulgaris and many benign (Hailey Hailey disease, Darier's disease and Grover's disease) and malignant (actinic keratoses and basal cell carcinoma) conditions may show this phenomenon.

Immunohistochemistry

The typical features of pemphigus vulgaris are the deposition of IgG and occasionally other classes of immunoglobulin and complement components in-

tercellularly in the epidermis (Fig. 11.1). The immunoreactants are found not only in involved skin, but also in non-involved skin. Indirect immunofluorescence detects a circulating antibody against the intercellular substance of the epidermis. Such circulating antibodies are characteristic for pemphigus, but may also be observed in AB blood mismatch reactions, trichophyton infections, morbilliform drug eruptions and severe burns patients. It has been suggested that the titre of the circulating pemphigus antibody correlates with the activity of the disease and this parameter is often used to monitor the response rates of patients with this disease (O'Loughlin et al 1978). Recent studies have shown, however, that pemphigus antibody titre and clinical disease only show a good correlation in about 50% of patients tested.

Pathogenesis

Pemphigus vulgaris is an autoimmune bullous dermatosis where the autoantibody or pemphigus antibody binds to the intercellular cement of the epidermis (Wolff & Schreiner 1971). This reaction

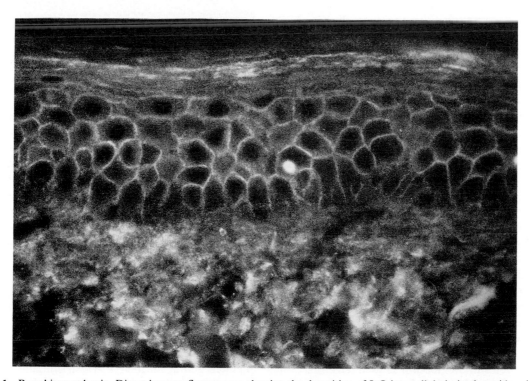

Fig. 11.1 Pemphigus vulgaris. Direct immunofluorescence showing the deposition of IgG intercellularly in the epidermis.

leads to the destruction of adhesion between keratinocytes and results in acantholysis. The initial findings of complement components and properdin in lesional skin of patients with pemphigus vulgaris suggested that the disease may be mediated through complement activation. Subsequent studies in vitro, however, have shown that acantholysis can be induced in the presence of complement-free serum (Schiltz & Michael 1976). Further studies have shown that acantholysis induced by the pemphigus antibody can be inhibited by soya bean trypsin inhibitor (Farb et al 1978). It would thus appear that the pemphigus antibody activates tissue proteinases which dissolve the intercellular cement and lead to acantholysis.

Bullous pemphigoid

Bullous pemphigoid is a blistering skin disease of old age with a peak incidence in the sixth and seventh decades. It is characterized by an initially pruritic eruption which becomes bullous, the blisters occurring on either normal or erythematous skin. Mucous membranes are rarely affected being involved in only about 10% of cases.

Histopathology

Two variants are recognized on histopathology: infiltrate rich and infiltrate poor. In both variants the predominant cell present in the dermal infiltrate is the eosinophil. There does not seem to be any correlaton between histopathological and clinical pictures in these variants. The primary abnormality seen in bullous pemphigoid is a subepidermal split with the presence of eosinophils in the blister cavity. At the ultrastructural level, the split in bullous pemphigoid is in the lamina lucida.

Immunohistochemistry

Characteristically there is a linear deposition of immunoglobulin and complement components, particularly C3, along the basement membrane zone in both affected and unaffected skin (Fig. 11.2). The immunoglobulin is generally IgG, but occasionally IgM and IgA may be found. Using

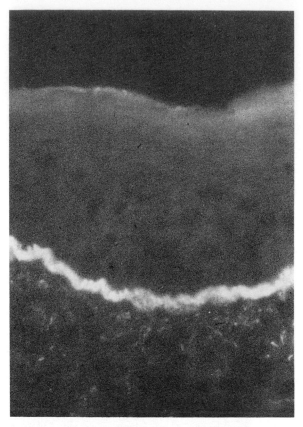

Fig. 11.2 Bullous pemphigoid. Direct immunofluorescence showing the deposition of IgG in a linear band along the basement membrane zone.

immunoelectromicroscopy, the immunoglobulin has been found to be deposited in the lamina lucida of the basement membrane zone. The majority of patients with bullous pemphigoid have a circulating IgG against the basement membrane zone of their own skin. This bullous pemphigoid antibody is present in about 75–80% of patients. It has been suggested that patients who do not show this circulating antibody have a higher incidence of internal neoplasm than age-matched controls (Hodge et al 1981).

Pathogenesis

Using the indirect C3 test the circulating bullous pemphigoid antibody has been shown to be a complement fixing antibody. It was initially assumed that this autoimmune disease was mediated by complement fixation and the damage caused by neutrophils recruited into the area by C3a and C5a. In sequential studies of developing lesions, however, an early change is seen in the distribution and granulation of mast cells. Also, the predominant cell present in the infiltrate is the eosinophil rather than the neutrophil. It has therefore been postulated that in bullous pemphigoid the antibody binds to the pemphigoid antigen present in the lamina lucida, fixes complement with the generation of C3a and C5a which activate mast cells, and it is the production of eosinophil chemotactic factor (ECF-A) by the mast cells that recruits eosinophils into the area, where release of their basic proteins causes the damage to the lamina lucida and the generation of the blister (Wintrobe et al 1977).

Bullous pemphigoid antigen has been isolated as a 31 000 mol.wt protein (Diaz et al 1977). This antigen is produced by the basal keratinocyte which can be identified in epidermal cell suspension by staining with bullous pemphigoid antibody which is seen as a cap on these cells.

Herpes gestationis

Herpes gestationis is a blistering disease which occurs generally during pregnancy, but may occur in the puerperium or with the onset of the first menstrual cycle following pregnancy. It is commonest in the third trimester of pregnancy and occurs in 1 in 10 000–30 000 pregnancies. There is no associated morbidity to the mother, but an increased mortality has been suggested in the infant (Kolodny 1969). The eruption may recur in subsequent pregnancies. The eruption often starts as a polycyclic pruritic eruption around the umbilicus which spreads peripherally and eventually gives rise to a bullous disease which can affect any part of the body.

Histopathology

As with bullous pemphigoid, herpes gestationis is a subepidermal blistering disease and the cleft is seen below the epidermis. In the early stages the dermal infiltrate is very sparse and generally only seen in established bullae, where it is predominantly eosinophilic. Electron microscopy of early and evolving lesions shows that the first abnormality occurs in the basal keratinocytes with damage to the cytoplasmic membrane and umbilication of cytoplasmic contents through defects in the cytoplasmic membrane (Harrington & Bleeker 1979).

Immunohistochemistry

In the majority of patients with herpes gestationis, the only abnormal finding is a linear band of C3 along the basement membrane zone. In 25% of patients a very weak band of IgG can also be identified along the basement membrane zone. All patients have a circulating factor, initially designated herpes gestationis or HG factor, which has now been identified as a low titre IgG which avidly binds complement. In general the HG factor is of such low titre that it can only be identified in 10–20% of patients by the normal indirect immunofluorescence technique. It can be identified in all patients, however, on the basis of its complement fixation using the indirect C3 test.

Pathogenesis

Unlike any of the other bullous diseases herpes gestationis appears to be mediated by a classical type II response. The binding of the autoantibody gives rise to complement-mediated damage to the basal keratinocytes and only later are eosinophils recruited into the area.

Cicatricial pemphigoid

This is a chronic blistering disease which primarily affects the mucous membranes. Oral and ocular involvement is common which ultimately leads to scarring due to repeated subepidermal bullae. However, in one variant, the Brunsting-Perry variant, there is an associated localized cutaneous vesicobullous eruption which usually occurs on the head and neck. The blisters recur repeatedly at the same site and give rise to severe scarring. The major morbidity with this disease is that of corneal damage and keratitis leading to blindness in up to 20% of patients.

Histopathology

Cicatricial pemphigoid is indistinguishable from bullous pemphigoid in showing a subepidermal blister. In electron microscopy the basal lamina shows evidence of replication (Susi & Shiklar 1971).

Immunohistochemistry

In lesional skin the findings in cicatricial pemphigoid are similar to those seen in bullous pemphigoid. Immunoglobulins, particularly IgG, are deposited in linear fashion along the basement membrane zone. C3 and in addition C1q, C4 and factor B are often detected. Only 10% of patients show circulating autoantibody against the basement membrane zone, but this is an IgG class immunoglobulin and may show organ specificity reacting with only mucosa or only skin, or species specificity reacting only with human rather than animal tissue and on occasions, idiotypic specificity reacting only with the patient's tissue and not with others.

Pathogenesis

The circulating antibody in cicatricial pemphigoid is a complement-fixing antibody and it is assumed that the damage in the basal lamina is mediated by complement fixation leading to infiltration by leucocytes. Why this disease has a predisposition to mucous membranes and why the cutaneous lesions in Brunsting-Perry disease occur repeatedly at the same site are questions still to be answered.

Dermatitis herpetiformis

This is an intensely pruritic vesicobullous eruption which occurs between the second and fourth decades of life. The eruption has a predilection for certain sites, occurring over bony prominences, elbows and knees, lower back and over the neck. In patients with the classical form of dermatitis herpetiformis there is invariably an associated gluten enteropathy. This may be manifest by an abnormal jejunal biopsy showing partial or subtotal villous atrophy (Weinstein 1974).

Histopathology

The pathognomonic feature of dermatitis herpetiformis is of a neutrophil microabcess in the dermal papillae. Initially oedema and a mild infiltration by neutrophils is seen. Following this a more intense neutrophil microabcess forms with fibrin accumulation in the papillary tips and subepidermal cleft begins. The blisters are initially multilocular but the vesicles coalesce leading to a single bulla.

At electron microscopy the basal lamina shows extensive damage and may be completely absent in advanced cases.

Immunohistochemistry

Both normal and uninvolved skin of patients with dermatitis herpetiformis show a unique im-

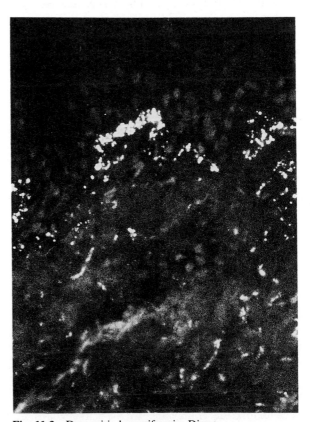

Fig. 11.3 Dermatitis herpetiformis. Direct immunofluorescence showing granular deposition of IgA in the papillary tips.

munohistochemical pattern. This is of a granular deposition of IgA in the papillary tips (Fig. 11.3). This form of the disease is associated with gluten enteropathy. A smaller percentage of patients with clinical dermatitis herpetiformis show a linear deposition of IgA along the basement membrane zone. These patients differ clinically from patients with the classical granular papillary tip IgA, not only in their association with gluten enteropathy, but also in their response to treatment.

Classical dermatitis herpetiformis responds well to dapsone and other sulphones while those showing a linear IgA deposition respond to steroids. Because of this difference, the linear IgA dermatitis herpetiformis patients are often grouped together with the chronic bullous disease of childhood under the term 'linear IgA disease'.

In indirect immunofluorescence, 16% of patients with classical dermatitis herpetiformis have circulating antireticulin IgG and in patients with linear IgA disease there is often a circulating IgA antibasement membrane zone antibody.

Pathogenesis

The pathogenesis of dermatitis herpetiformis is yet to be elucidated. The finding of IgA and complement components in the dermal papillae suggests that the deposition of IgA may lead to complement fixation via the alternative pathway, which will lead to the accumulation of neutrophils in the dermis and the damage to the basal lamina which is seen histologically. The close association between dermatitis herpetiformis and gluten enteropathy suggests the possibility of gluten being involved in the pathogenesis of the skin disease.

Immunoelectron microscopy has shown that the IgA in the granular form of dermatitis herpetiformis is located to dermal microfibrillar bundles in the papillary dermis (Yaoita & Katz 1976). It is possible, therefore, that the circulating antireticulin antibodies seen in patients with dermatitis herpetiformis, which have been shown to crossreact with gluten, may deposit in these microfibrillar bundles and that IgA will bind to this complex. Further studies, however, have failed to demonstrate gluten associated with the IgA deposition in the papillary dermis.

USE OF IMMUNOHISTOCHEMISTRY IN THE INVESTIGATION OF NON-AUTOIMMUNE-BULLOUS DERMATOSES

In various bullous dermatoses the level of the split within the skin is of enormous diagnostic value. This is well demonstrated in the group of diseases classed as *epidermolysis bullosa*. These diseases are linked by an abnormal fragility of the skin which manifests itself as blistering of the skin even after minor trauma. The three main variants of the genetic forms of epidermolysis bullosa — the dominant simplex form, the recessive junctional form and the dominant or recessive dystrophic forms, can now be differentiated by electron microscopy on the level at which the blister forms. In the simplex form this occurs through the basal keratinocytes; in the junctional form there is an absence of the subbasal plate of the hemidesmosome within the lamina lucida and the cleft occurs through the lamina lucida. In the dystrophic forms the cleft forms below the basal lamina, due to defects in the anchoring fibrils.

In the absence of electron microscopy, immunohistochemistry can be used to differentiate the three main forms of epidermolysis bullosa by examining a newly induced lesion. Using two markers of different parts of the basement membrane zone, i.e. antibodies against type 4 collagen for the basement membrane and bullous pemphigoid antibody for the lamina lucida, clefts at different levels in the skin can be identified and the three different forms of epidermolysis bullosa distinguished. In the simplex form both type 4 collagen and bullous pemphigoid antigen will be present in the floor of the blister (Fig. 11.4). In the junctional form of epidermolysis bullosa the cleft occurs through the lamina lucida and type 4 collagen will be found on the floor of the blister and bullous pemphigoid antigen will be found present in both the floor and the roof of the blister (Fig. 11.5). In the dystrophic forms of epidermolysis bullosa the cleft occurs beneath the basement membrane and type 4 collagen will therefore be found both in the floor and the roof of the blister, with bullous pemphigoid antigen only within the roof of the blister.

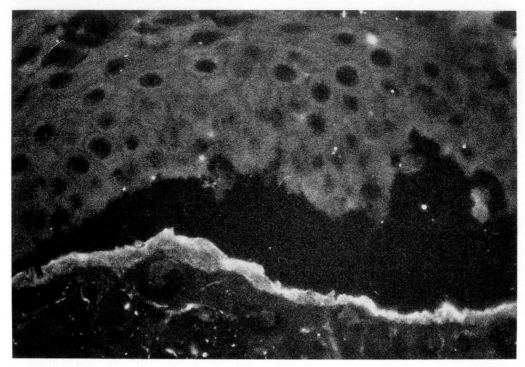

Fig. 11.4 Epidermolysis bullosa simplex — bullous pemphigoid antigen detected by serum from a patient with bullous pemphigoid is present on the floor of the blister.

More specific antibodies may also be used on unblistered skin biopsies. The antibody GB3 (Seralab) which binds to a component of the normal basement membrane, shows no affinity for the basement membrane in the severe forms of junctional epidermolysis bullosa (Heagerty et al 1986a). In the rarer, milder cases, in which sub-basal plates of hemidesmosomes may be demonstrable by electron microscopy, GB3 shows reduced rather than absent staining. Prenatal diagnosis of severe junctional epidermolysis bullosa is possible not only by electron microscopy, but also using the antibody GB3 on cryostat sections of a fetal skin biopsy taken at 16–18 weeks gestation (Heagerty et al 1986b).

In the dystrophic forms of epidermolysis bullosa, the anchoring fibrils, which attach the basement membrane to the collagen of the papillary dermis, are defective or rudimentary. Their presence in the normal can be detected by the use of an antibody to collagen type VII (monoclonal antibody LH7.2). In the recessive form of dystrophic epidermolysis bullosa there is grossly reduced binding of LH7.2, while the dominant form is characterized in most patients by a weaker than normal reaction, although some may be difficult to differentiate from the localized recessive form.

Lupus erythematosus

Lupus erythematosus is a chronic idiopathic disease which, in the skin, may be of two types. The first, chronic discoid lupus erythematosus, is generally restricted only to the skin and patients with this disorder do not show the autoantibodies and antinuclear factor characteristic of the systemic form of the disease. Skin lesions are erythematous and scaling and frequently heal to leave areas of scarring. When the condition involves hairy areas this leads to scarring alopecia. The second form of the disease is associated with systemic lupus erythematosus. Occasionally lesions found in systemic lupus erythematosus

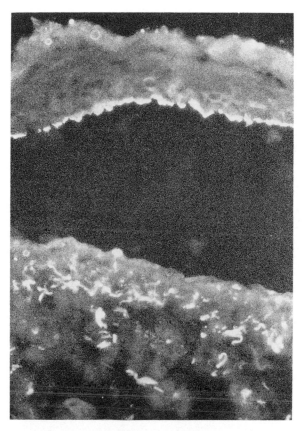

Fig. 11.5 Junctional epidermolysis bullosa — bullous pemphigoid antigen is detected in the roof and floor of the blister.

are identical to those of chronic dyscoid lupus erythematosus. More often, butterfly erythema is present in the malar area of the face.

Immunohistochemistry

Immunohistochemistry is important in lupus erythematosus for two reasons: the first is for diagnosis and the second is in differentiating the chronic discoid lupus erythematosus restricted to the skin from discoid lupus erythematosus associated with systemic lupus erythematosus. In lesional skin in both variants the immunocytochemical characteristics are identical and this consists of deposition of immunoglobulin, usually IgM or IgG, along the base membrane zone as a coarse granular layer. This corresponds histologically with the thickened basement membrane zone and PAS positive band which is

particularly prominent in chronic discoid lupus erythematosus. In addition to the immunoglobulins, complement components, in particular C3 may be found.

If normal skin is taken from patients with lupus erythematosus and examined for deposition of immunoreactants along the basement membrane zone, chronic discoid lupus erythematosus shows no reactivity, but 40–60% of patients with systemic lupus erythematosus show a positive immunoglobulin band along the basement membrane zone. This is useful in differentiating the localized from the systemic forms of the disease.

Vasculitis

The immunologically mediated vascular disease is the result of a type III response with deposition of immune complexes within blood vessels. Complement fixation and the recruitment of polymorphonuclear leucocytes into the vessel results in damage to the vessel and occlusion. This leads to the characteristic histological picture of leucocytoclasis — the presence of nuclear debris from polymorphonuclear leucocytes floating free between the infiltrating cells.

Timing of the biopsies is of importance in the cutaneous vasculitides. Studies have shown that if biopsies are taken more than 12 hours after the onset of the eruption the inflammatory infiltrate causes destruction and clearance of the immunoreactants.

Immunohistochemistry

In early cases of leucocytoclastic vasculitis immunoglobulins, particularly IgM and IgG, and complement components, particularly C3, may be found deposited within blood vessel walls. With time and further damage to the blood vessel, fibrin is then deposited within the blood vessel wall and can be identified using an anti-fibrin antibody.

Identification of different cell types in the skin

A variety of techniques can be used to identify different cell populations in the skin. These include enzyme histochemistry, polyclonal antibodies against cell surface determinants and monoclonal antibodies against different subpopulations of cells. This is of importance not only for the diagnosis of certain diseases but also in the investigation of the pathogenesis of skin disease. However, care must always be exercised in interpreting results using these techniques. Skin biopsies remove tissue at a single time point in a dynamic situation, and the mere presence of certain cells will not necessarily indicate a single pathogenetic mechanism. Indeed, the end result of the vast majority of cutaneous reactions is the accumulation of a cellular infiltrate with a predominance of helper T lymphocytes.

Where T lymphocytes are present Langerhans cells will also accumulate and it is impossible to interpret the finding of a predominant T cell infiltrate with Langerhans cells present, without information as to the stage of evolution of the eruption.

T-lymphocyte

The skin is now generally regarded as a T-cell organ of the body and within the milieu of the skin T-cells may undergo a certain degree of maturation (Rubenfeld et al 1981, Chu et al 1982) and development. The epidermis elaborates a number of cytokines that can influence both the maturation and the function of T lymphocytes and other mononuclear cells. Of particular importance is the epidermally derived lymphocyte differentiating factor (ELDIF), which is a poorly characterized factor or series of factors responsible for the in vitro differentiation of T-cells (Nicholas et al 1985).

A T-cell chemotactic factor produced by basal keratinocytes has recently been identified. Human keratinocytes have been shown to elaborate interleukin 1 (Luger et al 1981).

Cutaneous T-cell lymphoma (CTCL)

Cutaneous T-cell lymphoma is now regarded as a malignancy of helper T-cells which originates or predominantly affects the skin (Kung et al 1980). The major problem with this disease is its early diagnosis, as clinically CTCL in its early stages is often mistaken for a form of eczema or psoriasis and the correct clinical diagnosis may not be made until quite an advanced stage. Histopathology is extremely useful in this disease and the characteristic dense dermoepidermal lymphohistiocytic infiltrate containing numerous hyperchromatic cells with infiltration of the epidermis and formation of Pautrier microabcesses is easily recognized. However, even in patients with clinically typical CTCL, the histological picture may be difficult to interpret with only a mild infiltrate of mixed cellularity and little or no epidermotrophism. Better diagnostic aids are therefore essential, particularly as the disease starts in the skin and early aggressive treatment directed to the skin could potentially be curative. A large number of techniques have been used to try to augment histopathology in establishing a diagnosis of CTCL. This has included electron microscopy, enzyme histochemistry and immunohistochemistry.

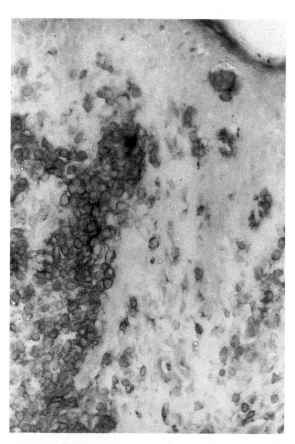

Fig 11.6 Cutaneous T cell lymphoma. CD4 positive cells stained with OKT4 are seen within the dermis and also in the epidermis, forming Pautrier microabscesses.

The identification of the Sezary cell as a specific morphological entity was initially very encouraging (Lutzner & Jordan 1968), but subsequent studies have demonstrated typical Sezary cells in a large number of benign dermatoses (Flaxman et al 1971). Similarly, enzyme histochemistry demonstrated that the vast majority of cells present within the infiltrate of CTCL were T-cells, as identified by a dot-like reaction with non-specific esterase stains (Chu et al 1981), but this did not allow differentiation from benign dermatoses such as lichen planus and even eczema.

Polyclonal antibodies against T-cells demonstrated the T-cell nature of the dermal infiltrate and allowed differentiation from B-cell lymphomas, but could not be used to differentiate benign from malignant infiltrates (Chu & Mac-Donald 1979). A large study on 92 patients with CTCL showed that monoclonal antibodies were of no greater value than polyclonal antibodies or even enzyme histochemistry in establishing a diagnosis. The majority of the cells present in CTCL were shown to be helper T-cells expressing CD3 and CD4 molecules (Fig. 11.6), but suppressor T-cells expressing the CD8 molecule were also present. Only in a minority of patients could the diagnosis be established by the demonstration of modulation of the CD3 molecule by the abnormal cells, or the simultaneous expression of both CD4 and CD8 molecules by the cells. In the vast majority of patients, however, monoclonal antibodies against conventional T-cell subpopulations could not differentiate CTCL from benign dermatoses (Chu et al 1984).

However, one monoclonal antibody, BE2, produced using leukaemic cells from a patient with CTCL as the antigen, will bind to tumour-associated antigens present on the surface of CTCL cells and will also bind to some EBV transformed B-cell lines, some long-term T-cell lines and T-cells that are infected with HTLV-1 (Berger et al 1982). BE2 does not react with normal or reactive T-cells. Using this antibody it has been possible to diagnose CTCL in patients by examination of peripheral blood mononuclear cells and also tissue biopsies from involved skin (Fig. 11.7).

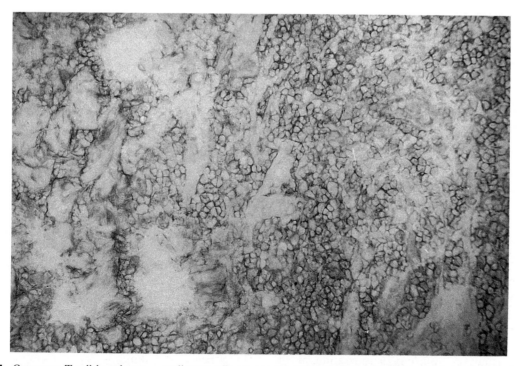

Fig. 11.7 Cutaneous T cell lymphoma — malignant cells are recognized by labelling with BE2, a monoclonal antibody against tumour-associated antigen on the lymphoma cells.

Graft versus host disease

Graft versus host disease is a constellation of clinicopathological findings which indicate an immunological attack on the host tissue by grafted cells. The theoretical requirements for GVHD are that the host must be immunosuppressed, that the host must have received foreign, allogeneic immunocompetent cells in sufficient quantity to constitute a graft and that there must be histocompatible differences between the graft and host cells. GVHD is seen in four clinical situations: 1) in utero, where the fetus has a congenital cellular immunodeficiency and maternal lymphocytes are able to transverse the placenta and induce GVHD within the developing fetus; 2) in newborns and infants with a primary cellular immunodeficiency following therapeutic blood transfusion or even transfusion of blood products; 3) in patients who are immunosuppressed due to chemotherapeutic treatment of malignancy in whom whole blood or blood products are given therapeutically and 4) after a lymphoid therapeutic graft for bone marrow aplasia, cellular immunodeficiency or leukaemia (Saurat 1981).

The final category of GVHD is the commonest seen in clinical practice and with the expansion of bone marrow transplantation schemes, numbers of patients with cutaneous GVHD will almost certainly increase.

In man the main targets of graft-versus-host disease are the skin, the gastrointestinal tract and the liver. The reaction can be acute or chronic and presents in a range of severities. The pathogenesis of graft-versus-host disease in the skin is still controversial.

The histopathology of acute cutaneous graft versus host disease has been described by Lerner et al (1974) who suggested a 0–4 histological grading system. Grade 1 consisted of focal or diffuse vacuolar degeneration of epidermal basal keratinocytes; grade 2 showed in addition to basal vacuolar degeneration a focal or diffuse spongiosis and dyskeratosis or eosinophilic degeneration of epidermal cells (Civatte bodies); Grade 3 was characterized by the findings seen in grade 2 with the addition of clefts and spaces with acantholysis and epidermolysis resulting in separation of the dermoepidermal junction. Grade 4 was seen where there was frank loss of epidermis.

In acute graft-versus-host disease a frequent finding is that of lymphocytes adjacent to dyskeratotic keratinocytes in the epidermis, a phenomenon called 'satellite cell necrosis'.

Chronic GVHD occurs three or more months after bone marrow transplantation in up to 25% of patients. The skin manifestations usually begin as erythematous or violaceous macules or papules which are asymptomatic and generalized. Over days or weeks the individual lesions evolve into a desquamative hyperpigmented eczematous process. Less frequently a generalized lichenoid eruption may be present, which can be indistinguishable from lichen planus, and a sclerotic variant is also well recognized, which resembles systemic sclerosis with progressive ulceration and contractures of the extremities.

In GVHD, immunofluorescence studies of the skin have generally been negative, but IgM deposition along the dermal/epidermal junction has been reported (Tsoi et al 1978). In GVHD examination of the cellular infiltrate may be of some value in establishing a diagnosis. In general there is a depletion of CD1 positive Langerhans cells in the epidermis (Perreault et al 1984). Some authors have suggested that the fall in Langerhans cell numbers is proportional to the severity of acute GVHD, a finding not supported by other studies. The infiltrate, although mixed, is predominantly CD8 positive (Lampert et al 1982), which is unlike most of the reactive skin disorders where the T-cell population is dominated by the CD4 positive cells.

Langerhans cells

The Langerhans cell is an immigrant population of cells in the epidermis being derived from a bone marrow precursor cell. Attention was drawn to the immunological function of this cell by study of its surface markers. See Table 11.1.

It is now regarded as the skin's accessory cell with potent antigen presenting function and other accessory cell functions (Braathen & Thorsby 1980). The bone marrow progenitor cell has not been characterized and it is unclear what signals attract it to the skin and allow maturation within the milieu of the skin.

The most specific marker for Langerhans cell is the Birbeck granule, which is a trilaminate struc-

Table 11.1

Aureophilic	Surface ATP-ase
Birbeck granule	Fc IgG receptors
C3b receptors	HLADR
CD1 complex	S100 protein
CD4 complex (activated)	1L2 receptor (activated)
α-Mannosidase	Peanut agglutinin (only for cells of LCH)

ture found within the cytoplasm of the cells. Elegant studies using immunogold electron microscopy have demonstrated that this structure can develop as the result of receptor-specific endocytosis when antibodies against the CD1 and class II antigens are used (Hanou et al 1986). The function of these structures, however, remains obscure. Of the cell surface markers the CD1 complex remains the most useful for the identification of these cells (Fig. 11.8), as the only other cell type that naturally expresses this antigen is the cortical thymocyte. Antibodies against this marker, however, require the use of fresh frozen tissue and retrospective analysis of paraffin embedded tissue is therefore impossible.

Of the other markers of Langerhans cells par-

ticular note should be taken of the CD4 complex, which was initially thought to be specific for helper-inducer T-cells (Reinherz et al 1980), but has now been found on a number of different cells of the monocyte macrophage series (McMillan et al 1986). In monocyte macrophages this antigen does seem to be an activation marker as resting Langerhans cells express only very small amounts of CD4 complex while activated Langerhans cells strongly express this antigen.

Enzyme histochemistry for surface ATP-ase and α-mannosidase are still very useful techniques in identifying Langerhans cells in cryostat sections. A useful marker for identifying Langerhans cells in paraffin sections is S100 protein, which does, however, tend to be a very non-specific marker

Fig. 11.8 Normal human skin stained with OKT6. Dendritic CD1 positive Langerhans cells are seen in the suprabasilar epidermis.

and within the skin will identify melanocytes and neural tissue as well as Langerhans cells. Other related cells, such as interdigitating reticulum cells which may also be present in cutaneous inflammatory reactions, may also be labelled with antibodies to S100 protein.

Langerhans cell histiocytosis (histiocytosis X)

This is a condition which predominantly affects children and has a wide variety of manifestations. The main organs involved are bone, skin, pituitary, lung and the reticuloendothelial and

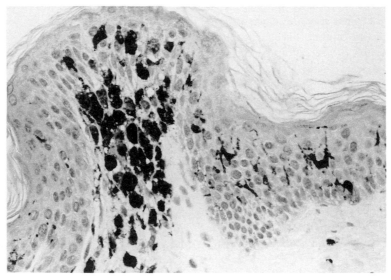

Fig. 11.9 Langerhans cell histiocytosis. Paraffin section stained for S100 protein using an immunoperoxidase technique. Normal Langerhans cells, LCH cells and melanocytes stain for S100 protein.

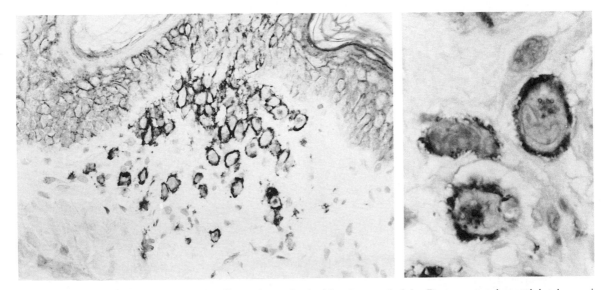

Fig. 11.10 Langerhans cell histiocytosis. Paraffin section stained with peanut agglutinin. Dense paranuclear staining is seen in the LCH cells, but normal Langerhans cells show only diffuse cytoplasmic labelling if any at all. Insert shows detail of the LCH cell staining.

haemopoietic systems (see also Ch. 20). The characteristic cell seen in lesions is the Langerhans cell histiocytosis (LCH) cell. This is a large eosinophilic histiocytic cell which, in the skin lesions, occurs in focal collections in the papillary dermis. The nuclei are often bean-shaped and the cells look histologically benign.

Unlike macrophages, LCH cells do not stain for non-specific esterase or for α-1-antichymotrypsin (Favara & Jaffe 1987). Using criteria defined by the Histiocyte Society (Chu et al 1987b), a 'diagnosis' is attained in the context of a characteristic lesion when the LCH cells have at least two of the following: positive stains for S100 protein (Fig. 11.9) (Wanatabe et al 1983), ATP-ase (Beckstead et al 1984), intense α-D-mannosidase

activity (Elleder et al 1977) and dense cell surface and paranuclear staining with peanut agglutinin (Fig. 11.10) (Jaffe 1984). A 'definite diagnosis' establishes the presence of Birbeck granules on electron microscopy (Basset & Turiaf 1965) or expression of the CD1 antigen (Chollet et al 1982), in the context of a histopathological picture consistent with LCH. It is important to state here that normal Langerhans cells may be present in high numbers in many benign dermatoses and if we relied totally on markers of Langerhans cells to establish a diagnosis of LCH, we would misdiagnose a large number of conditions incorrectly as LCH. The markers are essential in confirming a diagnosis of LCH, but can only do this if the histopathology is consistent with this diagnosis.

REFERENCES

Basset F, Turiaf J 1965 Identification par la microscopie electronique de particuler de nature probablement virale dans les lesions granulomateuses d'une histiocytose X pulmonaire. Comptes rendus hebdomadaires des séances de l'Académie des Sciences 261: 3701

Beckstead J H, Wood G S, Turner R R 1984 Histiocytosis X cells and Langerhans cells: Enzyme histochemical and immunologic similarities. Human Pathology 15: 826

Berger C L, Morrison S, Chu A et al 1982 Diagnosis of cutaneous T cell lymphoma by use of monoclonal antibodies reactive with tumour-associated antigens. Journal of Clinical Investigation 70: 1205

Braathen L R, Thorsby E 1980 Studies of human epidermal Langerhans cells. 1. Allo-activating and antigen presenting capacity. Scandinavian Journal of Immunology 11: 401

Chollet S, Duornovo P, Richard M S et al 1982 Reactivity of histiocytosis X cells with monoclonal anti-T6 antibody. New England Journal of Medicine 307: 685

Chu A C, MacDonald D M 1979 Identification in situ of T lymphocytes in dermal and epidermal infiltrates of mycosis fungoides. British Journal of Dermatology 100: 177

Chu A C, Morgan E, MacDonald D M 1981 Evaluation of acid alpha naphthylacetate esterase activity as a marker of T lymphocytes in benign and neoplastic cutaneous infiltrates. British Journal of Dermatology 104: 31

Chu A C, Berger C L, Patterson J 1982 Thymopoetin-like substance in human skin. Lancet ii: 766

Chu A C, Berger C, Morris J, Edelson R 1987a Induction of an immature T-cell phenotype in malignant helper T cells by cocultivation with epidermal cell cultures. Journal of Investigative Dermatology 89: 358

Chu A C, Patterson J, Berger C, Vonderhied E, Edelson R 1984 In situ study of T cell subpopulations in cutaneous T cell — lymphoma diagnostic criteria. Cancer 54: 2414

Chu T, D'Angio G J, Favara B, Ladisch S, Nesbit M, Pritchard J 1987b Histiocytosis syndromes in children. Lancet i: 208

Cordell J L, Falini B, Erber W N et al 1984 Immunoenzymic staining of monoclonal antibodies using immune complexes of alkaline phosphatase and monoclonal anti-alkaline phosphatase (APAAP complexes). Journal of Histochemistry & Cytochemistry 32: 219

Diaz C A, Calvanico N J, Tomasi T B, Jordan R E 1977 Bullous pemphigoid antigen: isolation from normal human skin. Journal of Immunology 18: 455

Elleder M, Povysil C, Rozkovscova J, Cihula J 1977 α-D-mannosidase activity in histiocytosis X. Virchows Archiv B Cell Pathology 26: 139–143

Farb E M, Dykes R, Lazarus G S 1978 Anti-epidermal cell surface pemphigus antibody detaches viable epidermal cells from culture plates by activation of proteinaze. Proceedings of the National Academy of Sciences of the USA 75: 459

Favara B E, Jaffe R 1987 Pathology of Langerhans cell histiocytosis. In: Osband M E, Pochedly C (eds) Histiocytosis X. Hematology/Oncology Clinics of North America. W B Saunders, Philadelphia, p 75

Flaxman B A, Zelazny G, Van Scott E J 1971 Non-specificity of characteristic cells in mycosis fungoides. Archives of Dermatology 104: 141

Hanou D, Gothelf Y, Fabre M et al 1986 Internalisation of T6 (CD1) antigen in a subset of human cord blood mononuclear cells expressing T6 surface antigen. Journal of Investigative Dermatology 87: 143

Harrington C I, Bleehan S S 1979 Herpes gestationis: immunopathological and ultrastructural studies. British Journal of Dermatology 100: 389

Heagerty A H M, Kennedy A R, Eady RAJ et al 1986a GB3 monoclonal antibody for diagnosis of junctional epidermolysis bullosa. Lancet i: 860

Heagerty A H M, Kennedy A R, Gunner D B, Eady RAJ 1986b Rapid prenatal diagnosis and exclusion of epidermolysis bullosa using novel antibody probes. Journal of Investigative Dermatology 86: 603–605

Hodge L, Marsden R A, Black M M, Bhogal B, Corbett M F 1981 Bullous pemphigoid: the frequency of mucosal involvement and concurrent malignancy related to indirect immunofluorescence findings. British Journal of Dermatology 105: 65

Jaffe R 1984 Case 5. Histiocytosis X involving the skin. Pediatric Pathology 2: 329

Kolodny R C 1969 Herpes gestationis: a new assessment of incidence, diagnosis and fetal prognosis. American Journal of Obstetrics & Gynaecology 104: 39

Kung P C, Berger C L, Goldstein G, Logerfo P, Edelson R 1980 Cutaneous T cell lymphoma. Characterisation by monoclonal antibodies. Blood 57: 261

Lampert I A, Janossy G, Suitters A J 1982 Immunological analysis of the skin in graft-versus-host disease. Clinical and Experimental Immunology 50: 123

Lerner K G, Kao G F, Storb R, Buckner C D, Clift R A, Thomas E D 1974 Histopathology of graft-vs-host reaction (GVHR) in human recipients of marrow from HL-A matched sibling donors. Transplantation Proceedings 6: 367–371

Luger T A, Stadler B M, Oppenheim J J et al 1981 A thymocyte activity factor produced by a murine keratinocyte cell line. Journal of Immunology 127: 1493

Lutzner M A, Jordan H W 1968 The ultrastructure of an abnormal cell in Sezary's syndrome. Blood 31: 719

McMillan M, Humphrey G B, Stoneking L et al 1986 Analysis of histiocytosis X infiltrates with monoclonal antibodies directed against cells of histiocytic, lymphoid and myeloid lineage. Clinical Immunology & Immunopathology 38: 395

Nicholas J F, Darlenne M, Faure M et al 1985 Production of a lymphocyte differentiation factor (ELDIF) by cultured human epidermal cells. Immunology 9: 6568

O'Loughlin S, Goldman G C, Provost T T 1978 Fate of pemphiguus antibodies following successful therapy. Archives of Dermatology 114: 1769

Perreault D, Pelletier M, Landry D, Gyger M 1984 Study of Langerhans cells after allogeneic bone marrow transplantation. Blood 63: 807

Reinherz E L, Kung P C, Goldstein G, Levey R, Schlossman S 1980 Discrete stages of human intrathymic differentiation: Analysis of normal thymocytes and leukaemic lymphoblasts of T cell lineage. Proceedings of the National Academy of Sciences of the USA 77: 1588

Rubenfeld M R, Silverstone A E, Knowles D M et al 1981 Induction of lymphocyte differentiation by epidermal cultures. Journal of Investigative Dermatology 77: 221

Saurat J H 1981 Cutaneous manifestations of graft-versus-host disease. International Journal of Dermatology 20: 249

Schlitz J R, Michael B 1976 Production of epidermal acanthosis in normal human skin in vitro by the IgG fraction from pemphigus serum. Journal of Investigative Dermatology 67: 254

Susi F R, Shiklar G 1971 Histochemistry and fine structure of oral lesions of mucous membrane pemphigoid. Archives of Dermatology 104: 244

Tsoi M-S, Storb R, Jones E et al 1978 Deposition of IgM and complement at the dermoepidermal junction in acute and chronic cutaneous graft-vs-host disease in man. Journal of Immunology 120: 1485

Wanatabe S, Nakajima T, Shimasato Y, Sato Y, Shimizv K 1983 Malignant histiocytosis and Letterer-Siwe disease. Neoplasm of T zone histiocyte with S100 protein. Cancer 51: 1412

Weinstein W M 1974 Latent celiac sprue. Gastroenterology 66: 489

Wintrobe B G, Soter N A, Mihm M C, Goetzel E, Austin K 1977 Functional and morphologic evidence of mast cell involvement in bullous pemphigoid. Clinical Research 25: 288

Wolff K, Schreiner E 1971 Ultrastructural localisation of pemphigus autoantibodies within the epidermis. Nature 229: 59

Yaoita H, Katz S I 1976 Immunoelectron microscopic localisation of IgA in skin of patients with dermatitis herpetiformis. Journal of Investigative Dermatology 67: 502

12. Gastrointestinal carcinoma and precursor lesions

M. Isabel Filipe

INTRODUCTION

Gastric and colorectal carcinomas are one of the major causes of cancer deaths in the world. In the advanced stage the prognosis is poor, with a 5-year survival rate after surgery of approximately 10% for both sites. In contrast, if detected at an early stage the prognosis is excellent with a 5-year survival rate, for both sites, of greater than 90% (Nagayo 1986, Morson et al 1977).

The fundamental problems facing the pathologist in everyday diagnosis are:

1. Definition of premalignant lesions; discrimination between reactive and neoplastic.
2. Assessment of malignant potential; the risk of a lesion to evolve to carcinoma.
3. Assessment of prognosis; the potential of a tumour to metastasize.
4. Identification of the primary site of metastatic carcinoma.

In answering these questions lies the ability to select patients at risk, establish the management of these patients in terms of follow-up, predict the aggressiveness of a tumour and devise therapeutic strategies. There is no definite answer as yet.

A number of markers have been developed and tested as potentially useful adjuncts to morphology for the establishment of clinical-pathological criteria or as independent indicators of malignant transformation and clinical course. The better known, and more recent and promising areas include:

Mucins   Antigens
Lectins   DNA
Oncogenes and markers of proliferative activity.

All investigations can be applied to paraffin sections; for methods see Appendices; for abbreviations of lectins see Chapter 2.

OESOPHAGUS AND STOMACH

Barrett's oesophagus

This condition predisposes to the development of oesophageal adenocarcinoma and is generally considered as a premalignant lesion. There is still much debate about the magnitude of the cancer risk in these patients. The prevalence reported in the literature varies between 0% and 47% (average 10%) and the incidence is calculated at 30- to 40-fold. One study found a 125-fold increase compared to the general population (Hamilton 1989, Hammeeteman et al 1989). The prognosis of advanced oesophageal carcinoma is extremely poor. Follow-up surveillance with biopsy and search for histological and histochemical indicators of cancer risk are mandatory.

The lining epithelium in Barrett's oesophagus is complex and three types are identified: fundic, cardiac and intestinal (specialized columnar type), either singly or in combination (Paull et al 1976), and this is reflected in the mixture of neutral, sialo- and sulphomucins secreted. Intestinal metaplasia is common and the presence of sulphomucins in both columnar and goblet cells is not infrequent. There is evidence, from prospective studies, that dysplasia and carcinoma in Barrett's arise more frequently in intestinal-type epithelium. Its presence in a biopsy should be considered a risk factor and warrant endoscopic surveillance of the patients (Hammeeteman et al

1989). Whether, in Barrett's oesophagus, the presence of sulphated material and, in particular, intestinal metaplasia type III has the selective relationship with malignancy as described in the stomach is uncertain (Rothery et al 1986).

Stomach

The workload of routine gastric pathology consists largely of the diagnosis of non-ulcer chronic gastritis, peptic ulcer and malignancy. The majority of cases pose no diagnostic difficulties. However, there are important clinical implications for the management of the patient in terms of follow-up and therapeutic decisions.

In routine diagnosis the pathologist recognizes different types of chronic gastritis (Correa 1980), intestinal metaplasia (Filipe & Jass 1986), and grades of cellular atypia and dysplasia (Ming et al 1984).

At least three distinct aetiopathological types of chronic gastritis are now recognized (Correa 1980):

1. 'Autoimmume' gastritis involves the body of the stomach and is associated with pernicious anaemia, intestinal metaplasia and an increased risk of malignancy.
2. 'Hypersecretory' antral gastritis is related to duodenal ulcer, and shows no association with intestinal metaplasia or carcinoma.
3. 'Environmental' type involves the incisura angularis and the antrum, can progress to intestinal metaplasia and is associated with the intestinal type of gastric carcinoma and peptic ulcer.

Since the reports on *Campylobacter pylori* by Warren and Marshall in 1983, a growing number of studies have shown association between *C. pylori* and chronic active gastritis, gastric and duodenal ulcers, increasing with age in the 'normal' population (Goodwin et al 1986, Wyatt & Dixon 1988). Whether *C. pylori* is a commensal, an opportunistic, or a primary pathogen is still a matter for debate. The organism is now known as *Helicobacter pylori* (Drumm et al 1990).

This spiral bacterium is readily identified in histological sections using a cresyl violet, Warthin and Starry silver stain, Gimenez or modified Giemsa. The organisms can be eradicated by bismuth compounds (DeNol), with remission of the symptoms and restoration of normal mucosal histology (Rokkas et al 1988).

The majority of gastric carcinomas, particularly the 'intestinal' (Laurén 1965) or 'expansive' type (Ming 1977), which predominates in high-risk populations, are preceded by a prolonged precancerous stage, and there is evidence for a progressive change from chronic atrophic gastritis through intestinal metaplasia (IM), and dysplasia, to carcinoma (Correa 1988).

The questions to be addressed are:

1. How to interpret a biopsy showing chronic atrophic gastritis or intestinal metaplasia, or mild forms of dysplasia? How to differentiate a reactive from a preneoplastic phenotype?
2. How to assess the magnitude of cancer risk in variants of intestinal metaplasia and grades of dysplasia?
3. Diagnosis of primary gastric carcinoma.
4. How to predict the clinical course of carcinoma?

Identification of precursors of gastric carcinoma (GCa)

There is a distinction between precursor and precancerous lesions. Precursor lesions 'precede carcinoma chronologically, but do not inevitably evolve to malignancy'. Precancerous lesions are 'obligate antecedents of invasive carcinoma' (Correa 1982). Chronic atrophic gastritis and intestinal metaplasia are precursors of gastric carcinoma. They can either regress or progress. Adenomas and high-grade dysplasia are precancerous lesions, with well recognized malignant potential and easy to define histologically. They are not common in the stomach and high-grade dysplasia often coexists with carcinoma (Ming et al 1984, Morson et al 1985).

In contrast, intestinal metaplasia is easy to define and its incidence is higher in high-risk as compared with low-risk gastric cancer areas (Filipe & Jass 1986). However, its occurrence in both benign and malignant conditions has limited its potential use as a predictive marker of cancer risk.

Morphology and markers of cell differentiation and cell kinetics have revealed a variety of phenotypic changes in preneoplastic gastric epithelium which may indicate subpopulations with increased cancer risk. These phenotypes have been identified by altered mucin secretion, loss of digestive enzymes, expression of oncofetal and differentiation antigens and oncogenes, and increased proliferative activity (Correa 1988, Filipe 1988).

Intestinal metaplasia is the most common change in the progression from superficial gastritis to chronic atrophic gastritis. As the process evolves, intestinal metaplasia (IM) changes from a mature well differentiated 'small intestinal' type (complete, type I IM) to immature forms mimicking 'colonic' epithelium (incomplete type III IM). Mucin histochemistry identifies the three main types of intestinal metaplasia (Filipe & Jass 1986):

1. Type I (complete) IM; regular architecture, straight crypts lined by mature absorptive cells (non-mucous secreting), goblet cells producing sialomucins and occasional sulphomucins (Fig. 12.1).

2. Type II IM (incomplete); slightly distorted crypts, few or no absorptive cells, columnar cells in variable number and degree of differentiation contain a mixture of neutral and sialomucins but no sulphated material, goblet cells as in type I IM.

3. Type III IM (incomplete); crypts are tortuous, cell immaturity is more marked, columnar mucous cells secrete predominantly sulphomucins, goblet cells are as in types I and II (Fig. 12.2).

In goblet cells sulphomucin may be present regardless of the type of intestinal metaplasia and is not a criterion for type III IM.

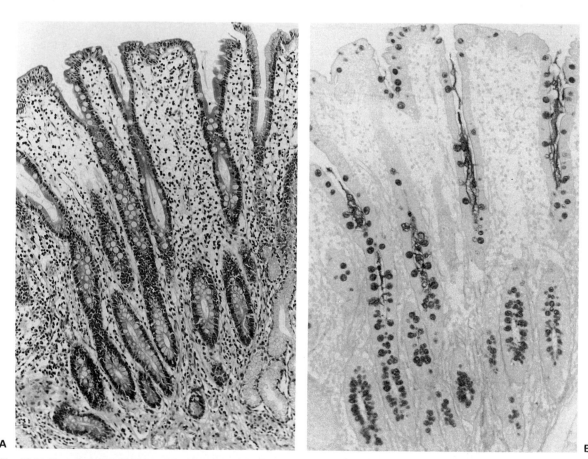

Fig. 12.1 Intestinal metaplasia type I: **A** straight crypts lined by mature absorptive and goblet cells; **B** on HID/AB stain, absorptive cells are non-secreting and goblet cells secrete sialomucins (grey). (**A** H & E, **B** HID/AB)

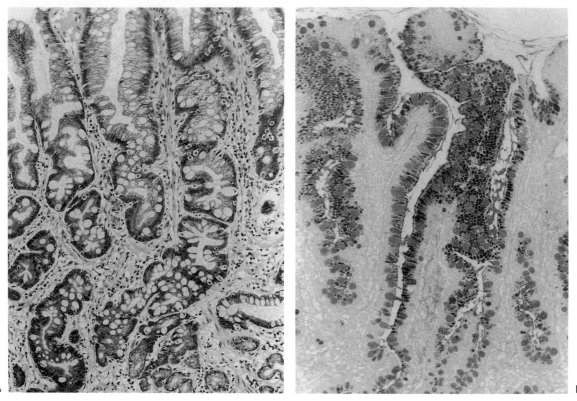

Fig. 12.2 Intestinal metaplasia type III: **A** the crypts are tortuous and lined by columnar mucous cells and interspersed goblet cells; **B** on HID/AB stain, columnar cells secrete predominantly sulphomucins (black) and goblet cells contain a mixture of sialo- (grey) and/or sulphomucins (dark grey/black). (**A** H & E, **B** HID/AB)

Our own data on gastric biopsies shows that type I IM is the most common (approximately 70%) and is the prevalent type in all conditions. Type II IM shows an equal prevalence in chronic gastritis, gastric ulcer and carcinoma. In contrast, type III IM is seen in approximately 10% of the intestinal metaplasia positive biopsies. However, it is seen in only 5% of benign conditions but is present in as many as 36% of carcinomas (Fig. 12.3). It does seem that type III intestinal metaplasia is more common in carcinoma, whilst the relationship with carcinoma is weaker for the non-sulphated types I and II (Filipe et al 1985b, Silva & Filipe 1986). Similar results have been described by others (Rothery & Day 1985, Huang et al 1986).

It is important to note that intestinal metaplasia is patchy and sampling error may account for the low yield of type III IM (36%) in biopsy specimens from cases of gastric carcinoma, compared with a high prevalence (77–80%) in mucosa adjacent to early gastric carcinoma and microcarcinomas described in gastrectomy specimens (Fig. 12.4) (Matsukura et al 1980, Hirota et al 1984). Furthermore, it should be pointed out that intestinal metaplasia and type III IM in particular are more common in 'intestinal' (IGCa) than in the 'diffuse' (DGCa) types of gastric carcinoma (67% and 9% for IGCa and DGCa respectively) (Fig. 12.4) (Matsukura et al 1980, Hirota et al 1984, Silva & Filipe 1986). The prevalence of type III IM in biopsies is higher in areas of greater cancer risk (unpublished). Transitional forms do exist between these main types of intestinal metaplasia suggesting a dynamic process, in which one form may evolve to another or regress (Silva et al 1990).

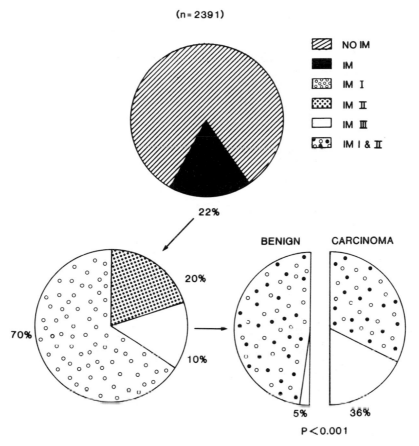

(n = 2391)

NO IM

IM

IM I

IM II

IM III

IM I & II

22%

20%

70%

10%

BENIGN   CARCINOMA

5%   36%

P < 0.001

Fig. 12.3 Intestinal metaplasia phenotypes: incidence and distribution in gastric biopsies. (Reproduced by courtesy of Elsevier Science Publ.)

In normal gastric mucosa sulphomucins may also be detected in the base of antral and pyloric glands, within cysts and in traces in mucous neck cells. Increased secretion of sulphomucin is noted in the neck region of the hyperplastic epithelium adjacent to carcinoma, in dysplastic glands and in a proportion of gastric carcinomas. Higher detectable amounts of sulphated glycoproteins in gastric mucus have been described in gastric carcinoma patients (Kakei et al 1984)

The concept that types of intestinal metaplasia with different morphological and mucin profiles, may have a different risk of gastric malignancy has support from early reports on intestinal marker enzymes and recent studies on antigenic expression, proliferative activity, and DNA analysis.

Enzymes

Heterogeneity of intestinal metaplasia has long been recognized by intestinal enzymes profiles, and classified into 'complete' and 'incomplete' types, which broadly correspond to types I and III referred to above. Complete intestinal metaplasia shows sucrase, trehalase, aminopeptidase (AMP) and alkaline phosphatase (ALP) activities. In incomplete IM, prevalent in intestinal type gastric carcinoma (IGCa) (80%) (Matsukura et al 1980), various activities are reduced or absent. However, these techniques, except for alkaline phosphatase, require frozen tissue which precludes their use in routine diagnosis.

Pepsinogens PGI and II are present in normal gastric epithelium (PGI in chief cells and neck

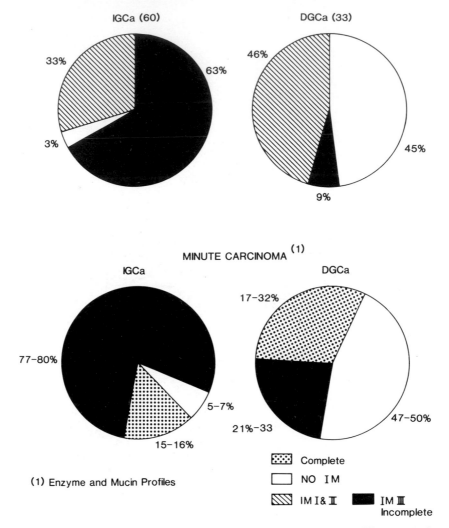

Fig. 12.4 Intestinal metaplasia types in gastrectomy specimens: incidence in intestinal and diffuse types of gastric carcinoma.

cells; PGII abundant in antral and cardiac glands). Their levels in serum can be used to monitor atrophic gastritis. Both pepsinogens are detected by immunoperoxidase techniques in IGCa and DGCa, but are absent in areas of IM either in carcinoma fields or benign conditions. Patchy positive staining, particularly with PGII, may be found in dysplasia (Bushy-Earle et al 1986). Again, these enzymes do not appear to be of use in the diagnosis of premalignancy, but high serum levels of PGII may indicate gastric cancer.

Antigens

Mucus-associated antigens, particularly fetal glycoproteins, can reappear in the course of malignant transformation (Table 12.1). Large intestinal mucin antigen (LIMA), derived from normal adult colonic epithelium, is present in fetal colon but not in normal gastric epithelium (Ma et al 1982a). The antigen is expressed in the columnar cells in type I IM (59%), type II IM (85%) and in all cases of type III IM and dysplasia, in association with gastric carcinoma (Filipe et al 1988a). It may be

Table 12.1 Fetal and carbohydrate related antigens in precursor lesions of gastric carcinoma

	N	SCG	CAG	IM	D	Ca field	CG
SFT	0	10	38	50	86		
CA19–9	0		18	61–100			
LIMA							
b	0	2	21.4	18 64 93	92	84	21.4
g	0	0	—	59 85 100	100		

Note: A gradual increase in antigenic expression from normal (N) through superficial (SCG) and atrophic gastritis (CAG) to grades of intestinal metaplasia (IM I–III) and dysplasia (D). Expression of LIMA is observed in greater number of biopsies with chronic gastritis (CG) from carcinoma compared to non-carcinoma patients (84% and 21.4% respectively).
Key: Values expressed as % of total. b = biopsies, g = gastrectomy specimens, SFT = 2nd trimester fetal antigen, CA19-9 = sialylated Lewis[a] determinant, LIMA = large intestinal mucin antigen.

detected in the histologically 'normal' mucosa adjacent to carcinoma, but rarely in benign conditions, and then only in foci of intestinal metaplasia in chronic atrophic gastritis. Only 6 out of 271 non-metaplastic biopsies exhibited LIMA (2%). These findings are similar to those observed by Bara's group (Nardelli et al 1983) using the intestinal mucin antigens M_3SI, M_3D and M_3C expressed in goblet cells from the adult small intestine, duodenum and colon respectively. Intestinal metaplasia of any type adjacent to GCa expresses M_3C + M_3D + M_3SI, but in benign conditions M_3C is not detected. The reappearance of fetal phenotypes in the gastric epithelium has been further illustrated by the expression of second trimester fetal antigen (SFT), with increasing frequency as the disease progresses from superficial gastritis (10%), to chronic atrophic gastritis (38%), intestinal metaplasia (50%) and dysplasia (86%) (Higgins et al 1984). Similar findings were obtained with blood-group-related antigens, particularly the sialylated Lewis[a] determinant CA19-9, which is present in fetal stomach, absent in the normal adult stomach, but is expressed in gastric carcinoma. This antigen is expressed in antral gastric biopsies related to the severity of the disease, from absent in the normal, to mild atrophy (18%), to grade III atrophy, with intestinal metaplasia (61%) and in all cases of marked atrophy with extensive metaplasia (Sipponen & Lindgren 1986). A good correlation between the fetal antigen YFA and increased proliferative activity has been shown in chronic

atrophic gastritis and intestinal metaplasia (Lipkin et al 1985). BD-5 is a mucus-associated antibody obtained from human gastric carcinoma cell line Kato III, expressed in normal and neoplastic cells of human intestine, but not gastric epithelium. It is detected in intestinal metaplasia of any type and dysplasia arising in areas of metaplasia, and in the majority of GCa (83%), independent of histological type and stage (Fiocca et al 1988). It does not appear to discriminate between types of intestinal metaplasia and is not expressed in all dysplasias.

Data described above support the progressive change from chronic atrophic to intestinal metaplasia, dysplasia and carcinoma (IGCa), and also illustrates an association between certain types of IM and IGCa.

Cell kinetics

In the past, tritiated thymidine uptake studies have demonstrated increased proliferation of epithelial cells (labelling index, LI) related to progression of disease from normal (LI 5.2%), to chronic atrophic gastritis (LI 10.6%) and chronic atrophic gastritis with extensive intestinal metaplasia (LI 20.5%). Furthermore, LI of intestinal metaplasia in patients with gastric carcinoma was much higher than in non-cancer patients (35.8% versus 13%) (Lehnert & Deschner 1986). However, the technique is not applicable to routine diagnosis. The development of a monoclonal antibody to bromodeoxyuridine (BrdU), which is incorporated into DNA in the

S-phase, the possibility of its use in vitro and its demonstration by immunohistochemical methods on routinely processed tissue, opens great possibilities (Wilson et al 1985, Sugihara et al 1986). In the normal antral epithelium, BrdU positive cells are localized in the neck region, extending upwards and downwards in the 'normal' epithelium adjacent to GCa, in the lower crypt in intestinal metaplasia and irregularly distributed in carcinoma (Danova et al 1988) (Fig. 12.5). Data is, however, still scarce and preliminary.

The quantitative evaluation of nucleolar organizer regions (NORs) revealed by a silver stain (AgNORs) on routine paraffin sections (Ploton et al 1986) may help to discriminate 'borderline' lesions (Rosa et al 1990a, Suarez et al 1989). NORs are loops of rDNA transcribing to rRNA and their numbers appear to relate to cell proliferation (Sahl 1982, Rosa et al 1996). Distinguishing between 'borderline' lesions is still essentially subjective and the point at which a particular lesion graduates to the next or regresses to normality remains uncertain. AgNORs could assist accurate definition of these lesions. Our recent data indicates that the mean number of AgNORs per nucleus increases with the degree of cellular atypia from normal to metaplasia and dysplasia, being higher in carcinoma, particularly the intestinal type (Rosa et al 1990a) (Figs. 12.6 and 12.7). Significantly higher AgNORs values are observed in carcinomas compared with other lesions except dysplasia and in dysplasia compared with type I IM, but it did not differ from type III IM. Of interest is the observation that AgNOR values of histologically normal mucosa adjacent to carcinoma are significantly higher than normal controls. Although the method is not practical for diagnostic purposes as it is time consuming, it may be useful as an adjunct to morphological criteria for better definition of precursor lesions of gastric carcinoma. The spread of AgNOR counts in intestinal metaplasia reinforces the concept that this lesion is a heterogenous entity reflecting a dynamic and continuous process.

DNA analysis

Few studies have used DNA ploidy to identify precancerous gastric lesions. The size and focality

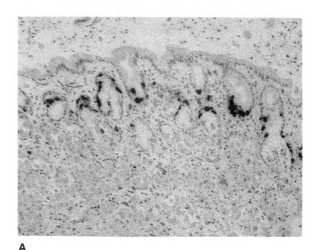

A

B

Fig. 12.5 Bromodeoxyuridine (BrdU) in vitro uptake in gastric mucosa. **A** Normal distribution of BrdU in the proliferative zone in the neck region; **B** intestinal metaplasia adjacent to carcinoma: cells in S-phase are distributed throughout the length of the metaplastic gland, indicating expansion of the proliferative zone. (Immunoperoxidase method; H & E counterstain See Appendix 2)

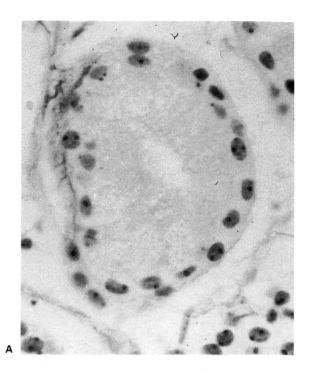

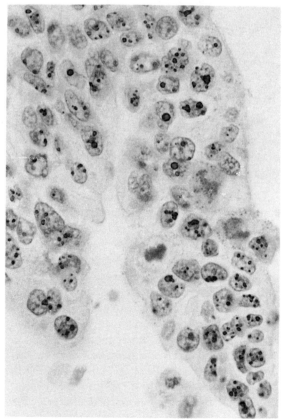

of the lesion is a limiting factor in flow cytometric measurements. The use of 40 μm sections from routinely processed paraffin tissue blocks makes it feasible to isolate the relevant preneoplastic and neoplastic areas by micromanipulation under a dissecting microscope. The selected foci are then processed for flow cytometry and/or the cell suspension used to prepare smears on slides, for fluorometry or Feulgen-stained for static microdensitometry (Quirke & Dyson 1986, Sowa et al 1988). Flow cytometry is quantitative and precise. Microdensitometry is based on DNA patterns of low and high ploidy, and can provide information on the occurrence of polyploid populations which may not be detectable by flow cytometry. Both have been applied in numerous studies of DNA content of preneoplastic lesions and carcinoma of the large bowel (see p. 187). DNA ploidy of gastric carcinomas (discussed below) has been extensively investigated, particularly by Japanese workers. Data on premalignant lesions of the gastric mucosa is scarce. Diploid values do not exclude malignancy as a proportion of GCa are diploid (27–45%) (McCartney et al 1986, Ballantyne et al 1987, Filipe et al 1990). On the other hand, the finding of aneuploid populations indicates malignant transformation, and analysis of the cell cycle may reveal subpopulations with different proliferative patterns, which deviate from normal and signal early malignant change. Aneuploidy had been found among dysplasias (Teodori et al 1984, McCartney et al 1986, Rubio & Kato 1988) and increased proliferative index (PI) were noted in the non-involved mucosa adjacent to aneuploid carcinomas; the PI of intestinal metaplasia was higher in stomachs with aneuploid carcinomas compared with those with diploid tumours (Odegaard et al 1987a,b). In Barrett's oesophagus, aneuploidy and/or increased G2/tetraploid fractions were detected in patients with severe dysplasia with or without associated carcinoma,

Fig. 12.6 Nucleolar organizer regions (NORs) can be identified by a silver staining technique (AgNORs) and shown as argyrophilic intranuclear black dots. **A** In normal gastric epithelium most cells have none, one or two AgNORs per nucleus. **B** In contrast, all cells in intestinal type gastric carcinoma contain multiple dots of AgNORs activity. (Silver staining, Appendix 2)

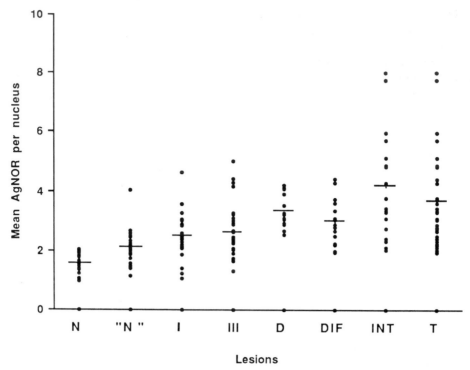

Fig. 12.7 Nucleolar organizer regions in the normal gastric epithelium, gastric carcinoma and its precursor lesions. Key: N = normal, "N" = normal mucosa adjacent to carcinoma, I and III = intestinal metaplasia types I and III, D = dysplasia, DIF and INT = diffuse and intestinal types of gastric carcinoma, T = all carcinomas.
AgNORs were counted in 100 randomly selected cells/lesion using ×100 oil immersion lens to a maximum magnification of ×1000. Each dot = 1 case.

but not in those diagnosed as 'indefinite' for dysplasia according to Riddell's classification (Riddell et al 1983), or metaplasia (Levine et al 1989).

Diagnosis of primary gastric carcinoma

The large majority of gastric malignancies are adenocarcinomas and the diagnosis can often be made with certainty on H & E staining. The most common areas of diagnostic difficulty are:

1. Severe dysplasia versus differentiated early gastric carcinoma (EGCa) of tubular pattern.

2. Dissociated malignant cells of undifferentiated and signet-ring type versus non-neoplastic mononuclear cells, especially macrophages in the lamina propria.

3. Overdiagnosis in relation to ulceration or erosion.

4. Undifferentiated carcinomas need to be distinguished from lymphomas, neuroendocrine tumours and, though uncommon, metastatic tumours need to be considered.

Cytological atypia and architectural pattern are similar in severe dysplasia and EGCa, but it is invasion of the lamina propria that is diagnostic of early carcinoma. Multiple sections may help and immunostaining with CAM5.2 may reveal budding or small collections of malignant epithelial cells into the lamina propria. It is not possible to diagnose EGCa from biopsy material because invasion beyond the submucosa can only be excluded by examination of a resected specimen.

The dissociated malignant cells in the lamina propria can be identified by diastase-PAS/AB pH 2.5 mucin staining, but additional criteria are needed. D-PAS alone is of little help as macrophages are positive and some carcinomas are negative for neutral mucin. The presence of intracellular AB pH 2.5 positive mucin globules is very suggestive of carcinoma. Nuclear pleomorphism and the presence of mitosis allows a certain diagnosis of malignancy to be made.

Distorted glands and cellular atypia in areas of ulceration and granulation tissue, with bizarre endothelial cells, histiocytes and fibroblasts, raises difficulty in the interpretation, which can be facilitated by immunostaining with CAM5.2 and with other markers such as factor VIII antigen, or UEA1 to identify the vessels and endothelial cells.

Poorly differentiated carcinomas can be differentiated from lymphomas on H & E staining (see Ch. 19). In small fragmented biopsies the interpretation is more difficult and the accuracy of diagnosis can, in most cases, only be achieved by immunohistochemical use of a panel of monoclonal antibodies, CAM5.2, CEA, and LCA (leucocyte common antigen).

In our department chromogranin is used as a first step in the identification or confirmation of a suspected neuroendocrine tumour in the stomach, followed by further characterization with the appropriate neuroendocrine markers (see Ch. 29).

The identification of the site of origin of a metastatic carcinoma in lymph nodes or distant organs is of the utmost importance. The approach to the problem should be stepwise and in most cases requires a combination of markers. There is no specific marker for gastric carcinoma, but an origin from the gastrointestinal tract, stomach, colon or pancreas can be strongly suggested. The panel of markers currently used in the diagnosis of metastatic gastric and colonic carcinoma is discussed at the end of this chapter (see also Pancreas, Ch. 27).

Prognostic indicators

The most accurate prognostic indicator in gastric carcinoma is stage. Detection of early stage gastric carcinoma (EGCa) offers an excellent prognosis. In large series from Japan, survival rate of patients with mucosal and submucosal carcinoma were 97.6% and 91.8% respectively at 5 years after surgery, and 87.3% and 75% respectively at 15 years after surgery (Nagayo 1986). The concept of EGCa is defined as an early stage in the evolution of all gastric carcinomas when the tumour is confined to the mucosa (intramucosal) and submucosa (submucosal) and the disease is potentially curable by surgical resection regardless of the presence of lymph node metastases. The behaviour of EGCa may be predicted from their growth patterns and classified in three main types: 'small mucosal', the superficial spreading 'super' and the penetrating type 'PEN'. A subtype 'PENA' with an expanded growth and destruction of the muscularis mucosa, though the least frequent (10%), is associated with worse prognosis (Kodama et al 1983). DNA analysis confirmed that 86% of PENA tumours have a high ploidy value compared with the other types, suggesting an aggressive behaviour (Inokuchi et al 1983). It does seem that EGCa falls into two broad categories: a slow-growing lesion which extends widely in the mucosa, for a long period, before invading the wall; and a more rapidly progressive tumour that is often small but which invades deeply at an early stage. If these observations are confirmed, then particular attention should be directed to early diagnosis of the latter type. Histochemical/immunohistochemical techniques are of no help at present. Other morphological parameters used to assess prognosis, such as histological type (intestinal versus diffuse) (Laurén 1965), grade (poorly versus well differentiated), and growth (expanded versus infiltrating) (Ming 1977), do not seem to hold when DNA ploidy values are considered. In EGCa, DNA ploidy seems to be an independent predictor of recurrence, metastatic potential and survival (Inokuchi et al 1983, Kamegawa et al 1986), including studies in minute (< 5 mm) carcinomas and biopsies (Korenaga et al 1986a). The DNA values in biopsies were shown to correlate with those detected in the subsequently resected tumour (Korenaga et al 1986b). If these observations are confirmed, DNA analysis could be an important investigation in preoperative biopsies in determining the management of the patient.

In advanced gastric carcinoma, the relationship between ploidy, the morphological parameters referred to above, and survival, differs between groups of workers and its value as an independent prognostic factor has not been proved (McCartney et al 1986, Ballantyne et al 1987, Filipe et al 1990). In our series of 122 patients with gastric carcinoma, tumour stage was the only prognostic factor. Neither DNA nor % S-phase values had any effect on survival. However patients with higher DNA index tumours showed a tendency to worse survival (Filipe et al 1990). However in a recent study of 226 patients with advanced GCa DNA ploidy using microspectrophotometry proved to be a major independant prognostic factor (Korenaga et al 1988).

There are important aspects to be considered which account for the different results obtained in various studies: a) the methodology (flow cytometry versus microdensitometry); b) formalin fixation of tissues, both the interval before fixation and duration of fixation; and c) tissue sampling to minimize intratumoral differences, which can occur in as much as 40% of gastric carcinomas (in contrast to 7.4% of colorectal carcinomas) (Sasaki et al 1988), and analysis of lymph nodes, which in a proportion of cases show different ploidy values from the tumour of origin. Our studies confirm that results can be greatly affected by these variables.

LARGE BOWEL

Normal

Mucins

Goblet cells in the normal adult colorectal epithelium secrete neutral and acid mucins, both sulphated and non-sulphated (sialomucins). Acid mucins with high sulphate content are predominant and both N-and O-acylated sialic acids including mono-, di- and tri-O-acyl derivatives are represented. Goblet cell mucin composition varies in the level of the crypts and in different segments of the colon (Filipe & Branfoot 1976, Filipe 1989).

Deviations from the 'normal mucus pattern' have been described in colonic epithelium in a variety of pathological conditions. This abnormal secretion is characterized histochemically by higher sialomucin/sulphomucin content, increased N-/O-acylated sialic acid ratios and altered proportion of various sugars (Filipe 1989). The suggestion that these qualitative and quantitative mucin changes could be related to malignant risk (Filipe & Branfoot 1976, Filipe 1989), have not been supported by others (Williams 1985).

Lectins

In the normal colonic epithelium, the lectin-binding profiles of the goblet cell mucin also vary both within the crypt and in different regions of the colon (Jacobs & Huber 1985, Bresalier et al 1985). A battery of lectins has been tested in the normal colonic epithelium and binding sites for DBA, WGA, SBA, LCA, ConA, RCA and UEA1 have been detected. No binding site for PNA (can be present in the golgi region), VVA and GSA-II have been found (Filipe 1989). Following neuraminidase digestion, a positive reaction with PNA and VVA is often seen in the goblet cell mucin. There is a decreasing gradient of frequency and intensity of binding of UEA1 and RCA1 from right to left colon, particularly marked for UEA1 binding, which is greatly reduced or absent in the left colon and rectum. Of the most commonly investigated, only PNA and UEA1 appear to provide potential useful diagnostic information (Filipe 1989).

Indicators of malignant potential in precancerous lesions

Adenomas

Adenomas are important precursors of colorectal carcinoma and yet only a minority will progress to invasive carcinoma (Muto et al 1975a). The criteria for predicting risk of malignancy of an individual adenoma is based on growth pattern (villous or tubular), degree of epithelial dysplasia (low or high grade) and size, the worse prognosis being associated with the larger, villous adenomas showing areas of high-grade dysplasia.

These are useful criteria but are insufficient, and other markers are necessary as indicators of malignant potential.

Table 12.2   O-acylated sialomucin content (PB/KOH/PAS effect) in adenomas, carcinomas and their adjacent 'transitional' mucosa

Staining intensity	TR	Adenomas				Carcinomas
		I	II	III	IV	
3+	34	10	25	17	0	4
2+	23	2	22	9	0	1
+/0	2	2	1	11	11	29
Totals	59	14	48	27	11	34

TR = transitional mucosa, I–III = degrees of dysplasia — mild, moderate and severe, IV = carcinoma in situ.

Adenomas secrete a mixture of neutral, sialo- and sulphomucins, with no clear relationship between the broader types of mucin and the morphological prognostic parameters (Filipe & Branfoot 1976, Greaves et al 1984, Lapertosa et al 1988, Filipe 1989). However, O-acylated sialomucins, present in normal colonic epithelium and adenomas, are markedly reduced or absent in areas of high-grade dysplasia and in villous adenomas, and more common in synchronous lesions in carcinoma patients (Table 12.2) (Greaves et al 1984, Muto et al 1985b). This feature is not seen in inflammatory polyps, but metaplastic polyps with increased size may reveal reduced O-acylation (Jass et al 1984). The sharp loss of O-acylation and the significant changes in the normal ratios of the various O-acylated sialic acids, particularly the tri-O-acylated variants described in association with malignancy (Rogers et al 1978, Greaves et al 1984, Muto et al 1985b, Hutchins et al 1988) could be of help in assessing malignant risk.

Lectin binding with PNA is detected in adenomas, increasing in frequency and intensity with the severity of dysplasia and the size of the polyp. Affinity for PNA is also more frequent in villous and synchronous adenomas, than in tubular and incidental lesions. A shift of PNA binding sites from goblet cell mucin to the apical cytoplasm and glycocalyx of the columnar cells accompanies the degree of dysplasia. The expression of PNA, however, does not provide additional useful prognostic information (Cooper & Reuter 1983, Filipe 1989). PNA binding is also found in metaplastic polyps and in the 'transitional' mucosa adjacent to carcinoma (Filipe 1989). No further information has been gained from other lectin-binding profiles tested so far (Campo et al 1988).

Enhanced expression of EMA, and CEA in particular appears to be related to the severity of dysplasia but shows no significant increase with onset of malignancy. The variations in staining intensity and patterns for both EMA and CEA, in neoplastic and non-neoplastic lesions, for instance metaplastic polyps and 'Transitional' mucosa, confirm their lack of specificity and sensitivity required for early detection of malignancy at tissue level (Rognum et al 1982b, Greaves et al 1984, Lapertosa et al 1988, Zamcheck et al 1988).

Changes in both intensity of staining and intracellular distribution of IgA and SC (secretory component) appear to be related to the degree of cellular atypia, being more pronounced in severe dysplastic areas within adenomas, apparently associated with loss of mucus secretion and may reflect functional immaturity. Loss of secretory component (SC) has been suggested as a diagnostic aid for malignant change, but this has not been confirmed (Isaacson 1982, Rognum et al 1982c, Lapertosa et al 1988).

DNA ploidy has been reported in a percentage of adenomas (12–35%), but whether aneuploidy is related to the grade of dysplasia is inconclusive (Goh & Jass 1986, Quirke et al 1988). In some adenomas with high-grade dysplasia, aneuploidy and high ploidy profiles similar to those observed in carcinoma have been described and may signal 'malignant transformation', as compared with other diploid adenomas (Hamada et al 1988). Of interest is the finding of a higher percentage of DNA aneuploidy in adenomas from patients with a family history of colorectal carcinoma compared to those with negative family history (78% and

20% respectively) (Sciallero et al 1988). In familial polyposis coli aneuploidy can be seen in small-sized adenomas (Quirke et al 1988).

DNA ploidy in adenomas may be an important independent indicator of malignant potential and may identify patients at risk of developing carcinoma, though its value is still debatable.

Dysplasia in ulcerative colitis

Patients with ulcerative colitis have an increased risk of developing carcinoma. The incidence of colorectal carcinoma in patients with long-standing and extensive disease is the subject of some debate, the most recent study calculated a cumulative risk of 7.2% at 20 years and 16.5% at 30 years (Gyde et al 1988). Screening for the development of malignancy involves continued follow-up with biopsies. The criteria for risk are clinical (extent of disease and duration) and morphological (the presence of dysplasia). Reliance on dysplasia may not, in a few cases, predict the development of carcinoma (Filipe et al 1985a, Thomas et al 1989). On the other hand, most patients with UC do not develop colorectal carcinoma. The management of these patients in terms of endoscopic surveillance and surgery depends on the assessment of malignant risk. Dysplasia is far from ideal as a 'marker' of increased cancer risk: it is patchy, difficult to interpret (Riddell et al 1983, Dixon et al 1988) and may be absent in the presence of carcinoma (Filipe et al 1985a, Thomas et al 1989). Thus additional functional criteria have been sought.

As mentioned above, quantitative and qualitative mucous secretion have been described in association with colorectal malignancy (Filipe & Branfoot 1976, Filipe 1989). Our findings indicate that the predominance of sialomucins and the loss of O-acylated sialic acids appear to be related to dysplasia and malignant risk (in the absence of dysplasia) in patients with ulcerative colitis independent of inflammation (Ehsanullah et al 1982, Ehsanullah et al 1985). Similar changes have been found in the absence of dysplasia in biopsies preceding the development of carcinoma (Filipe et al 1985a). These observations have gained support from various groups (Fozard et al 1987a, Allen et al 1988, Agawa et al 1988), but others deny its value (Jass et al 1986b).

Abnormal lectin binding patterns, particularly with PNA, are found in ulcerative colitis, and related to dysplasia and long-standing disease (Pihl et al 1985, Ahren et al 1987, Fozard et al 1987a, Cooper et al 1989, Boland et al 1984). The value of a positive staining for PNA in predicting the onset of malignant change has not been confirmed (Jass et al 1986b). However, it is interesting to note that increasing frequency of PNA binding is seen in non-involved, non-inflamed mucosa in patients with ulcerative colitis associated with carcinoma (see below). A similar pattern of binding is observed for UEA1 (Yonezawa et al 1982).

Other markers including IgA and SC and CEA (Rognum et al 1982a, Allen et al 1985) show variable results, and have not been successful in defining increased cancer risk in the individual patient.

Abnormal expression of mucous associated antigens (not available commercially) have also been described (Filipe et al 1988b), but their diagnostic use in assessing malignant risk needs to be further tested. The same problem occurs with other putative markers.

Another approach in assessing malignant risk in ulcerative colitis is the quantitative estimation of cellular DNA content by either flow cytometry (Hammarberg et al 1984, Fozard et al 1987b, Lofberg et al 1987) or microdensitometry on Feulgen-stained preparations (Cuvelier et al 1987). Data from various groups differ. In some studies DNA aneuploidy was more commonly found in long-standing disease, irrespective of the presence or absence of carcinoma (67% and 42% aneuploidy in carcinoma and non-carcinoma patients respectively), not related to dysplasia (21% and 15% aneuploidy in dysplasia and non-dysplasia respectively), and absent in over 50% of non-cancerous samples from carcinoma patients (Hammarberg et al 1984). Others have found an increased prevalence of DNA aneuploidy in dysplasia and also in a small proportion of histologically normal or inflamed biopsies (Lofberg et al 1987, Melville et al 1988).

Whether DNA aneuploidy signals an early

preneoplastic change and is a useful independent indicator of cancer risk is still debatable.

The non-neoplastic mucosa in carcinoma patients

Carcinogenesis is a stepwise process from normal through dysplasia to carcinoma, thus the identification of abnormal phenotypes in morphologically non-neoplastic mucosa within a putative carcinoma field may indicate increased individual susceptibility to malignant change.

The questions to be addressed are: is the normal mucosa from which dysplasia/adenoma arise normal? Is there an environment in which colorectal carcinomas are more likely to develop?

Increased frequency of abnormal patterns of mucin secretion, of lectin binding and antigenic expression, similar to those described in carcinomas and dysplasia/adenoma have been found in:

1. The so-called transitional mucosa (TM) adjacent to colorectal carcinoma and in foci at a distance from the tumour (Filipe & Branfoot 1974, Lapertosa et al 1984, Filipe 1989).

2. Between polyps in familial polyposis coli (Filipe et al 1980, Muto et al 1985b).

3. Ulcerative colitis complicated with carcinoma (Filipe et al 1985, Fozard et al 1987a, Hanby et al 1988).

The mucin changes are essentially lower neutral/sialomucin, higher sialo/sulphomucin and lower O/N-acylated sialic acids ratios (Dawson et al 1978, Rogers et al 1978, Greaves et al 1984, Reid et al 1985, Sugihara & Jass 1986, Hutchins et al 1988). Levels of tri-O-acylated sialic acids have been shown to decrease in relation to malignancy: 31.7, 1.8 and 0.6 in normal epithelium, 'normal' areas in carcinoma patients and carcinoma respectively (Hutchins et al 1988). This has been suggested to be related with a genetic defect (Sugihara & Jass 1986, Hutchins et al 1988).

Mucus-associated antigens expressed in carcinoma but not in normal mucosa, such as M1 antigen (Bara et al 1984) and SIMA (Ma et al 1982b) show strong positive immunostaining in 'transitional' mucosa (Ma et al 1982b, Bara et al 1984) and ulcerative colitis (Filipe et al 1988b). Blood group antigens of ABH type, present in the fetal colon, lost in the adult and re-expressed in malignancy, in particular the extended 'Le^y' antigens (KH-1 and CC-2), are detected in 'transitional' mucosa in a pattern closely resembling that of the adjacent tumour (Kim et al 1986, Yuan et al 1987). SSEA-1, a carbohydrate differentiation antigen [Gal-β-1 $\rightarrow$ 4 (Fuc-1 $\rightarrow$ 3) GlcNAc] defined by a monoclonal antibody raised against F9 murine embryonal carcinoma cells (Shi et al 1984), is not found in normal colonic epithelium but is expressed in colorectal malignancy. It has been detected with increasing frequency in the non-involved mucosa in familial polyposis coli and 'transitional' mucosa (Hara et al 1987).

Indicators of prognosis in carcinoma

The most important prognostic factor in colorectal carcinoma is the clinical and pathological Dukes' staging (Dukes 1932). Additional features related to prognosis include degree of tumour differentiation, growth pattern and lymphocytic infiltrate (Jass et al 1986a). However, these classifications have limitations: a) they are based an postoperative findings, so preoperative decisions on biopsy cannot be taken; b) they fail to define the tumour potential for growth; c) they fail to provide stage-independent information to predict recurrence and therapeutic-resistant tumour, features which will determine the management of the patient in terms of conservative versus radical surgery and adjuvant therapy.

A number of markers have been used, in addition to clinical morphological criteria, to increase the accuracy of prognostic assessment of colorectal carcinoma.

The better tested are CEA, CA 19-9, IgA and SC (Rognum et al 1982b,c, Arends et al 1984a,b, Wiggers et al 1986, Quentineier et al 1987, Lapertosa et al 1988, Wiggers et al 1988) and mucin patterns (Lapertosa et al 1988), but none provide a definite independent answer, though some can contribute to prognosis together with tumour grading and staging. Tumours which are negative for CEA and SC, are positive for CA 19-9 and serotonin, secrete predominantly sialomucin with loss of O-acylation and are aneuploid appear to have worse prognosis (Wiggers et al 1988). However, these observations may be biased as the majority of antigens are cell differentiation markers and are lost in poorly differentiated tumours.

The measurement of DNA ploidy may add important prognostic information to that obtained by staging. It has the advantages of being precise, rapid and quantitative as opposed to the subjective assessment of staining reactions, and can be available in preoperative biopsies. Abnormal DNA content (aneuploidy) has been reported in colorectal carcinoma in 39–82% of the patients. Most studies reported a significant or borderline relationship between poor survival and/or shorter disease-free interval in patients with aneuploid carcinomas compared with diploid tumours (Goh & Jass 1986, Enblad et al 1987, Schutte et al 1987). This has not been confirmed by others (Melamed et al 1985, Finan et al 1986). In most studies DNA ploidy was not related to stage, but whether it is an independent prognostic indicator is debatable (Tribukait et al 1983, Rognum et al 1987, Quirke et al 1987, Scott et al 1987). However, subgroups in advanced Dukes' C cases have been identified with distinct survival rates related to DNA ploidy (Rognum et al 1987). Local recurrence and vascular invasion appear to be more common in aneuploid tumours (Scott et al 1987).

Estimation of cellular DNA content, despite uncertainty regarding its independent status in predicting the clinical course of the disease, appears to be particularly useful in advanced cases when decisions on management are critical for the patient's survival.

GASTRIC AND COLONIC NEOPLASIA

Identification of origin of metastatic carcinoma

The approach to this problem is stepwise:

1. Identification of tissue of origin (in undifferentiated tumours): epithelial, lymphoid, soft tissue, neuroendocrine.

2. If epithelial, whether squamous or glandular origin.

3. If adenocarcinoma — which organ?

For steps 1 and 2 specific markers exist, and in the majority of cases diagnostic accuracy can be achieved with a panel of antibodies which identify the various categories of tumours. The markers we use as 1st step are CAM5.2, and LCA (leucocyte common antigen), followed, if appropriate, by vimentin, S-100 protein and chromogranin (Fig. 12.8), (see Chs. 4, 7, 19 and 29).

Diagnostic accuracy in step 3 is more difficult. With a few exceptions, no single marker specific for a certain type of tumour exists.

In recent years a large volume of publications has appeared on newly isolated antigens expressed in the GI tract, but mostly their value in routine diagnosis has not been fully investigated. Some mucin-associated antigens, such as M3C (Nardelli et al 1983), LIMA (Filipe et al 1988a) and BD-5 (Fiocca et al 1988) are promising but are not commercially available.

CEA remains the best marker for GI tract tumours. A negative reaction for CEA almost always rules out gastrointestinal malignancy, though poorly differentiated tumours can be negative (Arends et al 1984b, Quentineier et at 1987, Wiggers et al 1988). Conversely, a CEA-positive reaction strongly suggests GI tract origin of the tumour. Other malignant tumours from pancreas, lung, ovary and endometrium may show variable CEA staining. This limitation has been reduced with the isolation of monoclonal antibodies which recognize specific epitopes on the CEA antigen of colonic adenocarcinomas. A very recent addition is a monoclonal antibody, RWP1.1, produced against human pancreatic carcinoma cell line RWP1, which recognizes a unique epitope present on the CEA molecule. RWP1.1 has a very restricted specificity for primary and metastatic adenocarcinomas of the stomach, colon and pancreas (and colonic adenomas) but does not react with most adenocarcinomas from other sites (1/10 for breast and 1/6 for ovary), or other types of tumours. It does not react with any type of normal tissue (Kahn et at 1989).

For a correct diagnosis, clinical and morphological data have to be considered to narrow the likely sites of origin of the tumour, in addition to a selective use of a panel of monoclonal antibodies described in the appropriate chapters in this book.

FUTURE PROSPECTS: ONCOGENES

Recent developments in molecular biology, enabling the detection of subtle genetic abnormalities in transforming cells, have opened a new exciting era in the understanding of carcinogenesis.

The change from a normal cell to a cancer cell is a multistep process in which two main groups of genes, oncogenes and onco-suppressor genes, are implicated by their role in the regulation of cell growth and differentiation.

Oncogenes arise as an altered form of a normal cellular gene (the proto-oncogene), which can occur by a) transduction, b) insertional mutagenesis, c) chromosome translocation, d) mutation, e) amplification.

The oncogenes most investigated in human carcinomas are: a) *erb* oncoprotein, membrane location, homologous to epidermal growth factor receptor; b) *ras* oncoprotein, located in the internal part of the cytoplasmic membrane and implicated in signal transduction; c) *myc*, located in the nucleus, are phosphoproteins and bind to DNA (Weinberg 1985, Knudson 1986, Spandidos 1986, Spandidos & Anderson 1989).

The *ras* family of oncogenes, N-*ras*, Ki-*ras* and H-*ras*, the most widely investigated in both human and experimental tumours, are shown to be frequently activated in a variety of neoplasms and specific point mutations in the activated alleles at positions 12, 13, and 61 are not infrequent (Barbacid 1987, Guerrero & Pellicer 1987, Bos 1988). A high prevalence of *ras* oncogene mutations, around 50%, has been demonstrated in colonic adenocarcinomas and large adenomas compared with the frequency of 9% in small adenomas (Bos et al 1987, Forrester et al 1987) in most pancreatic carcinomas (93–95%) (Smit et al 1988, Almoguera et al 1988) but rare or none in gastric and oesophageal carcinomas (Nishida et al 1987, Hollstein et al 1988).

The development of monoclonal antibodies to oncogene products and their application in routinely prepared tissues may prove to be useful tools in a) diagnosis as 'tumour markers', for screening high-risk cancer patients and follow-up; b) in prognosis as predictors of tumour behaviour and determining therapeutic regimes.

Immunohistochemical detection of oncogene expression on routine sections refer particularly to colonic carcinoma and adenomas, but information on gastric carcinoma and its precursor stages is scanty.

Most studies refer to *ras* oncogenes. The results are controversial and inconsistent, due perhaps to different methods of raising antibodies and preparation of tissues (Ohuchi 1987, Michelassi et al 1987). There is also the problem of interpretation of data (Spandidos & Kerr 1984, Thor et al 1984, Williams et al 1985, Kerr et al 1986, Chesa et al 1987, Fujita et al 1987).

The same uncertainties and variable results are found in the scarce data published on the use of c-*myc* oncogene product (Evan et al 1985, Stewart et al 1986, Allum et al 1987, Ciclitira et al 1987, Jones et al 1987, Sikora et al 1987, Sundaresan et al 1987, Erisman et al 1988, Kate et al 1989).

A number of questions need to be answered at molecular level before the use of immunohistochemistry in routines sections can be exploited for diagnosis. To investigate the exact role of oncogenes in human carcinogenesis, it will be necessary to replay the sequential combination of molecular events involved in initiation and progression of each human malignancy.

A powerful new technique, the polymerase chain reaction (PCR), which can selectively amplify a single molecule of template DNA several million-fold in a few hours, has revolutionized molecular biology and genetics. The PCR method has greatly simplified the analysis of DNA sequences and significantly enhanced the sensitivity of detection of oncogenes (White et al 1989).

Of great importance in histological diagnosis is the fact that the automated PCR method can use material from 5 μm paraffin sections or smears (Almoguera et al 1988, Kumar & Barbacid 1988).

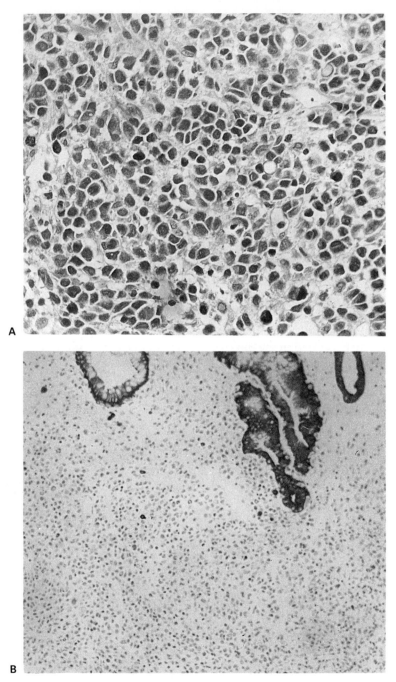

Fig. 12.8 A Gastric biopsy: undifferentiated malignant neoplasia (H & E)? tissue and site of origin. **B** CAM5.2 immunostaining shows cytokeratin expression in normal gastric epithelium. Tumour cells are negative. (Haematoxylin counterstain)

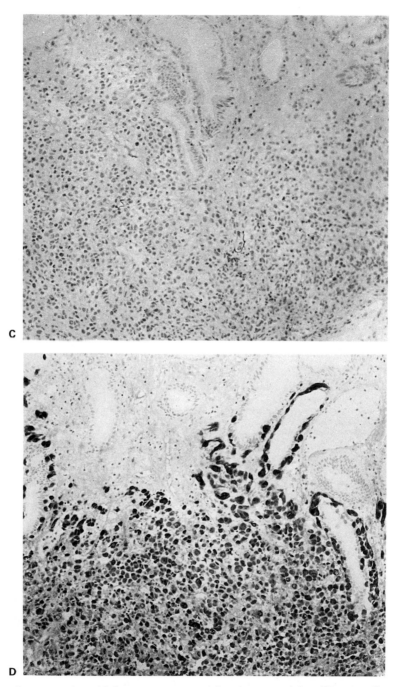

Fig. 12.8 C Tumour cells are negative with leucocyte common antigen immunostaining. (Haematoxylin counterstain.)
D S-100 protein is strongly expressed in the tumour cells. The normal gastric epithelium is negative.
Final diagnosis: metastatic malignant melanoma.

REFERENCES

Agawa S, Muto T, Morioka Y 1988 Mucin abnormality of colonic mucosa in ulcerative colitis associated with carcinoma and/or dysplasia. Diseases of the Colon and Rectum 31: 387–389

Ahren D J, Warren G H, Gruner L J et al 1987 Search for a specific marker of mucosal dysplasia in chronic ulcerative colitis. Gastroenterology 93: 1346–1355

Allen D C, Connolly N S, Biggart J D 1988 Mucin profiles in ulcerative colitis with dysplasia and carcinoma. Histopathology 13: 413–424

Allen D C, Biggart J D, Orchin J C, Foster H 1985 An immunoperoxidase study of epithelial marker antigens in ulcerative colitis with dysplasia and carcinoma. Journal of Clinical Pathology 38: 18–29

Allum W H, Newbold K M, MacDonald F, Russell B 1987 Evaluation of p62c-myc in benign and malignant gastric epithelia. British Journal of Cancer 56: 785–786

Almoguera C, Shibata D, Forrester K et al 1988 Most human carcinomas of the exocrine pancreas contain mutant C-K-ras genes. Cell 53: 549–554

Arends J W, Wiggers T, Thijs C T et al 1984a The value of secretory component (SC) immunoreactivity in diagnosis and prognosis of colorectal carcinoma. American Journal of Clinical Pathology 82: 267–274

Arends J W, Wiggers T, Verstijnen C et al 1984b Gastrointestinal cancer-associated antigen (GICA) immunoreactivity in colorectal carcinoma in relation to patient survival. International Journal of Cancer 34: 193–196

Ballantyne K C, James P D, Robins R A, Baldwin R W, Hardcastle J D 1987 Flow cytometric analysis of the DNA content of gastric cancer. British Journal of Cancer 56: 52–54

Bara J, Andre J, Gautier R, Burtin P 1984 Abnormal pattern of mucus-associated M1 antigens in histologically normal mucosa adjacent to colonic adenocarcinoma. Cancer Research 44: 4040–4045

Barbacid M 1987 ras Genes. Annual Review of Biochemistry 56: 779–827

Boland C R, Lance P, Levin B et al 1984 Abnormal goblet cell glycoconjugates in rectal biopsies associated with an increased risk of neoplasia in patients with ulcerative colitis. Early results of a prospective study. Gut 25: 1364–1371

Borkje B, Hostmark J, Skagen D W et al 1987 Flow cytometry of biopsy specimens from ulcerative colitis, colorectal adenomas and carcinomas. Scandinavian Journal of Gastroenterology 22: 1231–1237

Bos J L 1988 The ras gene family and human carcinogenesis. Mutation Research 195: 255–271

Bos J L, Fearon E R, Hamilton S R et al 1987 Prevalence of ras gene mutations in human colorectal cancers. Nature 327: 293–297

Bresalier R S, Boland C R, Kim Y S 1985 Regional differences in normal and cancer associated glycoconjugates of the human colon. Journal of the National Cancer Institute 75: 249–260

Bushy-Earle R M C, Williams A R W, Piris J 1986 Pepsinogens in gastric carcinoma. Human Pathology 17: 1031–1035

Campo E, Condom E, Palacin A et al 1988 Lectin binding patterns in normal and neoplastic mucosa. A study of Dolichos biflorus agglutinin, peanut agglutinin and wheat germ agglutinin. Diseases of the Colon and Rectum 31: 892–899

Chesa P G, Rettig W J, Melamed M R, Old L J, Niman H L 1987 Expression of p21 ras in normal and malignant human tissue: lack of association with proliferation and malignancy. Proceedings of the National Academy of Sciences of the USA 84: 3234–3238

Ciclitira P J, MaCartney J C, Evan G 1987 Expression of c-myc in non-malignant and pre-malignant gastrointestinal disorders. Journal of Pathology 151: 293–296

Cooper H S, Reuter V R 1983 Peanut lectin binding sites in polyps of colon and rectum. Adenomas, hyperplastic polyps and adenomas within situ carcinoma. Laboratory Investigation 49: 655–661

Cooper H S, Farano P, Coapman R A 1989 Peanut lectin binding sites in colons of patients with ulcerative colitis. Archives of Pathology and Laboratory Medicine 111: 270–275

Correa P 1980 The epidemiology and pathogenesis of chronic gastritis. Three aetiological entities. In: Van der Reis L (ed.) Frontiers of gastrointestinal research. Karger, Basel, p 98–108

Correa P 1982 Precursors of gastric and oesophageal cancer. Cancer 50: 2554–2565

Correa P 1988 A human model of gastric carcinogenesis. Cancer Research 48: 3554–3560

Cuvelier C A, Morson B C, Roels H J 1987 The DNA content in cancer and dysplasia in chronic ulcerative colitis. Histopathology 11: 927–939

Danova M, Riccardi A, Brugnatelli S et al 1988 In vivo bromodeoxyuridine incorporation in human gastric cancer: a study on formalin-fixed and paraffin embedded sections. Histochemical Journal 20: 125–130

Dawson P A, Patel J, Filipe M I 1978 Variations in sialomucins in the mucosa of the large intestine in malignancy: a quantimet and statistical analysis. Histochemical Journal 10: 559–572

Dixon M F, Brown L J R, Gilmour H M et al 1988 Observer space variation in the assessment of dysplasia in ulcerative colitis. Histopathology 13: 385–397

Drumm B, Perez-Perez G I, Blaser M J Sherman P M 1990 Intrafamilial clustering of helicobacter pylori infection. New England Journal of Medicine 322: 359–363

Dukes C E 1932 The classification of cancer of the rectum. Journal of Pathology and Bacteriology 35: 323–332

Ehsanullah M, Filipe M I, Gazzard B 1982 Mucin secretion in inflammatory bowel disease: correlation with disease activity and dysplasia. Gut 23: 485–489

Ehsanullah M, Nauton-Morgan M, Filipe M I, Gazzard B 1985 Sialomucins in the assessment of dysplasia and cancer-risk patients with ulcerative colitis treated with colectomy and ileo-rectal anastomosis. Histopathology 9: 223–235

Enblad P, Glimelius B, Bengtsson A et al 1987. The prognostic significance of DNA content in carcinoma of the rectum and rectosigmoid. Acta Chirurgica Scandinavica 153: 453–458

Erisman M D, Litwin S, Keidan R D et al 1988 Non-correlation of the expression of the c-myc oncogene in colorectal carcinoma with recurrence of disease or patient survival. Cancer Research 48: 1350–1355

Evan G, Lewis G K, Ramsay G, Bishop J M 1985 Isolation of monoclonal antibodies specific for human and mouse c-myc proto-oncogene products. Molecular Cell Biology 5: 3610–3616

Filipe M I, Branfoot A C 1974 Abnormal patterns of mucous secretion in apparently normal mucosa of large intestine with carcinoma. Cancer 34: 282–290

Filipe M I, Branfoot A C 1976 Mucin histochemistry of the colon. In: Morson B C (ed) Current topics in pathology. Springer Verlag, Heidelberg, p 143–178

Filipe M I, Mughal S, Bussey H J 1980 Patterns of mucous secretion in the colonic epithelium in familial polyposis. Investigative and Cell Pathology 3: 329–343

Filipe M I, Edwards M R, Ehsanullah M 1985a A prospective study of dysplasia and carcinoma in the rectal biopsies and rectal stump of eight patients following ileorectal anastomosis in ulcerative colitis. Histopathology 9: 1139–1153

Filipe M I, Potet F, Bogomoletz W V et al 1985b Incomplete sulphomucin-secreting intestinal metaplasia for gastric cancer. Preliminary data from a prospective study from three centres. Gut 26: 1319–1326

Filipe M I, Jass J R 1986 Intestinal metaplasia subtypes and cancer risk. In: Filipe M I, Jass J R (eds) Gastric carcinoma. Edinburgh, Churchill Livingstone, p 87–115

Filipe M I 1988 Intestinal metaplasia in the histogenesis of gastric carcinoma. In: Reed P I, Hill M J (eds) Gastric carcinogenesis. Elsevier Science, Amsterdam

Filipe M I, Barbatis C, Sandey A, Ma J 1988a Expression of intestinal mucin antigens in the gastric epithelium and its relationship with malignancy. Human Pathology 19: 19–26

Filipe M I, Sandey A, Ma J 1988b Intestinal mucin antigens in ulcerative colitis and their relationship with malignancy. Human Pathology 19: 671–681

Filipe M I 1989 The histochemistry of intestinal mucins. Changes in disease. In: Whitehead R (ed) Gastrointestinal and oesophageal pathology. Edinburgh, Churchill Livingstone, p 65–89

Filipe M I, Rosa J, Sanday A, Imrie P, Ormerod M, Morris R 1990 Is DNA ploidy and proliferative activity of prognostic value in advanced gastric carcinoma? Gut. Submitted for publication

Finan P J, Quirke P, Dixon M F et al 1986 Is DNA aneuploidy a good prognostic indicator in patients with colorectal cancer? British Journal of Cancer. 54: 327–330

Fiocca R, Villani L, Tenti P et al 1988 Widespread expression of intestinal markers in gastric carcinoma: a light and electron microscopy study using BD-5 monoclonal antibody. Journal of Clinical Pathology 4: 178–187

Forrester K, Almoguera C, Han K et al 1987 Detection of high incidence of K-ras oncogenes during human colon tumorigenesis. Nature 327: 198–303

Fozard J B J, Dixon M F, Axon A T R, Giles G R 1987a Lectin and mucin histochemistry as an aid to cancer surveillance in ulcerative colitis. Histopathology 11: 385–394

Fozard J B J, Quirke P, Dixon M F et al 1987b DNA aneuploidy in ulcerative colitis. Gut 26: 1414–1418

Fujita K, Ohuchi N, Yao T et al 1987 Frequent overexpression but not activation by point mutation, of ras genes in primary human gastric cancer. Gastroenterology 93: 1339–45

Goh H S, Jass J R 1986. DNA content and the adenoma-carcinoma sequence in the colorectum. Journal of Clinical Pathology 39: 387–392

Goodwin C S, Armstrong J A, Marshall B J 1986 *Campylobacter pyloridis*, gastritis and peptic ulceration. Review. Journal of Clinical Pathology 39: 353–365.

Greaves P, Filipe M I, Abbas S, Ormerod M G 1984 Sialomucins and carcinoembryonic antigen in the evolution of colorectal cancer. Histopathology 8: 825–834

Guerrero I, Pellicer A 1987 Mutational activation of oncogenes in animal model systems of carcinogenesis. Mutation Research 185: 293–308

Gyde S N, Prior P, Allan R N et al 1988 Colorectal cancer in ulcerative colitis: a cohort study of primary referrals from three centres. Gut 29: 206–217

Hamada S, Itoh R, Fujita S 1988 DNA distribution pattern of the so-called severe dysplasias and small carcinomas of the colon and rectum and its possible significance in the tumour progression. Cancer 61: 1555–1562

Hamilton S R 1989 Adenocarcinoma in Barrett's oesophagus. In: Whitehead R (ed) Gastrointestinal and oesophageal pathology. Churchill Livingstone, Edinburgh, ch. 38, p 683

Hammarberg C, Slezack P, Tribukait B 1984 Early detection of malignancy in ulcerative colitis. A flow-cytometric study. Cancer 53: 291–295

Hammeeteman W, Tytgat G N Y, Houthoff H J, Tweel J G Van Den 1989 Barrett's oesophagus: Development of dysplasia and adenocarcinoma. Gastroenterology 96: 1249–1256

Hara A, Watanabe M, Kodaira S et al 1987 The expression of stage-specific embryonic antigen 1 in non-cancerous coloretal epithelial of familial polyposis coli. Disease of the Colon and Rectum 30: 440 443

Higgins P J, Correa P, Cuello C, Lipkin M 1984 Fetal antigens in the precursor stages of gastric cancer. Oncology 41: 73–76

Hirota T, Okada T, Itabashi M et al 1984 Significance of intestinal metaplasia as a precancerous condition of the stomach. In: Ming S-C (ed) Precursors of gastric cancer. Praeger, New York, p 179–193

Hollstein M C, Smits A M, Galinana C et al 1988 Amplification of epidermal growth factor receptor gene but no evidence of ras mutation in primary human oesophageal cancers. Cancer Research 48: 5119–5123

Huang C, Xu J, Huang J, Meng X 1986 Sulphomucin colonic type intestinal metaplasia and carcinoma of the stomach. Cancer 57: 1370–1375

Hutching J T, Reading C L, Giavazzi R et al 1988 Distribution of mono-, di-, and tri-O-acetylated sialic acids in normal and neoplastic colon. Cancer Research 48: 483–489

Inokuchi K, Kodama Y, Sasaki O, Kamegawa T, Okamura T 1983 Differentiation of growth patterns of early gastric carcinoma determined by cytophotometric DNA analysis. Cancer 51: 1138–1141

Isaacson P 1982 Immunoperoxidase study of secretory immunoglobulin system in colonic neoplasia. Journal of Clinical Pathology 34: 14–25

Jacobs L R, Huber P W 1985 Regional distribution and alteration to lectin binding to colorectal mucin in mucosal biopsies from controls and subjects with inflammatory bowel disease. Journal of Clinical Investigation 75: 112–118

Jass J R, Filipe M I, Abbas S et al 1984 A morphological and histochemical study of metaplastic polyp of the colorectum. Cancer 53: 510–515

Jass J R, Atkin W S, Cuzick J et al 1986a The grading of rectal cancer: historical perspectives and a multivariate analysis of 447 cases. Histopathology 10: 437–459

Jass J R, England J, Miller K 1986b Value of mucin histochemistry in follow-up surveillance of patients with long standing ulcerative colitis. Journal of Clinical Pathology 39: 393–398

Jones D J, Ghosh A K, Moore M, Shofield P F 1987 A critical appraisal of the immunohistochemical detection of the c-myc oncogene product in colorectal cancer. British Journal of Cancer 56: 779–783

Kahn H J, Yeger H, Loftus R, Goldrosen M H 1989 Monoclonal antibodies to human pancreatic carcinoma cell line recognises gastrointestinal neoplasms. American Journal of Pathology 134: 641–649

Kakei M, Ohara S, Ishibara K, Goso K, Okabe H, Hotta K 1984 Sulfated glycoprotein biosynthesis in human gastric mucosal biopsies. Digestion 30: 59–64

Kamegawa T, Okamura T, Sugimechi K, Inokuchi K 1986 Preoperative detection of a highly malignant type of early gastric carcinoma by cytophotometric DNA analysis. Japanese Journal of Surgery 16: 169–174

Kate J T, Eidelman S, Bosman F T, Damjanov I 1989 Expression of c-myc proto-oncogene in normal human intestinal epithelium. Journal of Histochemistry and Cytochemistry 37: 541–545

Kerr I B, Spandidos D A, Finlay I G et al 1986 The relation of ras family oncogene expression to conventional staging criteria and clinical outcome in colorectal carcinoma. British Journal of Cancer 53: 231–235

Kim Y S, Yuan M, Itzkowitz S H et al 1986 Expression of Le^y and extended Le^y blood group-related antigens in human malignant, premalignant and non-malignant colonic tissues. Cancer Research 46: 5985–5992

Knudson A G 1986 Genetics of human cancer. Annual Review of Genetics 20: 231–251

Kodama Y, Inokuchi K, Soejima K, Matsusaka T, Okamura T 1983 Growth pattern and prognosis in early gastric carcinoma: Superficially spreading and penetrating growth types. Cancer 51: 320–326

Korenaga D, Mori M, Okamura T, Sugimachi K, Enjoji M 1986a DNA ploidy in clinical malignant gastric lesions less than 5 mm in diameter. Cancer 58: 2542–2545

Korenaga D, Haraguchi M, Okamura T et al 1986b Consistency of DNA ploidy between primary and recurrent gastric carcinomas. Cancer Research 46: 1544–1547

Korenaga D, Okamura T, Saito A, Baba H, Sugimachi K 1988 DNA ploidy is closely linked to tumour invasion, lymph node metastasis and prognosis in clinical gastric cancer. Cancer 62: 309–313

Kumar R, Barbacid M 1989 Oncogene detection at the single cell level. Oncogenes 3: 647–651

Lapertosa G, Fulcheri E, Acquarone M, Filipe M I 1984 Mucin profiles in the mucosa adjacent to large bowel non-adenocarcinoma neoplasia. Histopathology 8: 805–811

Lapertosa G, Barachini P, Abbas S et al 1988 Tissue evaluation of epithelial and functional markers of cell differentiation and mucins in colonic malignancy: assessment of diagnostic and prognostic value. Pathologia 80: 145–157

Laurén P 1965 The two histological main types of gastric carcinoma: diffuse and so-called intestinal-type carcinoma: an attempt at a histoclinical classification. Acta Microbiologica et Immunologica Scandinavica 64: 31–49

Lehnert T, Deschner E E 1986 Cell kinetics of gastric cancer and precancer. In: Filipe M I, Jass J R (eds) Gastric carcinoma. Churchill Livingstone, Edinburgh, p 45, 67

Levin D S, Reid B J, Haggitt R C, Rubin C E, Rabinovitch P S 1989 Correlation of ultrastructural aberrations with dysplasia and flow cytometric abnormalities in Barrett's epithelium. Gastroenterology 96: 355–367

Lipkin M, Correa P, Mikol I B et al 1985 Proliferative and antigenic modifications in human epithelial cells in chronic atrophic gastritis. Journal of the National Cancer Institute 75: 613–619

Lofberg R, Trikutait B, Ost A et al 1987 Flow cytometric DNA analysis in longstanding ulcerative colitis: a method of prediction of dysplasia and carcinoma development? Gut 28: 1100–1106

Ma J, deBoer W G R, Nayman J 1982a Intestinal mucinous substances in gastric intestinal metaplasia and carcinoma studied by immunflourescence. Cancer 49: 1664–1667

Ma J, deBoer W G R M, Nayman J 1982b The presence of oncofetal antigens in large bowel carcinoma. Australia and New Zealand Journal of Surgery 52: 30–34

McCartney J C, Camplejohn R S, Powell G 1986 DNA flow cytometry of histological material from human gastric cancer. Journal of Pathology 148: 273–277

McKenzie J K, Purnell D S, Shamsuddin A M 1987 Expression of carcinoembryonic antigen, T-antigen and oncogene products as markers of neoplastic and preneoplastic colonic mucosa. Human Pathology 18: 1282–1286

Matsukura N, Suzuki K, Kawachi T et al 1980 Distribution of marker enzymes and mucin in intestinal metaplasia in human stomachs and relation of complete and incomplete types of intestinal metaplasia to minute gastric carcinomas. Journal of the National Cancer Institute 65: 231–240

Melamed M R, Enker W E, Banner P et al 1985 Flow cytometry of colorectal carcinoma with three-year follow-up Disease of the Colon and Rectum 29: 184–186

Melville D M, Jass J R, Shepherd N A et al 1988 Dysplasia and deoxyribonucleic acid aneuploidy in the assessment of precancerous changes in chronic ulcerative colitis. Observer variation and correlations. Gastroenterology 95: 668–675

Michelassi F, Leuthner S, Lubienski M et al 1987 Ras oncogene p21 levels parallel malignant potential of different human colonic benign conditions. Archives of Surgery 122: 1414–1416

Ming S-C 1977 Gastric carcinoma: a pathological classification. Cancer 39: 2475–2485

Ming S-C, Bajtai A, Correa P et al 1984 Gastric dysplasia. Significance and pathologic criteria. Cancer 54: 1794–1801

Morson B C, Bussey H J, Samoorian S 1977 Policy of local excision for early cancer of the colorectum. Gut 18: 1045–1050

Morson B C, Jass J R, Sobin L H 1985 Precancerous lesions of the gastrointestinal tract. Bailliere Tindall, London

Muto T, Bussey J R, Morson B C 1975a The evolution of cancer of the colon and rectum. Cancer 36: 2251–2270

Muto T, Kamiya J, Sawada T et al 1985b Mucin abnormality of colonic mucosa in patients with familial polyposis coli. A new tool for early detection of the carrier? Diseases of the Colon and Rectum 28: 147–148

Nagayo T 1986 Histogenesis and precursors of human gastric cancer. Research and practice. Springer-Verlag, Heidelberg

Nardelli J, Bara J, Rosa B, Burtin P 1983 Intestinal metaplasia and carcinoma of the human stomachs: an immunohistochemical study. Journal of Histochemistry and Cytochemistry 31: 361–375

Nishida J, Kobayashi Y, Hirrai H, Takaku F 1987 A point mutation at codon 13 of the N-ras oncogene in a human stomach cancer. Biochemical and Biophysical Research Communications 146: 247–252

Odegaard S, Hostmark J, Skagen D W, Schrumpf E, Laerum O D 1987a Flow cytometric DNA studies in normal human gastric mucosa, gastritis and resected stomachs. Scandinavian Journal of Gastroenterology 22: 750–756

Odegaard S, Hostmark J, Skagen D W, Schrumpf E, Laerum O D 1987b Flow cytometric DNA studies in human gastric cancer and polyps. Scandinavian Journal of Gastroenterology 22: 1270–1276

Ohuchi N 1987 Enhanced expression of C-Ha-ras p21 in human stomach adenocarcinomas defined by immunoassays using monoclonal antibodies and in-situ hybridization. Cancer Research 47: 1413–1421

Paull A, Trier J S, Dalton M D, Camp R C, Loeb M P, Goyal R K 1976 The histological spectrum of Barrett's oesophagus. New England Journal of Medicine 265: 476–480

Pihl E, Peura A, Johnson W R et al 1985 T-antigen expression by Peanut agglutinin staining relates to mucosal dysplasia in ulcerative colitis. Diseases of the Colon and Rectum 28: 11–17

Ploton D, Menager M, Jeannesson P, Himber G, Pigeon F, Adnet J J 1986 Improvement in the staining and in the visualization of the argyrophilic proteins of the nucleolar organising regions at the optical level. Histochemical Journal 18: 5–14

Quirke P, Dyson J E D 1986 Flow cytometry: Methodology and applications in pathology. Journal of Pathology 149: 79–87

Quirke P, Dixon M F, Clayden A D et al 1987 Prognostic significance of DNA aneuploidy and cell proliferation in rectal adenocarcinomas. Journal of Pathology 151: 285–291

Quirke P, Dixon M F, Day D W et al 1988 DNA aneuploidy and cell proliferation in familial adenomatous polyposis. Gut 29: 603–607

Quentineier A, Moller P, Schwarz V et al 1987 Carcinoembryonic antigen, CA19-9, and CA125 in normal and carcinomatous human colorectal tissue. Cancer 60: 2261–2266

Reid P E, Owen D A, Dunn E L et al 1985 Chemical and histochemical studies of normal and diseased gastrointestinal tract. III Changes in the histochemical and chemical properties of the epithelial glycoproteins in the mucosa close to colonic tumours. Histochemical Journal 17: 171–181

Riddell R H, Goldman H, Ransohoff D F et al 1983 Dysplasia in inflammatory bowel disease: Standardised classification with provisional clinical applications. Human Pathology 14: 931–968

Rogers C M, Cooke K B, Filipe M I 1978 Sialic acids of human large bowel mucosa: O-acetylated variants in normal and malignant states. Gut 19: 587–592

Rognum T O, Elgjo K, Fausa O, Brandtzaeg P 1982a Immunohistochemical evaluation of carcinoembryonic antigen, secretory component and epithelial IgA in ulcerative colitis with dysplasia. Gut 23: 123–133

Rognum T O, Fausa O, Brandtzaeg P 1982b Immunocytochemical evaluation of carcinoembryonic antigen, secretory component and epithelial IgA in tubular and villous large-bowel adenomas with different grades of dysplasia. Scandinavian Journal of Gastroenterology 17: 341–348

Rognum T O, Thorud E, Elgjo et al 1982c Large bowel carcinomas with different ploidies, related to secretory component, IgA and CEA epithelium and plasma. British Journal of Cancer 45: 921–934

Rognum T O, Thorud E, Lund E 1987 Survival of large bowel carcinoma patients with different DNA ploidy. British Journal of Cancer 56: 633–636

Rokkas T, Pursey C, Uzoechina E et al 1988 Non-ulcer dyspepsia and short-term De-Nol therapy: a placebo controlled trial with particular reference to the role of Campylobacter pylori. Gut 29: 1386–1391

Rosa J, Mehta A, Filipe M I 1990a Nucleolar organizer regions in gastric carcinoma and its precursors stages. Histopathology 16: 265–269

Rosa J, Mehta A, Filipe M I 1990b Nucleolar organizer regions, proliferative activity and DNA index in gastric carcinoma. Histopathology. In press

Rothery G A, Day D W 1985 Intestinal metaplasia in endoscopic biopsy specimens of gastric mucosa. Journal of Clinical Pathology 38: 613–621

Rothery G A, Patterson J E, Stoddard C J, Day D W 1986 Histological and histochemical changes in the columnar lined (Barrett's) oesophagus. Gut 27: 1062–1068

Rubio C, Kato Y 1988 DNA profiles in mitotic cells from gastric adenomas. American Journal of Pathology 130: 485–488

Sahl A 1982 The nucleolus and nucleolar chromosomes. In: Jordon E G, Cullis C A (eds) The nucleolus. Cambridge University Press, Cambridge, p 1–24

Sasaki K, Hashimoto T, Kawachino K, Takahashi M 1988 Intratumoral regional differences in DNA ploidy of gastrointestinal carcinomas. Cancer 62: 2569–2575

Schutte B, Reynder M M J, Wiggers T et al 1987 Retrospective analysis of the prognostic significance of DNA content and proliferative activity in large bowel carcinoma. Cancer Research 47: 5494–5496

Sciallero S, Bruno S, Divinci A et al 1988. Flow cytometric DNA ploidy in colorectal adenomas and family history of colorectal cancer. Cancer 61: 114–120

Scott N A, Rainwater L M, Wieand H S et al 1987 The relative prognostic value of flow cytometric DNA analysis and conventional clinopathologic criteria in patients with operable rectal carcinoma. Diseases of Colon and Rectum 30: 513–520

Shi Z R, McIntyre L J, Knowles B et al 1984 Expression of carbohydrate differentiation antigen, stage-specific embryonic 1, in human colonic carcinoma. Cancer Research 44: 1142–1147

Sikora K, Chan S, Evan G et al 1987 C-myc oncogene expression in colorectal cancer. Cancer 59: 1289–1295

Silva S, Filipe M I 1986 Intestinal metaplasia and its variants in the gastric mucosa of Portuguese subjects: a comparative analysis of biopsy and gastrectomy material. Human Pathology 17: 988–995

Silva S, Filipe M I, Pinho A 1990 Variants of intestinal metaplasia in the evolution of chronic atrophic gastritis and gastric ulcer. A follow-up study. Gut In press

Sipponen P, Lindgren J 1986 Sialylated Lewis determinant CA19-9 in benign and malignant gastric tissue. Acta

Pathologica et Microbiologica Scandinavia. C Immunology 94: 305–311

Smit V T H B M, Boot A J M, Smits A M M et al 1988 KRAS colon 12 mutations occur very frequently in pancreatic adenocarcinomas. Nucleic Acids Research 16: 7773–7782

Sowa M, Yoshino H, Kato Y, Nishimura M, Kamino K, Umeyama K 1988 An analysis of DNA ploidy patterns of gastric cancer. Cancer 62: 1325–1330

Spandidos D A 1986 A unified theory for the development of cancer. Bioscience Reports 6: 691–708

Spandidos D A, Anderson M L M 1989 Oncogenes and onco-suppressor genes: their involvement in cancer. Review. Journal of Pathology 157: 1–10

Spandidos D A, Kerr I B 1984 Elevated expression of the human ras oncogene family in premalignant and malignant tumours of the colorectum. British Journal of Cancer 49: 681–688

Stewart J, Evan G, Watson J R, Sikora K 1986 Detection of the c-myc oncogene product in colonic polyps and carcinomas. British Journal of Cancer 53: 1–6

Suarez V, Newman J, Hiley C, Crocker J, Collins M 1989 The value of NOR numbers on neoplastic and non-neoplastic epithelium of the stomach. Histopathology 14: 61–66

Sugihara H, Hattori T, Fukuda M 1986 Immunohistochemical detetion of bromodeoxyuridine in formalin-fixed tissues. Histochemistry 85: 193–195

Sugihara K, Jass J R 1986 Colorectal goblet cell sialomucin heterogeneity: in relation to malignant disease. Journal of Clinical Pathology 39: 1088–1095

Sundaresan V, Forgas I C, Wight D G D et al 1987 Abnormal distribution of C-myc oncogene product in familial adenomatous polyposis. Journal of Clinical Pathology 40: 1274–1281

Teodori L, Capurso L, Cordelli E et al 1984 Cytometrically determined relative DNA content as an indicator of neoplasia in gastric lesions. Cytometry 5: 63–70

Thomas D M, Filipe M I, Smedley F H 1989 Dysplasia and carcinoma in the rectal stump of total colitics who have undergone colectomy and ileo-rectal anastomosis. Histopathology 14: 289–298

Thor A, Horan-Hand P, Wunderlich D et al 1984 Monoclonal antibodies define differential ras expression in malignant and benign colonic disease. Nature 311: 562–565

Tribukait B, Hammarberg C, Rubio C 1983 Ploidy and proliferation patterns in colorectal adenocarcinomas related to Dukes' classification and to histopathological differentiation: a flow cytometric DNA study. Acta Pathologica et Microbiologica Scandinavia (A) 91: 89–95

Warren J R, Marshall B J 1983 Unidentified curved bacilli on gastric epithelium in active chronic gastritis. Lancet i: 1273–1275

Weinberg R A 1985 The action of oncogene in the cytoplasm and nucleus. Science 230: 770–776

White T J, Arnheim N, Erlich H A 1989 The polymerase chain reaction. Trends in Genetics 5: 185–189

Wiggers T, Arends J W, Verstijnen C et al 1986 Prognostic significance of CEA immunoreactivity patterns in large bowel carcinoma tissue. British Journal of Cancer 54: 409–414

Wiggers T, Arends J W, Schutte B et al 1988 A multivariate analysis of pathologic prognostic indicators in large bowel cancer. Cancer 61: 386–395

Williams A R W, Piris J, Spandidos D A, Wyllie A H 1985 Immunohistochemical detection of the ras oncogene p21 product in an experimental tumour and in human colorectal neoplasms. British Journal of Cancer 52: 687–693

Williams G T 1985 Transitional mucosa of the large intestine. Commentary. Histopathology 9: 1237–1243

Wilson G D, NcNally N J, Dunphy E, Karcher H, Pfragner R 1985 The labelling index of human and mouse tumours assessed by bromodeoxyuridine staining in vitro and in vivo and flow cytometry. Cytometry 6: 641–647

Wyatt J I, Dixon M F 1988 Chronic gastritis — a pathogenetic approach. Review. Journal of Pathology 154: 113–124

Yonezawa S, Nakamura T, Tanaka S, Sato E 1982 Glycoconjugate with Ulex europaeus aggulatinin-binding sites in normal mucosa, adenoma and carcinoma of the human large bowel. Journal of the National Cancer Institute 69: 777–785

Yuan M, Itzkowitz S H, Ferral L D et al 1987 Expression of Lewis[x] and sialylated Lewis[x] antigens in human colorectal polyps. Journal of the National Cancer Institute 78: 479–488

Zamcheck N, Liu P, Thomas P, Steel G 1988 Search for useful biomarkers of pre- and early malignant colonic tumours. In: Steele G, Burt R W, Winawer S J, Karr J P (eds) Progress in clinical and biological research, vol 279. Alan R Liss Inc, New York, p 251–275

13. Malabsorption

Z. Lojda

INTRODUCTION

Appropriate nutrition depends on the adequate intake of food, its effective splitting by digestion and proper absorption and utilization of breakdown products. A deficiency in one or more of these processes leads to manifestations of malnutrition. In clinical terminology the set of symptoms which occur either regularly or occasionally in diseases with disorders of digestion, absorption, secretion and motility of the small intestine is designated as malabsorption syndrome (MS). Its classification is not easy (Jeffries et al 1969, Lojda et al 1971, Morson & Dawson 1979, Piris 1989): the enterocytes of the small intestine are responsible not only for absorption but also for the terminal digestion of nutrients, and the pathophysiological mechanisms leading to the manifestation of MS appear in many cases in various combinations. In every individual case the diagnosis is achieved on recognition of the nosological unit which causes the malabsorption symptomatology. The information on whether the malabsorption is caused by a defect in the enterocytes (primary MS) or outside them (secondary MS) is of basic importance and can often be resolved by intestinal biopsy. A close collaboration between clinician and pathologist is essential to yield the most useful information.

Diagnostic assessments based on purely morphological analysis of biopsies are limited and are summarized by Morson & Dawson (1979) and Piris (1989). Histochemistry has broadened them substantially (Lojda et al 1971, Lojda 1974, 1976a, 1981, 1984) and an outline of this approach is given in this chapter.

BIOPSY OF THE SMALL INTESTINAL MUCOSA

Specimens of the jejunal mucosa are usually taken with the aid of a suction biopsy capsule. The capsule is positioned just beyond the duodenojejunal flexure and its position checked by fluoroscopy. Several types of biopsy capsules are available. The Crosby-Kugler capsule, which is the most widely used, enables the biopsy of a larger specimen from one site (about 4 x 7 mm, 20–25 mg), while with other instruments several smaller biopsies from different sites can be obtained. Biopsies taken by endoscopy are small and often damaged. Biopsies are taken from fasting patients in the morning (to preclude the influence of the circadian changes of enzyme activities). In adult patients and older children, biopsies can be taken in the outpatient department. In babies and smaller children a short stay in the hospital is necessary. The value of the information obtained by biopsy far surpasses the difficulties and potential hazards of the intervention. Of decisive significance are biopsies in diffuse pathological states. In focal lesions a negative result does not exclude a suspected disorder. Duodenal biopsies are not suitable for the diagnosis of MS, because of a proximodistal gradient of enzyme activities (particularly of disaccharidases) in the enterocytes from the duodenum to the jejunum (Lojda 1976a, Lojda et al 1979b). For example, in alactasia the biopsy must be taken from the jejunum since there is sometimes no activity in the normal duodenum (Lojda 1974, 1976a). On the other hand, when a deficiency of enteropeptidase (enterokinase) is suspected, the duodenal mucosa should be examined, because in man the activity of this enzyme is normally low in the proximal jejunum (Lojda 1976a).

PROCESSING AND EVALUATION OF THE BIOPSY

The biopsy is carefully spread mucosal surface upwards on filter paper, a gelatine foil, or plastic micromesh. The dissecting microscope appearances are not as important as formerly contended and observation of unfixed specimens can be omitted or be kept as short as possible to prevent autolytic changes. Small biopsy specimens are quenched directly in light petroleum chilled in an acetone- or ethanol-dry ice mixture or in liquid nitrogen (see Lojda et al 1979a). If the specimen is large enough, it can be divided into two or more pieces for freezing and for routine paraffin sections or other special fixations. A portion may be taken for electron microscopy.

Good cryostat sections enable morphological evaluation so that the preparation of paraffin wax or plastic sections may be considered unnecessary.

Cryostat sections are cut perpendicularly to the mucosal surface and examined by the methods outlined below.

Important methods

Haematoxylin-eosin
PAS
Lactase (indigogenic, azo-dye, or GO-PO-DAB)
Trehalase (GO-PO-DAB)
Sucrase (GO-PO-DAB)
A fat stain (Oil red O, Fettrot 7B)
DPP IV (Gly-Pro-MNA)

Additional useful methods

Acid phosphatase (azo-dye, Gomori)
ATPase (Wachstein-Meisel)
Alkaline phosphatase (BCIP-NBT)
Enteropeptidase (trypsinogen procedure)
Tryptase (CBZ-Ala-Ala-Lys-MNA or
D-Val-Leu-Arg-MNA)
γ-Glutamyl transpeptidase (γ-Glu-MNA, Gly-Gly, FBB)
Immunoglobins A, M and G
Surface markers of lymphocytes
Amyloid

For details of methods see Appendix.

In haematoxylin and eosin-stained sections, the regularity of the surface, size and shape of the villi as well as depth of crypts are assessed. Other features easily recognized in these preparations are the enhanced basophilia of enterocytes, increased mitotic activity in crypts and increased number of intraepithelial lymphocytes, features which are associated with crypt hyperplasia (e.g. coeliac disease). The intraepithelial lymphocytes (belonging mainly to the OKT 8-positive subgroup of T-lymphocytes) can be easily identified by the reaction for ATPase (Lojda 1976b, Fig. 13.1). Crypt enterocytes with enlarged nuclei are seen in vitamin B_{12} deficiency. Greater numbers of goblet cells are found in patients with coeliac disease, protein intolerance, and cystic fibrosis. Paneth cells can likewise be observed in H & E-stained preparations and their granules which can be demonstrated by the reaction for tryptophan also display strong γ-glutamyl transpeptidase activity. At present they have no great significance in the diagnosis of the MS. The same holds true for endocrine cells which are therefore not considered here.

In the lamina propria hyperaemia and oedema in the apical portion of villi may be artefacts induced by the suction biopsy. Assessment of the cellularity is an important parameter. Plasma cells are revealed by the methyl green-pyronin reaction and, if required, their immunoglobulins can be identified by immunohistochemical methods (Fig. 13.2). Macrophage activity can be demonstrated by the reaction for acid phosphatase. Neutrophils are identified by their elastase-like activity best demonstrated by the reaction with N-acetyl-alanine-l-naphthyl ester and hexazonium-pararosaniline (Lojda 1981). The method using naphthol AS-D-chloroacetate and hexazonium-pararosaniline is inferior. In both methods chymase activity of mast cells coreacts. Neutrophils and eosinophils are selectively shown by the reaction for myeloperoxidase (using N-phenyl-p-phenylene-diamine and α-naphthol; Lojda et al 1979a). Eosinophils can be selectively shown by the reaction for tryptophan. Mast cells can be selectively demonstrated even in cryostat sections of snap-frozen samples due to their tryptase activity best shown by CBZ-Ala-Ala-Lys-

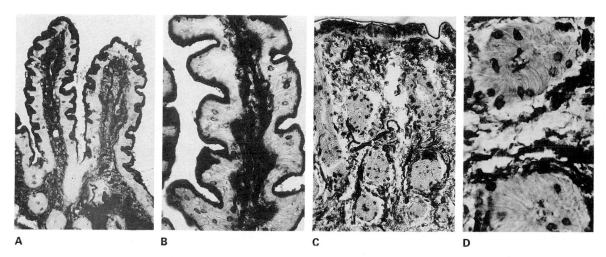

A **B** **C** **D**

Fig. 13.1 ATPase demonstrated with the Wachstein-Meisel method. **A** and **B** Normal jejunal mucosa, **C** and **D** jejunal mucosa of a patient with overt coeliac disease. In addition to a strong positive reaction in the brush border and in the lamina propria there is a remarkable positivity in the cell membranes of epithelial lymphocytes, the number of which is almost doubled in the patient with coeliac disease. **A** and **C** ×80, **B** ×160, **D** ×240

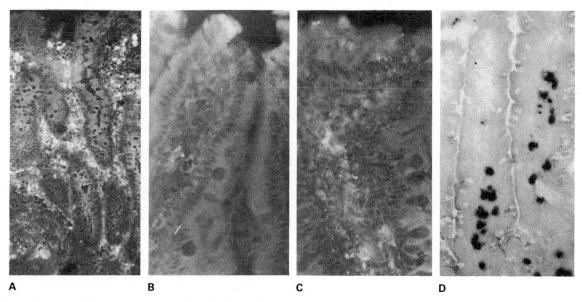

A **B** **C** **D**

Fig. 13.2 Immunofluorescent demonstration of IgA (**A** and **B**) and IgG (**C**) in the jejunal mucosa of a patient with coeliac disease (**A** and **C**) and with IgA deficiency (**B**). Many IgA positive plasmocytes in the propria (**A**). Only few IgA-positive plasmocytes in **B**. IgG-positive plasmocytes well visible in (**C**). **D** Tryptase activity in mast cells demonstrated with CBZ-Ala-Ala-Lys-MNA and Fast blue B. **C** Mast cells in the propria of the normal jejunal mucosa are selectively demonstrated. **A** ×100, **B** and **C** ×350, **D** ×110

4-methoxy-2-naphthylamide and fast blue B (Lojda et al 1987b, Fig. 13.2D see Appendix 5). D-Val-Leu-Arg-MNA introduced for the demonstration of mast cells by Garrett et al (1982) can also be used. Lymphocyte subtypes can be identified by immunohistochemical reactions for surface markers (Ch. 19) and the reaction for dipeptidyl peptidase IV (DPP IV) (see Appendix 5) depicts a subgroup of T-lymphocytes (belonging mainly to OKT 4-positive subset) which have a relation to interleukin 2 (Lojda 1985, Fig. 13.6B). In the normal lamina propria capillary endothelium and the brush border of enterocytes also reacts (Fig. 13.6A). A decreased DPP IV activity in the lamina propria points to its involvement in the pathological process.

In H & E-stained preparations, dilated lymphatic vessels can be recognized. When the dilation is caused by an obstruction of the lymph flow the detection of lipids using Sudan dyes (preferably Fettrot 7B or Oil Red O) can be useful. In these states lipids (both intra- and extracellular) are found in the lamina propria of fasting patients (Fig. 13.8B). The finding of numerous lipid droplets in enterocytes covering normal villi points to a-β-lipoproteinaemia or hypo-β-lipoproteinaemia.

The assessment of changes of activities of brush border enzymes of enterocytes plays the most important role in the diagnosis of the MS. The demonstration of disaccharidase activities is of fundamental importance. In routine practice the histochemical demonstration of these enzymes is more convenient than the determination of disaccharidase activities in homogenates of the jejunal mucosa. It requires much less material (the demonstration of one disaccharidase can be performed using one to three sections 10 μm thick) and enables, at the same time, the assessment of the morphological pattern. Although a quantitative determination of enzyme activities by microdensitometry or computer-assisted image analysis is possible (in the latter case a quantitative assessment of the surface area capable of cleaving of the respective disaccharide, represents the most precise approach) a semiquantitative evaluation is sufficient. For routine practice the assessments of lactase, sucrase and trehalase suffice. The method of choice for the demonstration of lactase activity is the indigogenic procedure with 5 Br-4 Cl-3-indolyl-β-D-fucoside (Lojda & Kraml 1971, cf. Lojda et al 1979a)(see Appendix 5)(Fig. 13.3,

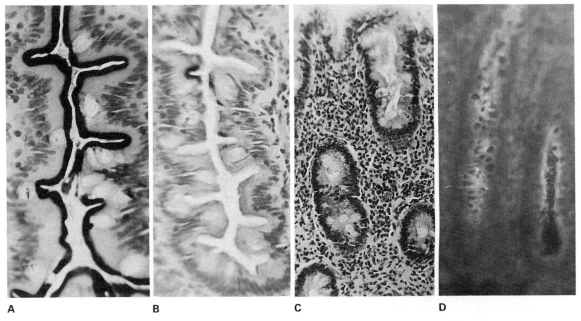

A   **B**   **C**   **D**

Fig. 13.3 Lactase activity demonstrated with indigogenic method of Lojda and Kraml (**A**, **B** and **C**). Lactase immunoreactivity demonstrated with guinea pig immunoglobulins against rat lactase and rabbit immunoglobulins against guinea pig immunoglobulins conjugated with fluorescein isothiocyanate (**D**). **A** Normal mucosa, strong reaction in the brush border. **B** Jejunal mucosa of a patient with isolated lactase deficiency. Two enterocyte populations can be seen: in one no reaction is apparent, in the other some reaction is present. **C** Jejunal mucosa with total villous atrophy in a patient with celiac disease. The reaction is completely negative. **D** Lactase immunoreactivity in the same patient. Positive reaction in the brush border is preserved. **A**, **B** ×490, **C** ×160, **D** ×210

13.8C). If this substrate is not available the azo-coupling reaction with 1-naphthyl-β-D-glucoside and hexazonium-pararosaniline can be used (Lojda 1975, Appendix 5). Lactase activity may also be demonstrated using the natural substrate lactose in the coupled glucose oxidase-peroxidase-diaminobenzidine method (GO-PO-DAB), using the longer incubation time of 24 hours (see Appendix 5). These reactions enable the demonstration of lactase deficiency. For this purpose, the demonstration of catalytic activity of lactase and not of its immunoreactivity is necessary (this demonstrates also catalytically inactive precursors, see Lojda et al 1985, Fig. 13.3D).

Lactase is the most sensitive indicator of the degree of differentiation and injury of enterocytes. In pathological processes its activity is first to be affected and the last to be restored. The second in respect to sensitivity is trehalase. Its activity is demonstrated with its natural substrate (trehalose) in the coupled glucose oxidase-peroxidase-diaminobenzidine (GO-PO-DAB) method (Lojda 1972, Lojda et al 1979a) (Appendix 5) (Fig. 13.4). Sucrase appears the most resistant of disaccharidases to injury. It can be demonstrated either with its natural substrate (sucrose) using the GO-PO-DAB method (Lojda 1972, Lojda et al 1979a) or with synthetic substrates 6-Br-2-naph-thyl-α-D-glucoside employed in the simultaneous azocoupling method with hexazonium-pararosaniline (Lojda 1965, Lojda et al 1979a) (Appendix 5) (Fig. 13.5 B,D) or 5-Br-4-Cl-3-indolyl-α-D-glucoside employed in a similar method as fucoside derivative in the demonstration of lactase (Lojda et al 1987a) (Appendix 5) (Fig. 13.5 A,C). Although synthetic substrates are also cleaved by glucoamylase (Lojda et al 1973, 1987a) sucrase-isomaltase deficiencies can well be

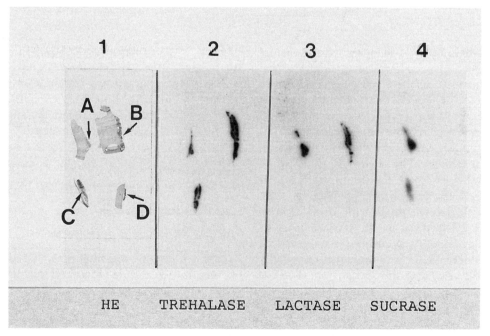

Fig. 13.4 Disaccharidases. GO-PO-DAB method with natural substrates. Macrophotograph. Four separate jejunal biopsies (**A,B,C,D**) are shown. Slide 1 is stained with HE to show the morphology of the jejunal biopsies which have been mounted on gelatine blocks. Slides 2,3 and 4, with the same set of biopsies in the same order, have been incubated as described in Appendix 5, with the gels containing trehalose (slide 2), lactose (slide 3) and sucrose (slide 4). The density of staining is proportional to the enzyme activity. Biopsy **A** is normal, biopsy **B** is deficient in sucrase activity yet has normal trehalase and lactase activities, biopsy **C** is deficient in lactase only, biopsy **D** is from a patient with severe subtotal villous atrophy and is deficient in trehalase, lactase and sucrase.

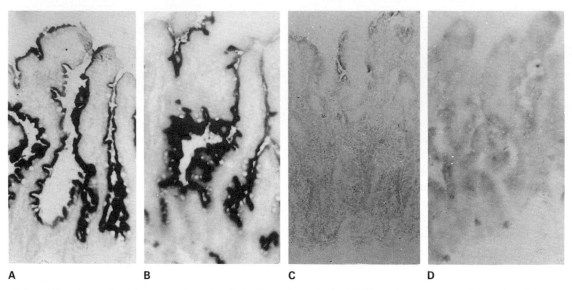

Fig. 13.5 α-Glucosidases (mainly sucrase-isomaltase). Indigogenic method with Kernechtrot counterstaining (**A** and **C**). Azo-coupling method with 6-Br-2-naphthyl-α-D-glucoside and hexazonium-pararosaniline (**B** and **D**). Normal jejunal mucosa (**A** and **B**), strong reaction in the brush border. Jejunal mucosa from a patient with sucrase-isomaltase deficiency (**C** and **D**); only traces of activity. **A,B,C** and **D** ×60.

detected by them because sucrase-isomaltase is the major α-glucosidase of human enterocytes. Isomaltase activity can be shown using palatinose as substrate in the GO-PO-DAB method with 24-hour incubation as for lactase.

Of the brush border proteases the demonstration of enteropeptidase (enterokinase) activity has a diagnostic meaning. The two-step reaction of Lojda & Mališ (1972; cf. Lojda et al 1979a), in which the activity of trypsin originating from trypsinogen is demonstrated with benzyloxycarbonyl-Gly-Gly-Arg-MNA or benzoyl-Arg-β-naphthylamide (BANA) as substrates and fast blue B as coupling agent, is recommended (Appendix 5) (Fig. 13.6 C,D). This method is more sensitive and less expensive than the method of Lojda & Gossrau (1983) using synthetic substrate Gly-Asp-Asp-Asp-Asp-Lys-2-naphthylamide and fast blue B, which enables a more precise localization.

For a better characterization of changes in enterocytes, particularly in the long-term follow-up of patients with MS (e.g. with coeliac disease), it is advisable to perform other reactions for activities of brush border proteases, e.g. γ-glutamyl transpeptidase (Fig. 13.7), brush border endopeptidase, DPP IV (Fig. 13.6), alanyl aminopeptidase

(Lojda 1981, 1985), particularly in connection with computer-assisted image analysis (Lojda 1988). Other parameters should also be studied (Lojda et al 1971, Lojda 1974). This, however, goes beyond the routine diagnosis and therefore will not be considered here further.

DIAGNOSTIC OUTLINE OF INDIVIDUAL FORMS OF MS IN ENTEROBIOPSIES

Isolated enzymopathies

Isolated disaccharidase deficiencies

These are either congenital or acquired and occur in childhood and in adults. Their incidence is race-dependent, being more frequent in individuals of African or Asian extraction. The morphological appearance of the mucosa is usually normal, or with only minor irregularities of villi, a somewhat larger number of goblet cells and a slightly increased cellularity in the lamina propria.

Lactase deficiency (hypolactasia, alactasia) is the most common. Histochemically three types are recognized. Most frequently lactase activity is not detectable (see Fig. 13.4). In others there is a general diffuse reduction of lactase activity. In the

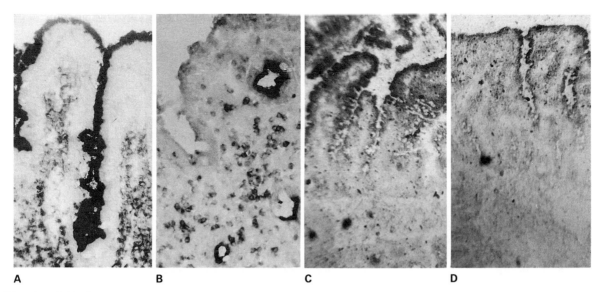

A   **B**   **C**   **D**

Fig. 13.6 DPP IV (**A** and **B**) and enteropeptidase (**C** and **D**). DPP IV was demonstrated with Gly-Pro-MNA and fast blue B according to Lojda, enteropeptidase with the trypsinogen procedure of Lojda and Mališ. **A** Normal jejunal mucosa, strong reaction in the brush border. In the lamina propria T-lymphocytes and endothelial cells of capillaries react. **B** Jejunal mucosa of a patient with overt coeliac disease: the reaction in the brush border is much weaker. In the lamina propria only T-lymphocytes (mostly helper cells) react. **C** Normal duodenal mucosa. The azo-dye is deposited in enterocytes, particularly in the apical portion of villi. **D** Duodenal mucosa of a patient with overt coeliac disease. The reaction is much weaker than in the preceding case. **A** and **B** ×250, **C** and **D** ×60.

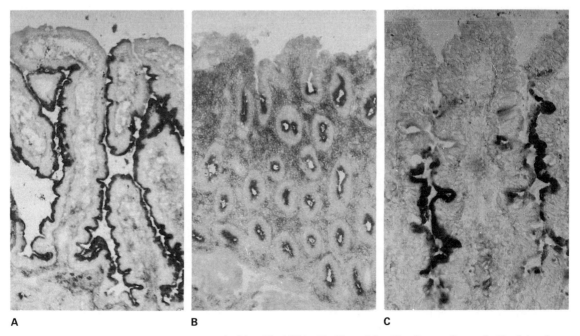

A   **B**   **C**

Fig. 13.7 γ-Glutamyl transpeptidase demonstrated with γ-Glu-MNA, Gly-Gly and fast Blue B according to Lojda. Jejunal mucosa of two patients with intractable diarrhoea (**A** and **C**). Strong reaction in the brush border of enterocytes in lower portion of otherwise normal villi. In the apical portions of some villi in **A** and all villi in **C** the reaction is negative. Distinct reaction in the basement membrane and in some cells of the propria. **B** Jejunal mucosa of a patient with overt coeliac disease. The reaction is positive in the brush border of crypt enterocytes. There is a remarkable increase of enzyme activity in the propria. **A** and **B** ×100, **C** ×250

third type there are two populations of enterocytes: one, which prevails numerically, is lactase-
negative, the other is lactase-positive (Fig. 13.3B).
It is not possible to decide by biopsy alone whether
the deficiency is congenital or acquired, although
the last type is usually acquired.

Isolated sucrase-isomaltase deficiency occurs much
less frequently. It is usually congenital. Sucrase
activity is present in traces only or may not be detectable (Figs 13.4, 13.5 C,D).

Isolated trehalase deficiency is the least frequent.
It is inborn and manifests itself after eating mush-
rooms. Trehalase activity is not detectable. A
diminished activity of trehalase (isolated or combined with hypolactasia) can be found in some
patients with cystic fibrosis whose jejunal mucosa
is otherwise normal.

Combined disaccharidase deficiencies

These secondary deficiencies occur particularly in
coeliac disease, tropical sprue, cow's milk protein
intolerance, Whipple's disease, radiation enteritis
and after administration of some drugs. In these
cases, however, changes are already seen in H &
E-stained preparations.

Enteropeptidase deficiency

Congenital enteropeptidase deficiency in children
is rare. Enzyme activity is completely missing.
Secondary deficiencies (decreased activity) occur
in some patients with partial or subtotal atrophy
of duodenal villi and in intractable diarrhoea of infancy. It is to be stressed that no judgment about
enteropeptidase activity can be made on the morphological appearance of the duodenal mucosa. In
many patients with coeliac disease displaying flat
mucosa the enzyme activity is within normal
limits. Conversely, in some patients with only
minor changes in the mucosa the activity can be
low (Lojda & Jodl 1974). Enteropeptidase activity
in patients with coeliac disease is generally significantly lower than in normal subjects (Lojda &
Gossrau 1983) (Fig. 13.6 C,D). It is suggested that
hypoproteinaemia and oedema in some patients
with protein malabsorption might be related to low
enteropeptidase activity.

A-β-lipoproteinaemia and hypo-β-lipoproteinaemia

These are rare disorders caused by defective synthesis of the apoprotein(s) in enterocytes necessary
for the formation of chylomicrons. Jejunal biopsy
usually shows no gross abnormality of the villi.
Enterocytes of fasting patients display a foamy appearance in sections stained with H & E. This is
due to neutral lipid droplets which can be
demonstrated by staining with Fettrot 7B. Lactase
activity may be low.

Coeliac disease (non-tropical sprue, coeliac sprue, gluten-induced enteropathy, gluten-sensitive enteropathy, primary malabsorption syndrome sensu stricto)

Changes in the jejunal mucosa depend on the stage
of the disease and its treatment. The characteristic
morphological features are the flattening of the
mucosal surface caused by changed villi, which are
short, broad and sometimes not visible at all (partial or subtotal villous atrophy). The enterocytes
on the surface facing the lumen are smaller and
crowded together so that the surface appears more
than one layer thick. There is deepening of crypts,
increased mitotic activity of enterocytes, which are
more basophilic than normal, increased number of
goblet cells, increased number of intraepithelial
lymphocytes, and thickening of the basement
membrane. When collagen is deposited between
the epithelium and capillaries (collagenic sprue)
there is a poor prognosis. In the lamina propria
oedema and increased cellularity consisting of
plasmocytes, lymphocytes, macrophages, neutrophils, eosinophils, and mast cells is found.

For the diagnosis of gluten-sensitive
enteropathy the villous pattern must improve after
a gluten-free diet and deteriorate after a gluten
challenge.

Although there is no diagnostic difficulty in sections stained with H & E, histochemistry can aid
in a more subtle assessment of the changes and of
the pattern after treatment. One cannot judge the
extent of biochemical changes and their improvement solely on the basis of morphological
evaluation. The evaluation of activities of brush
border enzymes is very helpful (see Fig. 13.4).

Lactase is most severely affected and in the acute phase its activity is virtually absent (Fig. 13.3C). Trehalase is similar in this respect and its activity is present in traces or may be lacking. The activity of sucrase is low and in some patients it cannot be demonstrated with natural substrate. Activities of brush border proteases are affected, particularly at the luminal surface (in decreasing order of involvement: brush border endopeptidase, γ-glutamyl-transpeptidase—Fig. 13.7B, DPP IV —Fig. 13.6B, alanylaminopeptidase). DPP IV enables a very rapid assessment of the involvement of the lamina propria. More serious pathology is indicated by a greater decrease of activity (Fig. 13.6B). In the lamina propria there is a significant increase of mast cells, IgE-containing cells, IgG-plasmocytes, and eosinophils in the luminal portion. Changes in deeper portions of the lamina propria are much less pronounced. The improvement of the given parameters after a gluten-free diet does not go parallel in all cases.

A variable expression of coeliac disease occurs also in patients with dermatitis herpetiformis Duhring.

Protein intolerance

A pattern closely resembling coeliac disease, which cannot be differentiated from it solely on the basis of the evaluation of a single jejunal biopsy, occurs in patients with an intolerance of cow's milk protein and soy bean protein. In these patients changes are usually of a somewhat milder degree.

Tropical sprue

Similar changes in the jejunal mucosa are found also in patients with tropical sprue (endemic sprue), which is endemic in south-western Asia, Indonesia, some parts of India, the Caribbean area, Nigeria and the Middle-East. It affects natives as well as immigrants and visitors. The disease can manifest itself in countries where it does not occur normally, in persons who return from endemic regions. Tropical sprue does not respond to gluten-free diet and responds well to antibiotics, folic acid and vitamin B_{12} treatment. No clear-cut distinction of tropical sprue from

coeliac disease is possible in biopsies (Lojda et al 1971, Lojda 1974). Lipids present in the basement membrane of the enterocytes, and the presence of brush border enzymes claimed by some to be characteristic for tropical sprue, are also found in many cases of coeliac disease.

Immunodeficiency disorders

MS can occur also in immunodeficiency disorders. In some of them the morphological pattern of the jejunal mucosa is normal with only a deficiency in a specific immunoglobulin (usually IgA)-producing plasma cell population (Fig. 13.2B). In these cases isolated alactasia occurs more frequently than in patients with a normal composition of plasma cells in the lamina propria. In some patients there are conspicuous morphological villous changes and a completely flattened mucosa may be found. About 1.5% of patients with coeliac disease responding to a gluten-free diet have IgA deficiency in the lamina propria, which is sometimes referred to as hypogammaglobulinaemic spruc. This deficiency can be demonstrated immunohistochemically. In these patients a compensatory increase in number of IgM plasmocytes is found.

Radiation enteritis and changes after cytotoxic agents

The lesions described in the preceding paragraphs belong to the crypt hyperplastic type. Lesions described in this paragraph are crypt hypoplastic. They occur after X-ray irradiation and after treatment with folic acid antagonists such as methotrexate. The cytoplasm of enterocytes does not display enhanced basophilia. Disaccharidase deficiencies of various degree are present and in the lamina propria a decrease of DPP IV positivity is remarkable.

Drug-induced changes

Other drugs such as chloramphenicol, aureomycin, terramycin, and pheninedione produce slight morphological changes. Hypolactasia or alactasia is very common. After neomycin treatment these changes are more pronounced and a hypotrehalasia can be demonstrated in addition to alactasia. A very impressive finding is the activation of macrophages in the apical region of villi well demonstrated with the acid phosphatase reaction. The overall cellularity of the lamina propria appears decreased. Severe changes with subtotal villous atrophy have been described after nefenamic acid, which return to normal after withdrawal of the drug.

Whipple's disease

This is a rare disorder usually classified under secondary malabsorption syndrome, in which the cause of malabsorption lies beyond the absorptive cells of the intestine. It is diagnosed very easily by the presence of sickle-form particle-containing cells (SPC-cells) in the lamina propria which are strongly PAS-positive (Fig. 13.8A). In the brush border of enterocytes no lactase is present. A significant decrease of trehalase and sucrase also occurs pointing to the participation of enterocytes in the malabsorption. After treatment with broad-spectrum antibiotics the number of SPC-cells

diminishes substantially. However, dilated lymph vessels and alactasia persist (Fig. 13.8C).

Cystic fibrosis

In some patients with cystic fibrosis the pattern in jejunal biopsies is completely normal. However, in more than 50% a decreased activity of trehalase, sometimes combined with hypolactasia, can be found even in cases in which there is no mucus hypersecretion apparent. In about one-half of patients the number of goblet cells is increased and PAS-positive mucus fills crypts and adheres to the surface of villi. In some patients the pattern of coeliac disease is found.

Efferent loop (bowel distal to a jejunostomy)

In evaluating changes in the mucosa in the efferent loop, it should be born in mind that all changes within 20 cm of the efferent loop have no diagnostic value. Alactasia is found sometimes in the mucosa beyond 20 cm.

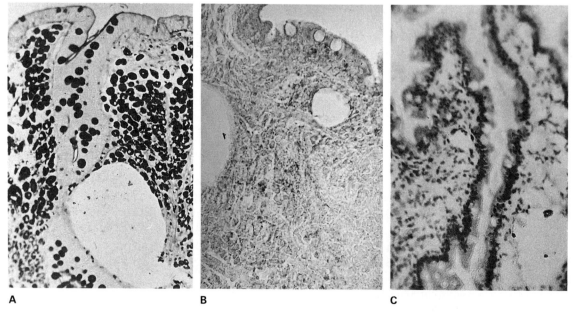

A   **B**   **C**

Fig. 13.8 Whipple's disease. **A** PAS reaction before the treatment with antibiotics. Strong positivity in SPC cells in the lamina propria. The goblet cells and brush border of enterocytes also show a positive reaction. Note a dilated lymphatic. **B** The reaction with Fettrot 7B reveals droplets of neutral lipids in enterocytes and in many cells of the lamina propria. **C** The same patient after the treatment with antibiotics. Although there is a remarkable improvement of the morphological pattern, lactase activity demonstrated with the indigogenic method (counterstained wath Kernechtrot) is lacking. Dilated lymphatics persist. **A** ×250, **B** and **C** ×140

Other disorders

No constant changes are found in the mucosa of patients with glucose-galactose malabsorption, in Hartnup disease, cystinuria, and other rare disturbances of the amino acid metabolism. Patchy, almost zero activities of some brush border enzymes found in enterocytes in the apical region of villi, particularly in cystinuria, are not constant. These changes cannot be considered diagnostic for these diseases because they occur also in some patients with intractable diarrhoea in which no disturbances of amino acid metabolism are found (Fig. 13.7A, C). In patients with Crohn's disease no constant changes are found. Alactasia or hypolactasia occurs very frequently.

Microvillus inclusion disease

Some infants with protracted diarrhoea have a defect of the brush border in which the microvilli appear involuted. At electron microscopy microvilli are present within the enterocytes. Routine microscopy shows a severe partial crypt hyperplastic villous atrophy and the presence of PAS patches, distinct from the goblet cells, in the apical portions of the enterocytes. The activity of alkaline phosphatase, which can be demonstrated even in paraffin sections after formalin fixation, is very much lowered in the brush border, and there are patches of activity in the apical portions of the enterocytes. This provides a quick and effective test in the diagnosis of microvillus inclusion disease for which electron microscopy had formerly to be used (Lake 1988). Disaccharidase activities are also depleted.

Amyloid

Demonstration of amyloid (best using Congo red with dichroism in polarized light) is performed when secondary amyloidosis is suspected, although a positive finding in jejunal biopsy is very rare.

REFERENCES

Garrett J R, Smith R E, Kidd A, Kyriacou K, Grabske R J 1982 Kallikrein-like activity in salivary glands using a new substrate, including preliminary secretory studies and observations on mast cells. Histochemical Journal 14: 967–979

Jeffries G H, Weser E, Sleisenger M H 1969 Malabsorption. Gastroenterology 56: 777–785

Lake B D 1988 Microvillus inclusion disease. Specific diagnostic features shown by alkaline phosphatase histochemistry. Journal of Clinical Pathology 41: 880–882

Lojda Z 1965 Some remarks concerning the histochemical detection of disaccharidases and glucosidases. Histochemie 5: 339–360

Lojda Z 1972 An improved method for the demonstration of disaccharidases with natural substrates. Histochemie 30: 277–280

Lojda Z 1974 Cytochemistry of enterocytes and of other cells in the mucous membrane of the small intestine. In: Smith D H (ed) Biomembranes. Intestinal absorption. Plenum Press, London, vol 4A, p 43–122

Lojda Z 1975 Suitability of the azocoupling reaction with 1-naphthyl-β-D-glucoside for the histochemical demonstration of lactase (lactase-β-glucosidase complex) in human enterobiopsies. Histochemistry 43: 349–353

Lojda Z 1976a Der Wert der Darmbiopsie für die Diagnostik und Beurteilung der Therapie des Malabsorptionssyndroms. Ergebnisse der experimentellen Medizin 27: 189–196

Lojda Z 1976b Cytochemical study on epithelial lymphocytes of the human jejunum. Sbornik lékařský 78: 263–268

Lojda Z 1981 Proteinases in pathology. Usefulness of histochemical methods. Journal of Histochemistry and Cytochemistry 29: 481–493

Lojda Z 1984 Die Bedeutung enzymhistochemischer Methoden bei der Differentialdiagnose des Malabsorptionssyndroms. Acta histochemica (Jena) Supplement 30: 189–199

Lojda Z 1985 The importance of protease histochemistry in pathology. Histochemical Journal 17: 1063–1089

Lojda Z 1988 Proteases of cell surface membrane. Achievements of the histochemical approach. Journal of Histochemistry and Cytochemistry 36: 889A

Lojda Z, Gossrau R 1983 Histochemical demonstration of enteropeptidase activity. New method with a synthetic substrate and its comparison with the trypsinogen procedure. Histochemistry 78: 251–270

Lojda Z, Jodl J 1974 Histochemical investigation of enterokinase in duodenal and jejunal biopsies of children with malabsorption syndrome. Československá Gastroenterologie a Výživa 28: 532–539

Lojda Z, Kraml J 1971 Indigogenic methods for glycosidases. III. An improved method with 4-Cl-5-Br-3-indolyl-β-D-fucoside and its application in studies of enzymes in the intestine, kidney and other tissues. Histochemie 25: 195–207

Lojda Z, Mališ F 1972 Histochemical demonstration of enterokinase. Histochemie 32: 23–29

Lojda Z, Gossrau R, Schiebler T H 1979a Enzyme histochemistry. A laboratory manual. Springer-Verlag, Berlin

Lojda Z, Kociánová J, Mařatka Z 1979b Histochemistry of the human duodenal mucosa with special reference to the gradient of activities of the brush border enzymes. Scandinavian Journal of Gastroenterology 14. Supplement 54: 7–13

Lojda Z, Frič P , Jodl J, Chmelik V 1971 Cytochemistry of the human jejunal mucosa in the norm and in malabsorption syndrome. Current Topics in Pathology 52: 1–63

Lojda Z, Kolínská J, Šmídová J, Gossrau R 1987a Are synthetic substrates suitable for the diagnosis of sucrase-isomaltase deficiencies in biopsies of jejunal mucosa of patients with malabsorption syndrome? Histochemical Journal 19: 615

Lojda Z, Kotalová R, Šmídová J, Ueberberg H 1987b Histochemical demonstration of proteases in mast cells of the jejunal mucosa of patients with malabsorption syndrome. Histochemical Journal 19: 601

Lojda Z, Slabý J, Kraml J, Kolínská J 1973 Synthetic substrates in the histochemical demonstration of intestinal disaccharidases. Histochemie 34: 361–369

Lojda Z, Šmídová J, Kolínská J, Kraml J 1984 A comparative study of lactase and sucrase-isomaltase activities and immunoreactivities in jejunal biopsies of patients sufferring from the malabsorption syndrome. Histochemical Journal 16: 373–376

Morson B C, Dawson I M P 1979 Gastrointestinal pathology, 2nd edn. Blackwell, London

Piris J 1989 Malabsorption and protein intolerance. In: Whitehead R (ed) Gastrointestinal and oesophageal pathology, Churchill Livingstone, Edinburgh, p 468–496

14. The diagnosis of Hirschsprung's disease and pseudo-obstruction

B. D. Lake

HIRSCHSPRUNG'S DISEASE

The diagnosis of Hirschsprung's disease by demonstration of the absence of ganglion cells in the distal portions of the large intestine is the standard method employed in most routine histopathology laboratories.

Full-thickness biopsies of rectum, performed under general anaesthetic, present few problems to the pathologist but have the disadvantage that such large biopsies may cause scarring and surgical complications in the subsequent pull-through procedure. Also, many patients present in the neonatal period, are often severely ill and a general anaesthetic is an unnecessary additional hazard. For these reasons suction rectal biopsies are now generally preferred and these may be taken without anaesthetic on the ward or as an outpatient procedure.

The density of ganglion cells varies in the last few centimetres of the rectum, and studies by Aldridge and Campbell (1968) emphasize the importance of knowing as precisely as possible the site of the biopsy in relation to the pectinate line. The diagnosis of 'not Hirschsprung's disease' is straightforward since the presence of ganglion cells excludes Hirschsprung's disease. However, as there is a normal hypoganglionic region extending 1–2 cm from the pectinate line, a minimum of 50 serial sections of an adequate biopsy must be examined before concluding no ganglion cells are present.

Acetylcholinesterase in the diagnosis of Hirschsprung's disease

The problems outlined above have led to other approaches to diagnosis. Early reports by Meier-Ruge and his colleagues (1972, 1974) suggested that the diagnosis could be made by assessment of the numbers of nerve fibres showing acetylcholinesterase activity in the mucosa of suction rectal biopsies. These claims were not upheld by all investigators, but our experience at The Hospital for Sick Children in London over a period of 16 years (1975–1990), and the experience of others with large series, has confirmed Meier-Ruge's observations. Provided the guidelines outlined below are followed and the biopsies are taken at known, carefully defined distances from the pectinate line, there should be no false-positives and no false-negatives (Lake et al 1989). The most usual causes of false positive or false negative results are a) lack of experience in interpretation, b) poorly prepared sections and c) poorly stained sections.

The method for demonstrating acetylcholinesterase-positive nerve fibres should be permanent and show the fibres readily with good contrast from the background. Koelle's modification of Gomori's method fades, particularly if mounted in synthetic media, and Karnovsky and Roots' direct colouring method sometimes has poor contrast. In 1972 Hanker, Anderson and Bloom described a method for ultrastructural studies in which the reaction product (copper ferrocyanide) of the Karnovsky and Roots' medium was made more electron dense by formation of osmium black.

Suction rectal biopsy diagnosis

Lake et al (1978) used Hanker's method for the light microscopic demonstration of acetylcholinesterase-positive nerve fibres in a series of suction rectal biopsies for the diagnosis of

Hirschsprung's disease, and this series has now been extended and includes biopsies from over 300 patients with Hirschsprung's disease. Biopsies are taken at a nominal 2 cm and 5 cm from the pectinate line. In practice, this means just above the pectinate line and as high as can reasonably be taken without risk of perforation. This allows the detection of the shortest of aganglionic segments. A third biopsy taken at an intermediate level is processed for routine microscopy. Cryostat sections of the suction rectal biopsies snap-frozen (supported on animal liver, thin wedges of 12% gelatin, or in OCT compound) are cut at 10 μm and air dried. Fixation in formal-calcium for 30 seconds is followed by a brief wash and incubation for 1 hour in the medium, as described in the Appendix. The intensifying agent is p-phenylene diamine (2HCl) instead of the original DAB. Although an inhibitor of pseudocholinesterase is added to the reaction medium, the inhibition of activity is not complete and smooth muscle cells have a diffuse reaction with accentuation at the edges of the cells, giving a network appearance. Tetraisopropylpyrophosphoramide (isoOMPA) is the usual inhibitor, but it may be difficult to obtain, and ethopropazine at the same concentration (10^{-4}) is a suitable alternative. If no inhibitor is added the pseudocholinesterase activity of the smooth muscle may mask the more subtle changes and could account for some claims of false negative

results. The inhibitor appears to have no effect on the esterases of the cells found in the lymphoreticular aggregates and these still give a strongly positive reaction.

In the normal rectum and descending colon only a few fine nerve fibres are demonstrated in the lamina propria and muscularis mucosae. If the biopsy has been taken low, i.e. close to the pectinate line, occasional nerve trunks and nerve fibres may be found in the submucosa, together with an apparent increase in numbers of blood vessels (Fenger 1988). In this region ganglion cells are scarce and on histological examination a false impression of Hirschsprung's disease may be obtained. A fine network of acetylcholinesterase-positive nerve fibres is present around blood vessels and this serves as an internal control for the method. Ganglion cells are sometimes heavily stained, but this is probably mainly due to the density of nerve fibres in the ganglia. Examination under high power reveals a stippled appearance on the neuronal cell surface, and nerves entering or leaving the ganglia may be seen. The acetylcholinesterase reaction is particularly helpful for the easy identification of the clusters of immature ganglion cells often found in the submucosa of the neonate (Fig. 14.1).

Hirschsprung's disease (Holschneider 1982) may be divided into three main groups. The first and most common group comprises those patients with

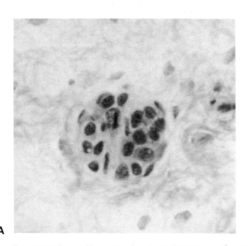 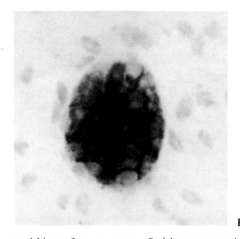

Fig. 14.1 'Immature' ganglion cells in the submucosa of a suction rectal biopsy from a neonate. Serial cryostat sections stained with H & E (**A**) and for acetylcholinesterase activity (**B**). The ganglion cells are more readily identified in acetylcholinesterase preparations. ×450

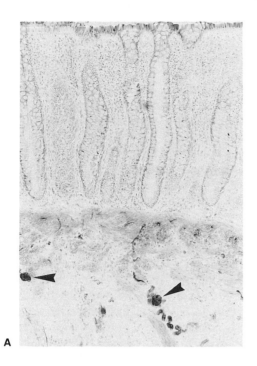

A

aganglionosis as far as the rectosigmiod junction (short segment disease). The second group of patients has aganglionosis extending beyond the rectosigmiod junction, but not involving the small bowel (long segment disease). The third group, which accounts for between 2% and 14% of all Hirschsprung's disease has aganglionosis extending into the small bowel, sometimes as far as the duodenum or stomach (total colonic aganglionosis).

In both short segment disease and long segment disease the appearance in the acetylcholinesterase preparations of suction rectal biopsies is the same. An increase in acetylcholinesterase-positive nerve fibres is present in the muscularis mucosae and this is accompanied by an absence of ganglion cells and an increase in nerve fibres and nerve trunks in the submucosa (Fig. 14.2). Increases in nerve fibres in the lamina propria are evident in older patients where the fibres are coarse, extend to the mucosal surface cells and may be seen running

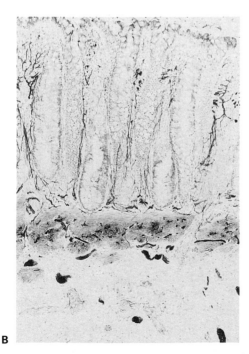

B

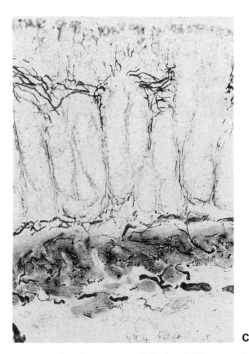

C

Fig. 14.2 Cryostat sections of suction rectal biopsies stained to demonstrate acetylcholinesterase activity in: **A** Normal, with a few fine nerve fibres in the muscularis mucosae, and submucosal ganglion cells (arrows). **B** Hirschsprung's disease. Increased numbers of nerve fibres are present in the lamina propria and muscularis mucosae, with small nerve trunks in the submucosa. **C** Hirschsprung's disease. In the older patient the increase in nerve fibres is marked and fibres may be seen running parallel to the mucosal surface

parallel to the mucosal surface (Fig. 14.2C). However, in the neonate the increase of nerve fibres in the lamina propria may not be apparent (Fig. 14.3) and this too has lead to claims of false negative results. It is clear, however, that there is a steady increase with age (Hersig et al 1979, de Brito & Maksoud 1987) and the age of the patient must be taken into consideration before a conclusion can be reached.

The increase of nerve fibres in the muscularis mucosae of patients with Hirschsprung's disease is present regardless of age and is found even at 30 weeks' gestation. Some authors subdivide short segment disease into short and ultrashort segment, reserving the latter for cases in which the aganglionic segment involves only the distal part of the rectum. There is no pathological justification for this subdivision. The changes are identical and the term causes confusion with the disorder achalasia of the internal sphincter also known as ultra short segment disease. The fact that in Hirschsprung's disease the length of the aganglionic segment is variable emphasizes the need for biopsies at several different levels as defined above.

In contrast, in total colonic aganglionosis the changes in the biopsy may be mild with no ap-parent increase in the number of nerve fibres in the muscularis mucosae and this can account for a false negative result. Submucosal nerve trunks can be increased, but this will require an adequate biopsy. Other cases of total colonic aganglionosis show changes in suction rectal biopsies which are identical with the more common short and long segment diseases.

The diagnosis of short and long segment Hirschsprung's disease may need to be made as a matter of clinical urgency, particularly in the neonate presenting with obstruction and at risk for enterocolitis. While, in other cases, the diagnosis is less urgent and can be on a 'next day' basis, the diagnosis in the neonate may need to be made within an hour so that theatre time can be booked. In this situation cryostat sections of snap-frozen rectal suction biopsies stained with haematoxylin and eosin are adequate to exclude Hirschsprung's disease if ganglion cells are present. If none are found, or their identity is not clear, a rapid acetyl-cholinesterase technique provides the necessary backup and a definitive diagnosis (see Fig. 14.3). Details of the rapid method adapted from Hedreen et al (1985), which can be accomplished within 30 minutes, can be found in Appendix 5. In spite of the improvement of staining time the method is

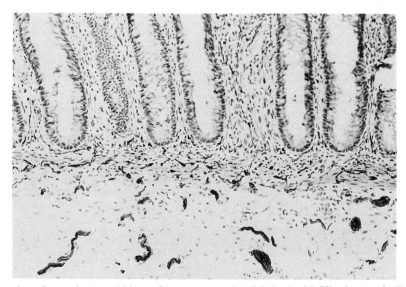

Fig. 14.3 Cryostat section of a suction rectal biopsy from a neonate (aged 3 days) with Hirschsprung's disease. Rapid acetylcholinesterase method. In the neonate the increase in nerve fibres is less prominent and is confined to the muscularis mucosae and submucosa. The pattern, although weaker than in the older cases, is pathognomonic and is readily distinguishable from the normal (see Fig. 14.2A).

still too long for use on the seromuscular biopsies taken during the pullthrough operation. A standard haematoxylin and eosin stain is adequate for the recognition of ganglion cells (neurons) or nerve trunks in the myenteric plexus, although some may prefer an additional rapid esterase stain which can be accomplished in about 5 minutes (Garrett & Howard 1969).

Other methods for the detection of nerves and neurons

The recognition of ganglion cells in suction rectal biopsies may pose problems for those who have limited experience in this area, but the acetylcholinesterase methods referred to in the preceding paragraphs will give a definitive diagnosis. The overall interpretation may be complicated in a few cases by the presence of extravasated blood in the lamina propria, giving rise to an appearance of nerve fibres. To overcome these difficulties a variety of monoclonal and polyclonal antibodies have been applied to cryostat sections and to routine sections. The application of immunocytochemical techniques adds an unnecessary complication, and does not improve the results of a properly performed acetylcholinesterase reaction. In many instances the increase in nerve fibres in Hirschsprung's disease is not shown by immunocytochemistry. In routine sections anti-S100 protein shows nerves and ganglia in which neurons can be recognized by their negativity. Of the other antibodies used PGP 9.5 appears to be the most useful in the diagnosis of Hirschsprung's disease (Mackenzie & Dixon 1987).

PSEUDO-OBSTRUCTION

Disorders causing symptoms of intestinal obstruction (termed pseudo-obstruction because no mechanical or anatomical cause can be found macroscopically) can be broadly divided into two main groups:

1. Disorders affecting nerves
2. Disorders affecting smooth muscle.

The diagnosis for both groups is based on an approach which should encompass routine histology,

histochemistry, electron microscopy and silver staining (Lake 1989).

Disorders affecting nerves

The normal mechanical function of the intestine depends on an integrated nerve supply interacting with the smooth muscle cells of the circular and longitudinal muscle coats. Derangement of the nerve supply can lead to pseudo-obstruction by the inco-ordinated muscle response to faulty stimulation. Increases or decreases in the neural population may be seen in some cases, but in others the evidence for a neural cause is based on motility studies, there being normal muscle structure with no evidence by microscopy of an abnormality of nerves. Proliferation of submucosal nerve trunks (Fig. 14.4), sometimes with ganglioneuromatous formations, is a feature of phaeochromocytoma and medullary carcinoma of the thyroid (as part of the multiple endocrine neoplasia syndrome) and, if found in a biopsy, the serum calcitonin and urinary VMA levels should be measured. Hyperplasia of the myenteric and submucosal plexuses, sometimes with submucosal ganglioneuromata, are found in *hyperganglionosis*, a condition which is also known as intestinal neuronal dysplasia (Fadda et al 1983, Garrett & Howard 1984). In the florid example there is, in suction rectal biopsies, an increase in fine nerve fibres (sometimes with a flattened ribbon-like appearance) accompanied by ganglion cells in the lamina propria, hyperplasia of the submucosal plexus and numerous nerves in the submucosa (Fig. 14.5). The muscularis mucosae may be normal. Some reports describe a decrease in adrenergic nerve fibres (as shown by FIF techniques) in patients with hyperganglionosis presenting in the neonatal period (Fadda et al 1983). The florid cases are readily recognizable in acetylcholinesterase preparations, but there is at present no accepted lower limit of abnormality or upper limit of normality, and slightly different criteria might account for differences in reported incidences in different countries. The presence of single ganglion cells in the lamina propria, although unusual, may be within normal limits and has indeed been seen in hypoganglionosis, and to consider a

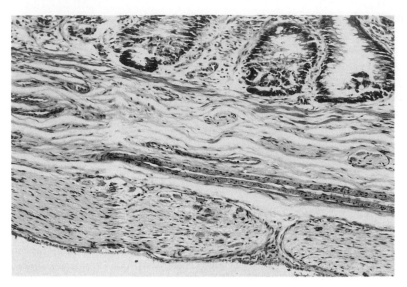

Fig. 14.4 Paraffin section of a suction rectal biopsy stained with HE. Numerous large nerve trunks containing ganglion cells (ganglioneuroma) in the submucosa of a patient presenting with obstruction, who later died with a medullary carcinoma of the thyroid.

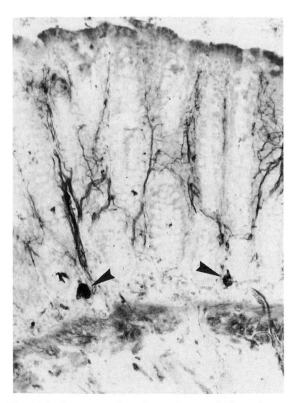

Fig. 14.5 Cryostat section of a suction rectal biopsy from a patient with hyperganglionosis (intestinal neuronal dysplasia); acetylcholinesterase. Increased numbers of nerve fibres are present in the lamina propria, together with ganglion cells (arrows). The muscularis mucosae is normal. Ganglion cells were present in the submucosa (not shown).

diagnosis of hyperganglionosis there should also be an increase of nerve fibres in the lamina propria and some indication of an excess of nerve fibres in the submucosa. Deeper biopsies, including the myenteric plexus, may be easier to interpret since the ganglia may be almost continuous, with projections into the circular muscle coat. Hyperganglionosis may also be present proximal to the aganglionic segment in Hirschsprung's disease and account for the residual symptoms in some patients who have had pull-through operations (Puri et al 1977, Fadda et al 1987).

In contrast, *hypoganglionosis* may be congenital or acquired. Routine sections of full thickness biopsies will generally be adequate for assessment of numbers of ganglion cells in the myenteric plexus. Suction biopsies cannot be used. Allowance should be made for any apparent reduction of normal numbers of neurons caused by dilatation of the bowel. Generally, very small ganglia with single scattered neurons indicates hypoganglionosis. However, normal variation has not been adequately documented and the criteria for the diagnosis of hypoganglionosis are vague. Acetylcholinesterase preparations are helpful in identifying neurons and size of ganglia (Fig. 14.6). Some patients also show a loss of nerve fibres within the circular and longitudinal muscle coats (Puri et al 1977b). Hypoganglionosis is a feature in a selected group of patients with chronic constipation (Howard et al 1984).

Silver staining has been used to demonstrate abnormalities of the argyrophil cells in the myenteric plexus (Smith 1982, Krishnamurthy & Schuffler 1987). It is important to know the age of the patient, since argyrophil cells (Fig. 14.7) may not be demonstrable in the normal until several weeks after birth, and absence at this age could still be normal. The disappearance of argyrophil cells in a neural disorder is probably a secondary phenomenon, but too few patients have had serial studies to confirm this suggestion. The technique is described in Appendix 2.

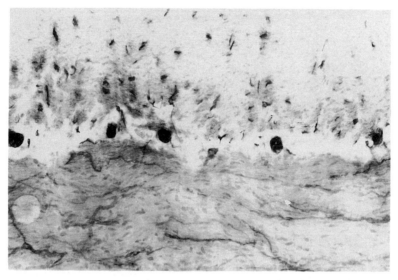

Fig. 14.6 Cryostat section of resected colon from a patient demonstrating hypoganglionosis. The numbers of ganglion cells and size of ganglia are decreased, but are readily visible in this acetylcholinesterase preparation.

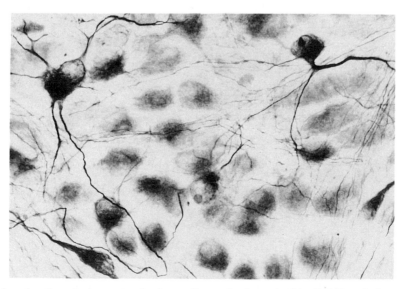

Fig. 14.7 Horizontal section through the myenteric plexus of normal colon stained by Smith's technique to show the argyrophilic neurons with argyrophilic processes. The argyrophobic neurons display no nerve processes (normal) in this preparation from an infant aged 5 months who died without evidence of bowel dysfunction.

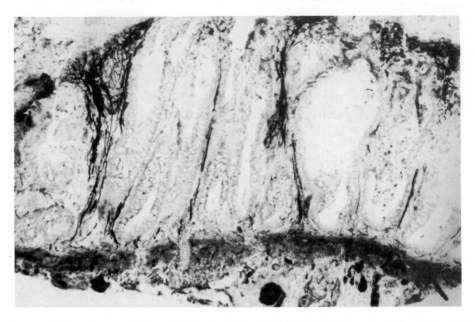

Fig. 14.8 Colonoscopy biopsy from an infant with allergic colitis. The histological picture is of ulcerative colitis, but there is a marked increase of thickened ribbon-like nerve fibres in the lamina propria shown by the acetylcholinesterase method. The muscularis mucosae shows no increase in nerve fibres and ganglion cells are present in the submucosa. A crypt abscess is present. ×90

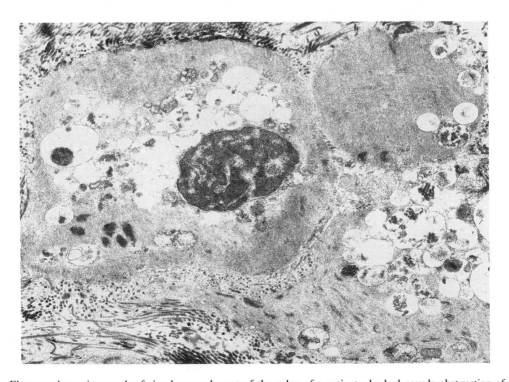

Fig. 14.9 Electron photomicrograph of circular muscle coat of the colon of a patient who had pseudo-obstruction of muscular origin. Degenerative changes and an increase of collagen fibres can be seen.

An increase in acetylcholinesterase-positive nerves in the lamina propria may be seen in some inflammatory bowel disease. The fibres may be coarse, but are generally ribbon-like and may be numerous, extending from the submucosa (Fig. 14.8). The muscularis mucosae is not involved. It is probable that the changes are secondary to the inflammatory process. A similar pattern in vasoactive intestinal polypeptide (VIP) — positive nerves in Crohn's disease has been reported (Bishop et al 1980). The increase in VIP-positive nerves was not confined to the inflamed mucosa, but was also present in the histologically uninvolved rectal mucosa.

Disorders affecting smooth muscle

In those disorders affecting nerves and ganglia, the muscle cells of the circular and longitudinal coats show no evidence of abnormality. A wide variety of changes of smooth muscle cells can be found in non-neural causes of pseudo-obstruction. Loss of, or destruction of, muscle cells is usually accompanied by replacement with connective tissue, and this is best demonstrated by staining with the picro-sirius technique (Appendix 2). Sometimes the muscle fibres show lytic changes, which are highlighted by an acid phosphatase reaction, but the most important investigation is electron microscopy, which will identify the organization of the myofilaments (Fig. 14.9) and define the nature of the muscle disruption (Puri et al 1983, Lake 1988, 1989). The cases so far are not numerous and many more need to be studied by the appropriate methods before any real classification can be made on pathological grounds. The many conditions which give rise to pseudo-obstruction are listed by Krishnamurthy and Schuffler (1987).

REFERENCES

Aldridge R T, Campbell P E 1968 Ganglion cell distribution in the normal rectum and anal canal. A basis for the diagnosis of Hirschsprung's disease by anorectal biopsy. Journal of Pediatric Surgery 3: 475–490

Bishop A, Polak J M, Bryant M G, Bloom S R, Hamilton S 1980 Abnormalities of vasoactive intestinal polypeptide containing nerves in Crohn's disease. Gastroenterology 79: 853–860

de Brito I A, Maksoud J G 1987 Evolution with age of the acetylcholinesterase activity in rectal suction biopsy in Hirschsprung's disease. Journal of Pediatric Surgery 22: 424–430

Fadda B, Maier W A, Meier-Ruge W, Scharli A, Daum R 1983 Neuronale intestinale Dysplasie. Eine kritische 10-Jahres-Analyse klinischer und bioptischer Diagnostik. Zeitschrift für Kinderchirurgie 38: 305–311

Fadda B, Pistor G, Meier-Ruge W, Hofmann von Kap-herr S, Muüntefering A, Espinoza R 1987 Symptoms, diagnosis and therapy of neuronal intestinal dysplasia masked by Hirschsprung's disease. Pediatric Surgery International 2: 76–80

Fenger C 1988 Histology of the anal canal. The American Journal of Surgical Pathology 12: 41–55

Garrett J R, Howard E R 1969 Histochemistry and the pathology of Hirschsprung's disease. Proceedings of the Royal Microscopical Society 4: 76–78

Garrett J R, Howard E R 1981 Myenteric plexus of the hind-gut: developmental abnormalities in humans and experimental studies. In: Development of the autonomic nervous system. Ciba Foundation Symposium 83. Pitman, London, p 326–354

Hanker J S, Anderson W A, Bloom F E 1972 Osmiophilic polymer generation: Catalysis by transition metal compounds in ultrastructural cytochemistry. Science 175: 991–993

Hedreen J C, Bacon S J, Price D L 1985 A modified histochemical technique to visualize acetylcholinesterase-containing axons. Journal of Histochemistry and Cytochemistry 33: 134–140

Hirsig J, Briner J, Rickham P P 1979 Problems in the diagnosis of Hirschsprung's disease due to false negative acetylcholinesterase reaction in suction biopsy in neonates. Zeitschrift fur Kinderchirurgie 26: 242–247

Holschneider A M (ed) 1982 Hirschsprung's disease. Hippokrates, Stuttgart

Howard E R, Garrett J R, Kidd A 1984 Constipation and congenital disorders of the myenteric plexus. Journal of the Royal Society of Medicine 77 supplement 3: 13–19

Krishnamurthy S, Schuffler M D 1987 Pathology of neuromuscular disorders of the small intestine and colon. Gastroenterology 93: 610–639

Lake B D 1988 Observations on the pathology of pseudo-obstruction. In: Milla P J (ed) Disorders of gastrointestinal motility in childhood. Wiley, Chichester

Lake B D 1989 Hirschsprung's disease and pseudo-Hirschsprung's disease. In: Whitehead R (ed) Gastrointestinal and oesophageal pathology. Churchill Livingstone, Edinburgh

Lake B D, Malone M T, Risdon R A 1989 Acetylcholinesterase in the diagnosis of Hirschsprung's disease, including a comment on intestinal neuronal dysplasia. Pediatric Pathology 9: 351–354

Lake B D, Puri P, Nixon H H, Claireaux A E 1978 Hirschsprung's disease. An appraisal of histochemically demonstrable acetylcholinesterase activity in suction rectal biopsy specimens as an aid to diagnosis. Archives of Pathology and Laboratory Medicine 102: 244–247

Mackenzie J M, Dixon M F 1987 An immunohistochemical study of the enteric neural plexi in Hirschsprung's disease. Histopathology 11: 1055–1066

Meier-Ruge W 1974 Hirschsprung's disease: its aetiology, pathogenesis and differential diagnosis. In: Current topics in pathology, vol 59. Springer, Berlin, p 1931–179

Meier-Ruge, W, Lutterbeck P M, Herzog B, Morger R, Schärli A 1972 Acetylcholinesterase activity in suction biopsies of the rectum in the diagnosis of Hirschsprung's disease. Journal of Pediatric Surgery 7: 11–17

Puri P, Lake B D, Nixon H H 1977b Adynamic bowel syndrome. Report of a case of disturbance of the cholinergic innervation. Gut 18: 754–759

Puri P, Lake B D, Nixon H H, Mishalany H, Claireaux A E 1977a Neuronal colonic dysplasia: An unusual association of Hirschsprung's disease. Journal of Pediatric Surgery 12: 681–685

Puri P, Lake B D, Gorman F, O'Donnell B, Nixon H H 1983 Megacystis-microcolon-intestinal hypoperistalsis syndrome. A visceral myopathy. Journal of Pediatric Surgery 18: 64–69

Smith B 1982 The neuropathology of pseudo-obstruction of the intestine. Scandinavian Journal of Gastroenterology (supplement) 71: 103–109

15. Histochemistry in diagnostic assessment of liver biopsy

B. C. Portmann

Besides H & E staining of traditional paraffin sections, the assessment of liver biopsy specimens often benefits from the use of special techniques. Collagen staining and/or reticulin silver impregnation, PAS after diastase digestion, Perls' reaction for iron and Shikata's orcein method are now routinely performed in many laboratories. In addition, more demanding histochemical methods and a range of antibodies, some of them monoclonal, are available to the pathologist for the detection of various tissue components. The potential information obtained from the use of these techniques in selected liver conditions are summarized in this chapter.

DEMONSTRATION OF VIRAL COMPONENTS

Hepatitis B virus (HBV)

The complete — 42 nm diameter — human HBV (Dane particle) consists of a core or nucleocapsid (HBcAg) containing the viral DNA molecule and surrounded by an outer envelope or hepatitis B surface antigen (HBsAg). Following infection the virus replicates within hepatocytes and both intact virions and excess protein coat material (HBsAg) are released into the blood. The latter is not infectious and may be the only material present in the serum and liver of a number of HBsAg carriers. Indication of active viral replication and infectivity of either blood products or liver tissue is ideally obtained by the demonstration of HBV-DNA in the serum. Alternatively, a third antigenic determinant — hepatitis B e antigen (HBeAg) — a soluble fraction of HBc, may be detectable in the serum and closely reflects active viral replication. The demonstration of HBc in the tissue also correlates with HBV-DNA and HBeAg seropositivity and therefore with active viral replication. These three viral components, but mainly HBsAg and HBcAg, may easily be stained in hepatocytes using commercially available reagents on fresh, or formalin fixed, paraffin embedded tissue (Gudat et al 1975, Huang 1975).

HBsAg material, owing to its content of disulphide bonds, is strongly stained by either Shikata's orcein (brown), aldehyde fuchsin (purple-violet) or Victoria blue (blue) methods in the form of cytoplasmic inclusions of various shapes (Shikata 1974, Deodhar et al 1975). The pattern of staining approximately corresponds to the less sensitive 'ground glass' transformation of hepatocyte cytoplasm observed on H & E stained sections and is identical to the more specific immunoperoxidase or immunofluorescent staining (Figs 15.1 and 15.2) (Portmann et al 1976). An exception is the so-called membranous HBs, a honeycomb-like pattern of cell membrane staining which is clearly demonstrated by immunostaining only (Ray et al 1976).

In contrast to HBsAg, HBcAg is preferentially localized in the nucleus and, to a lesser extent, in the cytoplasm of hepatocytes where it can be demonstrated by direct or indirect immunostaining with monospecific anti-HBc antisera (Fig. 15.3).

The availability of sensitive kits to detect HBV Ag and Ab in the serum and the uneven distribution of HBV components in the liver have somewhat reduced the need to demonstrate HBV in liver biopsy specimens. However, routine orcein

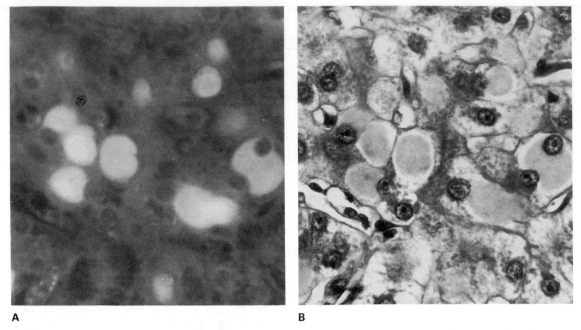

Fig. 15.1 Specific immunofluorescence for HBsAg (**A**) produces a similar pattern of staining as the ground glass change seen on H & E on the same liver section (**B**). ×450

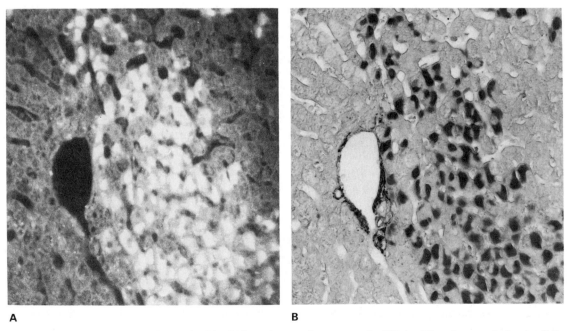

Fig. 15.2 Paraffin section of liver first stained by indirect immunofluorescence for HBsAg (**A**), subsequently by the Shikata's orcein method (**B**). An identical pattern of staining is achieved by both methods. ×160

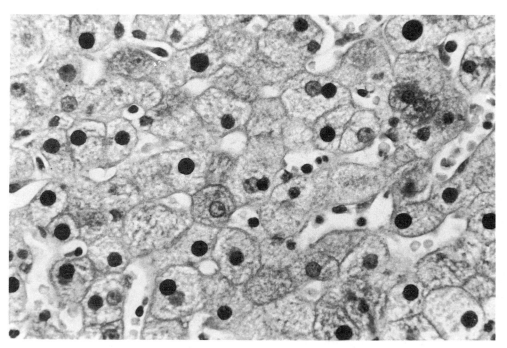

Fig. 15.3 Most of the hepatocyte nuclei show a positive reaction in this paraffin section stained by an indirect peroxidase method using monospecific anti-HBc antibodies. ×450

staining remains a rapid and simple method of tissue screening. Serum and tissue results may be at times discordant and *both* HBV serum profile *and* tissue expression are necessary investigations.

The patterns of distribution, relative proportion and degree of HBc and HBs staining in the liver reflect the efficacy of the host immune response in eliminating HBV and/or the degree of viral integration into the host genome. According to Bianchi et al (1987), four main patterns can be distinguished and summarized as follows:

1. In acute self-limited hepatitis the liver tissue is usually devoid of any stainable viral component.

2. Extensive HBc nuclear and HBs membranous staining in presence of a chronic, non-aggressive hepatitis is the pattern mostly observed in immunocompromised hosts, such as transplant recipients.

3. Prominent cytoplasmic, non-membranous HBs in absence of HBc staining and only minimal inflammatory changes is found in low-infectivity carriers in whom viral DNA may be largely in-

tegrated into the host genome.

4. Focal expression of nuclear HBc and membranous/cytoplasmic HBs reflects a partial immune insufficiency allowing both continuous tissue damage and viral replication with an inverse relationship between the two. This pattern may be associated with variable severity of chronic persistent or active hepatitis, the degree of the inflammatory activity being inversely proportional to the intensity of HBc staining.

Molecular hybridization techniques performed on extracts of liver tissue (Southern 1975, Bréchot et al 1980) have detected two different forms of HBV-DNA: episomal (free, low molecular weight single-stranded) and integrated into the host genome. HBV-DNA has also been detected in the tissue by in situ hybridization, in a predominant cytoplasmic localization independantly of the presence of HBcAg, suggesting non-encapsidated HBV-DNA (Burrell et al 1982, Rijntjes et al 1985). Recently the technique has been modified for electron microscopy allowing HBV-DNA localization at the ultrastructural level (Schirmacher et al 1988).

Hepatitis delta virus (HDV)

HDV is a defective RNA virus which can only replicate in the presence of HBV. HDV co- or superinfection is generally accepted as an aggravating factor leading to an increased severity of acute and chronic hepatitis B respectively (Rizzetto et al 1986). Similarly to HBcAg, HDVAg can be demonstrated preferentially in the nucleus and, to a lesser extent, in the cytoplasm of hepatocytes using human sera containing high-titre anti-HDV antibodies. The direct technique should be used because a secondary antihuman antiserum may produce false positive results when reacting with immunoglobulins, which are commonly found in hepatocyte nuclei. The anti-HDV antiserum must also be diluted adequately to abolish the weak anti-HBc activity invariably present in these human sera.

Cytomegalovirus (CMV)

CMV is a major cause of morbidity in allograft recipients and a potential risk factor for the development of chronic liver graft rejection (O'Grady et al 1988). The rapid detection of CMV in the liver is therefore critically important to allow early alteration in immunosuppression and antiviral therapy. Until now two different techniques have been successfully applied:

1. Immunostaining using monoclonal antibodies against early CMV antigens (Ten Napel et al 1984).
2. In situ hybridization using a commercially available biotin-labelled CMV-specific DNA probe (Naoumov et al 1988, Masih et al 1988).

Both techniques are sensitive and detect positive nuclear and, to a lesser extent, cytoplasmic CMV in cases which do not show obvious inclusions on H & E stained sections. The former, which preserves the morphology better, has the disadvantage of the inconsistency of commercial antibodies. The latter is dearer and technically more difficult, particularly the digestion and hybridization steps, which limits its use as a routine procedure.

COPPER AND COPPER-ASSOCIATED PROTEIN

Within hepatocytes copper may accumulate in two main forms:

1. A cytosolic fraction in which copper, bound to a low-molecular weight metallothionein, is highly soluble and not demonstrable histochemically.
2. A lysosomal fraction, less soluble, in which copper is bound to a sulphydryl-rich protein complex or copper-associated protein (CAP), of which a large portion is believed to be polymerized metallothionein.

CAP is strongly stained as coarse, dark brown granules by the Shikata's orcein method, the first step of which consists of copper release followed by oxidation of the exposed sulphydryl groups using an acidified permanganate solution. The intense granular staining is clearly different from the HBs material staining (Fig. 15.4A). An identical distribution of blue granules is obtained with the Victoria blue method. In the less soluble form copper may be directly stained by rubeanic acid (green/black) or p-dimethylaminobenzilidine rhodanine (red/orange) (Fig. 15.4C). The latter is preferred for giving the most consistent results and for being easier to distinguish from bile and iron pigments (Irons et al 1977). Discrepancy between CAP and copper staining may result from prolonged fixation and/or the use of acid fixatives, which removes copper but does not interfere with CAP staining.

Copper and CAP are normally found in fetal and newborn livers and therefore of little diagnostic assistance in this age group (Evans et al 1980). Their persistence in biopsy specimens obtained later, e.g. after surgical correction of extrahepatic biliary atresia or in cases of alpha-l-antitrypsin deficiency, is an indication of continuous disease activity. In adults, periportal CAP staining is a useful marker of biliary diseases (Salaspuro & Sipponen 1976), in particular primary biliary cirrhosis and primary sclerosing cholangitis in which

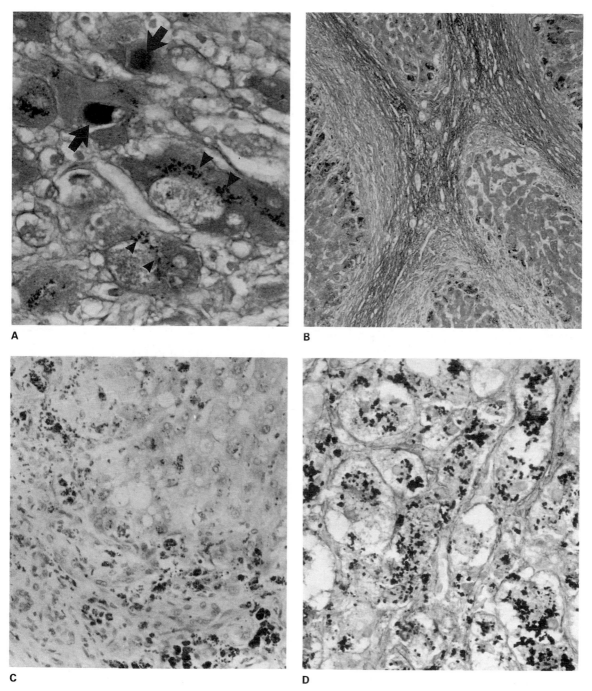

Fig. 15.4 Copper and CAP staining in paraffin sections of liver. **A** HBV associated cirrhosis; hepatocytes trapped within fibrosis variably contain HBsAg (arrows) or CAP (arrow heads) (orcein ×450). **B** Regular periseptal deposition of CAP in advanced primary biliary cirrhosis. Note the staining of elastic tissue corresponding to the more matured collagen of the fibrous septa (orcein ×140). **C** Rhodanine demonstration of copper within both hepatocytes and macrophages in Wilson's disease (×200). **D** Indian childhood cirrhosis. There is a widespread deposition of CAP in the form of coarse granules. (orcein ×450)

positive deposits are found from stage 1 onwards, whereas chronic non-biliary diseases, such as chronic active hepatitis, are negative in their precirrhotic stages (Guarascio et al 1983). When cirrhotic transformation has occurred, CAP deposition is a common finding and differences between cirrhosis of various aetiologies reside in the amount and pattern of deposition (Fig. 15.4). Heavy periseptal deposits regularly highlighting the parenchymal limiting plates are characteristic of a biliary cirrhosis (Fig. 15.4B). In cirrhosis due to other causes, deposits are usually not so abundant and more patchy in distribution, probably reflecting groups of hepatocytes trapped by fibrosis or parenchymal nodules disconnected from their normal biliary drainage (Fig. 15.4A).

In Wilson's disease, an autosomal recessive genetic disorder, copper accumulates massively in the liver and other organs. The basic defect is undetermined, but there seems to be an early lack of lysosomal CAP formation and a diminished biliary excretion of copper (Scheinberg & Sternlieb 1984). In the early stages of the disease, liver copper is mostly cytosolic and therefore not stainable, while tissue copper concentration measured by neutron activation gives the highest values. Later in the course of the disease, particularly when cirrhosis has set in, CAP appears in a random fashion, varying from periportal to widespread deposits in a proportion of the cirrhotic nodules only. Characteristic is the occurrence of coarse CAP deposits within both parenchymal and septal macrophages, suggesting a redistribution of CAP and copper following continuous hepatocellular damage (Fig. 15.4C) (Davies et al 1989). Indian childhood cirrhosis (ICC), an uncommon and as yet unexplained form of severe liver disease affecting infants and young children in India (Nayak & Ramalingswami 1975) is associated with gross excess of liver CAP and copper in a diffuse distribution (Portmann et al 1978, Tanner et al 1979) (Fig. 15.4D). Table 15.1 summarizes hepatic copper concentrations, morphological characteristics and diagnostic features in the various liver conditions discussed above.

ALPHA-1-ANTITRYPSIN (AAT)

AAT, the major circulating protein inhibitor (Pi) is an alpha-1-globulin produced in the liver prior

Table 15.1 Liver copper (Cu) concentration, CAP and Cu distribution and major diagnostic features of various liver conditions associated with excess Cu in liver

Disease	Mean hepatic Cu concentration (μg/g dry weight)*	Stainable Cu/CAP in tissue	Tissue distribution	Cells	Major diagnostic features
Early WD	983	0/0	0	0	Urinary Cu ↑ ↑ Caeruloplasmin ↓
Late WD	493	++/+++	Periportal/ random	HC/MA	Urinary Cu ↑ ↑ K-F ring Caeruloplasmin ↓
ICC	1832	+++/+++	Diffuse	HC/MA	Liver histology
PBC	411	with stage of	Periportal/	HC	AMA, APH ↑ ↑, IgM ↑
PSC	244	disease	periseptal	HC	ERCP, APH ↑ ↑, UC
CAH(cirrhotic)	40	(+)/+(+)	Patchy	HC	Autoantibodies/ HBV markers
Other cirrhosis	39	(+)/+(+)	Patchy	HC	Alcohol, iron load, variable
Neonate	435	++/++	Periportal	HC	0
Normal	31	0/0	0	0	0

*Data compiled from Evans et al 1980, Tanner et al 1979 and Vierling 1981.
WD = Wilson's disease, ICC = Indian childhood cirrhosis, PBC = primary biliary cirrhosis, PSC = primary sclerosing cholangitis, CAH = chronic active hepatitis, HC = hepatocyte, MA = macrophages, K-F = Kayser-Fleischer, AMA = anti-mitochondrial antibodies, APh = alkaline phosphatase, ERCP = endoscopic retrograde colangiopancreatography, UC = ulcerative colitis.

to its release into the serum. Several allelic variants of this polymorphic glycoprotein have been recognized and given alphabetic letters according to their relative mobility on acid starch gel. PiMM phenotype found in normal individuals is associated with a normal AAT serum level, whereas deficiency (regarded as less than 20% of normal activity) is usually found with the Z allele. A severe deficiency, such as observed in PiZZ individuals, carries a high risk of developing early panlobular emphysema and/or liver disease ranging from cholestasis in the neonate to chronic active hepatitis, childhood and adult cirrhosis and possibly primary liver cancer (Ishak & Sharp 1987).

In homo- or heterozygotes for the Z allele, a basic genetic defect leads to synthesis of an abnormal AAT glycoprotein which cannot be transported out of the hepatocytes and accumulates in the cisternae of the endoplasmic reticulum as an amorphous, moderately electron-dense material. Histologically the material is identifiable in the form of varying-sized eosinophilic globules in the cytoplasm of periportal hepatocytes; the globules are PAS positive after diastase digestion and immunoreactive to antisera to AAT (Fig. 15.5).

The size of the globules is an important consideration (Clausen et al 1984). Inclusions greater than 3 μm in diameter with a ring-shaped peripheral immunoreaction are pathognomonic of the Z allele. They are frequently found in liver specimens from PiZ-children and adults, but not in infants under 2 or 3 months of age. In early biopsies from PiZZ infants with neonatal jaundice, AAT is deposited in the form of dust-like granules

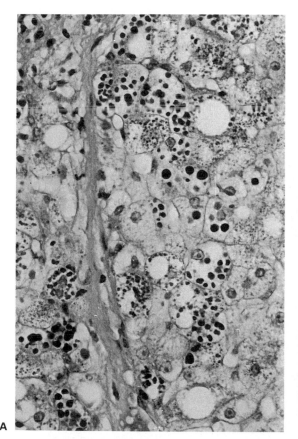
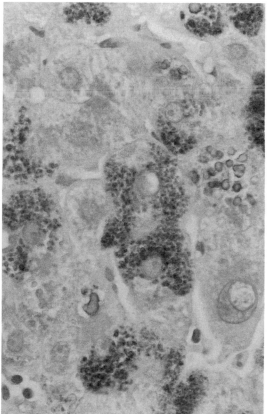

Fig. 15.5 AAT deficiency. Paraffin sections of liver. **A** Extensive deposition of darkly stained globules of varying size involves most of the hepatocytes on both sides of a thin fibrous septum (PAS after diastase ×240). **B** Stained to show AAT by the indirect PAP method. The larger globules exhibit a characteristically peripheral immunoreaction (×450).

indistinguishable from CAP, haemosiderin and bile pigment, which are invariably present. Even specific immunostaining does not allow a clear discrimination at this age. Positive results are obtained in other cholestatic diseases, suggesting that AAT might at times be poorly exported by the diseased hepatocyte. In infancy the diagnosis should therefore be excluded by serum Pi phenotyping, particularly in those cases whose histology mimics the changes of extrahepatic biliary atresia.

In children and adults with chronic cholestasis and/or cirrhosis, lipofuscin granules and CAP, both of which are PAS positive, are often present together with AAT. Confusion may be avoided by using the staining sequence 'digestion – Schmorl reaction – PAS', in which lipofuscin and CAP deposits stain blue leaving the AAT material bright red (Filipe & Lake 1983). However, this is now advantageously replaced by the more sensitive and specific immunohistochemical method.

CYTOKERATIN

In addition to their use in tumour pathology, antisera to various subtypes of intermediate filaments, in particular cytokeratins, may help in identifying tissue components (Moll et al 1982, Denk & Lackinger 1986). CAM 5.2 (Becton-Dickinson) stains both bile duct epithelium and the hepatocytes to a lesser and variable extent. The monoclonal antibody AE1 (Hybritech), which recognizes acidic low-molecular weight keratin polypeptides (Tseng et al 1982), selectively stains the bile duct epithelium, but not the normal hepatocyte. AE1 immunostaining is therefore useful in biopsy specimens from neonates to discriminate between apparent or genuine hypoplasia of the small interlobular bile ducts. In the adult it has an application in disorders associated with destruction and progressive loss of these structures, mainly hepatic allograft rejection or vanishing bile duct syndrome, graft-versus-host disease and primary biliary cirrhosis (Portmann & MacSween 1987). In the diseased liver, hepatocytes may show AE1 staining; Mallory's hyaline is particularly reactive and it has been claimed that a distinctive pattern of staining may help in distinguishing alcoholic from non-alcoholic liver disease (Ray 1987).

REYE'S SYNDROME

This syndrome, which predominantly affects

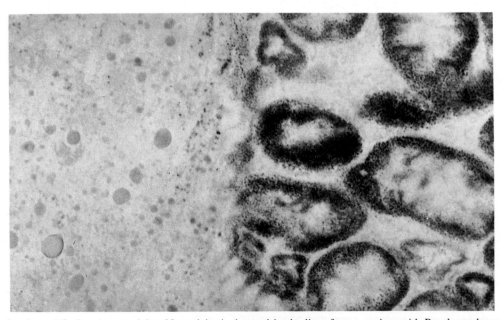

Fig. 15.6 Succinate dehydrogenase activity. No activity is detected in the liver from a patient with Reye's syndrome (left half of the field) when compared with the granular activity of a normal rat kidney (right half), used as both embedding support and control. Small to medium fat droplets can be seen in the liver. Cryostat section ×320

young children, combines an acute encephalopathy of unknown cause, hyperammonaemia and excess fat in the liver. The onset typically occurs during the recovery phase of a viral illness and various experimental and epidemiological studies suggest a reversible defect in mitochondria arising from an interaction of viral, toxic and host genetic factors (Crocker 1982).

In the liver, panlobular microvesicular fatty infiltration is associated with mild hepatocellular bile pigment and usually minimal portal inflammation. Occasional periportal necrosis may be present. Glycogen is depleted, unless hypoglycaemia has been vigourously treated by i.v. glucose solution. The marked mitochondrial disturbance leading to a defective urea cycle and high ammonia levels is corroborated by profound ultrastructural changes and by a complete loss of staining for succinate dehydrogenase activity (Fig. 15.6) in sections of snap frozen tissue. Other toxic conditions affecting mitochondria (e.g. mushroom poisoning) may also show a loss of SDH activity. In contrast, genetic defects affecting the urea cycle and resulting in high blood ammonia levels (deficiencies of carbamyl phosphate synthetase, ornithine carbamyl transferase, arginino succinate synthetase, or arginino-succinate lyase) are usually not associated with fat excess or loss of SDH activity (Filipe & Lake 1983). Fatty livers and Reye-like symptoms may occur in a variety of metabolic disorders (see Ch. 16), but in these there is no depletion of succinate dehydrogenase activity.

PORPHYRIAS

The porphyrias are disorders of the biosynthesis of haem leading to an excess of tissue deposition and excretion of porphyrin precursors. Porphyria

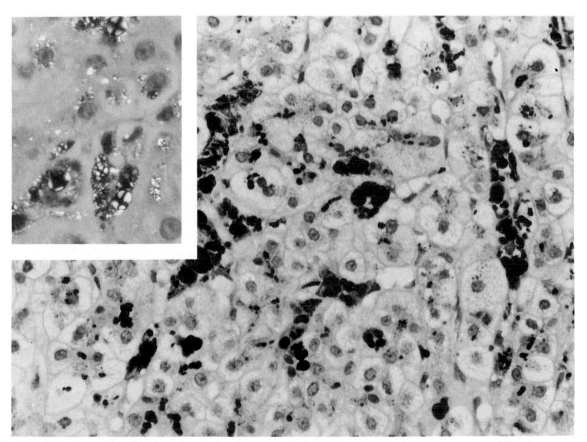

Fig. 15.7 Liver in erythropoietic protoporphyria. Paraffin section. Darkly stained deposits of protoporphyrin have accumulated in bile canaliculi, Kupffer cells and, to a lesser extent, hepatocytes. The larger deposits display a Maltese-cross birefringence (inset). H & E ×260

cutanea tarda (PTC) and erythropoietic protopor-phyria (EPP) affect the liver.

In PCT the uroporphyrin crystals are preserved in paraffin embedded tissue and can be visualized as needle-shaped crystals in unstained sections viewed under polarized light (Cortes et al 1980). The range of liver changes include variable haemosiderosis and fatty infiltration, spotty necrosis, chronic persistent and active hepatitis, periportal fibrosis and later cirrhosis and hepatocellular carcinoma in a proportion of the patients.

In EPP there is accumulation of protoporphyrin in the form of dark-brown deposits in hepatocytes, canaliculi and Kupffer cells (Fig. 15.7). By polarizing microscopy of either unstained or H & E stained paraffin sections, the protoporphyrin displays brilliant red birefringence with a Maltese-cross appearance in the canalicular casts (Klatskin & Bloomer 1974) (Fig. 15.7). Both protoporphyrin and uroporphyrin exhibit a red fluorescence when unstained paraffin sections or air-dried unfixed cryostat sections are examined under u.v. light using an iodine tungsten quartz lamp — 'blue light' (Bruguera et al 1976).

AMYLOIDOSIS

In primary amyloidosis, associated with im-munocyte dyscrasias, the deposited amyloid (AL type) shows staining with Congo red which is resistant to pretreatment with potassium perman-ganate. In contrast, Congo red staining of secondary amyloidosis (AA type), associated with familial Mediterranean fever, long-standing chronic infections and rheumatoid arthritis, is abolished by potassium permanganate pretreat-ment. This observation can also be confirmed by immunohistochemistry using specific antisera to amyloid AA. In the liver, secondary amyloidosis has a predominantly vascular pattern while primary amyloidosis shows sinusoidal deposition in addition (Looi & Sumithran 1988); however, this distinction is not always clear cut in practice.

TUMOURS MARKERS

Metastatic tumours in the liver are by far more common than primary cancers of the liver. Their diagnostic approach is similar to that applied in other sites. Mucins may be revealed by PAS-diastase or more specifically by Alcian blue (pH 2.5) in metastatic adenocarcinoma and chol-angiocarcinoma. Cholangiocarcinoma, in its peripheral form, has no distinctive features and its differentiation from metastatic adenocarcinoma is dependant upon exclusion of other primary sites.

Hepatocellular carcinoma (HCC) is the com-monest primary malignant tumour in adults. Its incidence shows a remarkable geographical vari-ation, being the highest in parts of Africa and Far East, and relatively uncommon in the western world. Most HCC arise in cirrhotic liver, more fre-quently caused by HBsAg-positive chronic hepatitis, alcohol, or haemochromatosis (Anthony 1987).

The majority of cases of HCC can readily be diagnosed in routine H & E sections owing to their growth pattern and cytology resembling that of the normal liver. The tumour cells may contain fat and glycogen. There may be evidence of bile secretion. Various cytoplasmic inclusions, such as Mallory's hyaline, PAS diastase resistant globules and 'ground-glass' bodies, are not uncommon (MacSween 1974, Peters 1976). Silver impreg-nation shows scanty or absent reticulin, often associated with an increased background staining. This pattern is useful to differentiate well-differentiated HCC from hyperplastic/dysplastic nodules and from hepatocellular adenoma.

In a proportion of cases other growth patterns, e.g. pseudoglandular, or cytological features such as clear or pleomorphic cells, and a poor degree of differentiation, may raise problems of diagnosis. In such situations various markers or enzyme staining patterns can be used to complement the mor-phological and clinical data, but none is specific on its own.

Alpha-fetoprotein (AFP)

One of the major fetal alpha-globulins, AFP, is detected in the 4-week-old fetus, where it is produced by the yolk sac, the mucosa of the gastrointestinal tract and the liver. From its peak concentration in the blood at 13–15 weeks' gestation, it rapidly falls thereafter to the normal range of 0–10 μg/1 2–5 weeks after birth.

In the adult, AFP is largely produced in HCC and yolk sac tumours of ovary and testis, and less frequently in other malignant tumours of endodermal or mesenchymal derivation. The less sensitive immunodiffusion assay, detecting levels in excess of 1000 μg/1, shows a great degree of specificity for these tumours. With more sensitive radio-immunoassays which detect levels down to 5 μg/1, an increased number of false-positives will be detected. These include other malignant tumours, with or without hepatic metastases, and various non-malignant situations associated with regeneration, such as untreated chronic active hepatitis and cirrhosis. As a guide, a value of 200 μg/l and above makes HCC highly likely, while 500 μg/l and over is almost pathognomonic for HCC, except for the theoretical possibility of AFP-producing gonadal tumours. Serial AFP estimations have an important role in screening patients with cirrhosis for the early detection of HCC, in follow-up and in monitoring the effects of therapy. In the tissue AFP may appear in the tumour cells of HCC as PAS-positive diastase-resistant globules similar to AAT. Immunohistochemical techniques on paraffin sections reveal granular or globular staining of a rather small number of cells (Thung et al 1979) in half the cases (Imoto et al 1985). Overall, this does not correlate with serum levels and the staining is much less rewarding in adult HCC than in hepatoblastomas (Schmidt et al 1985) (Fig. 15.8) or gonadal tumours.

Carcinoembryonic antigen (CEA)

This has been demonstrated in HCC tumour cells, in about 30% of cases (Imoto et al 1985). It has not been detected in normal hepatocytes, but is found in the apical cytoplasm and luminal surface of epithelial cells of bile ducts and ductules. CEA is not a specific marker of any particular neoplasm and therefore its diagnostic value is limited. However, in liver metastases a negative result virtually

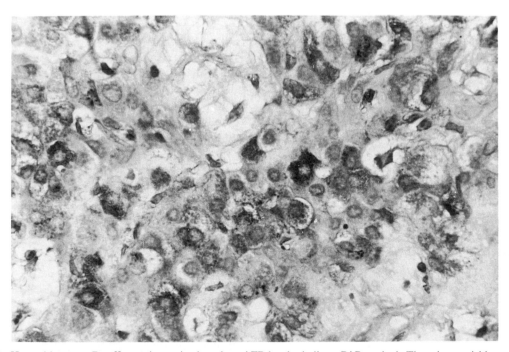

Fig. 15.8 Hepatoblastoma. Paraffin section stained to show AFP by the indirect PAP method. There is a variable granular staining of tumour cell cytoplasm. ×280

excludes a primary tumour in the GI tract (Gerber & Thung 1978).

Isoferritins

Acidic isoferritin, a third oncofetal protein, is apparently produced by HCC cells and normally in fetal liver (Kew & Dusheiko 1981). Ferritin has been demonstrated by immunohistochemical staining in up to 70% of HCC cases (Cohen et al 1984, Imoto et al 1985). In contrast, haemosiderin is invariably absent from HCC even on a background of haemochromatosis.

AAT and alpha-l-antichymotrypsin

These proteins are demonstrable in most cases of HCC in the form of fine cytoplasmic granules in tumour cells (Ordonez & Manning 1984). AAT is also present in the non-neoplastic hepatocytes adjacent to HCC, in some liver cell adenomas and in focal nodular hyperplasia (Palmer & Wolfe 1976).

HBV antigens

In most studies, both HBs and HBcAg, which were often present in large amounts in the non-tumorous liver, were not detected in HCC except for a few cells which may represent non-neoplastic liver cells trapped within the tumour (Anthony 1987). However, a study of 95 cases of HCC from Japan showed HBs in 11.6% of cases and HBcAg in 4.2% (Suzuki et al 1985). HBV-DNA has been shown integrated into the cell genome of HCC using cloned HBV-DNA probes and 'Southern blot' techniques (Shafritz & Kew 1981).

Other antigens

Other products demonstrated in HCC cells include fibrinogen and CAP, which are particularly abundant in fibrolamellar HCC (Teitelbaum et al 1985). This variant of HCC, composed of large mitochondrial-rich eosinophilic cells with a distinctive stroma of lamellar collagen, often occurs in non-cirrhotic liver of young patients and carries a better prognosis (Craig et al 1980). Recently, a detailed immunohistochemical study of hepatoblastoma using a wide range of antibodies to intermediate filaments and oncofetal proteins has brought some light as to the histogenesis of the various tumour components (Abenoza et al 1987).

Enzymes

A variety of liver enzyme activities, which can be demonstrated by histochemical techniques applied on fresh frozen cryostat sections, reveal changes in their amount and distribution between normal and neoplastic tissue (Gerber & Thung 1980).

In adenoma and non-neoplastic areas adjacent to HCC, a normal pattern is found, namely glucose-6-phosphatase (G6Pase) in the cytoplasm of hepatocytes throughout the lobules, adenosine triphosphatase (ATPase) in the bile canaliculi and γ-glutamyl transpeptidase (GGTase) in the bile canaliculi and in the epithelium of bile ducts and ductules. In contrast, HCC shows abnormal enzyme patterns in various combinations, the most common finding being decreased or absent G6Pase and ATPase and increased GGTase activities.

Prominent succinate dehydrogenase and NADH dehydrogenase activities reflect the increased numbers of mitochondria in the fibrolamellar variant of HCC.

REFERENCES

Abenoza P, Manivel J C, Wick M R, Hagen K, Dehner L P 1987 Hepatoblastoma. An immunohistochemical and ultrastructural study. Human Pathology 18: 1025–1035
Anthony P P 1987 Tumours and tumour-like lesions of the liver and biliary tract. In: MacSween R N M, Anthony P P, Scheuer P J (eds) Pathology of the liver, 2nd edn. Churchill Livingstone, Edinburgh, p 574–645
Bianchi L, Spichtin H P, Gudat F 1987 Chronic hepatitis. In: MacSween R N M, Anthony P P, Scheuer P J (eds) Pathology of the liver, 2nd edn. Churchill Livingstone,

Edinburgh, p 310–341
Bréchot C, Pourcel C, Louise A, Rain B, Tiollais P 1980 Presence of integrated hepatitis B virus DNA sequences in cellular DNA of human hepatocellular carcinoma. Nature 286: 533–535
Bruguera M, Esquerda J E, Mascaro J M, Pinol J 1976 Erythropoietic protoporphyria. A light, electron and polarization microscopical study of the liver in three patients. Archives of Pathology and Laboratory Medicine 100: 587–589

Burrell J B, Gowans E J, Jilbert A R, Lake J R, Marmion B P 1982 Hepatitis B virus DNA detection by in situ cytohybridization: Implications for viral replication strategy and pathogenesis of chronic hepatitis. Hepatology 2 Suppl: 85S–91S

Cohen C, Berson S D, Shulman G, Budgeon C R 1984 Immunohistochemical ferritin in hepatocellular carcinoma. Cancer 53: 1931–1935

Clausen P P, Lindskov J, Gad I et al 1984 The diagnostic value of α-1-antitrypsin globules in liver cells as a morphological marker of α-1-antitrypsin deficiency. Liver 4: 353–359

Cortes J M, Oliva H, Paradinas F J, Hernandez-Guio C 1980 The pathology of the liver in porphyria cutanea tarda. Histopathology 4: 471–485

Craig J R, Peters R L, Edmondson H A, Omata M 1980 Fibrolamellar carcinoma of the liver: a tumour of adolescents and young adults with distinctive clinicopathologic features. Cancer 46: 372–379

Crocker J F S 1982 Reye's syndrome. Seminars in Liver Disease 2: 340–352

Davies S E, Williams R, Portmann B 1989 Hepatic morphology and histochemistry of Wilson's disease presenting as fulminant hepatic failure: A study of 11 cases. Histopathology 15: 385–394

Denk H, Lackinger E 1986 Cytoskeleton in liver diseases. Seminars in Liver Disease 6: 199–211

Deodhar K P, Tapp E, Scheuer P J 1975 Orcein staining of hepatitis B antigen in paraffin sections of liver biopsies. Journal of Clinical Pathology 28: 66–70

Evans J, Newman S P, Sherlock S 1980 Observations on copper associated protein in childhood liver disease. Gut 21: 970–976

Filipe M I, Lake B D 1983 Liver: malignant tumours and chronic liver disease. In: Filipe M I, Lake B D (eds) Histochemistry in pathology. Churchill Livingstone, Edinburgh, p 150–159

Gerber M A, Thung S N 1978 Carcino-embryonic antigen in normal and diseased liver tissue. American Journal of Pathology 92: 671–680

Gerber M A, Thung S N 1980 Enzyme patterns in human hepatocellular carcinoma. American Journal of Pathology 98: 396–400

Guarascio P, Yentis F, Cevikbas U, Portmann B, Williams R 1983 Value of copper-associated protein in diagnostic assessment of liver biopsy. Journal of Clinical Pathology 36: 18–23

Gudat F, Bianchi L, Sonnabend W, Thiel G, Aenishaenslin W, Stalder G A 1975 Pattern of core and surface expression in liver tissue reflects state of specific immune response in hepatitis B. Laboratory Investigation 32: 1–9

Huang S N 1975 Immunohistochemical demonstration of hepatitis B core and surface antigens in paraffin sections. Laboratory Investigation 33: 88–95

Imoto M, Nishimura D, Fukuda Y, Sugiyama K, Kumada T, Nakamo S 1985 Immunohistochemical detection of α-fetoprotein, carcinoembryonic antigen, and ferritin in formalin-paraffin sections from hepatocellular carcinoma. American Journal of Gastroenterology 80: 902–906

Irons R D, Schenk E A, Lee J C K 1977 Cytochemical methods for copper semiquantitative screening procedure for identification of abnormal copper levels in liver. Archives of Pathology and Laboratory Medicine 101: 298–301

Ishak K G, Sharp H L 1987 Metabolic errors and liver disease. In: MacSween R N M, Anthony P P, Scheuer P J (eds) Pathology of the liver, 2nd edn. Churchill Livingstone, Edinburgh, p 99–180

Kew M C, Dusheiko G M 1981 Paraneoplastic manifestations of hepatocellular carcinoma. In: Berk P D, Chalmers T C (eds) Frontiers in liver disease. Thieme-Stratton, New York, p 305–319

Klatskin G, Bloomer J R 1974 Birefringence of hepatic pigment deposits in erythropoietic protoporphyria. Specificity and sensitivity of polarization microscopy in the identification of hepatic protoporphyrin deposits. Gastroenterology 67: 294–302

Looi L-M, Sumithran E 1988 Morphologic differences in the pattern of liver infiltration between systemic AL and AA amyloidosis. Human Pathology 19: 732–735

Masih A S, Linder J, Shaw B W et al 1988 Rapid identification of cytomegalovirus in liver allograft biopsies by in situ hybridization. The American Journal of Pathology 12: 362–367

McSween R N M 1974 A clinico-pathological review of 100 cases of primary malignant tumours of the liver. Journal of Clinical Pathology 27: 669–682

Moll R, Franke W W, Schiller D L 1982 The catalog of human cytokeratins: patterns of expression in normal epithelia, tumors and cultured cells. Cell 31: 11–24

Nayak N C, Ramalingaswami V 1975 Indian childhood cirrhosis. In: Popper H (ed) Clinics in gastroenterology, vol 4, no 2. W B Saunders, London, p 333–349

Naoumov N V, Alexander G J M, O'Grady J G, Aldis P, Portmann B, Williams R 1988 Rapid diagnosis of cytomegalovirus infection by in-situ hybridisation in liver graft. Lancet i: 1361–1364

O'Grady J G, Alexander G J M, Sutherland S et al 1988 Cytomegalovirus infection and donor/recipient HLA antigens: Interdependent co-factors in pathogenesis of vanishing bile duct syndrome after liver transplantation. Lancet ii: 302–305

Ordonez N G, Manning J T 1984 Comparison of alpha-1-antitrypsin in hepatocellular carcinoma. An immunoperoxidase study. American Journal of Gastroenterology 79: 959–963

Palmer P E, Wolfe H J 1976 Alpha-1-antitrypsin deposition in primary hepatocellular carcinoma. Archives of Pathology and Laboratory Medicine 100: 232–236

Peters R T 1976 Pathology of hepatocellular carcinoma. In: Okuda K, Peters R T (eds) Hepatocellular carcinoma. Wiley, New York, p 107–168

Portmann B, MacSween R N M 1987 Diseases of the intrahepatic bile ducts. In: MacSween R N M, Anthony P P, Scheuer P J (eds) Pathology of the liver, 2nd edn. Churchill Livingstone, Edinburgh, p 424–453

Portmann B, Tanner M S, Mowat A P, Williams R 1978 Orcein-positive liver deposits in Indian childhood cirrhosis. Lancet i: 1338–1340

Portmann B, Galbraith R M, Eddleston A L F, Zuckerman A J, Williams R 1976 Detection of HBsAg in fixed liver tissue — use of a modified immunofluorescent technique and comparison with histochemical methods. Gut 17: 1–9

Ray M B 1987 Distribution patterns of cytokeratin antigen determinants in alcoholic and nonalcoholic liver diseases. Human Pathology 18: 61–66

Ray M B, Desmet V J, Fevery J, De Groote J, Bradburne A F, Desmyter J 1976 Distribution patterns of hepatitis B surface antigen (HBsAg) in the liver of hepatitis patients. Journal of Clinical Pathology 29: 94–100

Rijntjes P J M, Van Ditzhuijsen Th J M, Van Loon A M, Van Haelst U J G M , Bronkhorst F B, Yap S H 1985 Hepatitis B virus DNA detected in formalin-fixed liver specimens and its relation to serologic markers and histopathologic features in chronic liver disease. The American Journal of Pathology 120: 411–418

Rizzetto M, Verme G, Gerin J L, Purcell R H 1986 Hepatitis delta virus disease. In: Popper H, Schaffner F (eds). Progress in Liver Diseases, vol VIII. Grune and Stratton, New York, p 417–431

Salaspuro M, Sipponen P 1976 Demonstration of an intracellular copper-binding protein by orcein staining in long-standing cholestatic liver disease. Gut 17: 787–790

Scheinberg I H, Sternlieb I 1984 Wilson's disease. W B Saunders, Philadelphia

Schmidt D, Harms D, Lang W 1985 Primary malignant hepatic tumours in childhood. Virchows Archiv A 407: 387–405

Shikata T, Uzawa T, Yoshiwara N, Akatsuka T, Yamazaki S 1974 Staining method of Australia antigen in paraffin sections. Detection of cytoplasmic inclusions bodies. Japanese Journal of Experimental Medicine 44: 25–36

Schirmacher P, Dienes H P, Worsdorfer M, Weber C, Hess G, Meyer zum Buschenfelde K H 1988 Demonstration of HBV-DNA at the ultrastructural level in the liver of patients with chronic hepatitis B. Investigation with in situ hybridization using biotinylated cDNA. Journal of Hepatology 7 (suppl 1): s75 Abstract 146

Southern E M 1975 Detection of specific sequences among DNA fragments separated by gel electrophoresis. Journal of Molecular Biology 98: 503–517

Suzuki K, Uchida T, Horiuchi R 1985 Localization of hepatitis B surface and core antigens in human hepatocellular carcinoma by immunoperoxidase methods: replication of complete virions of carcinoma cells. Cancer 56: 321–327

Shafritz D A, Kew M C 1981 Identification of integrated hepatitis B virus DNA sequences in human hepatocellular carcinoma. Hepatology 1: 1–8

Tanner M S, Portmann B, Mowat A P et al 1979 Increased hepatic copper concentration in Indian childhood cirrhosis. Lancet ii: 1203–1205

Teitelbaum D H, Tuttle S, Casey L C, Clausen K C 1985 Fibrolamellar carcinoma of the liver. Annals of Surgery 202: 36–41

Ten Napel C H H, Houthoff H J, The T H 1984 Cytomegalovirus hepatitis in normal and immune compromised hosts. Liver 4: 184–194

Thung S N, Gerber M A, Sarno E, Popper H 1979 Distribution of five antigens in hepatocellular carcinoma. Laboratory Investigation 41: 101–105

Tseng S C G, Jarvinen M J, Nelson W G, Huang J W, Woodcock-Mitchell J, Sun TT 1982 Correlation of specific keratins with different types of epithelial differentiation: monoclonal antibody studies. Cell 30: 361–372

Vierling J 1981 Copper metabolism in primary biliary cirrhosis. Seminars in Liver Disease 1: 293–308

16. Metabolic disorders of the liver

B. D. Lake

STORAGE DISORDERS AFFECTING THE LIVER

Storage disorders affect the liver in one of several ways. The enzyme defect may be manifest in the hepatic parenchymal cells alone, producing enlarged hepatocytes containing a stored product which is usually detectable in appropriately prepared sections. In extreme examples — phosphorylase kinase deficiency for instance — routine sections may appear to be composed of plant cells. In the disorders affecting the cells of the mononuclear phagocyte (reticuloendothelial) system the hepatic parenchymal cells may not be involved and the picture is then of infiltrating storage cells usually in the portal areas but also throughout the sinusoids. Gaucher's disease (all types) is an example of this category. In other situations hepatocytes are additionally involved and are enlarged and foamy with storage cells present in sinusoids and portal areas. In a few conditions — for example Farber's disease and later onset Niemann-Pick disease type C — routine sections show no abnormality even though marked changes are detectable, particularly in sections of snap-frozen tissue or by electron microscopy.

In all of these circumstances histochemical methods are important to discover whether a storage disorder is present and to define the nature of the disorder. As mentioned above, routine sections may show a storage disorder is present, but to define the type of disorder snap-frozen tissue is essential. In the absence of snap-frozen tissue, formalin-fixed frozen sections can provide much useful information, but with good communication between the clinician, surgeon and pathologist snap-frozen tissue should always be available.

GLYCOGEN STORAGE DISEASES

The glycogen storage diseases are a group of disorders (Scriver et al 1989) which result in the deposition of variable amounts of glycogen of differing structures. All require biochemical assay of the enzymes involved for final proof, but histochemical methods can go a long way towards the diagnosis and can eliminate those cases which do not fall into the glycogen storage group, thus saving the biochemists unnecessary assays. Routine histological examination will allow assessment of the liver architecture and degree of enlargement of the hepatocyte, but generally does not help in the diagnosis of glycogen storage disease. Where the diagnosis is suspected before the biopsy is undertaken only a small portion of tissue should be put into formalin, the majority being frozen for histochemistry and biochemical analysis.

The three commonly encountered types of glycogen storage disease affecting the liver are types 1, 3 and 6 with two or three minor types also being found on occasions. Table 16.1 lists the various types and their main clinical, enzymatic and pathological features. Some ten or more types of glycogen storage disease have been described in the literature, but with only one or two examples of the rarer types no clear histological or histochemical pattern has emerged, and they are not considered here. Figure 16.1 shows the pathways of glycogen metabolism.

Demonstration of glycogen

The glycogen found in the liver in the several types of glycogen storage disease varies in structure, and consequently in its solubility in aqueous

Table 16.1 Glycogen storage diseases

Type	Enzyme defect	Major clinical features	Major pathological features
0	Glycogen synthetase	Hepatomegaly, hypoglycaemia	Fatty liver, absent or grossly reduced glycogen
1A Von Gierke	Glucose-6-phosphatase	Hepatomegaly, hypoglycaemia, lactic acidosis	Fatty liver, renal proximal tubules contain much glycogen
1B	Glucose-6-phosphate translocase	As type 1A	As type 1A
1C	Phosphate/pyrophosphate transport protein		
2 infantile, Pompe	Acid α-1:4-glucosidase	Cardiomegaly, hypotonia, hepatosplenomegaly	Marked excess of glycogen in MP system, hepatocytes and cardiac and skeletal muscle
juvenile	Acid α-1:4-glucosidase	Muscular weakness	Gross vacuolar myopathy with excess glycogen
adult	Acid α-1:4-glucosidase	Muscular weakness, respiratory difficulty	Mild vacuolar myopathy with excess glycogen
3	Debranching enzyme (amylo-1:6-glucosidase)	Hepatomegaly, mild fasting hypoglycaemia	Marked excess glycogen in liver, portal fibrosis
4	Branching enzyme	Hepatosplenomegaly, jaundice	Cirrhosis, indigestible glycogen, bile stasis
5 McArdle	Myophosphorylase	Muscle weakness and cramps on exercise	Vacuolar myopathy with excess glycogen
6A⋆	Liver phosphorylase kinase	Hepatomegaly, very mild fasting hypoglycaemia	Marked excess glycogen in liver, slight fat accumulation
6B	Unclassifiable	Variable	Variable
7	Phosphofructokinase	Muscle weakness and cramps on exercise	Vacuolar myopathy with excess glycogen

⋆May also be referred to as type 6B or type 9 in the literature.

solvents also varies. Normal glycogen structure is found in types 1 and 6, a much more soluble limit dextrin in type 3, and an insoluble amylopectin in type 4. Much of the glycogen in type 2 is intralysosomal and in the β-particle form (monoparticulate) and is thus more soluble than the usual α-particles (rosettes) of glycogen. The glycogen in the muscle in types 5 and 7 has a normal structure. Routine fixation will extract variable amounts of glycogen and routine sections cannot be relied on to provide an accurate reflection of the glycogen content. Special fixatives, for example Bouin, Carnoy or alcohol, may provide a better guide to glycogen content, but this will waste valuable biopsy material which would be better used for cryostat sectioning and enzyme assay.

The protected PAS method (PAS celloidinized) on cryostat sections of snap-frozen tissue is the most accurate way of assessing the glycogen content. Normal glycogen levels range up to 6% (wet weight basis) and in this amount the hepatocyte appears full but not distorted. Very high contents of glycogen (up to 15%) are found in types 3 and 6A (Fig. 16.2). The glycogen in type 3 glycogen storage disease is very soluble and may have

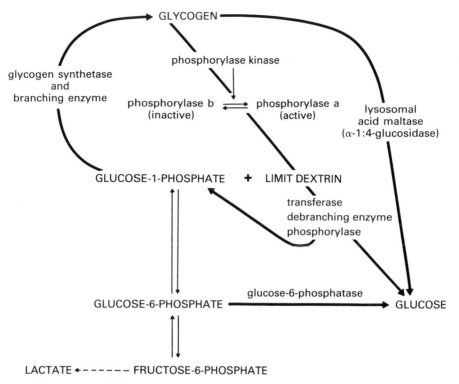

Fig. 16.1 Pathways of glycogen metabolism.

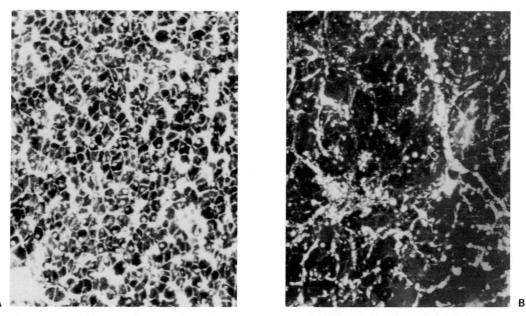

Fig. 16.2 Cryostat sections of liver stained with the protected PAS method (cell PAS) to demonstrate glycogen. A normal liver on the left contains 6% glycogen; on the right a biopsy from a patient with type 3 GSD contains 12% glycogen. Routine sections would show much less glycogen in the patient with type 3 GSD due to the greater solubility of the glycogen in this disorder. ×140

completely disappeared from routine sections. Some patients with type 1 glycogen storage disease also have a very high glycogen content, but more usually this is within the normal range and the pathology is dominated by marked fatty deposition. High glycogen contents (>7% wet wt) may also be found in the liver from diabetic patients, in organic acidemias and in infants who have been given intravenous dextrose feeds prior to surgery. The glycogen in all types (except type 4 and in an odd single-case report) is readily digestible through the protective celloidin coating by salivary amylase. The amylopectin in type 4 is not hydrolysed by amylase or diastase but is inexplicably attacked by pectinase preparations.

In types 5 and 7 the muscle glycogen may not be visibly in excess (see Ch. 10).

Pompe's disease (GSD2) shows glycogen within the Kupffer cells as well as the hepatocytes. Smooth muscle cells of the vasculature and endothelial cells also show increased amounts of glycogen. The heart is grossly affected, the whole of the cardiac muscle showing vacuolar change as does skeletal muscle. Deposits of an acid mucosubstance may also be shown in skeletal muscle and heart by the toluidine blue method of Haust & Landing (1961).

A few nuclei containing glycogen can usually be found in periportal areas in liver biopsies from most children. In types 1A and 1B glycogen storage disease the numbers of glycogenic nuclei are increased, becoming more noticeable with increasing age.

Glucose-6-phosphatase activity

Normal glucose-6-phosphatase activity is distributed throughout the hepatocyte with accentuated activity in the nuclear envelope. The reaction shows a gradient of activity from the terminal hepatic vein increasing towards the portal space and Zone 1 (Rappoport) of the acinus. An incubation time of 20 minutes is adequate, and is short enough not to show interference from non-specific phosphatases. In type 1A GSD no activity can be detected in hepatocytes (Fig. 16.3), but normal activity is shown in types 1B and 1C GSD because the cells are permeable in frozen sections and the glucose-6-phosphate translocase is not

necessary. Types 1B and 1C GSD can thus only be diagnosed by biochemical assay of glucose-6-phosphatase activity in intact microsomes prepared from fresh unfrozen liver tissue (Blair & Burchell 1988).

Glucose-6-phosphatase is also normally present in proximal renal tubular epithelium and in the epithelial cells of the small intestine. Jejunal biopsy may be used for diagnosis, but care should be taken since other conditions with normal jejunal morphology (fructose intolerance for example) show very low levels of activity, which may look like deficiency until compared with a deficient patient when subtle differences can be observed. The activity is present mainly in the supranuclear regions with a marked nuclear envelope activity. Alkaline phosphatase also hydrolyses glucose-6-phosphatase but its activity is confined to the brush border and causes no problems in interpretation. No excess glycogen is found in the enterocytes in type 1 GSD (Fig. 16.4).

The glucose-6-phosphatase activity in the hepatocytes of types 3 and 6A is reduced and often confined to the nuclear envelope and the periphery of the cell (see Fig. 16.3C), where the endoplasmic reticulum has been displaced by the glycogen deposition. The low activity should cause no difficulty since total absence is the hallmark of type 1 GSD and low activity is interesting but not diagnostic and is found in a variety of liver disorders.

Fructose intolerance can present as a neonatal hepatitis or in the older child with failure to thrive. The liver biopsy in the later onset shows gross fatty change with adequate glycogen, but the glucose-6-phosphatase activity of individual hepatocytes is often markedly increased.

Although activity of glucose-6-phosphatase can be shown in the islet cells of pancreas from several species, no such activity can be demonstrated in human pancreas.

Acid phosphatase activity

Acid phosphatase activity in the form of discrete particulate lysosomal staining pattern is found mainly in the pericanalicular regions of normal liver and in a stronger activity in Kupffer cells and histiocytes in the portal spaces. Normal or slightly

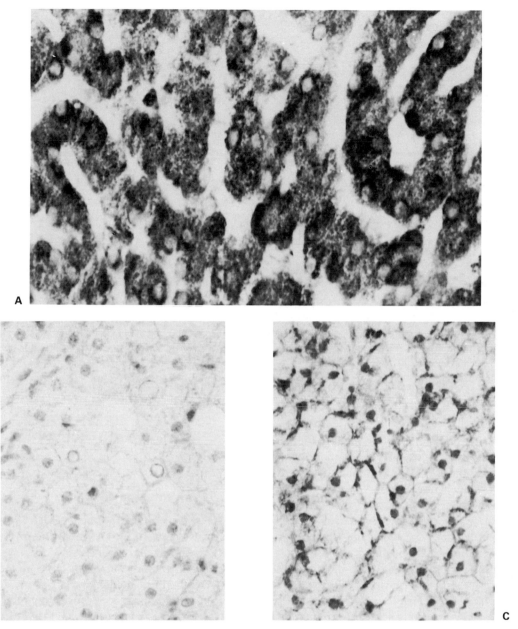

Fig. 16.3 Glucose-6-phosphatase activity in cryostat sections of: **A** Normal liver. Activity is present in the cytoplasm and in a perinuclear ring. (×600) **B** Type 1A GSD. No activity is detectable. (×450) **C** Type 6A GSD. Reduced activity is present and is confined to the periphery of the hepatocyte and around the nucleus. (×350) A light nuclear counterstain has been added to each preparation.

decreased activity is found in all types of GSD except type 2, where a marked change in distribution and increase in intensity is found in common with all lysosomal enzyme disorders. The discrete particulate pericanalicular activity is replaced by a strong activity throughout the hepatocyte. Enlarged Kupffer cells are also readily visible. Increased acid phosphatase activity is also found in the vacuoles in skeletal and cardiac muscle from patients with type 2 GSD.

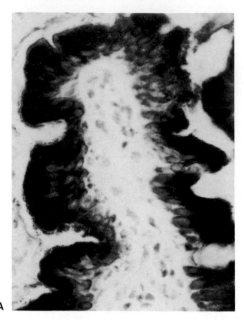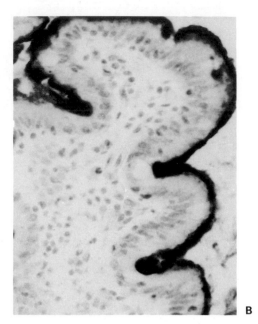

Fig. 16.4 Glucose-6-phosphatase activity in cryostat sections of: **A** Normal jejunum. Glucose-6-phosphatase is present throughout the enterocyte. Perinuclear activity is apparent. **B** Type 1A GSD. No glucose-6-phosphatase activity can be detected within the enterocyte. Alkaline phosphatase hydrolyases the substrate and its activity is present in the brush border. A light nuclear counterstain has been added. ×490

Phosphorylase activity

Of the many different modifications of the phosphorylase method, none is reliable for the detection of liver phosphorylase deficiency. Although the method will show low staining intensity in cases of biochemically proven phosphorylase deficiency, similar 'deficiency' is also found in phosphorylase kinase deficiency and in debranching enzyme deficiency. A good normal reaction is an indication of normal phosphorylase activity, but a low staining intensity is not of diagnostic significance. Methods for phosphorylase kinase are consequently also unreliable.

Phosphorylase methods for skeletal muscle by contrast are reliable and useful in the diagnosis of McArdle's disease where no activity can be detected in skeletal muscle fibres. Smooth muscle of vessel walls shows activity and acts as an internal control (see Ch. 10).

Fat

Fatty deposition in hepatocytes is the main feature of GSD 1. In type 6A there is a mild to moderate amount of fat, but in other types fat deposition is not found. The fat is mostly triglyceride and is well stained with oil red O. It shows no birefringence in polarized light.

Glycogen synthetase deficiency

In glycogen synthetase deficiency, for which only a few cases have been recorded, the glycogen content, as might be expected is very low or even absent, and there is abundant fat. The method for the histochemical demonstration of glycogen synthetase activity produces only a very low amount of reaction product in normal liver, and this coupled with the unpredictability of the glycogen-iodine colour of the phosphorylase reaction makes the detection of glycogen synthetase deficiency almost impossible by histochemical means. Abundant fat and very low or absent glycogen stores are also features of the liver in some fat oxidation defects, particularly of the medium chain acyl-CoA dehydrogenase deficiency. In fat oxidation defects, skeletal muscle is often involved in contrast with glycogen synthetase deficiency.

LIPID AND OTHER STORAGE DISEASES
(see Hers & van Hoof 1973, Scriver et al 1990)

Wolman's disease and cholesteryl ester storage disease

These two disorders appear identical in biochemical terms but they have quite different clinical courses and outcomes. Wolman's disease presents acutely in early infancy with failure to thrive, hepatosplenomegaly and adrenal calcification with death in the first year. Cholesteryl ester storage disease presents at any time with hepatomegaly and cases are often referred as a form of glycogen storage disease. Macroscopically the liver has an orange-yellow glistening appearance due to the very high fat content. There is portal fibrosis or cirrhosis and in cholesteryl ester storage disease regenerating nodules may be seen. The hepatic parenchymal cells are filled with lipid which stains strongly with oil red O and Sudan black. The Kupffer cells and foamy histiocytes in the sinusoids and portal areas are also strongly sudanophilic and in routine sections it is often possible to see that some of the lipid is in the form of a crystalline inclusion (see Ch. 4) particularly in Wolman's disease. In cholesteryl ester storage disease the foamy Kupffer cells occur in clusters and contain PAS positive lipopigment and lipid vacuoles (Fig. 16.5). Crystalline inclusions are rare. Examination of frozen or cryostat sections in polarized light also shows the crystalline lipid inclusions of cholesteryl esters and free fatty acids. Triclycerides show no birefringence. Stains for cholesterol (and esters) are strongly positive in hepatocytes and macrophages. It is not possible to differentiate between cholesterol and its esters by the use of digitonin unless the primary fixation includes digitonin. In sections (frozen or cryostat) the rate of formation of cholesterol digitonide is slower than the rate of solubility of the cholesterol in the solvent and thus although the test in Wolman's disease would appear to indicate the presence of cholesterol esters the same result would be obtained if cholesterol alone was present. Adams' method for triglyceride esters is positive in hepatocytes, but is not very sensitive. The macrophages and enlarged Kupffer cells contain free fatty acids which give a positive reaction with Holczinger's method, and Cain's Nile blue sulphate shows red hepatocytes (neutral fat) with purple macrophages indicating free fatty acids and neutral fat. The acid phosphatase reaction shows the macrophages with the crystalline and lipid inclusions outlined by the reaction product;

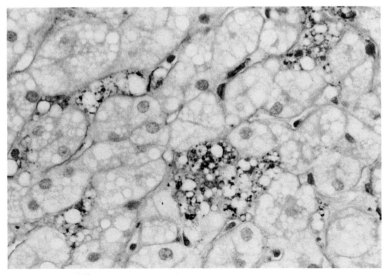

Fig. 16.5 Cholesteryl ester storage disease. Paraffin section, digested PAS. The clusters of histiocytes and Kupffer cells containing PAS-positive lipofuscin granules are more readily appreciated once the glycogen has been removed. Lipid vacuoles are present in hepatocytes and histiocytes, but no crystalline inclusions are visible.

hepatocytes also show a strong reaction. In frozen sections of formol-calcium/gum sucrose-processed tissue incubated in the presence of 10^{-4}M E600 to inhibit non-specific esterases, the absence of acid esterase is readily demonstrated with 1-naphthyl acetate as substrate. In normal liver, acid esterase can be shown mainly in Kupffer cells with lesser activity in hepatocytes (Lake & Patrick 1970) (Fig. 16.6). It should be noted that it is not possible to demonstrate the deficiency in cryostat sections of snap-frozen tissue since the activity of acid esterase in the normal is very low and is barely distinguishable from a deficiency under these conditions. Prefixation of a tissue block is essential. If only frozen tissue is available, the semi-permeable membrane technique may give useful results.

Gaucher's disease

Gaucher cells with their content of glucocerebroside are PAS-positive in routine and frozen sections and are only weakly sudanophilic. The cells are present as enlarged Kupffer cells in the sinusoids and in the portal areas, and are more numerous in the infantile form of the disease. Hepatocytes are not affected, and this is best demonstrated in acid phosphatase preparations where the Gaucher cells are strongly stained in contrast with the normal pericanalicular distribution in hepatocytes (Fig. 16.7). Other lysosomal storage disorders with visceral involvement affect hepatocytes and this is reflected by a strong diffuse acid phosphatase reaction. The histochemical methods for β-glucosidase (using the substituted indoxyl substrate, or the 6-bromo-2-naphthyl-β-glucoside or 1-naphthyl-β-glucosidase) are not sufficiently sensitive to discriminate between the isoenzymes of β-glucosidase and their differing pH optima, and thus cannot be used in the diagnosis of Gaucher's disease where only one isoenzyme is deficient.

Niemann-Pick disease types A and B

In contrast with Gaucher's disease the hepatocytes in Niemann-Pick disease types A and B are involved in the storage process. Large foamy Kupffer cells and hepatocytes are filled with sphingomyelin and cholesterol, and a variable

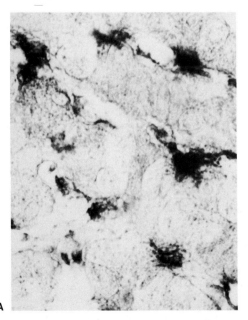
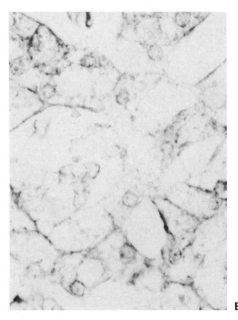

Fig. 16.6 Acid esterase activity in frozen sections of formol-calcium/gum sucrose fixed liver biopsies. On the left, a normal liver shows activity in Kupffer cells mainly, with some lesser activity in the hepatocytes. On the right no activity can be detected in a liver biopsy from a patient with Wolman's disease. ×950

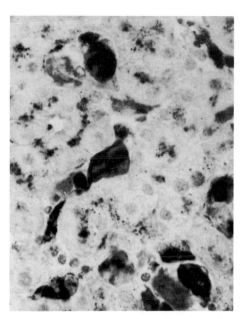

Fig. 16.7 Gaucher's disease, cryostat section of a liver biopsy stained to demonstrate acid phosphatase activity. Gaucher cells and Kupffer cells show strong activity while hepatocytes show normal pericanalicular activity. A light nuclear counterstain has been added. ×450

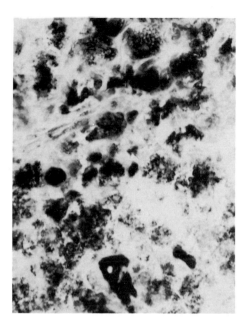

Fig. 16.8 Niemann-Pick disease type A; cryostat section of a liver biopsy stained to demonstrate acid phosphatase activity. Large foamy Niemann-Pick cells and hepatocytes show strong activity, in contrast with Gaucher's disease where hepatocytes are normal. A light nuclear counterstain has been added. ×470

amount of ganglioside. Sudan black and methods for cholesterol (Schultz, PAN) are positive, and in sections from which glycogen has been removed the foamy cells are weakly positive. The acid haematin method for sphingomyelin is not entirely reliable for two reasons. Firstly, liver normally contains sphingomyelin and secondly the stain appears to depend to a certain extent on the composition of the fatty acid side chain. In several instances the expected blue colour has not been found — instead the Niemann-Pick cells have been yellow-brown. The ferric haematoxylin method is better and more reliable and gives consistent results in those cases where the acid haematein reaction failed. After staining with Sudan black sphingomyelin often will show a red birefringence in polarized light. This is not a specific test but is an indication of an orderly lipid lamellar arrangement. The acid phosphatase reaction shows a characteristic foamy appearance of the hepatocytes and Niemann-Pick cells (Fig. 16.8) with the periphery of the vacuoles being outlined by the reaction product. There is no staining method available for the demonstration

of sphingomyelinase activity. Liver failure and cirrhosis is a cause of death in the adult type B Niemann-Pick disease.

Niemann-Pick type C

This condition, known in the past by a variety of names (Niemann-Pick type C, Niemann-Pick type D, subacute Niemann-Pick disease, juvenile dystonic lipidosis, neurovisceral storage disease with vertical supranuclear ophthalmoplegia, etc.), does not store sphingomyelin and there is no connection with classic Niemann-Pick disease (Neville et al 1973). Niemann-Pick disease type C presents in three main ways: with neonatal hepatitis and splenomegaly in 60% of cases, with hepatosplenomegaly and dementia in childhood, or with isolated splenomegaly. Routine sections of liver in those presenting early with neonatal hepatitis show changes of neonatal hepatitis with bile pigment retention and giant cell change. As in most hepatitic liver samples, PAS-positive macrophages are present but in Niemann-Pick type C

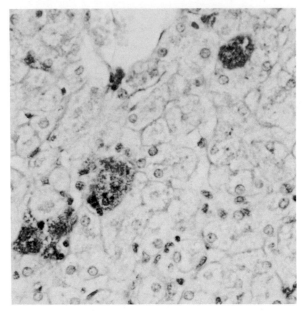

Fig. 16.9   Niemann-Pick disease type C. Routine section of a liver biopsy stained with PAS after amylase digestion. Four large storage cells and several Kupffer cells stain positively. Hepatocytes appear to be normal. ×360.

most Kupffer cells are enlarged and foamy and palely PAS-positive (Fig. 16.9). This change is· best seen after digestion of the glycogen, but may be difficult to distinguish from other causes of neonatal hepatitis even in cases where the diagnosis is known. In frozen or cryostat sections the foamy cells are sudanophilic. In cryostat sections, first coated with celloidin then digested with saliva and stained with PAS (cell. dig. PAS), the strong staining of the foamy storage cells and Kupffer cells is clearly seen. Feyrter's thionin shows rose-pink metachromasia in the storage cells and hepatocytes. The storage cells show a strong acid phosphatase reaction as do the hepatocytes, but these are normally quite strong in liver damage. The liver disease in most cases does not progress beyond a portal fibrosis, which remains quiet. Cirrhosis is very rare.

In those patients presenting with unexplained hepatosplenomegaly at any age, routine sections of liver appear quite normal. However, PAS after digestion of glycogen shows positive staining of storage cells and Kupffer cells, which are sudanophilic in frozen or cryostat sections. The

cell.dig. PAS method is also strong in these cells but is much reduced in fixed material, indicating a water-soluble component. Although the hepatocytes appear to be normal they show strong acid phosphatase activity (Fig. 16.10) indicating the lysosomal nature of the disorder which is confirmed by electron microscopy. A defect in cholesterol esterification has been identified in a large group of patients (Vanier et al 1988).

Farber's disease

Routine sections show a normal liver and although ceramide is present in excess it does not stain to any great extent with lipid dyes or methods. No storage cells are apparent, but as in Niemann-Pick type C the acid phosphatase reaction shows strong staining in hepatocytes.

Cystinosis

Although cystine is relatively insoluble, the volume of fluid used for fixation, in floating out and staining, is sufficient to ensure that most cystine is removed from sections of routinely processed tissue. All that remains is a foamy appearance in Kupffer cells resembling those found in a variety of other storage disorders. Snap-frozen tissue, sectioned in a cryostat and examined under polarized light, is required to show the crystal habit and will preserve all the cystine within the section. If preferred, the cryostat section can be stained with alcoholic basic fuchsin (1% basic fuchsin in 70% alcohol) to show nuclei and general structure. If the concentration of cystine is high enough, the Woollaston test (Patrick & Lake 1968) is positive. Under the influence of concentrated hydrochloric acid, cystine dissolves and reprecipitates as cystine hydrochloride with change of crystal shape (see Fig. 18.7).

Metachromatic leucodystrophy

Deposition of sulphatide is not seen in hepatocytes or Kupffer cells. However, sulphatides are found stored in bile duct epithelium where they show brown metachromasia with cresyl fast violet or toluidine blue in frozen or cryostat sections. In some cases the bile duct epithelium can be in-

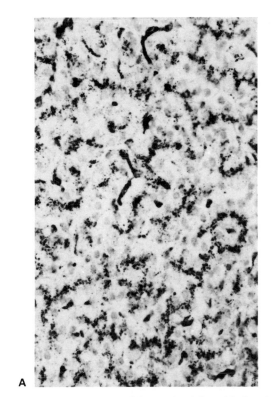
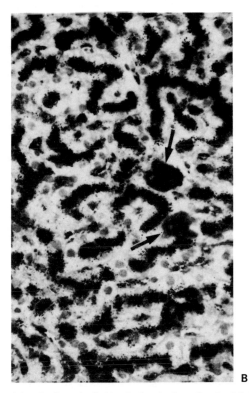

A
B

Fig. 16.10 Cryostat sections of liver stained for acid phosphatase activity (lead method) 30 min incubation. On the left is a normal liver showing Kupffer cells and pericanalicular activity in hepatocytes. On the right is a section of a liver biopsy from a patient with Niemann Pick disease type C showing intense activity in hepatocytes. Two storage cells are evident (arrows). Histologically the liver appeared normal. A light nuclear counterstain has been added. ×300

volved to such an extent that the gall bladder may show papillomatous change with marked sulphatide deposition in epithelium and in macrophages in the subepithelial connective tissue.

G_{M2}-gangliosidosis

There is usually no evidence of storage of ganglioside G_{M2} or its asialo derivative in Tay-Sachs disease and related disorders. In occasional cases, however, the Kupffer cells may show strong PAS-positivity with pale sudanophilia in frozen or cryostat sections.

G_{M1}-gangliosidosis

Type 1 infantile

There are differences in liver involvement in the two forms of G_{M1}-gangliosidosis. Type 1 (infantile)

shows marked vacuolation of hepatocytes and foamy enlarged Kupffer cells and appears at first sight rather like a fatty liver. No fat can be demonstrated and the vacuoles contain a water-soluble substance which has so far defied histochemical demonstration. The substance:

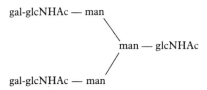

should be PAS-positive but its solubility appears to prevent its demonstration. Although the patients present as a mucopolysaccharidosis no mucopolysaccharide can be demonstrated in the vacuoles by any method.

At least one other storage product in liver has been identified containing sialic acid residues in addition to galactose and glucosamine. This suggests that it may be similar to the compound mentioned above and may account for the alcian blue-positive reaction in the enlarged foamy Kupffer cells seen in sections of routinely processed wax-embedded tissue. These cells are also PAS-positive.

The β-galactosidase deficiency can be detected reliably in cryostat sections of fresh tissue using the substituted indoxyl method (Lake 1974). Normal or unaffected tissue shows marked activity after overnight incubation, but no activity can be demonstrated in either hepatocytes or histiocytes in patients with G_{M1}-gangliosidosis of type 1 or type 2 (Fig. 16.11).

Type 2 juvenile

The general hepatic architecture is undisturbed and even though Kupffer cells are known to show storage by electron microscopy no firm evidence can be obtained by light microscopy. Occasional large foamy histiocytes, present in sinusoids, show a moderate PAS reaction in routine sections.

These large histiocytes show strong acid phosphatase activity in contrast with normal activity in hepatocytes and Kupffer cells. The demonstration of the enzyme defect by the indoxyl method mentioned above is reliable.

Mucopolysaccharidosis

The several types of mucopolysaccharidosis all show deposition of mucopolysaccharide in the liver. The type of mucopolysaccharide deposited varies from type to type, but since the histochemical methods are sufficiently imprecise for individual identification, only the presence of an acidic mucopolysaccharide can be shown. These substances are extremely water-soluble and are totally lost on fixation, and even using cryostat sections of snap-frozen specimens localization is poor. The only satisfactory method which is reasonably sensitive is that of Haust & Landing (1961) using fixation of the section in a tetrahydrofuran/acetone mixture. Staining of the mucopolysaccharide is based on the metachromasia induced with toluidine blue in a 25% acetone solution. Localization to individual cells is often impossible although the general im-

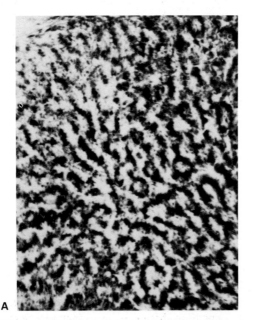

A   **B**

Fig. 16.11 Cryostat sections of liver stained for β-galactosidase activity. Neutral red has been added as a nuclear counterstain. On the left is a normal liver. On the right is a liver biopsy from a patient with G_{M1}-gangliosidosis type 2, in which no activity can be demonstrated. ×110

pression of hepatocyte or Kupffer cell storage is evident. Normal tissue shows no metachromasia, apart from the occasional mast cell, the granules of which are clearly defined.

In addition to the storage of mucopolysaccharide, foamy cells storing a ganglioside-like substance can be found in routine sections in portal areas and also scattered throughout the lobule. These cells are PAS-positive and resist amylase digestion. Their sialic acid content imparts colloidal iron positivity which becomes negative after sialidase digestion.

Fucosidosis

The storage of oligosaccharides, glycoproteins and glycolipids containing a terminal α-linked fucose is found in hepatocytes and foamy Kupffer cells. The epithelium of bile ducts is particularly affected. The substances involved are generally of low molecular weight and very soluble, and are thus difficult to demonstrate without adequate protection. In frozen sections the stored material is variably PAS-positive (or negative where it has been extracted) and is not sudanophilic. The variability of staining from one cell type to another is explained by the differing types of fucose-containing material stored. There are at least two clinical phenotypes of fucosidosis and the liver involvement may be different in the differing forms.

Mannosidosis

Routine sections of liver may appear entirely within normal limits, and no evidence of PAS positivity can be seen. The soluble nature of the mannose-containing storage material makes it difficult to detect (as in the peripheral blood lymphocytes). The substances involved are deposited in hepatocytes, Kupffer cells and bile duct epithelium. A reaction for acid phosphase activity should show the lysosomal nature of the disorder, and pick out the enlarged Kupffer cells. The methods for α-mannosidase activity are not sufficiently sensitive to distinguish between the various isoenzymes of α-mannosidase and do not allow deficient activity to be demonstrated.

Nephrosialidosis

Few cases of this variant of absent sialidase activity have been recorded (Le Sec et al 1978). The liver sinusoids are filled with foamy storage cells without apparent involvement of the hepatocytes (Fig. 16.12). The stored sialyl-compounds are likely to be extremely water-soluble and vigorous protective treatments (snap-frozen tissue, celloidinized sections or freeze drying) will be necessary.

Aspartylglucosaminuria

There is variable storage of aspartylglucosamine in hepatocytes and marked vacuolation of Kupffer cells. The substance is soluble and does not stain with any method. Lipofuscin deposits may be present in hepatocytes and Kupffer cells.

OTHER METABOLIC DISORDERS OF THE LIVER

A variety of other metabolic disorders affect the liver and have no particular histochemical features to assist in their identification. Tyrosinaemia and galactosemia are recognized on morphological grounds (MacSween et al 1987) backed up by biochemical findings. The pseudoductular change observed in both is not specific and is found also in Alpers disease (progressive neuronal degeneration of childhood) (Harding et al 1986), where fat is often a predominant feature and may occur in patches within cirrhotic lobules. Whenever extreme fatty change in a liver is observed in childhood, a metabolic cause should be suspected. The microvesicular change seen in Reye's syndrome is an indication of an acute phenomenon, whereas macroglobular fat is usually a feature of a chronic problem. Metabolic disorders of lipid metabolism (short- medium- or long-chain fatty acid oxidation defects, glutaric aciduria) and carbohydrate disorders (fructose intolerance, fructose 1–6 diphosphatase deficiency, GSD I) have macroglobular fat as a major feature.

The urea cycle defects have no particular morphological or histochemical features, except that with adequate planning (see Ch. 6) it should be possible to detect ornithine carbamyl transferase

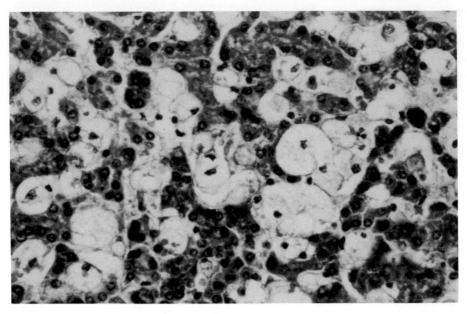

Fig. 16.12 Nephrosialidosis. Section of liver stained with H & E. Large foamy storage cells are present among apparently normal hepatocytes. ×390

deficiency in the male and the mosaic of activity (Lyonization) in the female, using the method described by Wareham et al (1983). Giant mitochondria (eosinophilic in H & E, also shown by the Nitro BT succinate dehydrogenase method) are found in the urea cycle defects, but they are non-specific and are seen in Wilson's disease, alcoholic liver disease and a range of other situations.

Alpha-1-antitrypsin deficiency (in which an abnormal alpha-1-antitrypsin accumulates in hepatocytes where it is synthesized) should be excluded in all cases of neonatal hepatitis. The characteristic periportal PAS-positive globules found in the older cases are not readily recognizable in the neonate since the globule size is small and PAS-positive material (lipofuscin, copper associated protein) is present in neonatal hepatitis. Amylase (or diastase) digestion, followed by a Schmorl reaction and finally with PAS, is a more specific sequence in which lipofuscin and copper associated protein are blue with red alpha-1-antitrypsin deposits. Antibody methods reduce the uncertainty, but some case are difficult as alpha-1-antitrypsin is an acute phase reactant and

will be present in the macrophages and Kupffer cells in neonatal hepatitis. Chapter 15 contains further details of liver disease.

Peroxisomal disorders

A number of disorders in which the activity of one or more peroxisomal enzymes is deficient have been recognized (Schutgens et al 1986, 1989, Monnens & Heymans 1987). Conditions regarded as peroxisomopathies, which encompass not only liver but brain, adrenal and skin pathologies, are:

Zellweger cerebrohepatorenal syndrome
Pseudo-Zellweger syndrome
Infantile Refsum's disease
Adult Refsum's disease
Hyperpipecolicacidaemia
Neonatal adrenoleucodystrophy
Pseudo-neonatal adrenoleucodystrophy
X-linked adrenoleucodystrophy
Adrenomyeloneuropathy
Rhizomelic chondroplasia punctata
Primary hyperoxaluria type 1
? Cerebrotendinous xanthomatosis

In Zellweger's cerebrohepatotrenal syndrome and infantile Refsum's disease peroxisomes are not detectable in the liver, while reduced numbers of smaller peroxisomes are found in neonatal adrenoleucodystrophy. In these three conditions a micronodular cirrhosis may be the only light microscopical abnormality in the liver. Paucity of bile ducts, giant cell hepatitis and deposition of iron are other features which may occur in the liver in Zellweger's cerebrohepatorenal syndrome, while glomerulocystic changes, abnormal C-shaped olivary nuclei and polymicrogyria are consistently found. Peroxisomes, unpredictably detected in cryostat sections, are best demonstrated by the method described by Roels & Goldfischer (1979) on lightly fixed tissue blocks. Since none can be found in Zellweger's cerebrohepatorenal syndrome or infantile Refsum's disease, it is important to run a normal control at the same time. Internal controls such as the haemoglobin peroxidase of RBSs and the myeloperoxidase of neutrophils are helpful indicators that the high pH catalase method has been applied correctly and that the tissue is in the correct state. The peroxisomes may be seen by light microscopy or by electron microscopy (Fig. 16.13).

'Pseudo' disorders, clinically simulating Zellweger's cerebrohepatorenal syndrome or neonatal adrenoleucodystrophy have been described (Goldfischer et al 1987, Pol-The et al 1988). Their chemistry and the nature of the structural changes of the peroxisome are different from the disorders they simulate.

Clearly the peroxisomopathies represent a new group of disorders which will grow as the complexities of the peroxisome become apparent. Approaches to the diagnosis of these disorders are in a state of flux and no clear conclusions can be drawn as to the best microscopical approach. At present, demonstration of the presence, absence, or abnormal shape of the peroxisome represents the baseline. Immunohistochemical methods may be of some help eventually (Litwin et al 1988), but final diagnosis is likely to remain a biochemical necessity.

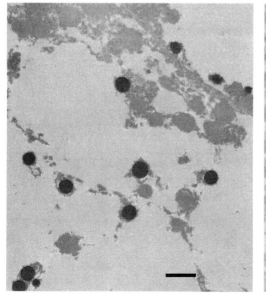 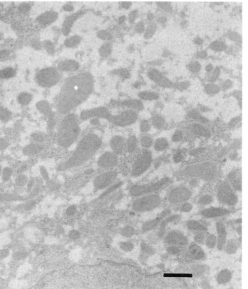

Fig. 16.13   Electron photomicrograph: liver after incubation in a medium to show the peroxidatic activity of catalase. Peroxisomes in normal liver (**A**) are electron dense and are readily distinguished from mitochondria which are unstained. In Zellweger's syndrome (**B**) no peroxisomes can be identified. Scale mark 1 μm. ×8000

REFERENCES

Blair J N R, Burchell A 1988 The mechanism of histone activation of the hepatic microsomal glucose-6-phosphatase system; a novel method to assay glucose-6-phosphatase activity. Biochemica et Biophysica Acta 964: 161–167

Goldfischer S, Collins J, Rapin I et al 1986 Pseudo-Zellweger syndrome; deficiencies in several peroxisomal oxidative activities. Journal of Pediatrics 108: 25–32

Harding B N, Egger J, Portmann B, Erdohazi M 1986 Progressive neuronal degeneration of childhood with liver disease. Brain 109: 181–206

Haust M D, Landing B H 1961 Histochemical studies in Hurler's disease. A new method for localization of acid mucopolysaccharide and an analysis of 'lead acetate' fixation. Journal of Histochemistry and Cytochemistry 9: 79–86

Hers H G, van Hoof F (eds) 1973 Lysosomes and storage diseases. Academic Press, New York

Lake B D 1974 An improved method for the detection of β-galactosidase activity, and its application to GM1-gangliosidosis and mucopolysaccharidosis. Histochemical Journal 6: 211–218

Lake B D, Patrick A D 1970 Wolman's disease. Deficiency of E-600 resistant acid esterase activity with storage of lipids in lysosomes. Journal of Pediatrics 76: 262–266

Le Sec G, Stanescu R, Lyon G 1978 Un nouveau type de sialidose avec atteinte renale: la nephrosialidose. II etude anatomique. Archives Francais de Pediatrie 76: 830–844

Litwin J A, Volkl A, Stachura J, Fahimi H D 1988 Detection of peroxisomes in human liver and kidney fixed with formalin and embedded in paraffin: the use of catalase and lipid β-oxidation enzymes as immunohistochemical markers. Histochemical Journal 20: 165–173

MacSween R N, Anthony P P, Scheuer P J (eds) 1987 Pathology of the liver. Churchill Livingstone, Edinburgh

Monnens L, Heymans H 1987 Peroxisomal disorders: clinical characterization. Journal of Inherited Metabolic Disease 10 (suppl 1): 23–32

Neville B G R, Lake B D, Stephens R, Sanders M 1973 A neurovisceral storage disease with vertical supranuclear ophthalmoplegia and its relationship to Niemann-Pick disease. A report of nine patients. Brain 96: 97–120

Patrick A D, Lake B D 1968 Cystinosis: electron microscopic evidence of lysosomal storage of cystine in lymph nodes. Journal of Clinical Pathology 21: 571–575

Poll-The B T, Roels F, Ogier H et al 1988 A new peroxisomal disorder with enlarged peroxisomes and a specific deficiency of acyl-CoA oxidase. American Journal of Human Genetics 42: 422–434

Roels F, Goldfischer S 1979 Cytochemistry of human catalase; the demonstration of hepatic and renal peroxisomes by a high temperature method. Journal of Histochemistry and Cytochemistry 27: 1471–1477

Schutgens R B H, Heymans H S A, Wanders R J A, van den Bosch H, Tager J M 1986 Peroxisomal disorders; a newly recognized group of genetic diseases. European Journal of Paediatrics 144: 430–440

Schutgens R B H, Schrakamp G, Wanders R J A, Heymans H S A, Tager J M, van den Bosch H 1989 Prenatal and perinatal diagnosis of peroxisomal disorders. Journal of Inherited Metabolic Disease 12 (suppl 1): 118–134

Scriver C R, Beandet A L, Sly W S, Valle D (eds) 1989 The metabolic basis of inherited disease, 6th edn. McGraw Hill, New York

Vanier M T, Wenger D A, Comly M E, Rousson R, Brady R O, Pentchev P 1988 Niemann-Pick disease group C; clinical variability and diagnosis based on defective cholesterol esterification. Clinical Genetics 33: 331–348

Wareham K A, Howell S, Williams D, Williams E D 1983 Studies of X-chromosome inactivation with an improved histochemical technique for ornithine carbamoyl transferase. Histochemical Journal 15: 363–371

17. Kidney: glomerular disease

R. A. Risdon

The identification and classification of the various patterns of glomerular disease recognized in renal biopsy specimens routinely requires both conventional light microscopy (LM) and immunochemical staining. Although in a minority of conditions electron microscopy (EM) is also essential for correct diagnosis, the major contribution of this technique has been that a better understanding of the different patterns of glomerular disease provided by EM has aided interpretation at the LM level. In addition, LM has been applied to sections of resin-embedded tissue processed by techniques originally developed for EM with vastly improved results in terms of histological clarity and resolution. This is principally related to better tissue preservation and to the thinner sections which can be cut from resin- compared with paraffin-embedded tissue. The introduction of beam-stable, hydrophilic acrylic resins such as LR White (Germain 1982) has allowed the use of a full range of conventional stains normally applied to paraffin sections, as well as the performance of EM on the same tissue blocks.

For LM, whether performed on paraffin or plastic sections, fixation of a renal needle biopsy specimen in 10% phosphate-buffered formalin or Dubosq-Brazil (alcoholic Bouin) is recommended. After processing, 1 μm plastic sections, or 2 μm paraffin sections are cut and stained by haematoxylin and eosin (HE), Masson's trichrome, elastic/van Gieson, periodic acid-Schiff (PAS) and periodic acid-methenamine silver (PAMS) (see Bancroft & Stevens 1982). In addition, stains to detect amyloid, such as Congo red, or Thioflavin T should be applied to thicker (5 μm) sections.

For EM, fixation of small fragments of tissue in glutaraldehyde and postfixation in osmium tetroxide gives consistently good results. After processing, semi-thin sections are cut and the block trimmed to include the glomeruli. Ultrathin sections are stained with uranyl acetate and lead citrate.

The standard method for immunochemical staining uses antisera labelled with fluoresccin applied to 4 μm cryostat sections of snap-frozen fresh tissue. More recently, the use of formalin-fixed paraffin sections digested with pronase to which peroxidase or fluorescein-labelled antisera are applied, has become more popular (Tubbs & Gephardt 1983, Sinclair et al 1981). Using peroxidase conjugates, not only can paraffin sections be employed, but a permanently stained preparation results. Antisera to human immunoglobulins (IgA, IgG, IgM) fibrinogen/fibrin and complement (Clq, C3) are applied using direct rather than indirect methods; in addition, antisera against kappa and lambda light chains may be used.

The histological classification of glomerulonephritis (GN) employed is that of the World Health Organization (see Churg & Sobin 1982) and the following conditions will be described here.

A. Minor glomerular abnormalities
B. Focal/segmental lesions
C. Diffuse glomerulonephritis
   1. Diffuse membranous GN
   2. Diffuse proliferative GN
      a. Mesangial proliferative GN
      b. Endocapillary proliferative GN

c. Mesangiocapillary GN
d. Crescentic GN
D. Glomerulonephritis of systemic diseases
1. Lupus nephritis
2. Nephritis of anaphylactoid purpura
3. Berger's (IgA) nephropathy
4. Goodpasture's syndrome
E. Glomerulonephritis in vascular diseases
1. Periarteritis nodosa
2. Wegener's granulomatosis
3. Thrombotic microangiopathy
F. Glomerular lesions in metabolic diseases
1. Diabetes mellitus
2. Amyloidosis
3. Glomerular lesions in dysproteinaemias

A. MINOR GLOMERULAR ABORMALITIES

This categorization refers to what was previously called minimal change disease. Patients are usually children with highly selective proteinuria without haematuria and a nephrotic syndrome that responds to corticosteroid therapy (Churg et al 1970). Classically, no significant alterations are seen in the glomeruli by LM, but with EM in untreated cases there is generalized fusion of podocyte foot processes and the epithelial cell cytoplasm is applied directly to the outer aspect of the glomerular basement membrane. In many cases minor glomerular changes are discerned by LM and these include slight focal hypercellularity in some peripheral mesangial areas, mild generalized mesangial expansion (up to twice normal) and focal splitting, wrinkling or thickening of tuft capillary walls best seen in silver stains (PAMS). Immunochemical staining is usually completely negative. Occasionally flecks of C3 or C1q and scanty mesangial deposits of IgM or IgG may be seen. The significance, if any, of these findings is debatable and may reflect some variability in antisera specificity.

B. FOCAL/ SEGMENTAL LESIONS

Focal segmental hyalinosis and sclerosis

Controversy still exists over whether or not this lesion represents a distinct entity (Habib 1973, Saint-Hillier et al 1975). Morphologically, it is characterized by focal segmental lesions involving only some glomeruli, the remainder showing no more than the minor lesions described in the previous section or a slight diffuse mesangial cell proliferation. Particularly early in the disease process, glomerular lesions selectively involve glomeruli in the deep cortex (Rich 1957). These lesions are characterized by segmental sclerosis of part of the glomerular tuft, often associated with deposits of hyaline material in the mesangium or capillary lumina, foam cells in the sclerosed segment and capsular adhesions with localized reactive proliferation of adjacent epithelial cells. These focal lesions frequently involve the glomerular hilus, but may first appear in peripheral lobules. Progression of the disease is marked by extension of the sclerotic process eventually to affect the whole tuft (global sclerosis) and by the involvement of an increasing number of glomeruli. Tubular atrophy is associated with the glomerular lesions from an early stage and its presence should alert the pathologist to make a careful search for focal sclerotic lesions which may be sparse and require the examination of serial sections.

Immunochemical staining is often completely negative but deposits of immunoglobulin (especially IgM) and C3 may be identified in the segmental sclerotic lesions and occasionally more diffusely in the mesangial areas.

One group of patients with focal segmental hyalinosis and sclerosis on renal biopsy present with persistent proteinuria or the nephrotic syndrome, which is resistant to steroid therapy, but may partially respond to cyclophosphamide. Haematuria and hypertension may also be features and the natural history of the disease is progression at a variable rate to chronic renal failure. Other patients with steroid-sensitive nephrotic syndrome and only minor glomerular abnormalities (minimal change disease) on initial biopsy may develop focal sclerotic lesions after years of relapsing nephrotic syndrome. In these patients the segmental lesions do not have the same sinister clinical implications. In addition, segmental sclerotic lesions morphologically similar to those described above may be seen in association with a variety of other glomerular lesions,

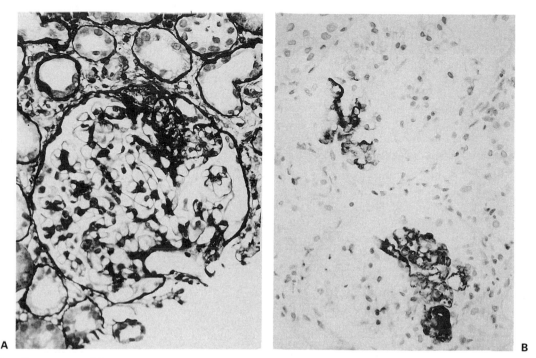

Fig. 17.1 Focal segmental hyalinosis and sclerosis. A Segmental sclerosis of part of the tuft. PAMS ×150. **B** IgM deposition in sclerosed portions of two tufts, sparing the remainder. Direct immunoperoxidase. ×100

such as diabetic nephropathy, hypertensive nephrosclerosis, Alport's syndrome, reflux nephropathy, and as a part of a scarring process in some forms of proliferative glomerulonephritis.

Focal glomerulonephritis

This category describes the involvement of some, but not all, glomeruli in a given biopsy by segmental lesions of varying morphology. These include segmental mesangial/endothelial cell proliferation, segmental tuft necrosis (often with thrombosis of the relevant tuft capillary loop), segmental sclerosis, capsular adhesion formation with associated localized reactive epithelial cell proliferation, or even frank crescent formation.

Such lesions occur in a variety of clinical contexts, particularly in association with systemic diseases such as lupus erythematosus, anaphylactoid purpura, Goodpasture's syndrome, Berger's disease, Wegener's granulomatous and periarteritis nodosa (see sections D and E) as well as in infec-

tive endocarditis and rheumatic fever. Focal glomerulonephritis may also occur in isolation (idiopathic form); it may precede diffuse proliferative glomerulonephritis and during resolution of proliferative glomeruloncphritis residual focal lesions may be apparent.

Consideration of the full clinical picture is necessary to place the morphological changes in focal glomerulonephritis into their proper context. Immunochemical staining is also vital in this respect and is discussed in sections D and E.

C. DIFFUSE GLOMERULONEPHRITIS

Diffuse membranous glomerulonephritis

This lesion is characterized at LM by diffuse thickening of glomerular capillary walls in most or all of the glomeruli in a biopsy. In silver stains (PAMS), short argyrophilic projections ('spikes'), sometimes with clubbed ends, extend from the epithelial aspect of the glomerular basement

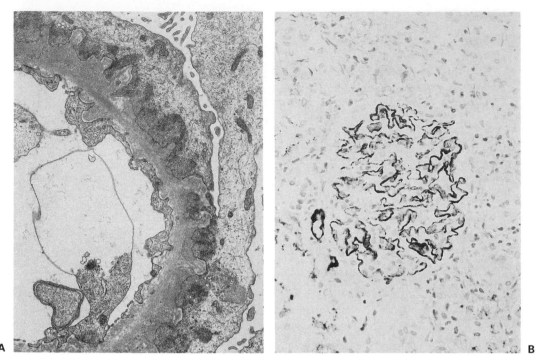

Fig. 17.2 Membranous GN. **A** Electron micrograph showing epithelial spikes and electron dense deposits on the epithelial aspect of the glomerular basement membrane. ×5000. **B** Granular IgG staining diffusely along the glomerular capillary walls. Direct immunoperoxidase. ×125

membrane. Immunochemical staining reveals granular deposits of IgG (and rarely of IgA and IgM) as well as C3 diffusely along the epithelial side of the GBM. By EM, both the spikes and the intervening deposits are clearly demonstrated. In resin sections stained with toluidine blue, the deposits but not the spikes are stained. Various stages in the evolution of membranous nephropathy are recognized in renal biopsies (Ehrenreich & Churg 1968). In the early stages, glomerular capillary wall thickening and spike formation are inconspicuous or absent, but deposits are clearly seen on immunochemical staining or EM. In the fully developed lesion the appearances are those described above and there may also be some diffuse mesangial expansion and rarely some mesangial cell proliferation. In the late stages the spikes tend to fuse and coalesce to form a greatly thickened, chain-like glomerular basement membrane with lucent areas. Deposits are hard to recognize at this stage by EM and the lucent areas may represent sites from which the deposits have been reabsorbed.

Clinically, patients present at any age with proteinuria and often the nephrotic syndrome, which may persist for many years before finally ending in renal failure. This progression is not invariable, however, and spontaneous resolution may occur in about 25 % of adults (Row et al 1975) and an even higher percentage of children (Habib et al 1973).

Diffuse proliferative glomerulonephritis

Diffuse mesangial proliferative glomerulonephritis

This variety of glomerulonephritis is recognized by a diffuse and uniform increase in mesangial cells, usually accompanied by an expansion of the mesangial areas in all or nearly all the glomeruli in a biopsy specimen. The glomerular capillaries remain patent and the capillary walls in the periphery of the glomerular lobules are unthickened.

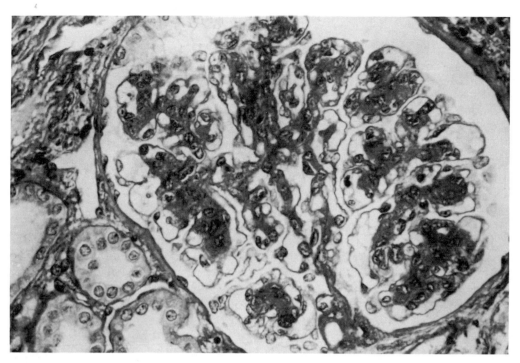

Fig. 17.3 Diffuse mesangial proliferative GN. Mesangial cell proliferation and increase in mesangial matrix. Peripheral tuft capillaries are patent and their walls are unthickened. Trichrome. ×250

The appearances may be seen in pure form or in association with other lesions, such as focal segmental hyalinosis and sclerosis or membranous glomerulonephritis (see appropriate sections).

Clinically, like focal glomeronephritis, diffuse mesangial proliferative glomerulonephritis may occur in systemic conditions such as lupus erythematosus, anaphylactoid purpura and Berger's disease (see section D) as well as in infective endocarditis and resolving diffuse endocapillary glomerulonephritis. Particularly in patients with lupus erythematosus and anaphylactoid purpura, superimposed focal segmental lesions, such as segmental necrosis and sclerosis, capsular adhesions and crescent formation, are not unusual. Knowledge of the clinical background and immunochemical staining are essential for placing the biopsy appearances in mesangial proliferative glomerulonephritis in their appropriate context, as discussed in sections D and E.

Some patients with diffuse mesangial proliferative glomerulonephritis have no accompanying systemic disease. These may present at any age (but often in childhood) with proteinuria, sometimes of nephrotic proportions, and frequently with gross or microscopic haematuria. The proteinuria is usually not responsive to steroid therapy and whilst spontaneous remission occurs in many, some progress to renal failure. In these patients without associated systemic disease, immunochemical staining may be negative but in some mesangial deposits of IgM and C3 are seen. These latter findings are sometimes described as IgM nephropathy (Tejani and Nicastri 1983)

Diffuse endocapillary proliferative glomerulonephritis

This term describes a diffuse and uniform proliferation of endocapillary (predominantly mesangial) cells in all glomeruli. It is distinguished from mesangial proliferative glomerulonephritis by the intensity of the proliferative change which is sufficient to narrow or occlude the glomerular capillary lumina. Neutrophil polmorphs may also infiltrate the glomerular tufts in varying numbers.

Classically this form of glomerulonephritis is seen in patients, particularly children, who present with an acute nephritic syndrome following a beta-haemolytic streptococcal infection, usually of serotypes 12, 4 or 1 ('acute poststreptococcal glomerulonephritis'). However, it is now recognized that a similar picture may occur after other infections (staphylococcal, pneumococcal or viral) and the less committed term 'acute post-infective glomerulonephritis' is now preferred. Indeed, preceding streptococcal infection is now increasingly rare (Meadow 1975). In addition, diffuse endocapillary proliferative glomerulonephritis may be seen in patients with systemic diseases such as lupus erythematosus, periarteritis nodosa, systemic lupus erythematosus and infective endocarditis.

Acute post-infective glomerulonephritis usually resolves completely in children, but in adults, not only is the initial disease usually more severe, but in a proportion there is progressive glomerular scarring leading to chronic renal failure.

In acute poststreptococcal glomerulonephritis EM may reveal large electron-dense deposits ('humps') on the epithelial aspect of the glomerular basement membrane in the early stages of the disease. These 'humps' are occasionally sufficiently large to be seen by LM, particularly in resin sections.

In acute postinfective glomerulonephritis the most constant immunochemical finding is the presence of granular C3 deposits along the glomerular capillary walls. IgG and occasionally IgA and IgM may also be found. Similar deposits are seen less frequently in the mesangium.

Diffuse mesangiocapillary proliferative glomerulonephritis

In this lesion, also termed membranoproliferative glomerulonephiritis, all or most of the glomeruli are enlarged and lobulated. There is diffuse, but variable, mesangial cell proliferation and mesangial expansion combined with thickening of the glomerular capillary walls. Infiltration of the glomerular tufts by monocytes and neutrophil polymorphs, as well as some degree of crescent formation, are occasional features. The lesion often evolves into so-called lobular glomerulonephritis in which accentuated tuft lobulation is associated with relatively acellular centrilobular hyalinization.

EM studies have demonstrated three varieties of mesangiocapillary glomerulonephritis of which two (types I and III) are probably variants of the same process, but the third (type II) is distinctly different (Levy et al 1979).

Types I and III are characterized by extension of the mesangial cell cytoplasm and matrix around the circumferences of peripheral glomerular capillary walls between the glomerular basement membrane and its lining endothelium (so-called 'mesangial interposition'). On its endothelial aspect the interposed mesangium is delineated by a second layer of basement membrane-like material, either condensed mesangial matrix or a second basement membrane formed by the displaced endothelial cells. This can be seen at the LM level in sections stained with PAMS as a second argyrophilic layer in the glomerular capillary wall (so-called 'double contouring' or 'tram-tracking'). In type 1, electron-dense deposits, some of which are large, are seen along the endothelial side of the glomerular basement membrane, in the mesangium and paramesangium, in interposed mesangium between the glomerular basement membrane and the endothelium and occasionally encroaching on the basement membrane itself. Less commonly, deposits occur on the epithelial side of the glomerular basement membrane. In type III, similar deposits are seen mainly in the disrupted lamina densa of the glomerular basement membrane.

Immunochemical staining reveals granular deposits along peripheral glomerular capillary walls and often in the mesangial areas. Three patterns may be distinguished. In the first, deposits of immunoglobulins, Clq and C3 outline the periphery of the tuft, but spare the mesangium. In the second, the appearances are similar, but deposits are also seen in the mesangium. In the third, deposits of C3 alone are seen along glomerular capillary walls and in the mesangium.

Clinically, types I and III mesangiocapillary glomerulonephritis occurs in older children and

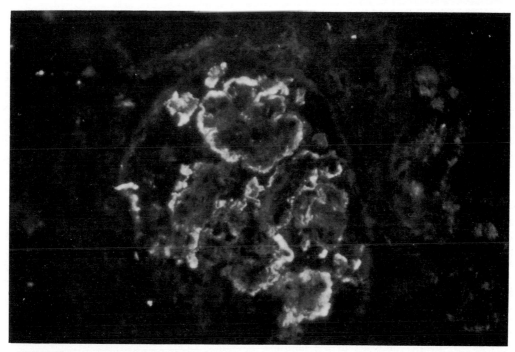

Fig. 17.4   Mesangiocapillary GN (type 1). Granular deposition of IgG outlining the periphery of the glomerular tuft. Immunofluorescence microscopy. ×250

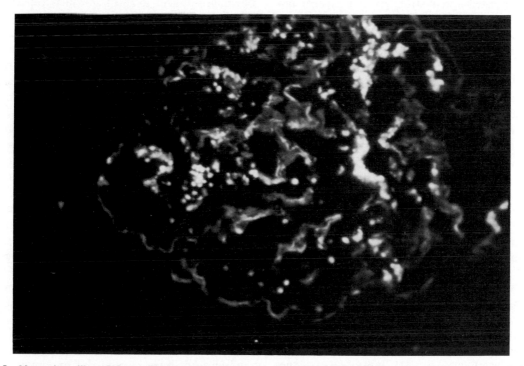

Fig. 17.5   Mesangiocapillary GN type II (dense deposit disease). Deposits of C3 as bright nodules in the mesangium. Less intense staining of the glomerular capillary walls. Immunofluorescence microscopy. ×250

young adults. The disease may present insidiously with proteinuria and microscopic haematuria developing into the nephrotic syndrome. Sometimes presentation is more abrupt, with an acute nephritic syndrome evolving to the nephrotic syndrome; hypertension and renal failure are common. Depression of serum levels of C3 and also of early components of complement (Clq, C4 and C2) is an important finding and develops in the majority of patients, even if not demonstrated at the onset. In addition, the serum contains a globulin called the C3 nephritic factor (C3 NeF),which is an indirect activator of C3. The significance of the complement disorders in mesangiocapillary glomerulonephritis is reviewed by Peters & Williams (1975). In type II mesangiocapillary glomerulonephritis (also known as dense deposit disease) (see Fig. 17.5) the thickening of the glomerular capillary walls is quite different from that seen in types I and III. By EM the striking change is the presence of highly electron-dense ribbon-like deposits in the lamina densa of the thickened glomerular basement membrane. These deposits often form long continuous or interrupted stretches and involve not only the glomerular basement membrane, but also that of Bowman's capsule and some tubules. Less electron-dense, discrete deposits may also be seen beneath the endothelium and sometimes as 'humps' on the epithelial aspect of the glomerular basement membrane (Habib et al 1975).

By LM the dense deposits stain intensely with PAS, but they are not argyrophilic so do not stain with PAMS. A useful pointer in the PAS stain is to follow the extension of the PAS-positive deposits as they form a continuous line from the glomerular basement membrane into that of Bowman's capsule. The poor staining of the glomerular basement membrane in PAMS preparations is sometimes mistakenly dismissed as being due to technically inferior staining. In the Masson trichrome, the deposits give a characteristic birefringent appearance,but they are best demonstrated at the LM level in resin-embedded sections in which they stain darkly with toluidine blue (Vargas et al 1976). Associated capsular crescent formation is more common in type II than in types I and III mesangiocapillary glomerulonephritis.

Immunochemical staining reveals positive staining with C3 alone. The antiserum stains strongly well-defined rounded deposits in the mesangium and less intensely the linear deposits in the glomerular and tubular basement membranes.

Clinically, the features are similar to those in types I and III. Serum complement levels are almost invariably depressed and this is due to a low C3 level, the earlier components remaining normal. The presence of C3 NeF in the serum is also a constant feature. Sometimes mesangiocapillary glomerulonephritis is associated with partial lipodystrophy and when this occurs the glomerulonephritis is of the type II (dense deposit) variety. Partial lipodystrophy is also associated with hypocomplementaemia which may precede the clinical onset of glomerulonephritis by several years.

Diffuse crescentic glomerulonephritis

This term is used to describe glomerulonephritides in which prominent cellular proliferation in Bowman's space, to form so-called capsular crescents, is a feature in the majority of glomeruli. Authorities vary on the exact criteria, some requiring crescents in more than 50% and others requiring them in more than 80% of glomeruli.

The crescents tend to compress the adjacent glomerular tufts and may obscure any underlying glomerular pathology. However, several varieties of glomerulonephritis — focal segmental proliferative glomerulonephritis associated with systemic diseases, and diffuse endocapillary or mesangiocapillary glomerulonephritis being the most frequent — may be associated with crescent formation. When this latter change affects a majority of glomeruli it is useful to designate the morphological changes appropriately, for example, diffuse mesangiocapillary glomerulonephritis (type II) with 80% capsular crescents.

The importance of crescentic glomerulonephritis lies in its usually poor prognosis and accounts for the older term of rapidly progressive glomerulonephritis. Accepting the usually grave implication of crescentic glomerulonephritis, the outlook and likely response to treatment depends partly on the nature of any underlying glomerular

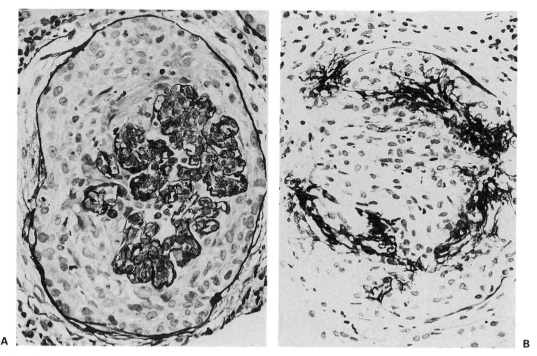

Fig. 17.6   Diffuse crescentic GN. **A** A cellular capsular crescent compressing the glomerular tuft. PAMS. ×160 **B** Fibrin deposition in the crescent but not in the glomerular tuft. Direct immunoperoxidase. ×160

disease and partly on the degree of organization of the crescents. These can be (a) cellular, i.e. composed of proliferating visceral and parietal epithelial cells with a varying admixture of macrophages, monocytes and sometimes neutrophil polymorphs, (b) fibrocellular, i.e. also containing basement membrane-like strands and collagen fibres, or (c) fibrous, i.e. composed largely of fibrous tissue with little or no cellular elements.

Often fibrin can be demonstrated, particularly in cellular crescents, either by special staining on LM or immunochemically. Breaks in the glomerular basement membrane in crescentic glomerulonephritis can normally be demonstrated by EM (Morita et al 1973) and it has been suggested that the crescents are a response to the consequent fibrin leakage into the urinary space.

The immunochemical findings in crescentic glomerulonephritis depend largely on the nature of the underlying glomerular disease, but sometimes the only abnormality is the presence of fibrin/fibrinogen in the crescents.

D. GLOMERULONEPHRITIS OF SYSTEMIC DISEASES

Lupus nephritis

Renal involvement in systemic lupus erythematosus is common and the manifestations, both clinically and on renal biopsy, range from very mild to severe. The renal biopsy appearances may be helpful both prognostically and in assessing the likely response to treatment. A wide variety of morphological changes are encountered on renal biopsy and these have been classified as follows (Churg & Sobin 1982).

Minor glomerular abnormalities are encountered relatively uncommonly since these patients have few clinical manifestations related to renal involvement and are thus seldom biopsied. By LM the glomeruli are normal, or show only very minor abnormalities such as those described in Section A.

However, electron-dense deposits in the glomeruli are usually demonstrated by EM and immunochemical staining reveals both

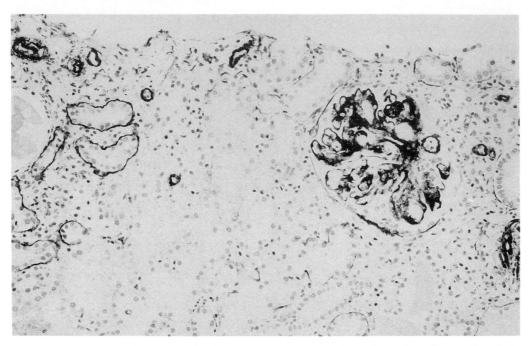

Fig. 17.7 Lupus nephritis. Granular deposits of IgG in the glomerular mesangium, along tuft capillary walls and along tubular basement membranes. Direct immunoperoxidase. ×80

immunoglobulin and complement deposition (see below).

Mesangial lupus nephritis at its mildest merges with minor abnormalities, but in general the changes are identical to those described under diffuse mesangial glomerulonephritis (see Section C, Mesangial proliferative GN). Clinical manifestations are mild (usually slight proteinuria and microscopic haematuria) and the histological changes may remain stable, regress spontaneously, or more often following steroid or immunosuppressive therapy, or progress to diffuse lupus nephritis (see below).

Focal lupus nephritis is characterized by focal and segmental proliferative change, often with focal necrosis and local thrombosis of tuft capillaries. Scarring of these lesions produces segmental tuft sclerosis, often associated with capsular adhesions and local reactive epithelial cell proliferation; sometimes frank capsular crescents are formed.

Isolated focal lupus nephritis is a mild disease with a similar clinical significance to pure mesangial lupus nephritis. However, the lesions of focal lupus nephritis are usually superimposed on another pattern (particularly mesangial lupus nephritis), when they indicate active progression of the disease towards diffuse lupus nephritis.

Diffuse lupus nephritis implies involvement of all or nearly all glomeruli and occurs in various forms, often in combination. These include diffuse endocapillary glomerulonephritis similar by LM to acute postinfective glomerulonephritis, a diffuse mesangiocapillary proliferative glomerulonephritis, diffuse crescentic glomerulonephritis and (most commonly) diffuse necrotizing and sclerosing glomerulonephritis. The latter is characterized by diffuse, mainly mesangial proliferative change combined with varying degrees of segmental necrosis and sclerosis. Typically the intensity of these changes varies from glomerulus to glomerulus and between lobules in an individual glomerulus. Necrosis is often associated with neutrophil polymorph infiltration and the presence of fragments of nuclear debris. These sometimes take the form of so-called 'haematoxyphil bodies', which are the tissue counterpart of LE cells.

Localized thickening of glomerular tuft capillary walls ('wire loops') are produced by large subendothelial deposits and similar deposits may occlude the capillary lumina ('hyaline thrombi'). Clinically, diffuse lupus nephritis is associated with heavy proteinuria, often with a nephrotic syndrome, and with haematuria, hypertension and renal insufficiency. The disease tends to be progressive and resistant to therapy.

Membranous lupus nephritis may occur in a 'pure' form resembling idiopathic diffuse membranous glomerulonephritis (see Section C, Diffuse membranous GN), or in combination with other varieties of lupus nephritis. Clinically, membranous lupus nephritis is associated with the nephrotic syndrome and haematuria which does not respond to steroid or immunosuppressive therapy. Whilst the clinical course may be fairly indolent, resembling that of idiopathic membranous nephritis, sometimes hypertension and renal failure develop rapidly. This transformation is more likely when proliferative or focal segmental lesions are superimposed.

Interstitial and vascular disease — Arteritis and tubulointerstitial disease may affect the kidney in lupus erythematosus, usually in association with severe glomerular lesions. Immunochemical staining may reveal deposition of immunoglobulins and complement along tubular basement membranes and in blood vessels as well as in the glomeruli.

Electron microscopy confirms the various changes seen by LM and demonstrates electron-dense deposits, which are often large and numerous. These are situated beneath the endothelium, on the epithelial aspect of the glomerular basement membrane as well as within the lamina densa and in the mesangium. These deposits are usually homogenous or granular, but occasionally they are organized into a whorled or 'fingerprint' pattern.

Immunochemical staining usually reveals deposits of IgA, IgG, IgM, C3, C1q, C4 and properdin in various patterns depending on the histological type. In focal lupus nephritis immunochemical staining of the glomeruli is diffuse and is not confined to glomeruli showing alterations by LM. Fibrinogen/fibrin is seen in necrotic segments and in capsular crescents.

Nephritis of anaphylactoid purpura

Glomerulonephritis accompanies Henoch-Schönlein (anaphylactoid) purpura in only a minority of patients (usually children) affected by the disease. When it does occur, renal involvement is usually mild and self-limiting. However, in a small proportion of cases, and usually when attacks of anaphylactoid purpura are repeated, renal disease may be progressive and lead to chronic renal failure.

Morphologically the appearances are of a diffuse mesangial proliferative or focal proliferative glomerulonephritis (see Sections B, Focal glomerulonephritis, and C, Mesangial proliferative GN) and often the two pictures are superimposed. Less commonly a diffuse endocapillary, mesangiocapillary or crescentic glomerulonephritis (see Section C, Endocapillary proliferative, Mesangiocapillary and Crescentic GN) are seen either separately or in combination. By EM, electron-dense deposits are found mainly in the mesangium, to a lesser extent beneath the endothelium, and rarely on the epithelial aspect of the glomerular basement membrane. Immunochemical staining reveals mainly mesangial deposits of IgA with lesser amounts of IgG and occasionally IgM. C3 is also present, but early complement components are not seen. Fibrinogen/fibrin is present in capsular crescents and frequently in the mesangium.

Berger's (IgA) nephropathy

The characteristic feature of this type of glomerulonephiritis, first described by Berger & Hinglais (1968), is the presence of diffuse mesangial deposition of IgA. Clinical manifestations of systemic disease such as lupus erythematosus and anaphylactoid purpura are lacking, but the disease may occur in patients with hepatic cirrhosis (Sinniah 1984) and mucin-secreting adenocarcinomas (Sinniah 1982). Despite the lack of clinical evidence of anaphylactoid purpura, the LM appearances and immunochemical findings in Berger's nephropathy are often very similar and some authorities regard IgA nephropathy as an undeveloped form of anaphylactoid purpura.

World-wide, IgA nephropathy is now recognized as perhaps the most common variety of glomerulonephritis and an important cause of chronic renal failure.

By LM, the characteristic pattern is of mild diffuse mesangial proliferative glomerulonephritis, accompanied by focal segmental lesions consisting of focally accentuated mesangial proliferation, focal sclerosis with capsular adhesions and occasionally necrosis. Other patterns include very mild changes and only some barely discernible increase in mesangial matrix, more severe changes with focal capsular crescent formation and rarely a diffuse endocapillary proliferative glomerulonephritis. By EM, electron-dense deposits are scattered throughout the mesangium, the largest deposits often occupying a site between the mesangium and the adjacent reflection of the glomerular basement membrane (paramesangial deposits). These paramesangial deposits can sometimes be seen by LM, particularly in resin sections.

Immunochemical staining invariably demonstrates diffuse mesangial IgA deposition (Fig. 17.8), although the amount varies from case to case. Lesser amounts of IgG and rarely of IgM are sometimes present. C3 deposition is a consistent finding, but Clq and C4 are absent. Deposits of IgA and C3 may also be found in skin and muscle biopsies at the dermal-epidermal junction and around small blood vessels.

Clinically, the disease is most often identified in children and young adults presenting with gross or microscopic haematuria and mild proteinuria, often precipitated by an upper respiratory tract infection. Less commonly the disease presents as an acute nephritic or nephrotic syndrome. Serum IgA levels are elevated in a proportion of patients. Over a period of years progression to hypertension and renal failure may occur. IgA deposits tend to recur in renal transplants.

Goodpasture's syndrome

This describes the rare combination of pulmonary haemorrhage with glomerulonephritis which is usually crescentic in type. By LM, the typical

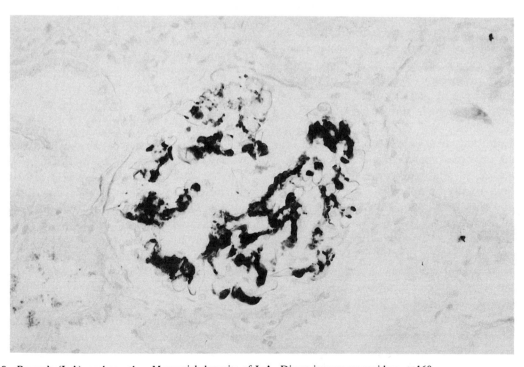

Fig. 17.8 Berger's (IgA) nephropathy. Mesangial deposits of IgA. Direct immunoperoxidase. ×160

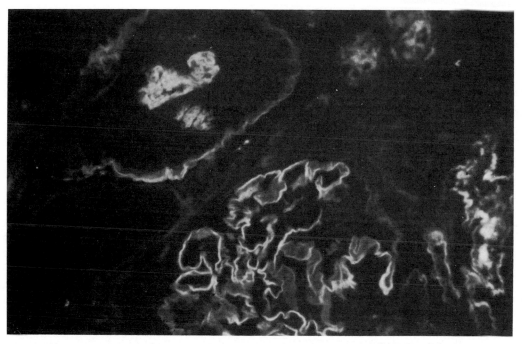

Fig. 17.9 Goodpasture's syndrome. Bright linear deposits of IgG along glomerular and tubular basement membranes. Immunofluorescence microscopy. ×250

finding is of a diffuse crescentic glomerulonephritis with large occlusive crescents compressing the glomerular tufts which show little other change. Less often focal segmental changes are seen with focal mesangial proliferation, focal tuft necrosis and thrombosis, and small crescents containing fibrin. The diagnostic feature seen on immunochemical staining is of diffuse linear staining of all the glomerular basement membranes by IgG and less constantly by C3. Characteristically, circulating antibodies against glomerular basement membrane can be demonstrated in the serum.

Clinically, the disease is most common in young male adults who present with pulmonary haemorrhage ranging in severity from mild blood-staining of the sputum to dramatic life-threatening bleeding from the lungs. This may be preceded by a mild 'flu-like' illness and in some cases there is a history of exposure to volatile hydrocarbons. Renal manifestations include haematuria and proteinuria of varying degree. Very rarely this form of glomerulonephritis may occur in the absence of pulmonary haemorrhage. Although rapid progression to renal failure is the rule, some

patients recover spontaneously, either temporarily or permanently.

E. GLOMERULONEPHRITIS IN VASCULAR DISEASES

Periarteritis nodosa

Renal involvement in periarteritis nodosa may be of the so-called macroscopic or microscopic forms. In the former, large muscular arteries are involved and the renal parenchymal alterations include infarcts of various sizes and ischaemic changes. Arteritic lesions are seldom recognized in small needle biopsies. In the microscopic form, a glomerulonephritis characterized by focal segmental thrombosis, sclerosis, necrosis, cellular proliferation and crescent formation is seen; this may evolve to a diffuse crescentic glomerulonephritis. Sometimes an acute necrotizing angiitis may be recognized in the biopsy specimen.

Immunochemical staining usually reveals only fibrinogen/fibrin in the necrotic lesions, in capsular crescents and sometimes in arterial walls. Rarely immunoglobulins such as IgG and C3 may be deposited in the tufts.

Wegener's granulomatosis

This disease, characterized by necrotizing lesions of the upper respiratory tract, mouth and middle ear, with pulmonary infiltrates and cavitation, may be accompanied by a microscopic periarteritis and renal lesions identical to those described in periarteritis nodosa. Occasionally granulomata may be recognized in the glomeruli or in the renal interstitium on renal biopsy.

Thrombotic microangiography

This describes the renal glomerular and arteriolar changes accompanying the haemolytic-uraemic syndrome. The latter usually affects young children, especially boys under the age of 4 years, but may also occur in adults. The disease usually follows an upper respiratory or gastrointestinal infection and it is characterized by vomiting, purpura, haematemesis and bloody diarrhoea. A haemolytic anaemia with fragmented erythrocytes in the peripheral blood smear occurs and the renal manifestations include haematuria, proteinuria and rapidly developing oliguria with uraemia. With proper supportive therapy, death or irreversible renal failure are rare in children but more common in adults.

The changes on renal biopsy are variable from case to case and between the glomeruli in an individual specimen. In some there may merely be marked congestion, or alternatively collapse of the glomerular tuft capillaries. Occasionally capsular crescents or total infarction may be evident. More commonly the tuft capillary walls are thickened and double contours are visible in PAS or PAMS stains. Fragmented red cells or fibrin thrombi may be seen in capillary lumina. Mild lesions may resolve completely, but more severe lesions lead to bland sclerosis of the tufts in which cells other than some parietal epithelial cells disappear, although the lobular structure of the contracted tuft is preserved and capsular adhesions are hardly ever seen.

Arterioles show fluffy material in the subendothelial space leading to luminal narrowing and sometimes luminal thrombosis occurs. Larger arterioles and small arteries may show mucoid and cellular intimal thickening with marked reduction in luminal diameter and microaneurysms of some afferent arterioles may be apparent.

EM shows separation of the endothelium from the glomerular basement membrane and accumulation of fluffy material in the space so formed. Within this material occasional fibrin strands and fragments of red blood cells may be visible. A newly formed basement membrane is often seen beneath the endothelial cells. Pale material also collects in the mesangium and the matrix may be disrupted causing mesangiolysis.

Immunochemical staining reveals fibrinogen/fibrin in the mesangium and in capillary walls in the early stages. Immunoglobulins and complement are less commonly observed.

F. GLOMERULAR LESIONS IN METABOLIC DISEASES

Diabetes mellitus

The development of glomerular disease in diabetes mellitus is associated with proteinuria which sometimes reaches nephrotic proportions; occasionally haematuria also occurs. Slow progression over a number of years to chronic renal failure develops and hypertension is frequently a feature. Diabetic glomerulosclerosis is much commoner in insulin-dependent juvenile diabetes than in maturity-onset disease.

Early in the disease glomerular enlargement without other abnormalities is seen. The principal glomerular changes that develop are related to a diffuse thickening of the glomerular basement membranes and widening of the mesangium without associated mesangial hypercellularity. On EM this is seen to be due to widening of the lamina densa and to an increase in mesangial matrix. The glomerular basement membrane may be thickened to between 5 and 10 times normal. In about 25% of cases, mesangial widening progresses to centrilobular rounded nodule formation, first described by Kimmelstiel & Wilson (1936). These nodules vary in size, affect mainly peripheral lobules and involve only a proportion of the tufts.

Hyaline arteriolar sclerosis is prominent in both afferent and efferent arterioles. Localized hyaline deposits containing fibrin derivatives and lipid (Salinas-Madrigal et al 1970) also form in the glomeruli ('fibrin cap' lesions) and in Bowman's capsule ('capsular drop' lesions).

Immunochemical staining is often negative, but sometimes faint linear staining of IgG and albumin is seen along glomerular and sometimes tubular basement membranes.

Amyloidosis

Renal involvement is frequent in systemic amyloidosis of both the AL (primary and myeloma-associated) as well as the AA (secondary) varieties (Cohen et al 1983) and is characterized clinically by heavy proteinuria or the nephrotic syndrome, leading to renal failure over one or two years.

Amyloid deposits are seen in both the glomerular mesangium and in capillary walls. The mesangial deposits form nodules of varying size which encroach on the capillary lumina. Capillary wall deposits form beneath the endothelium and involve the basement membrane. They occasionally extend as spike-like spicules from the epithelial aspect of the glomerular basement membranes. Although amyloid deposits generally fail to stain with PAMS, the epithelial spicules are positive and resemble somewhat the spikes in membranous glomerulonephritis. Amyloid deposits are demonstrated by LM by their metachromasia with crystal violet and toluidine blue, their reddish staining with Congo red (with dichroic-red and green-colouration under partly polarized light) and by their yellow-green staining with Thioflavin-T with ultraviolet illumination.

Amyloid is also deposited in arteries, arterioles and peritubular interstitial tissues. By EM, amyloid is composed of fine, criss-crossing, non-branching fibrils 7–10 nm in diameter. In spicules the fibrils lie at right angles to the glomerular basement membrane.

Immunochemical staining occasionally demonstrate Ig and C3 in association with the amyloid deposits, which themselves frequently autofluoresce. L amyloid stains for light chains and A amyloid with AA antisera.

Glomerular lesions in dysproteinaemias

Dysproteinaemia refers to quantitative or qualitative changes in gamma globulins, particularly immunoglobulins and their fragments. These changes are usually associated with neoplastic proliferation of cells of the B-lymphocyte/plasma cell series.

Myelomatosis

This is a monoclonal neoplastic proliferation of plasma cells, which produce IgG, IgA, IgD or IgE, or immunoglobulin fragments, principally kappa or lambda light chains. The latter may be excreted in the urine as 'Bence-Jones' protein. In the kidneys tubular damage is the most frequent change and glomerular lesions are uncommon. The tubules contain dense, sometimes multilayered casts composed of Bence-Jones protein, which sometimes elicit a multinucleate giant cell response. The casts sometimes stain for amyloid and can be demonstrated on immunochemical staining to contain kappa and lambda light chains. In a small proportion of cases amyloid deposits can be demonstrated in the glomeruli and occasionally mesangial nodules, like those in diabetic glomerulosclerosis, are seen. In about 15% of myelomatosis cases the changes of so-called 'light chain deposition disease' (Ganeval et al 1984) (Fig. 17.10) are encountered. By LM, the glomerular changes resemble those of diffuse mesangiocapillary glomerulonephritis. In resin sections and by EM, linear deposits are seen in the glomerular basement membrane. These are distinguished from dense deposit disease by their variable and more granular appearance, and by their situation toward the endothelial aspect of the glomerular basement membrane rather than centrally in the lamina densa. On immunochemical staining, the deposits are shown to contain a paraprotein corresponding to that in the circulation, usually a kappa chain, less frequently a lambda chain or a mixture, and rarely a heavy chain. These deposits also occur in the mesangium, in Bowman's capsule and in tubular basement membranes.

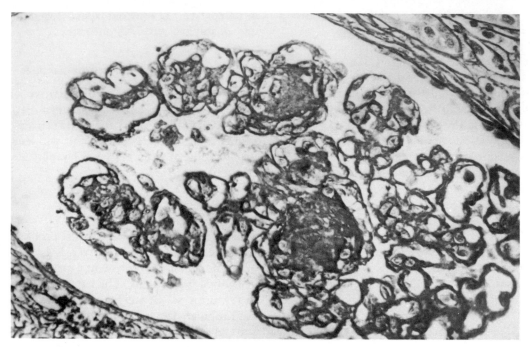

Fig. 17.10   Light chain deposition disease. Mesangial nodules and focal double contouring of glomerular capillary walls. Immunofluorescence microscopy revealed kappa light chain deposition. Trichrome. ×400

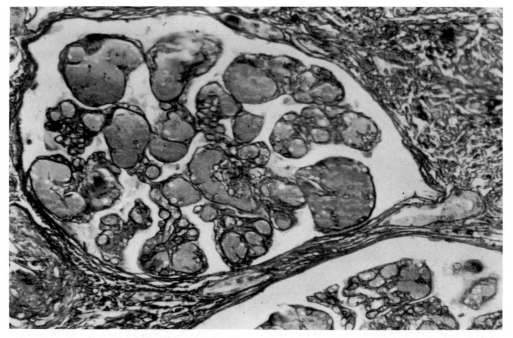

Fig. 17.11   Waldenstrom's macroglobulinaemia. Plugging of glomerular capillaries by deposit shown by immunochemical staining to contain IgM. PAMS. ×250

Waldenstrom's macroglobulinaemia

This is a proliferation of lymphoid cells producing a monoclonal elevation of IgM. The glomerular lesions are similar to those in myelomatosis, but in addition glomerular capillaries may be 'plugged' by hyaline thrombi which, on immunochemical staining, are rich in IgM.

Cryoglobulinaemia

Cryoglobulins are gamma globulins which precipitate on cooling; some are capable of damaging the kidney. Monoclonal and mixed cryoglobulinaemias are recognized and the latter are more likely to cause glomerular disease. The glomerular lesions encountered include a diffuse proliferative or mesangiocapillary glomerulonephritis, often with focal crescent formation. Subendothelial and intraluminal deposits are visible by LM, and EM shows them to have an organized arrangement composed of parallel arrays of fibrils and tubules resembling that in systemic lupus erythematosus.

Immunochemical staining shows the deposits to contain individual immunoglobulins or a mixture of usually IgG and IgM. Complement components are also generally present.

REFERENCES

Bancroft J D, Stevens A 1982 Theory and practice of histological techniques, 2nd edn. Churchill Livingstone, Edinburgh

Berger J, Hinglais N 1968 Les dépôts intercapillaries d'IgA — IgG. Journal d'Urologie (Paris) 74: 594

Churg J, Habib R, White R H R 1970 Pathology of the nephrotic syndrome in childhood. Lancet i: 1299

Churg J, Sobin L H 1982 Renal disease: classification and atlas of glomerular diseases. Igaku-Shoin Ltd, Tokyo

Cohen A S, Shirahama T, Sipe J D, Skinner M 1983 Amyloid proteins, precursors, mediator and enhancer. Laboratory Investigation 48: 1

Ehrenreich T, Churg J 1968 Pathology of membranous nephropathy. In : Sommers S C (ed) Pathology annual, vol 3, Appleton-Century-Crofts, New York, p 145

Gavenal D, Noël L H, Preud'homme J L, Droz D, Grünfeld J P 1984 Light-chain deposition disease: its relation with AL-type amyloidosis. Kidney International 26: 1

Gavenal D, Mignon F, Preud'homme J L et al. Visceral deposition of monoclonal light chains and immunoglobulins: a study of renal and immunopathologic abnormalities. In: Hamburger J, Grosnier J, Grünfeld J P, Maxwell M H (eds), Advances in nephrology, vol 2. Year Book Medical Publishers, Chicago, p 24

Germain J 1984 Resin embedding — the growth of resin histology. Optical and Electron Microscopy 8: 4

Habib R 1973 Editorial. Focal glomerular sclerosis. Kidney International 7: 204

Habib R, Kleinknecht C, Gubler M-C 1973 Extramembranous glomerulonephritis in children: report of 50 cases. Journal of Pediatrics 82: 754

Habib R, Gubler M C, Loirat C, Ben Maiz H, Levy M 1975 Dense deposit disease: a variant of membranoproliferative glomerulonephritis. Kidney International 7: 204

Kimmelstiel P, Wilson C 1936 Intercapillary lesions in glomeruli of kidney. American Journal of Pathology 12: 83

Levy M, Gubler M C, Habib R 1979 New concepts on membranoproliferative glomerulonephritis. In: Kincaid-Smith P, d'Apice A J, Atkins R C (eds) Progress in glomerulonephritis. John Wiley, New York, p 177

Meadow S R 1975 Post streptococcal nephritis — a rare disease. Archives of Disease in Childhood 50: 379

Morita T, Suzuki Y, Churg J 1973 Structure and development of the glomerular crescent. American Journal of Pathology 72: 349

Peters, D K, Williams D G 1975 Pathogenic mechanisms in glomerulonephritis. In: Jones N F (ed) Recent advances in renal disease. Churchill Livingstone, Edinburgh, vol 1, p 90

Rich A R 1957 A hitherto undescribed vulnerability of the juxtaglomerular glomeruli in the lipoid nephrosis. Bulletin of the Johns Hopkins Hospital 100: 173

Row P G, Cameron J S, Turner D R et al 1975 Membranous nephropathy. Long-term follow up and association with neoplasia. Quarterly Journal of Medicine 44: 207

Saint-Hillier Y, Morel-Maroger L, Woodrow D, Richet G 1975 Focal and segmental hyalinosis. Advances in Nephrology 5: 67

Salinas-Madrigal L, Pirani C L, Pollack V E 1970 Glomerular and vascular 'insudative' lesions of diabetic nephropathy: electron microscopic observations. American Journal of Pathology 59: 369

Sinclair R A, Burns Y, Dunhill M S 1981 Immunoperoxidase staining of formalin-fixed, paraffin-embedded human renal biopsies with comparison of the peroxidase-antiperoxidase (PAP) and indirect methods. Journal of Clinical Pathology 34: 859

Sinniah R 1982 Mucin secreting cancer with mesangial IgA deposits. Pathology 14: 303

Sinniah R 1984 Heterogeneous IgA glomerulonephropathy in liver cirrhosis. Histopathology 8: 947

Tejani A, Nicastri A 1983 Mesangial IgM nephropathy. Nephron 35: 1

Tubbs R R, Gephardt G N 1983 Use and interpretation of immunoperoxidase procedures. In: Glomerular disease. Churchill Livingstone, Edinburgh, p 11

Vargas R, Thomson K J, Wilson D et al 1976 Mesangiocapillary glomerulonephritis with dense 'deposits' in the basement membranes of the kidney. Clinical Nephrology 5: 73

18. The leukaemias

I. M. Hann   J. S. Earley   J. R. O'Donnell

INTRODUCTION

When a haematologist has to decide whether or not a patient has leukaemia, several considerations are helpful. A comprehensive clinical history and examination for signs of lymphadenopathy, hepatosplenomegaly and bone marrow failure is essential. The blood count usually gives away the diagnosis, but the most important factor is a good bone marrow aspirate. The Romanowsky stain is often sufficient to define the particular type of leukaemia; however, the cytochemical tests which are detailed here will frequently give additional and confirmatory information which allows the correct treatment to be instituted. We have chosen not to give details of every potential cytochemical test, but rather to concentrate on those which give diagnostically useful information.

Enzyme histochemistry has traditionally been applied to define the different types of leukaemia, but of the many enzyme techniques available relatively few are necessary. Methods for myeloperoxidase, 'non-specific esterase' (NSE), chloroacetate esterase (CAE), β-glucuronidase, acid phosphatase, neutrophil alkaline phosphatase (NAP), and α-naphthyl acetate esterase (ANAE) are those most commonly used. Ultrastructural demonstration of enzyme activity (Breton-Gorius et al 1981) may have been necessary in the past to define megakaryocytes and megakaryoblasts by virtue of their platelet peroxidase (PPO) activity, but this has been superseded in most cases by the availability of platelet antibodies (Choate et al 1988). However, the demonstration of PPO may be necessary in some cases of megakaryoblastic leukaemia where the cells are sufficiently immature and do not express the usual platelet markers.

In recent years monoclonal antibodies have been helpful in identifying cell types and lineages. Application of these antibodies, together with the immunocytochemical demonstration of TdT (terminal deoxynucleotidyl transferase), a marker of immature lymphocytes, Ia-like antigen, distribution of immunoglobulins, and platelet antibodies, allows immunological typing of the different leukaemias of childhood. Additional useful information can also be derived from chromosomal analysis.

Newer techniques which may prove useful include in situ hybridization to demonstrate *abl* and *sis* oncogenes in the myeloid leukaemias (Kemnitz et al 1987). This is an expanding area and many other applications are probable. The use of new simple dyes may help to eliminate the more complex techniques. Basic Blue 93 appears to be such a dye which mirrors the myeloperoxidase stain and has none of the disadvantages of fading, toxicity, or the need for relatively fresh samples common to the current methods (Kass 1988). Further evaluation will be necessary before this dye can be used routinely.

GENERAL CONSIDERATIONS

We have alluded above to the need to take into account a number of considerations before assigning a specific diagnosis of leukaemia to a patient. This is of extreme importance as nowadays therapy will be significantly affected by this process and patients could be exposed to relatively ineffective or unnecessarily toxic regimens. In the tables we have detailed those cytogenetic and immunological tests which significantly help in diagnosis, particularly in difficult cases. The process is often like piecing together a jigsaw and if a piece does not fit, careful reconsideration is necessary. The role of the clinical history and examination cannot be overemphasized. For instance, a young baby presenting with a high white cell count, fevers, gum hypertrophy, raised serum lysozyme and central nervous system disease may well have a monoblastic leukaemia.

Controls

When interpreting the results of cytochemistry it is important first to look at a control for each and every stain and then to examine the patients' slides for the presence of positive cells. For instance, when looking at the Sudan black stain always search for positive neutrophil or myelocyte granules as a good internal control. In acute lymphoblastic leukaemia (ALL) this may not always be easy as there are often very few residually normal cells. It may seem trite to remind people that the first premise must be to ensure that the stain has worked. However, experience tells us that this is not often adequately considered. In order to achieve the best results, marrows and bloods are best produced directly without anticoagulant, but if this is not feasible, a *short* period in EDTA does no significant harm. Samples should always be dealt with and examined fresh, because the stains may fade. Cytocentrifuge preparations are extremely dependant on technique and morphology must be interpreted with the greatest care, and, if possible, other methods of diagnosis should be used. For example, the detection of terminal deoxynucleotidyl transferase (TdT) in lymphoblasts by immunofluorescence in CSF from children with acute lymphoblastic leukaemia would confirm CNS relapse.

Trephines

Bone marrow trephine core samples are rarely of value in diagnosis of leukaemias. However, in rare cases of ALL, severe myelofibrosis (sometimes with marrow infarction) prevents aspiration of marrow. Also, frequently in acute megakaryoblastic leukaemia (FAB M7) fibrosis is an integral part of the disease, perhaps because of the production of platelet-derived growth factor. In this case paraffin sections of a trephine will show strong positivity with factor VIII antibody in megakaryoblasts and megakaryocytes (Koike 1984).

Cytochemistry

The prognostic significance of certain cytochemical stains has been examined in recent years and in childhood ALL greater positivity of the PAS reaction correlates with a better outcome (Hann et al 1979).

Classification

The French-American-British (FAB) classification (Bennett et al 1985) has helped both to classify leukaemias (see Tables 18.1 and 18.2) and sometimes to stratify therapy. In ALL of childhood, the FAB types L_1 and L_2 have independent prognostic significance, with L_2 cases doing worse. L_3 cases are usually mature B-cell ALL which require much more intensive therapeutic regimens (Palmer et al 1980).

Table 18.1 FAB classification*

Acute lymphoblastic leukaemia

L_1	75% or more of the cells are small, with high nucleocytoplasmic ratio, regular nuclear membane and non-prominent nucleoli
L_2	More than 25% of cells have prominent nucleoli and low nucleocytoplasmic ratio. More than 50% are usually large and nuclear membrane irregularity may be present
L_3	Large blast cells, intensely basophilic cytoplasm, multiple cytoplasmic vacuoles. Large nucleoli

Acute myeloid leukaemia (AML)

M1	Acute myeloblastic leukaemia without differentiation
M2	Acute myeloblastic leukaemia with differentiation
M3	Acute promyelocytic leukaemia
M4	Acute myelomonocytic leukaemia
M5	Acute monoblastic leukaemia
M6	Acute erythroid leukaemia
M7	Acute megakaryoblastic leukaemia

*For further details see Bennett J M et al 1985.

Table 18.2 Monoclonal antibodies useful in the immunological classification of leukaemia

CD* no.	Typical antibodies	Specificity
CD1	OKT6, NA1/34	Cortical thymocytes
CD2	OKT11	E rosette receptor
CD3	OKT3, UCHT1	Mature T cells
CD4	OKT4, Leu 3a	Helper/inducer T cell subset
CD5	OKT1, UCHT2	Pan T, subpopulation B cells
CD6	OKT12, MBG6	Mature T cells
CD7	WT1, Campath 2	Pan T, all T ALL
CD8	OKT8, UCHT4	Suppressor, cytotoxic T cell subset
CD9	BA2, FMC8	B cell, platelets, some monocytes
CD10	AL2, J5, BA3	CALLA (common ALL antigen)
CD13	MY7, MCS.2, WM15	Monocytes, myeloid cells
CD14	MY4, MO2, UCHM1	Monocytes
CD15	TG1, MCS.1	Granulocytes, myeloid cells
CD19	B4, HD37	Pan B
CD20	B1	Pan B, more mature B cells
CD22	RFB4	Pan B, more mature B cells and intracytoplasmic in very early B cells
CD33	MY9	Early myeloid cells, monocytes
CD34	BI-3C5, ICH-3	Immature lymphoid and myeloid cells
CD38	OKT10	Immature marrow cells, thymocytes, B blasts and plasma cells
	gpIIIA, J15, C16	Platelets, megakaryocytes, megakaryoblasts
	FA6152	Erythroid cell line

*CD = cluster of differentiation as defined at International Workshops on leucocyte differentiation (see Ch. 19).

Table 18.3 Cytochemistry of acute lymphoblastic leukaemia

Stain	Result
Myeloperoxidase	Negative
Sudan black	Negative
PAS	Positive in 80% of cases. Coarse granules and blocks
NSE	Negative
CAE	Negative
Acid phosphatase	Positive in 90% of T-cell cases — usually single polar block positivity

CYTOCHEMISTRY OF LYMPHOCYTIC LEUKAEMIAS

The majority of cases of ALL are PAS positive and Sudan black negative (Table 18.3), in contrast with the immunological typing results (see Table 18.4) where differences are detectable. In 90% of T-cell ALL cases there is a strong block of acid phosphatase activity corresponding to the Golgi apparatus. B-cell ALL is usually negative with PAS, Sudan black and for acid phosphatase activity, but shows positivity with Oil Red O. About one-fifth of common, null and pre-B ALL cases are negative with all stains and one has to rely on the immunophenotyping (Table 18.4) for precise diagnosis. Before the advent of the latter technique, some cases of ALL were misclassified as AML, particularly when L_2 FAB morphology was present. Very occasional cases of biphenotypic (ALL and AML) leukaemia are now recognized, especially in babies and in Philadelphia chromosome-positive acute leukaemias. Although this may be obvious on Romanowsky, PAS and Sudan black staining, immunophenotyping is essential for confirmation.

The chronic lymphocytic leukaemias (CLL)

These conditions differ in general from the acute leukaemias in having a higher white cell count and the leukaemic cells show greater differentiation with some characteristic distinguishing features, e.g. in hairy cell leukaemia (HCL), prolymphocytic leukaemia (PLL) and cutaneous T-cell lymphoma (CTCL) (see Tables 18.5 and 18.6). Although common disorders like B-CLL may often be diagnosed on a well-stained blood film, the proper identification of rarities like PLL and T-cell variants depends on immunophenotyping (Table 18.5). Conventional cytochemistry provides

Table 18.4 Immunological typing of leukaemia in childhood

Diagnosis		Antibody reactivity								
		TdT	Ia	Ig	CD19	CD10	CD13	CD14	FA6152 (erythroid)	gpIIIA (platelet)
ALL	T-cell	+	–	–	–	–	–	–	–	–
	B mature	–	+	Smg	+	–	–	–	–	–
	Common	+	+	–	+	+	–	–	–	–
	Pre-B	+	+	CIg	+	–	–	–	–	–
	'null'	+	+	–	+	–	–	–	–	–
AML	M1/2	–	+	–	–	–	+	–	–	–
	M3	–	–	–	–	–	+	–	–	–
	M4/5	–	+	–	–	–	+	+	–	–
	M6	–	–	–	–	–	–	–	+	–
	M	–	–	–	–	–	–	–	–	+

For details of the FAB classification, see Table 18.1.

SmIg = Surface marker immunoglobulin
CIg = Cytoplasmic immunoglobulin

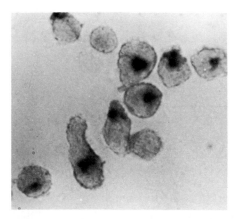

Fig. 18.1 Acid phosphatase in bone marrow lymphoblasts from a 2-year-old girl with ALL and mediastinal enlargement. The acid phosphatase has a compact paranuclear ('T' type) distribution. ×1300

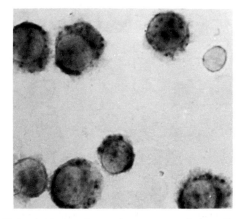

Fig. 18.3 Acid phosphatase, bone marrow, hairy cell leukaemia. The field contains six hairy cells with strong activity which was tartrate-resistant, and a normal lymphocyte. ×1000

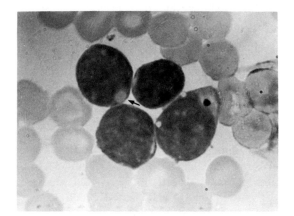

Fig. 18.2 PAS stain, bone marrow, Ph′ positive ALL. One lymphoblast contains a strongly positive block, another (arrow) a spherical inclusion staining with moderate intensity (compare with clear vacuoles in cell at right). ×1300

less clear-cut information than in the acute leukaemias, but typical patterns are seen in HCL where most (but not all) patients show tartrate-resistant acid phosphatase activity (isoenzyme 5) and in the rare CTCL (Sezary syndrome, mycosis fungoides) where the acid phosphatase and β-glucuronidase reactions are positive. Table 18.5 summarizes the commoner cell surface phenotypes in these disorders using cell markers and monoclonal antibodies. These methods are now within the scope of many laboratories with the introduction of the alkaline phosphatase/anti-alkaline phosphatase (APAAP) immunocytochemical technique (Erber et al 1986).

Table 18.5 Immunophenotype of chronic lymphocytic leukaemias

Markers	CLL B	T	HCL	PLL B	T	CTCL
B cell						
SmIg	+/−	0	+	++	0	0
m rosettes	++	0	+/−	0	0	0
CD 20 (B1)	+	0	+	+	0	0
HLA (DR)	+	0	+	+	0	0
CD5 (OKT1)	+	0	0	+/−	0	0
CD20 (B1)	+	0	+	+	0	0
T cell						
E-rosettes	0	+	0	0	+	+
CD2 (OKT11)	0	+	0	0	+	+/−
CD3 (OKT3)	0	+	0	0	+	+
CD4 (OKT4)	0	0	0	0	+	+
CD8 (OKT 8)	0	+	0	0	+/−	0
CD7 (WT 1)	0	+/−	0	0	+	0

CLL = chronic lymphocytic leukaemia, HCL = hairy cell leukaemia, PLL = prolymphocytic leukaemia, CTCL = cutaneous T-cell lymphocytosis/lymphoma.

Table 18.6 Cytochemistry of chronic lymphocytic leukaemias

Stain	CLL	HCL	PLL	CTCL
Peroxidase	0	0	0	0
Sudan black	0	0	0	0
Acid phosphatase	0 or ++ in T cell	++ tartrate resistant	0 or weak diffuse	+ or ++
β-Glucuronidase	+ granules, ++ block in T cell	0	0	+ or ++
PAS	+ or ++ granules	++ diffuse	0	+ or ++
ANAE*	0 or + granules, more in T cell	+ faint diffuse fluoride resistant	0 in B-cell ++ single block in T-cell	+ or ++ dots

*Alpha-naphthyl acetate esterase.

CYTOCHEMISTRY OF MYELOID LEUKAEMIAS (Figs 18.4–18.7)

The classification of this group is now complicated and dependent to some extent upon cytochemistry (see Tables 18.1 and 18.7). The essential criteria of the FAB classification are:

M 1 Myeloblastic without maturation, with 3% or more blasts positive for peroxidase or Sudan black.

M 2 Myeloblastic with maturation with >20% nucleated cells as myelocytes or more mature granulocytes.

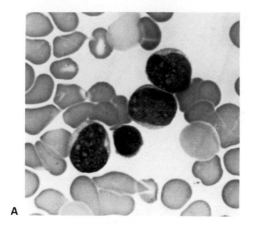

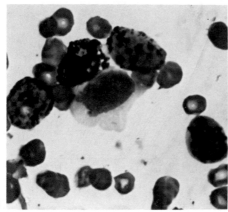

Fig. 18.4   May-Grünwald-Giemsa (**A**) and myeloperoxidase (**B**) stains, bone marrow, AML M1. The myeloperoxidase stain is strongly positive even though no granulation was evident in the Romanowsky stain. The field in both A and B contains a lymphoid cell. ×1000

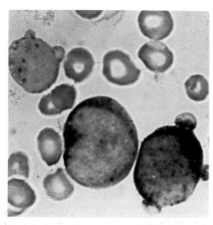

Fig. 18.5   β-Glucuronidase, bone marrow, AML M1. Two myeloblasts have a diffuse and granular reaction. A lymphocyte contains dots on a negative background. ×1300

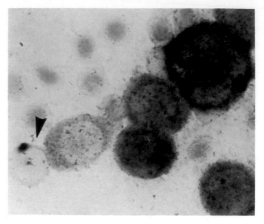

Fig. 18.6   α-Naphthyl acetate esterase, bone marrow, AML M5. The field contains four monocytes with positive reaction which was fluoride-sensitive, and a lymphocyte (arrow) with a single block. ×1000

M 3   Promyelocytic, hypergranular or microgranular.

M 4   Myelomonocytic with >20% monocytic forms and more than 10% myeloblasts.

M 5   Monoblastic with non-specific esterase positivity.

M 6   Erythroleukaemia, with 30% myeloblasts and either more than 50% erythroblasts or more than 30% bizarre erythroblasts.

M 7   Megakaryoblastic with positivity for platelet antibodies or factor VIII related antigen, and extreme myelofibrosis.

Chronic myeloid leukaemia (CML) is diagnosed on the basis of the clinical picture and a high total white cell count (usually 50–200 × 10^9/l). Typically, there is hyperplasia of the entire granulocytic series, without discontinuity in maturation, but with a left shift showing a low proportion of myeloblasts. There is often an eosinophilia and basophilia and the latter cells are frequently hypogranular. The bone marrow appearances are of extreme granulocytic hyperplasia and virtually no fat may be evident. There may also be eosinophilia, basophilia and sometimes megakaryocytic hyperplasia. Cytogenetics has made an enormous contribution to the understanding of this disease and the Philadelphia (Ph′) chromosome, classically involving a translocation from the long arm of chromosome 22 to

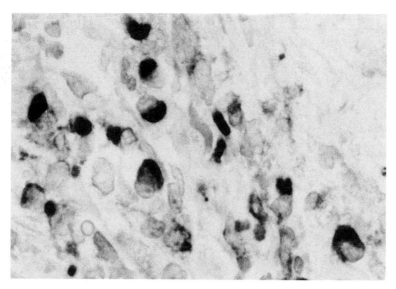

Fig. 18.7   Bone marrow trephine: paraffin section. Acute megakaryoblastic leukaemia (M7). The megakaryoblasts are numerous and can be identified immunocytochemically by anti-factor VIII-related antigen.

Table 18.7   Cytochemistry of acute myeloid leukaemias

	Myeloperoxidase	SB	Ac. phos	PAS	NSE	CAE
M1	+ or −	+ or −	0	0	0	+ or 0
M2	+	+	0	0	0	+
M3	++	++	0	0	0	++
M4	+	+	+ Diffuse	0 or mild diffuse	+	+
M5	0 or fine granules	0 or fine granules	+ Diffuse	0 or mild diffuse	++	0
M6	0	0	++ Unipolar	+ granular or diffuse	+	0
M7	0	0	+ Diffuse	+	+	0

SB = Sudan black, Ac. phos = acid phosphatase, PAS = periodic acid-Schiff, NSE = non-specific esterase, CAE = chloroacetate esterase.

chromosome 9 [t(9;22)], is present in over 90% of cases. More complex translocations and rearrangements involving other chromosomes may also occur (see Table 18.8). When CML enters the accelerated and/or acute transformation phase the Ph′ chromosome may persist, but frequently additional abnormalities, e.g. a double Ph′ chromosome, extra chromosome 8 or isochromosome for the long arm of number 17 (17q), occur. The enzyme TdT may become apparent before acute transformation, but is not sufficiently

specific a marker to distinguish those cases with a myeloblastic (about 70%) as opposed to a lymphoblastic transformation.

The neutrophil alkaline phosphatase (NAP) stain produces a low score in CML, but this is also seen in juvenile CML and often in AML. The diagnosis of juvenile CML depends on the usual clinical features of a myeloproliferative disorder in childhood together with absence of the Ph′ chromosome and often a raised level of haemoglobin F. Monosomy 7 is also associated

Table 18.8   Common chromosome changes in leukaemia

Type		Gains	Losses	Rearrangements
CML		+8	−7 (rare)	t(9;22), i(17q)
Blast crisis		+Ph′		
AML	M2	+8	−7, −5	t(8;21)
	M3			t(15;17)
	M4	+8	−7	inv(16)
	M5			t(11q) or del (11q)
CLL	B	+12		14q+
	T			inv(14)
ALL	Null	+12; +6	−9	t(9;22), t(4;11)
	B			t(8;14)
	Pre-B			t(1;19)
	T			t(11;14)

From: Rowley J D 1987 In: Nathan, Oski (eds) Hematology of infancy and childhood. Saunders, Philadelphia, ch 43

with chronic leukaemia in children and several familial varieties are described. Eosinophilia may occur in association with ALL as well as AML (especially M4) and CML. However, as with basophilia, this does not represent leukaemic proliferation of these cells but is presumably a reactive response. Another probable variant of CML is subacute or chronic myelomonocytic leukaemia (CMML). This is an uncommon disorder in those over 60 years and is regarded by some as forming a part of the spectrum of myelodysplastic (preleukaemic) conditions. The white cell count tends to be lower than in CML and the NSE stain may help in identification of the atypical monocytic cells. The NAP score is variable and the Ph′ chromosome is usually absent.

REFERENCES

Bennett J M, Catovsky D, Daniel M T 1985 Proposed revised criteria for the classification of acute myeloid leukaemia. Annals of Internal Medicine 103: 626–629

Breton-Gorius J, Gourdin M F, Reyes F 1981 Ultrastructure of the leukaemic cell. In: Catovsky D (ed) The leukaemic cell. Churchill Livingstone, Edinburgh, ch 4

Choate J J, Domenico D R, McGraw T P, Fareed J, Molnar Z, Schumacher H R 1988 Diagnosis of acute megakaryoblastic leukaemia by flow cytometry and immunoalkaline phosphatase techniques. Utilization of new monoclonal antibodies. American Journal of Clinical Pathology 89: 247–253

Erber W N, Mynheer L C, Mason D Y 1986 APAAP labelling of blood and bone marrow samples for phenotyping leukaemia. Lancet i: 761–765

Hann I M, Evans D I K, Palmer M K, Jones P M,

Haworth C 1979 The prognostic significance of morphology features in childhood ALL. Journal of Laboratory and Clinical Haematology 1: 215–225

Kass L 1988 Basic Blue 93. An alternative to the traditional myeloperoxidase stain. American Journal of Clinical Pathology 89: 195–202

Kemnitz J, Helmke M, Cohnert T R, Buhr T, Freund M 1987 In situ hybridization in the analysis of *abl* and *sis* oncogenes in myeloproliferative disorders. Human Pathology 18: 1302–1303

Koike T 1984 Megakaryoblastic leukaemia: the characterization and identification of megakaryoblasts. Blood 64: 683–692

Palmer M K, Hann I M, Jones P M, Evans D I K 1980 A score at diagnosis for predicting length of remission in childhood ALL. British Journal of Cancer 42: 841–849

19. Immunohistology of lymphoid tissue and lymphomas

A. J. Norton P. G. Isaacson

INTRODUCTION

In this chapter we review the immunohistological techniques which may be used to aid in the diagnosis of human lymphoproliferative disease. Since the publication of the first edition there have been considerable advances, both in the range of antisera generally available, and in the sensitivity of methods applicable to lymphoreticular immunopathology. Firstly, monoclonal antibodies to myriad leucocyte antigens are now accessible. The specific cell surface molecules recognized by the more important antibodies are known in many cases, and their roles at the cellular level are beginning to be understood. However, there had been increasing confusion arising from this wealth and diversity of antisera. Each reagent had its own unique designation and in many instances equivalent antibodies were not recognized as such. It has only been through the efforts of four International Workshops devoted to the leucocyte antigens that a standardized nomenclature for these antibodies has emerged, under the clusters of differentiation (CD) system (Bernard et al 1984, Reinherz et al 1986, McMichael et al 1987). It should not be forgotten, however, that even today good polyclonal antisera still have a considerable diagnostic role to play. Secondly, impressive refinements have been made in immunohistological methods and in the overall quality and consistency of commercial reagents. Not only have the methods been improved in relation to immunohistology on cryostat sections, but there has been much success in applying these methods to specially prepared tissue sections and, more importantly, routinely processed histological material. The latter has made the immunohistological diagnosis of lymphoma accessible to nearly every routine laboratory.

With the wide acceptance of these techniques come the inherent problems of adequate control material, quality assurance, reliability of commercial antisera, and above all else, the realization that interpretation of immunohistochemical staining is as much an art as any other area of histopathology.

It remains to be seen whether the increasing contributions being made by molecular biology and in-situ hybridization will supersede immunohistology in the diagnosis of lymphoreticular disease, or whether, as seems more likely, they will become complementary modalities.

PRACTICAL DETAILS OF LYMPHOMA HISTOLOGY

Antigens to be detected by immunohistology

Antibodies and the clusters of differentiation

Table 19.1 gives a complete list of the well characterized anti-leucocyte antibodies deserving of a cluster of differentiation (CD) as determined at the fourth most recent International Workshop on leucocyte antigens (Knapp et al 1989)

For an antibody to be assigned to one of the established clusters it must fulfil stringent selection criteria; it must bind to a range of 50 or more cell lines in a fashion identical to a reference clustered antibody, and it has to immunoprecipitate a molecule or molecules of the appropriate molecular weight. In addition, the test antibody should bind to cells transfected with the gene coding for the cluster's target antigen. To establish a new cluster there should be at least two

Table 19.1 Clustered monoclonal antibodies to leucocyte surface antigens grouped according to their major specificity. Cluster numbers are those given at the Fourth International Workshop.

CD Number	Antigen Molecular weight (kDa)	Cellular distribution and function	CD Number	Antigen Molecular weight (kDa)	Cellular distribution and function
T-cell related clusters			*B-cell related clusters*		
1a	49kDa	Cortical thymocytes, dendritic cells, Langerhans' cells	24	42kDa	B-cell subset. Granulocytes. Epithelia. Sialoglycoprotein
1b	45kDa	Cortical thymocytes, dendritic cells, Langerhans cells	37	40, 52kDa	B-cells. Weakly on macrophages and T-cells. Follicular dendritic cells
1c	43kDa	Cortical thymocytes, dendritic cells, Langerhans' cells, B-cell subset. All are linked to β_2 microglobulin. Possibly needed for antigen presentation or act as hormone receptors	38	45kDa	Follicle centre cells. Plasma cells T-cells.
			39	80kDa	Mantle and marginal zone B-cells. Macrophages. Endothelium. Activated T-cells
2	50kDa	Mature T-cells. Also known as LFA-2. Consists of an $\alpha\beta$ or $\gamma\delta$ heterodimer. Ligand for LFA-3. T-cell activation	40	50kDa	B-cells. Dendritic cells. Epithelia. Homology with nerve growth factor receptor. Ligand for IL-6. Phosphatase activity
3	22, 20kDa	T-cells. Complex closely associated with the T-cell antigen receptor. Most antibodies identify the ϵ chain. T-cell activation	72	39, 43kDa	B-cells. Subset of macrophages.
			73	69kDa	Mantle zone B-cells. T-cell subset. Dendritic cells. Epithelia. Ecto-5'-nucleotidase
4	55kDa	T helper/inducer cells and macrophages. Closely associated with TCR and CD3. Together act as ligand for HLA class II plus antigen. Complexed with protein tyrosine kinase, p58^{LcK}. Receptor for HIV 1	74	41, 35, 33kDa	B-cells. Macrophages. Epithelia. Invariant chain of HLA complex (sli), predominately intracellular
			w75	?53kDa	Follicle centre cells. B-blasts. T-cell subset
5	67kDa	Mature T-cells. B-cell subset. Increases pool of intracytoplasmic second messengers	76	67, 85kDa	Mantle zone B-cells. T-cell subset. Granulocytes. Epithelia
			77	?	Follicle centre cells. Epithelia Endothelium. Gb3.
6	90kDa	Mature T-cells. B-cell subset	w78	?67kDa	B-cells. Subset of macrophages. Epithelia.

Myeloid/macrophage related clusters

No.	kDa	Description
12		NOT A CLUSTER IN PRESENT USAGE
13	130, 150kDa	Granulocytes. Monocytes. Osteoclasts. Epithelia. Connective tissue. Aminopeptidase-N
14	55kDa	Monocytes. Kupffer cells. Granulocytes. Dendritic cells. Possible growth factor receptor
15	105–200kDa	Granulocytes. Dendritic cells. Epithelia. Lacto-N-fucopentaosyl III (Hapten X)
16	50–70kDa	Granulocytes. Subset of macrophages. Low affinity $Fc\gamma$ receptor III
32	40kDa	Granulocytes. B-cells. Monocytes. Platelets. Endothelium. $Fc\gamma$ receptor II
33	67kDa	Early myeloid cells. Macrophages. Homology with myelin associated protein
34	107–120kDa	Myeloid subset. Endothelium. Leukosialylin family
35	220kDa	Follicular dendritic cells. B-cell subset. Erythrocytes. Granulocytes. Glomeruli. C3b receptor
64	75kDa	Monocytes. Macrophages. $Fc\gamma$ receptor I
w65	23, 41, 63kDa	Granulocytes. Monocytes. Fucoganglioside
66	180–200kDa	Granulocytes. Subset of macrophages. Granulocyte phosphoprotein
67	100kDa	Granulocytes. Subset of macrophages.
68	110kDa	Macrophages. Granulocytes. B-cell subset

No.	kDa	Description
7	40kDa	Thymocytes and mature T-cells. Earliest antigen to appear on T-cells. Receptor for IgM
8	32kDa	Suppressor/cytotoxic T-cells. Splenic sinusoidal cells. Closely associated with TCR and CD3. Together act as ligand for HLA class 1 plus antigen. Complexed with protein tyrosine kinase, $p58^{Lck}$
27	55kDa	Activated T-cells. Plasma cells
28	44kDa	T-cell subset. Stabilizes key lymphokine mRNAs during activation (IL-2, γIFN, GMCSF, TNFα)
w60	14, 54, 56kDa	T-cell subset. Platelets. Melanoma cells. Disialoganglioside

B-cell related clusters

No.	kDa	Description
10	95, 100kDa	Follicle centre cells. B and some T lymphoblasts. Renal tubules. Also know as common ALL antigen (CALLA). Neutral endopeptidase
19	95kDa	B-cells. Follicular dendritic cells. Present very early in ontogeny. Induces B-cell differentiation
20	37, 35kDa	B-cells. Phosphoprotein. Calcium ion channel
21	140kDa	B-cells. Follicular dendritic cells. C3d receptor (CR2). Receptor for Epstein Barr virus
22	135kDa	B-cells. Present very early in ontogeny. Increases pool of cytoplasmic second messengers
23	45, 50kDa	Two forms of low affinity $Fc\epsilon$ receptor II: (a), B-cell subset, follicular dendritic cells. (b), monocytes, eosinophils, T-cells and B-cell subsets

Table 19.1 Cont'd

Platelet related clusters

CD Number	Antigen Molecular weight (kDa)	Cellular distribution and function
29	180kDa	Platelets. T-cell subset. Many non-lymphoid tissues. VLA (very late activation antigen) β4-chain
31	130, 140kDa	Platelets. Granulocytes. Monocytes. Endothelium. Glycoprotein IIa
36	85kDa	Platelets. Monocytes. Endothelial cells. Thrombospondin/collagen receptor. Glycoprotein IV
41a	140, 95kDa	Platelets. Megakaryocytes. Fibrinogen/von Willebrand factor/collagen receptor. Glycoprotein IIb/IIIa intact complex
41b	140kDa	Platelets. Megakaryocytes. Glycoprotein IIb
42a	23kDa	Platelets. Megakaryocytes. von Willebrand factor receptor. Glycoprotein IX
42b	150kDa	Platelets. Megakaryocytes. Glycoprotein Ib
w49b, d, f	167, 135, 120kDa	Platelets. Many leucocytes. Collagen, fibronectin and laminin receptors respectively. VLA α2, α4, α6 chains respectively

Non-lineage and NK cell related antigens

CD Number	Antigen Molecular weight (kDa)	Cellular distribution and function
11b	170kDa	Granulocytes. Monocytes. C3bi receptor α chain
11c	150kDa	Granulocytes. Macrophages. α chain of p 150, 95 antigen
18	95kDa	Many white cells. See CD11 entry. Common β chain
43	95kDa	T-cells. Granulocytes. Early and activated B-cells. Megakaryocytes
44	80, 95kDa	All leucocytes. Epithelia. In (a) and In (b) blood group antigens
45	180, 190, 205, 220kDa	All leucocytes. Tyrosine phosphatase. Probably dephosphorylates the T-cell receptor complex and thus antagonizes p58^{Lck}. Leucocyte common antigen
45RA, B, 0		As above but showing restricted binding to distinct exons of the molecule. Different exons are present in different cell lineages
46	66, 56kDa	Wide tissue expression. Membrane co-factor protein. Associated with HLA class I. Complement clearance

CD	Molecular weight	Expression / comments
47	47, 53kDa	Wide tissue expression. Associated with the rhesus group system
48	41kDa	Wide leucocyte expression
w50	108, 140kDa	Wide leucocyte expression
51	120kDa	Platelets. Many leucocytes. Vitronectin receptor α chain
w52	25–30kDa	Leucocytes and epithelia. Campath-1 antigen
53	32, 40kDa	Wide leucocyte expression except platelets and erythrocytes
54	85kDa	Wide tissue expression. ICAM-1 antigen. Ligand for LFA-1. Rhinovirus receptor
55	73kDa	Wide tissue expression. Decay accelerating factor. Complement clearance
56	135, 220kDa	NK cells. Some monocytes. Neuroectodermal cells
57	110kDa	NK cells. T-cell and B-cell subsets. Some monocytes. Neuroectodermal cells. Homology with myelin associated glycoprotein
58	45, 60kDa	Wide leucocyte expression. LFA-3. Ligand for CD2
59	18, 20kDa	Wide tissue expression
61	114kDa	Platelets. Glycoprotein IIIa. Vitronectin receptor β chain
62	150kDa	Platelets. α-granule protein. PADGEM (GMP-140)
63	53kDa	Platelets. Activation related protein (gp-53)

Activation antigens

CD	Molecular weight	Expression / comments
25	55kDa	Activated cells. Macrophages. IL-2 receptor α chain
26	150, 200kDa	Activated T-cells. Bile canaliculi. Dipeptidyl-peptidase IV (DPP IV)
30	120kDa	Activated cells
69	28, 34kDA	Activated cells. 'Activation Inducer Molecule' (AIM)
w70	?	Activated cells. Very late activation antigen
71	90kDa	Activated cells. Transferrin receptor

Non-lineage and NK cell related antigens

CD	Molecular weight	Expression / comments
11a	180kDa	Most leucocytes. LFA-1 α chain. Ligand for ICAM-1 and ICAM-2

completely identical antibodies submitted from different laboratories.

Perhaps one of the most intriguing aspects of monoclonal antibodies is their recognition of small, discrete epitopes. On some larger molecular species there may be many potential unique antibody-binding sites. Thus antibodies of the same cluster may recognize quite distinct epitopes. For practical purposes this phenomenon poses very few problems, but it is noteworthy that the different antibodies of the leucocyte common family (CD45 and CD45RA, B, 0) operate precisely on this principle. Leucocyte common antigen is a useful illustration of the subtleties of epitopic variation (See Table 19.2). Conventional leucocyte common (CD45) recognizes an invariant portion of the molecule, whereas CD45R (restricted) antibodies bind to discrete differentially expressed exons (Thomas & Lefrançois 1988, Norton & Isaacson 1989a).

There are many useful reagents still looking for their cluster, but, in general, clustered antibodies are to be preferred because of their more thorough characterization.

Lastly, and perhaps surprisingly, the clustering process only applies to leucocyte surface antigens. Clearly there are a host of useful target molecules for monoclonal antibodies which are exclusively resident in the cytoplasm or nucleus. Indeed, many macrophage-specific markers, and markers effective in fixed and routinely processed tissue fall into this category.

Antibodies to fixation resistant antigens

Many monoclonal antibodies detect antigens which are unable to withstand formalin fixation and paraffin wax embedding. Of necessity these are restricted for use on cryostat sections, or on tissues which have been very lightly fixed. A range of antibodies is now available to fixation-resistant leucocyte antigens and these are enumerated in Table 19.2. A few of the monoclonal antibodies fall into established clusters, but many recognize cytoplasmic antigens, which, by definition, cannot be clustered, and included in the table there are several very useful polyclonal reagents. A significant problem shared by many of these antibodies is their lack of absolute T or B lineage

specificity. Thus they need to be used in a panel by a pathologist conversant with their normal range of reactivity and aware of the potential diagnostic pitfalls.

Preparation of tissue samples

If lymphoma immunopathology is to be effectively performed then lymph node and allied biopsies should be received unfixed in the histopathology department with the minimum of delay. If tissue has to be transported over long distances, or if any lengthy delay is envisaged, then the specimen should be immersed in tissue transport medium (Appendix 1). For larger specimens, for example a gastric resection for lymphoma, the specimen can be wrapped in towels soaked in normal saline. Delays of greater than 12 hours should be avoided as this will seriously impair the morphology, especially if the specimen is unrefrigerated. Antigenicity, on the other hand, is retained by many tissues for up to 36 hours postmortem (Pallesen & Knudsen 1985). On receipt of the specimen it should be described in the usual manner and then be cut into 4 mm thick slices with a fresh clean scalpel or new razor blade. Slices should be taken at right angles to the long axis of the lymph node such that both cortex and medulla appear in sections. Spleens should be similarly sliced in their entirety as thinly as possible.

Imprint preparations

One cut surface of the tissue should be gently touched onto half a dozen clean glass slides. One of these imprint preparations may be stained by a routine haematological stain for a rapid cytological diagnosis, and the rest can be air dried for any enzyme cytochemical staining, if required.

Frozen tissue

The amount of tissue to be frozen largely depends upon the local needs of the department. For immunophenotyping a single slice is sufficient, but more can be taken if genotypic or other studies are to be performed. The method of freezing is unimportant as long as it is rapid (see Appendix 1).

It is not necessary to use any embedding compound at this stage unless the specimen is very small. It is notable that some commercial embedding compounds have endonuclease activity and can digest the cellular DNA, rendering further genetic studies difficult.

Storage of frozen tissue

Frozen tissue should be stored tightly wrapped in aluminium foil in small containers. A liquid nitrogen tank or a -70°C freezer are both suitable for long-term storage. Keeping the tissues at a higher temperature, in say a domestic freezer, results in morphological deterioration with loss of antigenicity over time.

Preparation and fixation of sections

For details of section preparation and fixation see Chapter 3 and Appendix 6. For fixation, acetone for 30 minutes is used in the authors' laboratory.

Fixed embedded tissue

The standard fixative in most histopathology laboratories is unbuffered formol saline and the slightly acid pH favours staining with most of the antibodies listed in Table 19.2. Buffered formalin is preferred in some centres; however, a number of antibodies, notably UCHL1, perform suboptimally with this fixative. The rest of the antibodies listed in Table 19.2 perform adequately provided the tissue is not overfixed and a sensitive detection system is employed. Tissue should remain in formalin for no longer than 18 hours before processing and embedding. Longer fixation will result in the need for lengthy predigestion of tissue sections before immunostaining. It is worth mentioning that some of the fixation – resistant antigens can even be detected in infarcted lymphoid tissues (Norton & Isaacson 1988).

Special fixation protocols

In an effort to improve antigen preservation in fixed material many special fixatives and processing protocols have been devised.

At its simplest, the use of acid formalin (5% acetic acid in formol saline — Curran & Jones 1978) can improve immunoglobulin staining in paraffin sections. However, the improvement achieved is minimal compared to unbuffered formalin fixation, when a sensitive technique such as the immunogold silver (IGSS) (Holgate et al 1983), streptavidin biotin peroxidase complex (strept-ABC) (Guesdon et al 1979), or alkaline phosphatase antialkaline phosphatase (APAAP) (Mason 1985) methods are used.

Precipitating fixatives such as Bouin's fluid, formol sublimate and B5 give excellent cytological preservation and improve staining with some of the antibodies listed in Table 19.2. LN1 and LN2, for example, give greatly enhanced staining (Epstein et al 1984). However, there are exceptions. UCHL1 performs very badly in Bouin's fixed material, giving a diffuse overall pattern of tissue binding (Norton et al 1986) . Furthermore, antibodies requiring proteolytic digestion can give disappointing results, especially the anti-immunoglobulins. None of the antibodies restricted to cryostat sections is effective in tissue fixed in any of these media.

In an effort to get wide-ranging immunoreactivity in paraffin sections other radical fixation protocols have been devised (Stein et al 1985, Holgate et al 1986, Sato et al 1986, Delsol et al 1989, Islam et al 1988), which are of limited value in routine laboratories.

Proteolytic digestion

In the immunohistological assessment of routinely fixed and processed biopsies, pretreatment of tissue sections with proteolytic enzymes to unmask antigens has become an established procedure (Huang et al 1976). Not all monoclonal antibodies require a digestion step, and indeed for some reagents it may abolish staining altogether. As a general rule most polyclonal antibodies need a predigestion step, whereas the individual needs of monoclonal antibodies can vary.

The proteolytic enzyme chosen may also affect the staining result, especially with regard to the tissue distribution of the antigen. The enzyme trypsin is widely used and the artifacts produced by it are well described. On the other hand,

Table 19.2 Antibodies effective in paraffin sections.

Antibody	Monoclonal or Polyclonal	CD number	Reactivity in normal Tissues and Comment
2B1 (DAKO-LC)	Monoclonal	45	Conventional leucocyte common antigen. Leucocyte common has variable exons at the N terminus which can be alternately spliced out. All exons A, B and C are seen in this isoform. Present on all white cells except very early lymphoblasts and mature myeloid cells
PD7/26/16 (DAKO-LC)	Monoclonal	45RB	Constituent of DAKO-LC. Restricted binding to exon B of leucocyte common antigen. Similar to CD45 but binds to fewer T-cells
4KB5 (DAKO-4KB5) F8-11-13 MB1	Monoclonal	45RA	Restricted binding to exon A of leucocyte common antigen. Present on B-cells and naive T-cells. The latter are a minor (<50%) population in reactive lymphoid tissues
UCHL1 DAKO-UCHL1	Monoclonal	45R0	Restricted binding to leucocyte common antigen lacking either A, B or C exons. Present on memory T-cells, mature myeloid cells and macrophages. Some pre-plasma cells are also stained
βF1	Monoclonal	none	Invariate portion of T-cell receptor β-chain. Most T-cells, i.e. those with αβ T-cell receptors
Poly-CD3	Polyclonal	3 related	Fixation resistant epitopes on the ε chain of the CD3, T-cell receptor associated molecule. T-cells
MT-21	Monoclonal	w60	T-cells and B-cell subset Macrophages
MAC 387 (DAKO-MAC 387) anti-L1 / S22	Monoclonal / Polyclonal	none	Recognizes the cystic fibrosis antigen. Present on macrophages, myeloid cells and squamous epithelia
KP1 DAKO-KP1)	Monoclonal	68	Macrophages, plasmacytoid T-cells and myeloid cells
E11 To5 (DAKO-C3bR)	Monoclonal	35	C3b receptor (CR1). Present on follicle dendritic cells, macrophages and some B-cells
T09 C3D-1 (DAKO-M1) 3C4 anti-LeuM1	Monoclonal	15	Hapten X. Present on myeloid cells, some dendritic cells and epithelia

Antibody	Type	CD	Comments
MT2	Monoclonal	45R(?)	Restricted leucocyte common antigen. Possibly binds to exon C. Present on memory T-cells and on all but germinal centre B-cells
Ki-B3	Monoclonal	45R	Possible leucocyte common restricted antibody. Present on all but marginal zone B-cells and a proportion of follicle centre cells. T-cells negative
MT1 DFT-1 L60	Monoclonal	43	Primitive B-cells and pre-plasma cells. All T-cells, myeloid cells and macrophages
L26 (DAKO-L26)	Monoclonal	none	Pan B-cell antibody. Possible cytoplasmic precursor of CD20
MB2	Monoclonal	none	Present on B-cells, some macrophages, endothelium and various epithelia
HD66 CRIS-4 DBA44	Monoclonal	76	Mantle zone B-cells, T-cell subset and endothelium
LN1 HH2 OKB4 DNA7 DNA8	Monoclonal	w75	Present on mature B-cells, subset of T-cells, macrophages and some epithelia
LN2 MB3 DND53	Monoclonal	74	Cytoplasmic invariant chain of HLA complex. Present on B-cells, macrophages, endothelium and some epithelia

Antibody	Type	CD	Comments
Y2/51	Monoclonal	61	Platelet glycoprotein IIIa. Present on platelets and megakaryocytes
NP57 (DAKO-elastase)	Monoclonal	none	Granulocyte elastase. Present in granulocytes
MPO-7 (DAKO-MPO)	Monoclonal	none	Myeloperoxidase. Granulocytes and monocytes
Ret40f	Monoclonal	none	Glycophorin C. Present in erythrocytes and precursors
LN3 TAL-1B5 (DAKO-HLA-DR/Alpha)	Monoclonal	N/A	HLA-DR $\alpha\beta$ and β chains respectively. B-cells, macrophages and activated T-cells. Many non-lymphoid tissues
Ber-H2 (DAKO-Ber-H2)	Monoclonal	30	Activated T and B-cells. Some activated macrophages
Anti-Tac	Monoclonal	25	Activated T and B-cells
Anti-immunoglobulin heavy and light chains	Polyclonal	N/A	B-cells and plasma cells
Anti-α_1-antitrypsin	Polyclonal	N/A	Macrophages, granulocytes and many non-lymphoid tissues.
Anti-α_1 anti-chymotrypsin	Polyclonal	N/A	Macrophages, granulocytes and many non-lymphoid tissues
Anti-lysozyme	Polyclonal	N/A	Macrophages and myeloid cells. Some non-lymphoid tissues

pronase has its advocates, but it can produce quite different staining patterns, especially with immunoglobulin and anti-protease antibodies. In trypsinized sections anti-immunoglobulins give immunoreactivity in the region of the perinuclear space and also produce a focal flare of cytoplasmic staining, presumed to be in the golgi region (Isaacson et al 1980). However, with pronase-digested tissue the staining result differs, tending to give a rather diffuse or overall granular result in the cytoplasm alone. The latter is, of course, acceptable in plasma cells, but is very difficult to interpret in other cell types. Predigestion with neuraminidase can profoundly alter the tissue distribution of antibody binding, especially if the antigen is largely carbohydrate in composition. CD15 antibodies on neuraminidase-treated sections bind to nearly all white cells and not just the granulocytes, as is seen in trypsinized material (Hsu et al 1986). Neuraminidase may even abolish tissue binding by antibodies recognizing heavily sialylated antigens such as CD43. Because of these variations in staining with different enzymes we will only discuss staining in relation to trypsin predigestion in the rest of the chapter.

The duration of proteolytic digestion may also influence the staining achieved. It may not always be possible to gauge the correct digestion time at first, except to say that with shorter fixation times, the shorter is the digestion period required. In order to optimize staining, paired digestion times of 5 and 10 minutes are desirable. Overdigestion will result in sections falling off the slides. Lastly, a word of warning with regard to kappa and lambda immunoglobulin light chain staining; only slides digested for the same length of time should ever be compared one with another to assess light chain restriction.

Immunohistological labelling procedures

Choice of detection system

For monoclonal antibodies on cryostat sections an indirect immunoperoxidase method is quite adequate. If greater sensitivity is needed, then an avidin-biotin peroxidase complex method can be used.

For paraffin sections it is imperative to use a very sensitive detection system, otherwise it is a waste of time performing immunoglobulin staining. Peroxidase antiperoxidase and indirect immunoperoxidase are too insensitive and have largely been superseded by the avidin-biotin peroxidase complex and streptavidin-biotin peroxidase complex methods which give excellent results with any of the antibodies listed in Table 19.2.

Choice of antibodies for frozen tissue

From Table 19.1 it can be seen that, despite the clustering process, there is an unwieldy range of antibodies available. Since the selection of a rational panel from this list has to be governed by local needs and interests, we will confine ourselves to the selection of a minimum useful diagnostic panel.

Most modern lymphoma classification schemes make a distinction between tumours derived from T- and B-lymphocytes and inclusion of relevant antibodies is clearly necessary (Stansfeld et al 1988). The selection of pan-B cell antibodies is relatively simple as the vast majority of such tumours will express the CD20 or CD22 antigens. For T-cell tumours, however, the choice is somewhat more vexed. For reasons to be discussed in the section covering lymphoma diagnosis, a much wider panel is desirable; pan-T cell antibodies CD3, CD5 and CD7 should be combined with T-cell subset-related markers CD1a, CD4 and CD8. In the assessment of B-cell clonality, anti-kappa and lambda light-chain monoclonals are pivotal. Anti-heavy chain antibodies can be limited to IgM and IgD, as heavy chain restriction is not a feature of lymphoma and these two reagents help to highlight different B-cell compartments (Fig. 19.1). Clearly, not all cells in a lymph node are B- or T-cells. Thus antibodies to macrophages, such as CD68 or CD11c, and follicular dendritic cells (CD35 or R4/23) will yield valuable additional information.

There are of course many useful, yet still unclustered antibodies which can be added to the panel. The R4/23 antibody mentioned above is one such reagent, recognizing follicular dendritic reticulum cells. Another antibody of interest is Ki67, which identifies a nuclear antigen found in cells in the G_1, S, G_2 and M phases of the cell cycle and can give a very good indication of the proliferative activity of a given tumour. Assignment of a Ki67 score to a lymphoma has been shown to provide useful prognostic information in a number of published series (Gerdes et al 1984, Schrape et al 1987).

Choice of antibodies for paraffin sections

It could be argued that in many cases immunocytochemistry on frozen tissue is unnecessary as similar information can be obtained by using the antibodies from Table 19.2 on fixed material. However, it can be seen from Table 19.2 that there are not insignificant lineage infidelities to be found with many of the reagents to fixation-resistant antigens. These problems are not insurmountable and the added advantages of being able to use routine paraffin sections far outweigh any demerits. By selecting a suitable antibody panel a reliable assessment of phenotype can be made in the vast majority of cases (Norton & Isaacson 1989a and 1989b). A reasonable diagnostic panel should include a conventional leucocyte common antibody (CD45), B-cell related antibodies (L26 and MB2), T-cell related antibodies (UCHL1 and MT1) and a macrophage marker, either KP1 or Mac 387. Polyclonal anti-immunoglobulin light chains and anti-IgM will enable light chain restriction to be shown provided a good detection system is in use. There are two further points to be made. Firstly, it may have been noticed that polyclonal CD3 and βF-1 were not mentioned as desirable in the T-cell panel. Although both of these antisera represent major advances, they are presently of limited availability, or too costly for routine use (Mason et al 1988, Francis et al 1988). Secondly, in a busy laboratory, and for ease of diagnosis, a very satisfactory half-

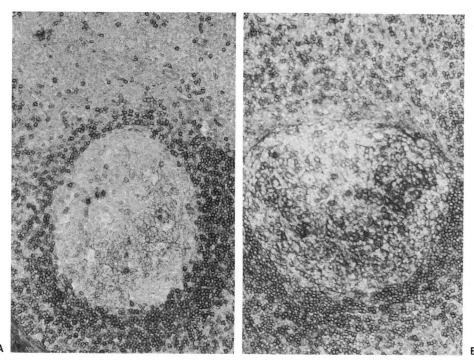

Fig. 19.1 Cryostat sections of reactive lymph node showing a B-cell follicle stained with monoclonal antibodies to (**A**) IgM and (**B**) IgD. Note the IgM staining of follicular dendritic cells and mantle zone B-cells, while IgD staining is confined to the mantle zone. Immunoperoxidase. ×150

way measure can be adopted with frozen and paraffin section phenotyping. In this way the frozen immunophenotyping is reserved only for those cases in which an ambiguous phenotype emerges in the paraffin sections. We must stress that this last option is only viable if a large paraffin-reactive antibody panel is used in conjunction with immunoglobulin staining.

Specificity controls

The technical measures for controls are detailed in Chapter 3. However, a number of points should be made. Firstly, nearly every antibody used in the diagnosis of lymphoma will stain some cells whatever the condition, since reactive B- and T-lymphocytes are a common component of all lymphoreticular tumours. Thus, if no staining at all is observed in a test section, then the procedure should be repeated. This applies equally to immunoglobulin staining, where a minor population of reactive plasma cells are nearly always present. Secondly, in order to control for staining conditions and to eliminate mix-ups with antibodies, the inclusion of a piece of control reactive tonsil on the same slide as the test section is very helpful. This last measure also acts as an easy reference for the normal distribution of binding for the test antibody.

PRACTICAL APPLICATIONS

Interpretation of staining results

Cryostat sections

Before embarking on lengthy cryostat section staining, the pathologist must be familiar with the normal distribution of the antigens being sought. Failure to comply with this simple requirement will undoubtedly result in misdiagnoses.

The antibodies listed in Table 19.1 largely recognize molecules situated on or within the plasma membrane. Some antibodies will also identify cytoplasmic precursors of these surface antigens, for example CD22 and CD3. Pure cytoplasmic staining with either of these two antibodies is usually restricted to primitive lymphocytes and

their derived tumours, and is difficult to see except in cytospin preparations. Indeed, the most commonly observed positive result is a ring of staining around the cell body, presumably plasma membrane. Ki67 is a notable exception in this respect as it binds to discrete areas of the cell nucleus, probably the nucleoli. Interpretation of immunoglobulin staining can be difficult. These normal plasma constituents can give marked background non-specific binding to connective tissue. It is only by placing the kappa and lambda stained sections side by side on the same microscope stage that a direct and reliable comparison of the staining intensities can be made (Fig. 19.2). With anti-immunoglobulins, more than any other reagents, it is imperative to check the control sections on each staining run, and to adjust the antibody titre if the light chain staining intensities differ.

Paraffin sections

Because of the superior cytomorphology, the precise cellular distribution of antibody binding is readily seen in paraffin sections. This is important since non-specific tissue binding sometimes occurs, especially with polyclonal antisera, and recognition of specific staining patterns can help to overcome this difficulty. Significant deviations from an accepted pattern of staining should be interpreted with caution.

CD45 and CD45R (A, B, 0) all bind to the plasma membrane of lymphoid cells. In some cases of large cell lymphoma, coexistent surface and focal cytoplasmic (golgi?) staining are seen (Kurtin & Pinkus 1985). Diffuse cytoplasmic staining with nearly all the CD45 group has been described in a wide range of tumours, including carcinomas, as well as occurring as a non-specific phenomenon (Warnke & Rouse 1985). It is therefore inadvisable to interpret diffuse cytoplasmic staining as a positive result. CD45 isoforms are modulated in both B- and T-cells during both maturation and activation and it is not entirely surprising that occasional aberrantly stained cells may be seen in any one tumour. For practical purposes, the consensus staining should be noted and other non-CD45 antibodies will improve the accuracy of the diag-

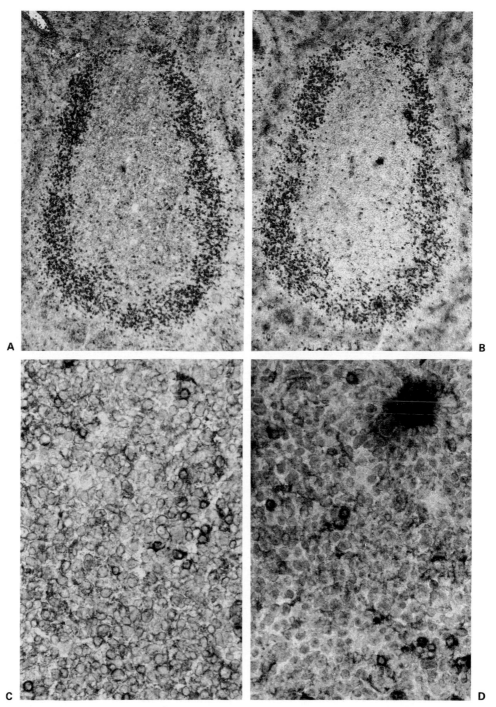

Fig. 19.2   Cryostat sections of a centroblastic/centrocytic, follicular lymphoma stained with monoclonal antibodies to (**A**) kappa and (**B**) lambda light chains. The mantle zone is polytypic while the follicle centre shows kappa light chain restriction. Immunoperoxidase.   ×60

At higher magnification the surface kappa (**C**) and lambda staining (**D**) is more clearly seen. Note that isolated reactive B-cells within the neoplastic follicle centre express lambda light chain. Immunoperoxidase. ×380.

nostic panel. This is doubly necessary as CD45RA and CD45R0 (UCHL1) antigens show reciprocality of expression irrespective of cell lineage.

Other antibodies identifying predominately surface moieties are CD43, KiB3, βF-1 and L26. The last antibody also identifies a cytoplasmic antigen, which is difficult to demonstrate in paraffin sections. Again, it is better to only take regard of surface staining with these reagents.

MB2 recognizes a cytoplasmic antigen in immunoreactive cells. This can present minor problems in interpretation if the staining is only weak, as the result may be indistinguishable from endogenous peroxidase activity.

CDw75 antibodies, such as LN1, present problems in interpretation. The pattern of binding to B-lymphocytes has two components; a weak surface staining and a stronger cytoplasmic dot. If the surface pattern alone is interpreted then LN1 is probably a very good marker of activated B-cells. If the cytoplasmic positivity is included then the antibody identifies a significant number of T-cell derived tumours. This is compounded by surface staining with LN1 being only very weak in formalin-fixed tissues.

The CD74 antibodies, including LN2, have a staining distribution similar to the CDw75 reagents. The antigen is predominately intracellular with very weak surface expression. Again T-cell lymphomas may have cytoplasmic activity, but surface staining in these tumours is very uncommon.

CD15 and CD30 antibodies also have a dual surface and cytoplasmic pattern of binding, but with both of these reagents there is no significant difference in cell types stained in relation to the pattern observed.

Most antibodies to cellular enzymes give predominant intracytoplasmic binding. These include Mac387, KPl (CD68) and the antiproteases, $\alpha_1$anti-trypsin and $\alpha_1$anti-chymotrypsin, and anti-lysozyme. The last three polyclonal antibodies have caused endless difficulties and debate regarding the correct interpretation of staining. In our view a diffuse cytoplasmic immunoreactivity is almost always due to passive diffusion or uptake of antigen by effete or phagocytic cells (Isaacson & Wright 1979, Mason & Biberfeld 1980). A reliable positive result requires a distinct staining pattern;

coarse cytoplasmic granules clustering close to the cell nucleus. The choice of fixative is important in this regard, Bouin's fluid and mercuric fixatives being unsuitable due to the need for high titres of antiserum with attendant background staining.

The interpretation of immunoglobulin staining in paraffin sections can be difficult. Control tonsil sections should always be run in parallel, allowing the relative intensities of kappa and lambda staining to be adjusted and to check that the correct degree of sensitivity has been reached. Ideally there should be strong staining of both the follicle centre and mantle zone cells, and not just plasma cells! The distinctive features and reasons for non-specific staining have been well-described previously (Isaacson & Wright 1979, Isaacson et al 1980, Mason & Biberfeld 1980, Norton & Isaacson 1987). In trypsinized immunoperoxidase-stained sections a positive result should give perinuclear space staining and in larger cells a focal cytoplasmic dot, possibly golgi in origin (Figs 19.3 and 19.4). True surface staining is rarely seen, but is possibly observed when using a multilayered APAAP or IGSS staining method. The golden rules when comparing kappa and lambda test slides are as follows. Firstly, if no reactive cells are seen on either slide, or if the control sections are unsuitable, then the staining has to be repeated. Secondly, only compare slides digested for the same length of time and those stained in the same laboratory run. Thirdly, compare like section with like section, and area for area. The last is a very important point as the optimal staining may be found only in some parts of an unevenly fixed specimen. Lastly, always ask yourself if the observed staining is giving a structured pattern or whether it is only due to non-specific uptake or over digestion?

Immunocytochemistry in the diagnosis of lymphoma

Immunocytochemistry initially played only a peripheral role in the classification of lymphoreticular tumours and it is only in recent years that there has been a shift from purely morphological considerations to incorporating the immunophenotype into classification systems (Stansfeld et al 1988). Early expectations were that

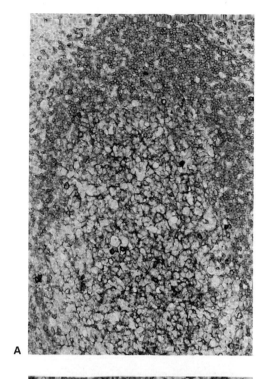

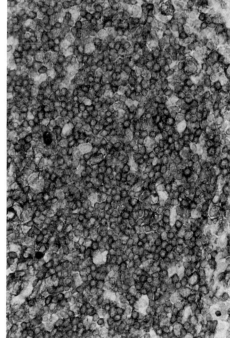

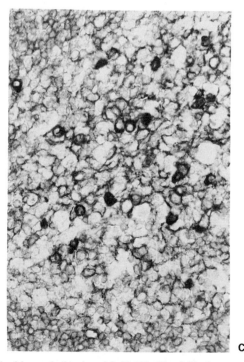

well-defined tumour entities would have a corresponding immunological fingerprint (Stein et al 1984). However, with greater experience it would seem that this is not entirely the case. Discrimination between T- and B-cell tumours is relatively easy and B-cell clonality can be inferred from immunoglobulin light chain restriction in most cases. T-cell clonality is still very difficult to determine in tissue sections. Some antibodies to T-cell receptor families do exist and can be used to infer clonality (Clark et al 1986). However, there is a need for at least 20 such reagents to cover the whole T-cell receptor beta chain repertoire, and this has yet to be achieved. It is in the application of very large antibody panels on cryostat sections that the lack of pure immunological boundaries between distinct subtypes of lymphoma has become increasingly evident (Schuurman et al 1987). Nevertheless a proportion of low-grade non-Hodgkin's lymphomas of B- and, to a lesser extent, T-cell types have a common phenotype which is expressed in the majority of cases. These

Fig. 19.3 Paraffin section of a reactive tonsil showing a follicle stained with a polyclonal anti-IgM (**A**). The follicular dendritic cell network is strongly stained and there are scattered follicle centre cells which are positive. The mantle zone is strongly positive. **B** High power detail of mantle zone showing perinuclear space staining of all cells. **C** Detail of follicle centre showing staining of follicular dendritic cell processes and scattered cytoplasmic staining of centroblasts and centrocytes. Immunoperoxidase. **A** ×60, **B** and **C** ×380

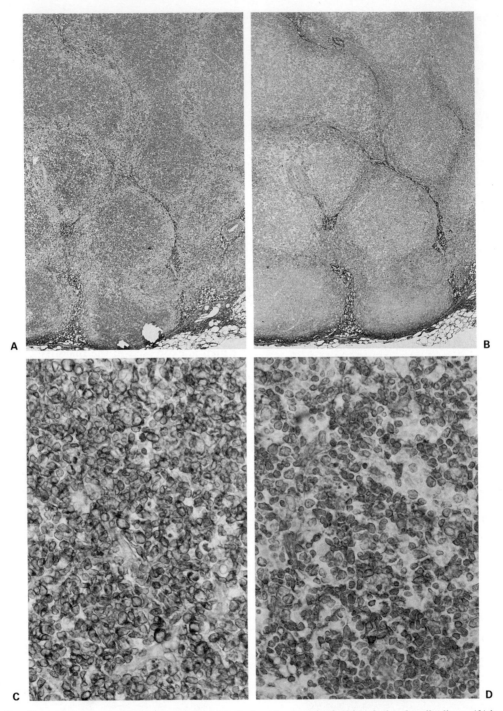

Fig. 19.4 Paraffin sections of centroblastic/centrocytic follicular lymphoma stained with polyclonal antibodies to (**A**) kappa and (**B**) lambda light chains. There is kappa light chain restriction. Immunoperoxidase. ×25

Higher power detail of a follicle centre stained for (**C**) kappa and (**D**) lambda light chains. Perinuclear space staining of centroblasts and centrocytes is evident, with focal cytoplasmic expression. Immunoperoxidase. ×380

immunophenotypes are listed in Table 19.3 and for the interested reader the following references take the controversy further (Stein et al 1984, Schuurman et al 1987, Smith et al 1989). Thus the expression of such a commonly occurring phenotype helps to confirm a diagnosis, whereas the absence of such a pattern does not necessarily exclude the possibility. A more fundamental change in emphasis of late has been the possibility that the expression of patently abnormal phenotypes is a hallmark of malignancy (Mason 1987). This is seen especially in T-cell lymphomas and in some high-grade B-cell tumours. Here one sees loss of expected antigens, or the gain of aberrant ones. Some T-cell tumours will have shed as many as half a dozen pan-T or T-cell subset markers (Picker et al 1987). Whilst diagnostically useful, it does necessitate a much larger panel of T-cell than B-cell reagents. The loss of the CD7 T-cell antigen alone should be interpreted cautiously as this may also disappear in some inflammatory skin diseases (Ralfkiaer 1989). While these remarks largely concern immunophenotyping in cryostat sections, the same applies to paraffin phenotyping (Norton & Isaacson 1989b). Table 19.4 lists the commonly found immunophenotypes of non-Hodgkin's lymphoma in paraffin sections. In high-grade T- and B-cell tumours the reliability of the antibodies steadily falls off, necessitating the use of a large diagnostic

Table 19.3 Immunophenotypes of non-Hodgkin's lymphomas using antibodies effective only on cryostat sections. Lymphomas are classified by the updated Kiel classification. T-cell lymphoma types are specified as these show minor variations in frozen sections.

| Lymphoma type | Typical immunophenotype (CD) | | | | | | | | | | | | | |
| | B-related clusters | | | | | T-related clusters | | | | | | Macrophage clusters | Activation clusters | |
	10	20	22	23	Ig	1a	3	5	7	4	8	11c	25	30
B-lymphocytic	−	+	+	+	M&D	−	−	+	−	−	−	−	−	−
Hairy cell leukaemia	−	+	+	+	GorA	−	−	−	−	−	−	+	+	−
Lymphoplasmacytoid/cytic	−	+	+	−	MorG	−	−	+	−	−	−	−	−	−
Centroblastic/centrocytic	+	+	+	+	MorG	−	−	−	−	−	−	−	−	−
Centrocytic	−	+	+	−	M&D	−	−	+	−	−	−	−	−	−
Centroblastic and B-Immunoblastic	−	+	+	+	any	−	−	−	−	−	−	−	±	−
B-lymphoblastic	+	+	+	+	M	−	−	−	−	−	−	−	−	−
T-lymphocytic	−	−	−	−	−	−	+	+	+	+>−	−>+	−	−	−
Mycosis fungoides	−	−	−	−	−	−	+	+	±	+		−	−	−
Angioimmunoblastic T-cell	−	−	−	−	−	−	+	+	±	+>−	−>+	−	−	−
Lennert's lymphoma	−	−	−	−	−	−	+	+	±	+	−	−	−	−
T-zone lymphoma	−	−	−	−	−	−	+	+	±	+>−	−>+	−	±	−
Pleomorphic medium/large cell T	−	−	−	−	−	−	±	±	±	+>−	−>±	−	±	±
Large cell anaplastic T	−	−	−	−	−	−	±	±	±	−>±	−>±	−	+	+
T-lymphoblastic	−	−	−	−	−	+	+	±	+	+or−	+or−[S]	−	−	−

Key to Table 19.3
+ Majority of cases positive (nearly 100%).
± Most cases positive but can be patchy.
+>− More cases positive than negative.
−>+ More cases negative than positive.
+or−[S] CD4 and CD8 antigen expression positive, negative or both positive.

Table 19.4 Immunophenotypes of non-Hodgkin's lymphomas using monoclonal and polyclonal antibodies effective in routinely processed tissue sections. Tumours are classified by the updated Kiel system. For simplicity low grade T and high grade T cell lymphomas are grouped together as they show little individual variation in phenotype.

Tumour type	Antibody or CD designation if known													
	CD45	CD45RA	CD45R0	MT2	KiB3	CD43	CDw75	CD74	L26	MB2	CD15	CD30	CD68	α1AT
B-lymphocytic	+	+	−	+	+	+	p	+	+	+	−	−	−	−
Hairy cell leukaemia	+	+	−	+	+	−	−	+	+	+	−	−	+	−
Lymphoplasma-cytoid/cytic	+	+	−/±	+	+	±	PB	+	+	+	−	−	−	−
Centroblastic/centrocytic	+	+	−	45%	±	−	+	+	+	+	−	−	−	−
Centrocytic	+	+	−	+	+	±	−	+	+	+	−	−	−	−
Plasmacytic	−/±	−	−/±	−	−	−/±	±	−	−	−	−	+	−	−
Centroblastic and B-immunoblastic	+	+/−	−/+	+/−	+/−	−/+	+	+	+	+/−	−/+	−/+	−	−
B-lymphoblastic	+/−	+/−	−	+	−	+	−	+	+	+	−/+	−/+	−	−
T-lymphoblastic	+/−	−	+/−	−	+/−	+	−	−	−	−/+	−/+	−	−	
Low grade T-cell	+	−	+	−	−	+	−	−	−	−	−	−	−	
High grade T-cell	+/−	−/+	+/−	−/+	−/+	+/−	−/+	−/+	−	−/+	−/+	−/+	−	−/+
Large cell anaplastic	+/−	−/±	+/−	−	−	+/−	−/+	+/−	−	−	−/+	+	−	+/−

Key to Table 9.4
    + Majority of cases positive (nearly 100%)
    ± Most cases positive, but patchy distribution
    − Most cases negative
    +/− Either completely positive or negative
P Proliferation centres alone positive
B Blast cells positive, rest of tumour negative

panel. Again the expression of a bizarre phenotype is almost pathognomonic of lymphoma. In both cryostat and paraffin section immunohistology, the histopathologist must be familiar with the normal phenotype and cellular distributions of different lymphoid subsets in order to avoid misdiagnosis. For example, the expression of CD43 (MT1) and CD45R0 (UCHL1) by preplasma cells is a potentially serious diagnostic pitfall unless an immunoglobulin stain is performed concurrently.

Another aspect of lymphoma immunocytochemistry is the use of non-lymphoid markers, such as anti-keratins and anti-epithelial membrane antigen (EMA). At a fundamental level in combination with CD45 antibodies they help to distinguish between carcinomas and lymphomas

(Fig. 19.5). However, there are some pitfalls even here. EMA is known to be expressed by mature plasma cells, some large cell lymphomas and notably the large cell anaplastic lymphoma (Delsol et al 1984), a tumour which closely mimics anaplastic carcinoma (Fig. 19.6) Lymphoid tumours reactive with cytokeratins are very rare, but plasmacytomas are a notable exception (Wotherspoon et al 1989). Furthermore, there is a population of dendritic shaped cells in lymph nodes known to be immunoreactive with anti-cytokeratin reagents (Jack et al 1989). Another area where epithelial markers are diagnostically useful is in the demonstration of typical lymphoepithelial lesions in lymphomas of mucosal associated lymphoid tissues. These can be hard to

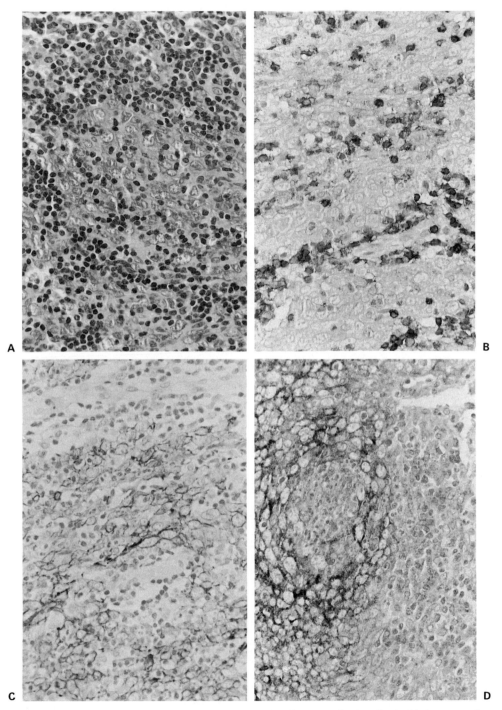

Fig. 19.5 Paraffin sections of a nasopharyngeal carcinoma metastatic to lymph node. **A** stained with haematoxylin and eosin, **B** with CD45, **C** for low molecular weight cytokeratin and **D** for epithelial membrane antigen. The undifferentiated epithelial cells are CD45 negative in contrast to accompanying reactive lymphoid cells. The tumour cells are both positive for cytokeratin and epithelial membrane antigen. H & E and immunoperoxidase. ×380

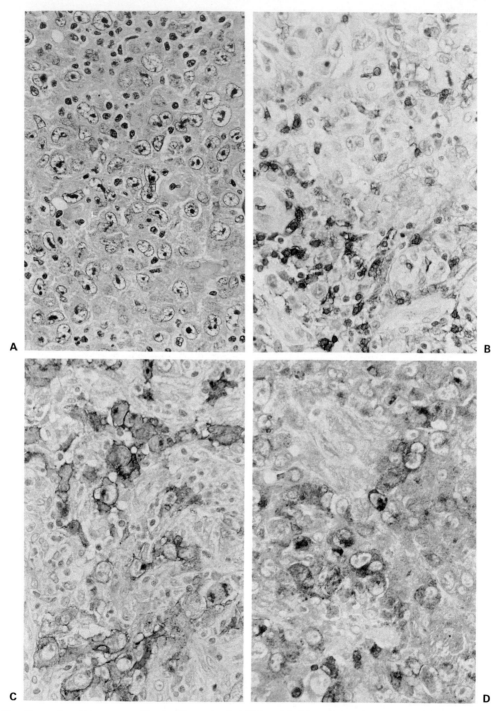

Fig. 19.6 Paraffin sections from a case of large cell anaplastic (Ki-1 positive) lymphoma. **A** stained with haematoxylin and eosin, **B** with CD45, **C** with CD30 and **D** for epithelial membrane antigen. The tumour cells are negative with CD45, show characteristic staining with CD30 and also stain for epithelial membrane antigen. H & E and immunoperoxidase. ×380

find by conventional light microscopy, but the use of an anti-keratin will highlight the disrupted intestinal glands (Fig. 19.7).

The discussion has largely evaded the question of distinguishing between Hodgkin's disease and non-Hodgkin's lymphoma. While it is relatively easy to define the latter tumours immunologically, the Reed-Sternberg cell has no characteristic profile. The closest phenotype is that of the large cell anaplastic lymphoma which expresses activation markers CD25, CD30, CD70, CD71 and HLA-DR (Stein et al 1985). T-cell and B-cell specific markers are only very rarely expressed, with a slight predominance of B-cell examples in the case of Hodgkin's disease. Curiously there is usually complete lack of the leucocyte common antigen (CD45). Two further points concerning Hodgkin's disease immunohistology remain. Firstly antibodies of the CD15 cluster, such as anti-Leu Ml, are not Reed-Sternberg cell specific as was first thought, with many T-cell lymphomas and some B-cell tumours being immunoreactive (Hall & d'Ardenne 1987). Secondly, in contrast to the other subtypes, nodular lymphocyte predominance Hodgkin's disease or nodular paragranuloma, has a unique immunophenotype. The Reed-Sternberg cell variants in this tumour always express B-cell markers and CD45 antigen and can be shown to contain J-chain (Stein et al 1986). It would seem likely, and especially when the singular natural history is considered, that nodular lymphocyte predominance Hodgkin's is a special type of B-cell proliferation, distinct from the rest of Hodgkin's disease.

In any chapter on the immunohistology of lymphoma, one will find the criteria for defining a true histiocytic, monocyte/macrophage-derived tumour. While these tumours do exist, their true incidence has been grossly overinflated in the past. Included amongst the older cases were large cell anaplastic lymphomas and large cell T- and large cell B-cell lymphomas (Isaacson 1985, Lennert et al 1987, Delsol et al 1988). Immunochemistry has been in part responsible for the decline of this diagnosis, but the major contribution has been made by gene rearrangement studies (Weiss et al 1985, Isaacson et al 1985). Expression of genuine monocyte/macrophage markers in either paraffin

or cryostat sections is clearly a prerequisite. Some cases will express S100 antigen and show features suggesting derivation from dendritic reticulum cells. The expression of $\alpha_1$anti-trypsin and $\alpha_1$antichymotrypsin, either in combination or alone, is regrettably no longer good enough since there are well documented cases of T-cell lymphomas expressing these antigens (Norton & Isaacson 1986). Lysozyme is confined to the cells of myeloid and monocyte/macrophage lineage, and is a more reliable marker. However, the staining pattern needs correct interpretation to avoid the pitfalls of passive diffusion or uptake of antigen discussed earlier. It is paradoxical that lysozyme was originally discarded as staining too few cases of histiocytic lymphoma, when in fact it was probably truly reflecting the incidence of a very rare disease.

Immunocytochemistry in the acquired immunodeficiency syndrome

Immunocytochemistry not only has a role to play in the classification of non-Hodgkin's lymphomas, but it may assist in the diagnosis of reactive lymphoid proliferations. The singular area in immunochemistry most useful for distinguishing between a reactive and neoplastic proliferation is the demonstration of immunoglobulin light-chain restriction. An alternative in laboratories lacking the expertise required for immunoglobulin detection is the antibody MT2, which does not stain reactive germinal centres but will react with 40–50% of cases of follicular lymphoma (Norton et al 1989c). These methods are, of course, limited to B-cell disorders, and currently there is nothing short of gene rearrangement studies to assist in the routine determination of T-cell clonality.

In the context of the acquired immunodeficiency syndrome there are a number of distinctive immunochemical features. The light microscopic changes of follicular hyperplasia, followed by involution and lymphocyte depletion, are all well described. In cryostat sections, fragmentation of germinal centres may be demonstrated using a CD35 antibody or R4/23 (Piris et al 1987). The follicle dendritic cells are swollen with clubbing or

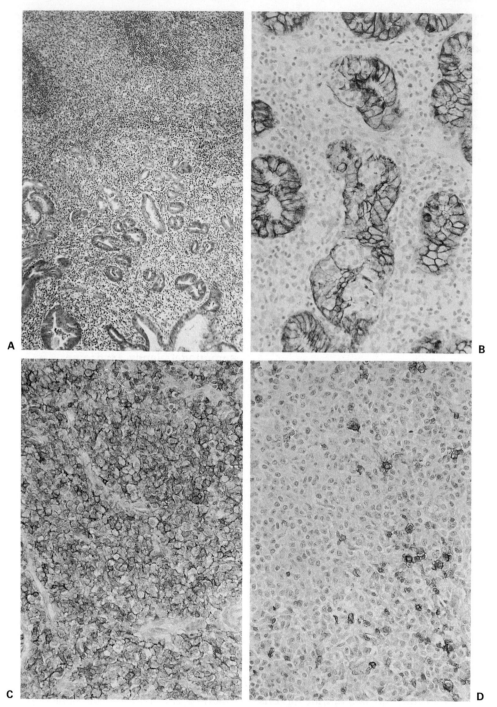

Fig. 19.7   A Paraffin section of a case of primary gastric lymphoma, low-grade B-cell type, arising in mucosal associated lymphoid tissue. Reactive follicles can be seen at the bottom of the illustration and lymphoepithelial lesions can be seen in the mucosa which is infiltrated by centrocyte-like cells. H & E. ×25 **B** Paraffin section stained with an anti-cytokeratin highlights the lymphoepithelial lesions. Staining of the mucosal infiltrate for **C**: B-cells using L26 and **D**: T-cells using UCHL1 (CD45R0). These stains confirm the B-cell nature of the tumour. Immunoperoxidase. ×250

matting of their cytoplasmic extensions. A later feature is the loss of CD4 positive T-cells, with an increase of the CD4:CD8 ratio to between 1:5 and 1:10 (Baroni et al 1985). Where available, antibodies to the p24 and p18 HIV1 antigens may be used to localize virus within follicle dendritic cells, T-cells and endothelium (Baroni et al 1986). Unfortunately, most of these manoeuvres are restricted to cryostat sections with the attendant health and safety constraints. In paraffin sections, antibodies to viral antigens have been disappointing, although some success is reported in B5 fixed

tissues (Ward et al 1987). The follicular disintegration and dendritic cell disruption can be readily demonstrated in fixed material using CD35 reagents or anti-S100 protein (Carbone et al 1987) (Fig. 19.8). The lymphomas arising in AIDS may be adequately phenotyped in paraffin sections using the range of antibodies discussed above. In addition, antisera to infective agents such as cytomegalovirus, cryptosporidium, toxoplasma gondii and JC virus may help to characterize the many opportunistic infections so characteristic of the disease.

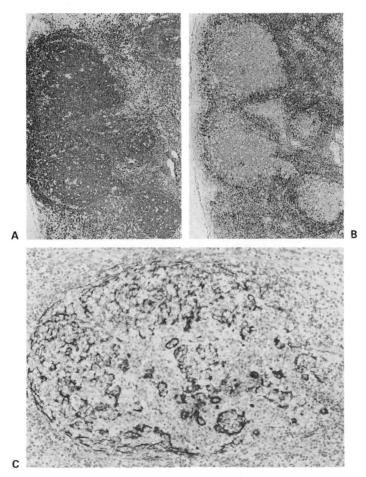

Fig. 19.8 Paraffin sections of lymph node from a patient with persistent generalized lymphadenopathy syndrome (HIV 1 positive) showing follicular hyperplasia stained: **A** for B-cells using L26 and **B** for T-cells using UCHL1 (CD45R0). Note the irregular follicles infiltrated by T-cells. Immunoperoxidase. ×20

Staining of paraffin sections with CD35 shows the follicular dendritic cells. Note the disintegration of the follicular dendritic cell network in the follicle centre. Immunoperoxidase. ×380

CONCLUSION

Immunochemistry has become a fundamental tool in the differential diagnostic armoury of the lymphoreticular pathologist. It is now clear that pure immunophenotypes do not exist for individual tumour types. More importantly, through the immunologists we are gaining insight into the functional significance of these leucocyte molecules. In the future we may find useful functional parameters of lymphoma behaviour, either through the expression of integrin and other leucocyte adhesion molecules, or the expression of tissue addressins, the leucocyte homing molecules. With the cytogenetics of lymphoma becoming more important, immunochemistry may help to detect abnormal gene products reflecting a karyotypic abnormality. Indeed, there is already one antibody which identifies the bcl-2 gene product as a consequence of the t(14;18) in follicular lymphomas (Ngan et al 1988). Although molecular biology has become one of the preeminent lines of research in lymphoreticular disease, immunocytochemistry will always play a pivotal role in primary diagnosis.

REFERENCES

Baroni C D, Pezzella F, Stoppacciaro A et al 1985 Systemic lymphadenopathy (LAS) in intravenous drug abusers. Histology, immunohistochemistry and electron microscopy: pathogenic correlations. Histopathology 9: 1275–1293

Baroni C D, Pezzella F, Mirolo M, Ruco L P, Rossi G B 1986 Immunohistochemical demonstration of p24 HTLV III major core protein in different cell types within lymph nodes from patients with lymphadenopathy syndrome (LAS). Histopathology 10: 5–13

Bernard A, Boumsell L, Dausset J, Milstein C, Schlossmann S F 1984 Leucocyte typing. Springer-Verlag, Heidelberg

Carbone A, Manconi R, Poletti A, Volpe R 1987 Immunostaining for S-100 protein: another suitable tool for analysing dendritic reticulum cells in AIDS-related lymphadenopathy. Journal of Pathology 151: 163–165

Clark D M, Hall P A, Boylston A W, Carrels S 1986 Antibodies to T-cell antigen receptor beta chain families detect monoclonal T-cell proliferation. Lancet ii: 835–836

Curran R C, Jones E L 1978 Hodgkin's disease: an immunohistochemical and histological study. Journal of Pathology 125: 39–48

Delsol G, Gatter K C, Stein H et al 1984 Human lymphoid cells express epithelial membrane antigen. Implications for diagnosis of human neoplasms. Lancet ii: 1124–1129

Delsol G, Al Saati T, Gatter K C et al 1988 Coexpression of epithelial membrane antigen (EMA), Ki-1, and interleukin-2 receptor by anaplastic large cell lymphomas. Diagnostic value in so-called malignant histiocytosis. American Journal of Pathology 130: 59–70

Delsol G, Chittal S, Caverivière P et al 1989 Immunohistochemical demonstration of leucocyte differentiation antigens on paraffin sections using a modified AMex (ModAMex) method. Histopathology 15: 461–472

Epstein A L, Marder R J, Winter J N, Fox R I 1984 Two new monoclonal antibodies (LN-1, LN-2) reactive in B5 formalin-fixed, paraffin embedded tissues with follicular center and mantle zone human B lymphocytes and derived tumors. Journal of Immunology 133: 1028–1036

Francis N D, Clark D M, Boylston A W 1988 Definitive identification of human T cells in formalin fixed paraffin wax embedded tissue. Journal of Clinical Pathology 41: 900–903

Gerdes J, Dallenbach F, Lennert K, Lemke H, Stein H 1984 Growth fractions in malignant non-Hodgkin's lymphomas as determined in situ with monoclonal antibody Ki67. Hematological Oncology 2: 365–371

Guesdon J L, Ternynck T, Avrameas S 1979 The use of avidin-biotin interaction in immunoenzymatic techniques. Journal of Histochemistry and Cytochemistry 27: 1131–1139

Hall P A, d'Ardenne A J 1987 Value of CD15 immunostaining in diagnosing Hodgkin's disease: a review of published literature. Journal of Clinical Pathology 40: 1298–1304

Holgate C S, Jackson P, Cowen P N, Bird C C 1983 Immunogold-silver staining: new method of immunostaining with enhanced sensitivity. Journal of Histochemistry and Cytochemistry 31: 938–944

Holgate C S, Jackson P, Pollard K, Lunny D, Bird C C 1986 Effect of fixation on T and B lymphocyte surface membrane antigen demonstration in paraffin processed tissue. Journal of Pathology 149: 293–300

Hsu S-M, Ho Y-S, Li P-J et al 1986 L&H variants of Reed-Sternberg cells express sialylated Leu M1 antigen. American Journal of Pathology 122: 199–203

Huang S, Minassian H, More J D 1976 Applications of immunofluorescent staining in paraffin sections improved by trypsin digestion. Laboratory Investigation 35: 383–391

Isaacson P G 1985 Histiocytic malignancy. Histopathology 9: 1007–1011

Isaacson P G, Wright D H 1979 Anomalous staining patterns in immunohistologic studies of malignant lymphoma. Journal of Histochemistry and Cytochemistry 27: 1197–1199

Isaacson P G, Wright D H, Judd M A, Jones D B, Payne S V 1980 The nature of the immunoglobulin containing cells in malignant lymphoma: An immunoperoxidase study. Journal of Histochemistry and Cytochemistry 28: 761–770

Isaacson P G, O'Connor N T J, Spencer J et al 1985 Malignant histiocytosis of the intestine: a T-cell lymphoma. Lancet ii: 688–691

Islam A, Archimbaud E, Henderson E S, Han T 1988

Glycol methacrylate (GMA) embedding for light microscopy. II immunohistochemical analysis of semithin sections of undecalcified marrow cores. Journal of Clinical Pathology 41: 892–896

Jack A S, Grigor I, O'Brien C J, McMeekin W, Lewis F, McNicol A M 1989 Association between CAM 5.2 and anti-CD1a reactivity in lymph nodes and gastrointestinal tract epithelium. Journal of Clinical Pathology 42: 271–274

Knapp W, Dörken B, Gilks W R et al (eds) 1989 Leucocyte typing IV. White cell differentiation antigens. Oxford University Press. Oxford

Kurtin P J, Pinkus G S 1985 Leukocyte common antigen — a diagnostic discriminant between hematopoietic and nonhematopoietic neoplasms in paraffin sections using monoclonal antibodies: Correlation with immunologic studies and ultrastructural localization. Human Pathology 16: 353–365

Lennert K, Feller A C, Radzun H J 1987 Malignant histiocytosis/histiocytic sarcoma and related neoplasms. Recent Advances in RES Research 24: 1–16

McMichael A J, Beverley P C L, Cobbold S et al 1987 Leucocyte typing III. white cell differentiation antigens. Oxford University Press, Oxford

Mason D Y 1985 Immunocytochemical labelling of monoclonal antibodies by the APAAP immunoalkaline phosphatase technique. In: Bullock G R, Petrusz P (eds) Techniques in immunocytochemistry. Academic Press, London, p 25–40

Mason D Y 1987 Editorial. A new look at lymphoma immunohistology. American Journal of Pathology 128: 1–4

Mason D Y, Biberfeld P 1980 Technical aspects of lymphoma immunohistology. Journal of Histochemistry and Cytochemistry 28: 731–745

Mason D Y, Krissansen G W, Davey F R, Crumpton M J, Gatter K C 1988 Antisera against epitopes resistant to denaturation on T3 (CD3) antigen can detect reactive and neoplastic T cells in paraffin embedded tissue biopsy specimens. Journal of Clinical Pathology 41: 121–127

Ngan B-Y, Chen-Levy Z, Weiss L M, Warnke R A, Cleary M L 1988 Expression in non-Hodgkin's lymphoma of the bcl-2 protein associated with the t(14;18) chromosomal translocation. New England Journal of Medicine 318: 1638–1644

Norton A J, Isaacson P G 1986 Immunocytochemical study of T-cell lymphomas using monoclonal and polyclonal antibodies effective in routinely fixed wax embedded tissues. Histopathology 10: 1243–1260

Norton A J, Isaacson P G 1987 Detailed phenotypic analysis of B-cell lymphoma using a panel of antibodies reactive in routinely fixed and wax-embedded tissue. American Journal of Pathology 128: 225–240

Norton A J, Isaacson P G 1988 Antigen preservation in infarcted lymphoid tissue. A novel approach to the infarcted lymph node using monoclonal antibodies effective in routinely processed tissues. American Journal of Surgical Pathology 12: 759–767

Norton A J, Isaacson P G 1989a Lymphoma phenotyping in formalin-fixed and paraffin wax-embedded tissues. I. Range of antibodies and staining patterns. Histopathology 14: 437–446

Norton A J, Isaacson P G 1989b Lymphoma phenotyping in formalin-fixed and paraffin wax-embedded tissues. II. Profiles of immunoreactivity in the various tumour types. Histopathology 14: 557–579

Norton A J, Rivas C, Isaacson P G 1989c A comparison

between monoclonal antibody MT2 and immunoglobulin staining in the differential diagnosis of follicular lymphoid proliferations in routinely fixed and wax embedded biopsies. American Journal of Pathology 134: 63–70

Norton A J, Ramsay A D, Smith S H, Beverley P C L, Isaacson P G 1986 Monoclonal antibody (UCHL1) that recognises normal and neoplastic T cells in routinely fixed tissues. Journal of Clinical Pathology 39: 399–405

Pallesen G, Knudsen L M 1985 Leucocyte antigens in post mortem tissues: their preservation and loss as demonstrated by monoclonal antibody immunohistological staining. Histopathology 9: 791–804

Picker L J, Weiss L M, Medeiros L J, Wood G S, Warnke R A 1987 Review article: immunophenotypic criteria for the diagnosis of non-Hodgkin's lymphoma. American Journal of Pathology 129: 434–440

Piris M A, Rivas C, Morente M, Rubio C, Martin C, Olivia H 1987 Persistent and generalized lymphadenopathy: a lesion of follicular dendritic cells? An immunohistologic and ultrastructural study. American Journal of Clinical Pathology 87: 716–724

Ralfkiaer E 1989 Reactivity of the workshop panel of CD7 antibodies in benign cutaneous conditions and known and suspected cutaneous T-cell lymphomas. Tissue Antigens. Histocompatibility and Immunogenetics 33: Abstract 110

Reinherz E L, Haynes B F, Nadler L M, Bernstein I D 1986 Leukocyte typing II. Springer-Verlag, New York

Sato Y, Mukai K, Watanabe S, Goto M, Shimosato Y 1986 The AMeX method. A simplified technique of tissue processing and paraffin embedding with improved preservation of antigens for immunostaining. American Journal of Pathology 125: 431–435

Schrape S, Jones D B, Wright D H 1987 A comparison of three methods for the determination of the growth fraction in non-Hodgkin's lymphomas. British Journal of Cancer 55: 283–286

Schuurman H-J, Van Baarlan J, Huppes W, Lam B W, Verdonck L F, Van Unnik J A M 1987 Immunophenotyping of non-Hodgkin's lymphoma. Lack of correlation between immunophenotype and cell morphology. American Journal of Pathology 129: 140–151

Smith J L, Jones D B, Bell A J, Wright D H 1989 Correlation between histology and immunophenotype in a series of 322 cases of non-Hodgkin's lymphoma. Hematological Oncology 7: 37–48

Stansfeld A G, Diebold J, Kapanci Y et al 1988 Updated Kiel classification for lymphomas. Lancet i: 292–293

Stein H, Gatter K, Asbahr H, Mason D Y 1985 Use of freeze-dried paraffin-embedded sections for immunohistologic staining with monoclonal antibodies. Laboratory Investigation 52: 676–683

Stein H, Lennert K, Feller A C, Mason D Y 1984 Immunohistological analysis of human lymphoma: correlation of histological and immunological categories. Advances in Cancer Research 42: 67–147

Stein H, Mason D Y, Gerdes J et al 1985 The expression of the Hodgkin's associated antigen Ki-1 in reactive and neoplastic lymphoid tissue. Evidence that Reed-Sternberg cells and histiocytic malignancies are derived from activated lymphoid cells. Blood 66: 848–858

Stein H, Hansmann M-L, Lennert K, Brandtzaeg P, Gatter K C, Mason D Y 1986 Reed-Sternberg cells and Hodgkin cells in lymphocyte-predominant Hodgkin's disease of nodular subtype contain J chain. American Journal of Pathology 86: 292–297

Thomas M L, Lefrançois L 1988 Differential expression of the leucocyte-common antigen family. Immunology Today 9: 320–326

Ward J M, O'Leary T J, Baskin G B et al 1987 Immunohistochemical localization of human and simian immunodeficiency viral antigens in fixed tissue sections. American Journal of Pathology 127: 199–205

Warnke R A, Rouse R V 1985 Limitations encountered in the application of tissue section immunodiagnosis to the study of lymphomas and related disorders. Human Pathology 16: 326–331

Weiss L M, Trela M J, Cleary M L, Turner R R, Warnke R A 1985 Frequent immunoglobulin and T-cell receptor gene rearrangements in 'histiocytic' neoplasms. American Journal of Pathology 121: 369–373

Wotherspoon A C, Norton A J, Isaacson P G 1989 Immunoreactive cytokeratins in plasmacytomas. Histopathology 14: 141–150

20. Blood, bone marrow, spleen and lymph nodes in metabolic disorders

B. D. Lake

PERIPHERAL BLOOD CELLS IN STORAGE DISORDERS

Many of the metabolic disorders express their enzyme defect in circulating blood cells and in some there is also morphological and histochemical evidence on which a diagnosis can be made (Hansen 1972, Lake 1989). In some of these disorders the evidence in stained blood films can only point in a particular direction but in others there can be sufficient information to make a definitive diagnosis. Correlation of the findings with clinical information is, as always, most important. The simple screening by examination of stained blood films can be of help in the exclusion of certain conditions and can save considerable time and expensive biochemical reagents. It has the added advantage that the techniques are available in the majority of laboratories in contrast with the specialized biochemical methods available only at a handful of laboratories.

In the majority of storage disorders the vacuolated lymphocytes occur with greater frequency in the tail of the film. Consequently, most haematologists will not find the cells in their routine examination, preferring the parts of the film more evenly spread. Thus, it is important carefully to examine not only the 'best' part of the film, but to pay particular attention to the tail when a storage disease is suspected.

Lymphocytes

To most people the presence or vacuolated lymphocytes is a relatively non-specific phenomenon; but when the size and distribution of the vacuoles and their staining characteristics are taken into account, together with the clinical presentation, a reasonably accurate diagnosis can be made. Table 20.1 lists a number of conditions and the features found in stained blood films. In the first instance, a routinely stained film (Giemsa, Wright, Leishman, etc.) prepared from a fingerprick sample or from anticoagulated blood should be examined. Further examination by the special methods indicated in Table 20.1 can be made on the spare films prepared at the initial sampling.

Some drugs, particularly those of an amphiphilic nature, are known to induce vacuolation of lymphocytes (Lullman et al 1973) and this should be borne in mind when examining blood films for evidence of a storage disorder.

In the mucopolysaccharidoses the inconstant Reilly bodies cause much confusion. Their presence is variable and their identity is not certain since in the original description it was not clear whether they occurred in circulating blood cells or in bone marrow. Thus the presence of absence of Reilly bodies in neutrophils is unhelpful in the differential diagnosis of the mucopolysaccharidoses. In contrast, Alder granules occur specifically and constantly in Maroteaux-Lamy disease (MPS 6), mucosulphatidosis and β-glucuronidase deficiency.

Table 20.1 is not an exhaustive list but includes conditions in which ancillary tests can be helpful. Other disorders in which large, well-defined lymphocytic vacuoles are found include mucolipidosis I (identical with sialidosis II), and Salla disease. Vacuolated lymphocytes are also mentioned variably in fucosidosis, the cherry red spot — myoclonus syndrome (sialidosis I) and in some patients with Niemann-Pick disease type C. In none of these is there any specific staining

Table 20.1 Histochemical detection of storage disorder in blood cells

Disorder	Lymphocytes (Giemsa stain)	Staining reaction of lymphocytic vacuoles	Histochemical enzyme method	Neutrophils	Other comments
Wolman's disease (including cholesteryl ester storage disease)	Small discrete vacuoles 1–6 per cell in most lymphocytes	ORO:SBB	Acid esterase negative (Lake 1971)	Usually not involved; some patients show marked lipid accumulation in neutrophils	Platelets also involved; foam cells in blood are exceedingly rare
Pompe's disease (and other variants of GSD II)	Small discrete vacuoles 1–6 per cell in majority of lymphocytes	Cell PAS + ve Dig. cell PAS −ve	Method for α-glucosidase not suitable or of sufficient sensivity	Not involved	Fewer lymphocytes are involved in the adult form
Niemann-Pick type A	Small discrete vacuoles 1–4 per cell in most lymphocytes	Contents not readily shown	No method available for sphingomyelinase	Not involved	No lymphocyte vacuoles are present in Niemann-Pick disease type B
Mannosidosis	Numerous small discrete vacuoles in some lymphocytes. Others contain several large, well-defined vacuoles	PAS cell, not digested. Not easily shown because of extreme solubility of mannosides	Method for α-mannosidase not selective enough or of sufficient sensitivity	Not involved	No metachromasia
I-cell disease (mucolipidosis II)	Numerous large bold vacuoles	Vacuoles appear empty	Defect not amenable to histochemical methods	Not involved	No metachromasia; Eosinophils have normal granules
G$_{MI}$-gangliosidosis type 1	Numerous large bold vacuoles in most lymphocytes	Vacuoles appear empty	β-galactosidase activity not detectable (Lake 1974)	No storage; β-galactosidase activity not detectable	Eosinophil granules are large, grey and sparse. Some residual or different β-galactosidase may be seen in eosinophils. No metachromasia. Platelets also β-galactosidase deficient
G$_{MI}$-gangliosidosis type 2	No vacuoles	—	ditto	ditto	
Juvenile Batten's disease	Several large bold vacuoles in up to 30% of lymphocytes in the tail of the film	Vacuoles appear empty	Defect not known	Not affected	EM may disclose some rare fingerprint bodies within the numerous vacuoles
Aspartyl-glucosaminuria	Several large bold vacuoles in some lymphocytes; not frequent. Reddish inclusions may be present in the vacuoles	Vacuoles appear empty	Defect not amenable to histochemical methods	Not affected	Lymphocyte vacuoles may be difficult to find

Table 20.1 (Cont'd)

Disorder	Lymphocytes (Giemsa stain)	Staining reaction of lymphocytic vacuoles	Histochemical enzyme method	Neutrophils	Other comments
Mucopoly-saccharidosis (MPS)					
MPS IH (Hurler)	Occasional vacuoles	Metachromatic inclusions present generally in <5% of lymphocytes	None available		Gasser cells may be seen (vacuolated lymphocytes with basophilic inclusions). These can be present in other MPS
MPS IS (Scheie)	Occasional vacuoles	Metachromatic inclusions in <5% of lymphocytes	None available		
MPS IH/S (Hurler-Scheie compound)	5–10% of cells contain vacuoles	Metachromatic inclusions in <5% of lymphocytes	None available		
MPS 2 (Hunter)	Occasional vacuoles	Metachromatic inclusions generally in <20% of lymphocytes	None available		
MPS 3 (Sanfilippo) types ABC & D	Occasional vacuoles	Metachromatic inclusions generally in >20% of lymphocytes	None available		Type A most common in UK. Type C most common in Scandinavia
MPS 4 Morquio type A	Not vacuolated	No metachromasia detectable	None available	Occasional, rare basophilic inclusions	
type B	Vacuoles present in significant number of lymphocytes	No metachromasia detectable			β-galactosidase activity is deficient
MPS 6 (Maroteaux-Lamy)	Vacuoles present in significant numbers of lymphocytes	Metachromatic inclusions present in lymphocytes	None available	Alder granulation present in all neutrophils. The granulation is metachromatic and shows birefringence (Haust & Landing method 1961) in contrast to toxic granulation	
MPS 7 β-glucuronidase deficiency	Occasional vacuoles		β-glucuronidase activity in lymphocytes and neutrophils is deficient	Alder granulation present, sometimes very coarse	

Table 20.1   (Cont'd)

Disorder	Lymphocytes (Giemsa stain)	Staining reaction of lymphocytic vacuoles	Histochemical enzyme method	Neutrophils	Other comments
Mucosulphatidosis			Sulphatases A, B and C. Not tested histochemically	Alder granulation present	Eosinophils few and may have abnormal granulation
Sialic acid storage disease	Numerous small vacuoles in many lymphocytes	Vacuoles appear empty	Defect not known	Not affected	Eosinophil granules large, grey and sparse

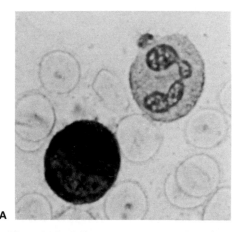
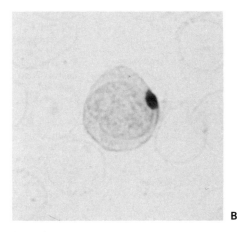

Fig. 20.1   Normal blood film stained to show acid esterase activity. No E600 has been added. In **A** a neutrophil is unreactive, while a monocyte shows strong 'non-specific' esterase activity serving as a control for the method. In **B** the normal lymphocyte (T type) shows a single strong spot of activity, not present in Wolman's disease. ×1800

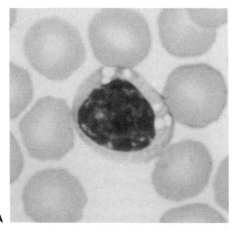
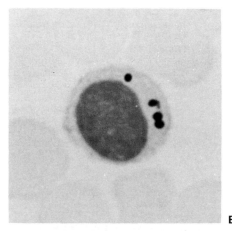

Fig. 20.2   Pompe's disease (glycogen storage disease type 2); lymphocytes in a blood film stained on the left with a routine haematological method (May-Grunwald-Giemsa) and on the right with the protected PAS method. The cytoplasmic vacuoles seen in the routine film are full of glycogen. ×2100

reactions of the contents of the vacuoles and suitable enzyme staining methods are not available.

Ultrastructural abnormalities of lymphocytes can be detected in infantile Batten's disease (deposits of GROD), late infantile Batten's disease (curvilinear bodies) (Lake 1981) and mucolipidosis IV (membranous cytoplasmic bodies) (Lake et al 1982) even though no abnormality can be detected by light microscopy.

Neutrophils

Abnormalities of the neutrophil are less common in metabolic disorders but occasionally vacuolated neutrophils are found in 'carnitine deficiency'. The contents of the vacoules stain with oil red O. Lipid inclusions in neutrophils may also be present in some cases of Wolman's disease. Care should be taken to ensure that the acute toxic changes producing vacuolation of neutrophils are not mistaken for either of the above disorders.

Myeloperoxidase deficiency is readily detected by any of the methods for demonstration of peroxidase activity. Although neutrophils normally contain much peroxidase activity they are negative in myeloperoxidase deficiency while eosinophils maintain their strong reaction.

In a screening programme (Kitahara et al 1981), one patient with a total deficiency was readily detected. However, a large number of partially deficient patients were also found. Their partial deficiency was not clearly defined by staining methods but was apparent in quantitative assays. Low myeloperoxidase activity may also be found in chronic myeloid leukaemia.

BONE MARROW IN STORAGE DISORDERS

In storage disorders affecting the reticuloendothelial (mononuclear phagocyte, MP) system, examination of bone marrow films is an important diagnostic procedure. The differential diagnosis in an infant, young child or adult with hepatosplenomegaly will include storage disorders, infection, portal hypertension or malignancy. Study of bone marrow films will exclude storage conditions if an adequate sample of marrow contains no storage cells. Films should be examined in preference to routine sections of wax-embedded trephine or clotted samples of bone marrow, because processing will extract the characteristic storage material and thus prevent an accurate diagnosis.

Although morphology of the affected histiocytes is best assessed in films stained by Giemsa or similar routine haematological methods, the exact nature of the storage cell can only be determined by further staining methods. It should be remembered that the presence of 'foam' cells in bone marrow is not necessarily indicative of a storage disorder since they may be acquired, as in the hyperlipidaemias and chronic myeloid leukaemia, or drug-induced.

In very general terms there are two types of 'foam' cell which may be found in bone marrow samples. The first is the Gaucher type of cell with its fibrillar appearance and the second is the Niemann-Pick type of cell which has foamy rather than fibrillar cytoplasm and is the type found in most storage disorders.

The storage cells are always present and may be seen in the marrow of fetuses affected with visceral storage disorders. It is the site of choice for early confirmation of the diagnosis in fetal material from termination of pregnancies shown to be affected on biochemical assay of chorionic villus samples or cultured amniotic fluid cells.

Gaucher's disease

The Gaucher cell is usually described as having cytoplasm 'like wrinkled tissue paper', and illustrations show this feature. However, in bone marrow films and tissue sections the characteristic appearance is not seen in every Gaucher cell, and an extended search may be necessary before the 'typical' Gaucher cell is found. The majority of the cells show features ranging from the typical form to almost vacuolar. The glucocerebroside stored within the Gaucher cell gives a mild PAS positivity and stains only pale grey with Sudan black. The fibrillar nature of the cell is best shown in marrow films which have been fixed before staining with Sudan black. The cells show strong acid phosphatase activity with β-glycerophosphate as substrate, and an acid phosphatase reaction will help to decide whether storage cells (or histiocytes) are present in a marrow film by examination with a low power objective (× 10) (Fig. 20.3).

Megakaryocytes show some acid phosphatase activity but this is often localized in one area of the cell and is finely punctate whereas histiocytes and storage cells show strong diffuse activity throughout the cell. Pseudo-Gaucher cells may occur in chronic myeloid leukacmia, thalassemia or in any condition where there is an overload on the MP system due to excessive turnover of blood cells. The numbers of pseudo-Gaucher cells are small and it should be possible to distinguish these 'acquired' storage conditions from Gaucher's disease by the range of morphology with some histiocytes still containing ingested WBCs.

All forms of Gaucher's disease have similar cells in the bone marrow and it is not possible to differentiate infantile from adult on morphology or staining characteristics. The histochemical method for the detection of β-glucosidase activity is not reliable for the diagnosis of Gaucher's disease because, of the several isoenzymes of β-glucosidase, only one is deficient, and with the limitations of the capture reaction, differentiation of pH optima is not possible. It should also be noted that patients with the adult form of Gaucher's disease have a greater incidence of malignancy with myeloma, myeloid leukaemia and lymphoma occurring in 10% of those who died in one series of 200 (Lee 1981).

Niemann-Pick disease types A and B

In the infantile form of Niemann-Pick disease (type A) the marrow cells are foamy with the vacuoles being reasonably uniform in each cell. Only rarely are there any densely-staining inclusions (nuclear debris) or ingested RBCs. Sphingomyelin is stored together with cholesterol and a variable amount of ganglioside, and this combination gives rise to a rather pale blue staining with Sudan black which in most instances shows red birefringence in polarized light, due to sphingomyelin. The ferric haematoxylin method for sphingomyelin (see Appendix 4) is reliable and effective on marrow films, giving a dense blue colour in sphingomyelin-containing cells. Stains for cholesterol are also positive (Schultz type or PAN). The variable ganglioside content results in

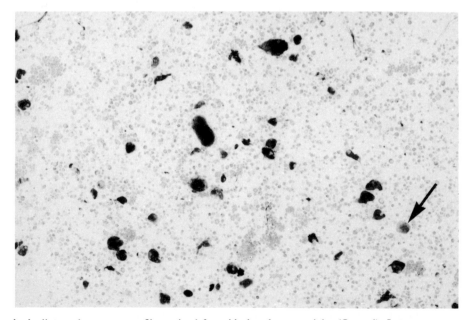

Fig. 20.3 Gaucher's disease; bone marrow film stained for acid phosphatase activity (Gomori). Low-power survey shows numerous storage cells with a variety of shapes and sizes. A megakaryocyte is present (arrow) with only low activity. A light nuclear counterstain has been added. ×90

a variable PAS reaction. The foamy nature of the Niemann-Pick cells is readily appreciated in an acid phosphatase reaction when the vacuoles are outlined by the reaction product in contrast with the punctuate reaction product localized between the lipid droplets in the acquired storage cells of the hyperlipidaemias.

The juvenile and adult forms of the disease can present from 18 months of age onwards (past the age at which neurological involvement is grossly evident in the infantile form) with hepatosplenomegaly, abnormal pulmonary function and coagulation defects. In the older patients the most striking storage cell in the marrow is the 'sea-blue histiocyte' which is profuse in number and whose cytoplasm is filled with blue stained granules of lipofuscin in routine haematological stains. The granules are autofluorescent, PAS-positive, stain grey-black with Sudan black and show acid phosphatase positivity. In a marrow with cells which match these criteria the most likely diagnosis is adult Niemann-Pick disease. Fewer cells, and those with sparse blue granules, represent the normally functioning histiocyte, perhaps in response to a variety of other conditions (Long et al 1977). In adult Niemann-Pick disease a few foamy storage cells may also be found.

Niemann-Pick disease type C

Crocker & Farber (1958) in their series of patients with Niemann-Pick disease included a number of patients from Nova Scotia and some with what has been called the subacute type of Niemann-Pick disease, later known as Niemann-Pick disease type D and type C respectively. The patients in these two subgroups do not show deficiency of sphingomyelinase, neither do they show accumulation of sphingomyelin in the liver or brain, both of which are grossly affected by storage. The clinical presentations vary widely from neonatal hepatitis to unexplained splenomegaly in an otherwise well child of up to 7 years. All develop ataxia, dementia and a vertical supranuclear opthalmoplegia later in the disease and it is this latter phenomenon which led to the term 'ophthalmoplegic lipidosis'. Many other terms have been used, causing some confusion (Lake 1984). At a symposium in Prague 1982, it was decided to retain the term Niemann-Pick disease group C, until the chemistry is clarified. The pathology is the same in all, the ultrastructure is distinct and different from other storage conditions (Neville et al 1973). A defect in cholesterol esterification has been identified (Vanier et al 1988).

Marrow films show an abundance of large foamy storage cells which may be multinucleate. There is variation in the sizes of the cytoplasmic vacuoles within the cell and it is common to find one or more small, densely-staining inclusions (probably erythroid debris) in the routine staining method especially in the older child. The cells do not stain with Sudan black but show PAS-positivity, the colour of which often seems to be a little redder than the colour of the other marrow cells. The acid phosphatase reaction is strong. In the older patients occasional sea-blue histiocytes may also be found.

Even in the absence of neurological deterioration, neuronal storage can be demonstrated in rectal biopsies several years before the onset of dementia and is present even at 34 weeks' gestation (Adam et al 1988).

It should be noted that the typical foam cells of Niemann-Pick disease type C (including scattered 'sea-blue histiocytes') may be present in marrow samples from heterozygotes, and in testing younger, apparently clinically unaffected, siblings a rectal suction biopsy should be examined to determine the presence (or absence) of neuronal storage (see Ch. 9) before the gloomy prognosis can be given.

G_{M1}-gangliosidosis

The storage cells present in G_{M1}-gangliosidosis are different in the two known types. In type 1 — presenting early and similarly to patients with Hurler's disease (MPS I) — the numerous storage cells are large and foamy. No inclusions are present and it is not possible to demonstrate any storage material within the vacuoles. The cells do, however, show acid phosphatase activity. In type 2 — presenting in the late infantile/juvenile age range but without features of a mucopolysaccharidosis and usually with hepatosplenomegaly — the storage cells are few in number. In the routine Giemsa stain the cells have a 'sky-blue' colour and resemble, to a limited extent, Gaucher cells in their morphology. They are PAS-positive, greyish with Sudan black and are positive for acid phosphatase activity. The activity of β-galactosidase is deficient in both types 1 and 2 and this deficiency is readily shown by the indoxyl method (Appendix 5).

Wolman's disease and cholesteryl ester storage disease

Both of these disorders are caused by a deficiency of acid esterase activity and should be regarded as the infantile and juvenile/adult forms of the same disease. The foamy storage cells present in the marrow contain 'neutral fat' in the vacuoles and stain strongly with oil red O or Sudan black (Fig. 20.4). The neutral fat is a mixture of triglyceride and cholesteryl esters, and stains to demonstrate cholesterol (free or as ester) are also positive. Cain's Nile blue method imparts a deep blue/purple colour to these cells due to their content of free fatty-acids — though why free fatty acids originate in macrophages deficient in acid-esterase activity is not clear. No other storage disorder has cells with this characteristic staining reaction with Nile blue. The acid phosphatase reaction is positive.

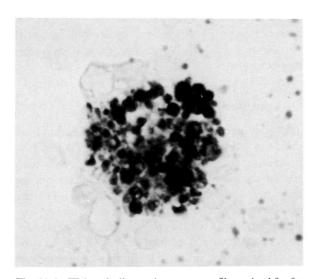

Fig. 20.4 Wolman's disease; bone marrow film stained for fat with oil red O. A large foamy cell is filled with neutral fat and stains strongly. A very similar appearance is found with Nile blue sulphate. ×1200

Although acid esterase activity is easily demonstrated in blood films without the addition of the non-specific esterase inhibitor E600, its addition is necessary in marrow films because of the widespread distribution of many fatty acid ester hydrolases which would otherwise interfere. In the infantile and juvenile presentations the cells are numerous but in those presenting in the adult age range the storage cells are less common and not readily recognisable in the routine Giemsa stain. The cells may be large and angular and contain neutral fat, the fat vacuoles being diffusely scattered through the cytoplasm. The appearance may be difficult to differentiate from acquired neutral fat storage seen in the hyperlipidaemias.

Cystinosis

The large macrophages containing the cystine crystals are quite fragile and in taking the aspirate and in making the films the cells can become ruptured leaving scattered cystine crystals over the slide. Thus the films should be made with great care. Although cystine is relatively insoluble, aqueous reagents should be avoided since the large volumes used relative to the small concentration of cystine will dissolve some or all of the crystals leaving at best a distorted crystal shape. In routine Giemsa-stained films scattered vacuolated macrophages may be found containing the occasional rounded crystal. The cystine crystals are adequately preserved in films fixed in ethanol, and stained for 5 minutes in 1% basic fuchsin in 70% ethanol before dehydration, clearing and mounting. Examination with polarized light will show the characteristic rectangular-shaped birefringent crystals. The hexagonal form (end on) shows no birefringence. The concentration of cystine in individual cells is insufficient for chemical confirmation and the Woollaston test (see Fig. 20.7) will be negative. A more sensitive but impermanent technique for detection of cystine in marrow aspirates is to place one drop of the aspirate on a slide, add a cover slip and view in polarized light. Many more macrophages can be seen with this technique. It should be borne in mind that dirt, dust and glass chips are also birefringent and the crystal habit should be carefully considered before making the diagnosis.

Mucopolysaccharidoses

Diagnosis of the mucopolysaccharidoses by examination of marrow aspirates is not a usual exercise since the diagnosis is more easily and accurately made on blood and urine samples. However, in some cases, particularly in the Sanfilippo types (MPS 3) where the patients do not always have the characteristic physical features, a bone marrow aspirate may be taken in the investigation of hepatosplenomegaly. In routine Giemsa-stained preparations overt storage cells are not seen in MPS 3, but around the marrow fragments occasional histiocytes with scattered fine basophilic granules can be found. The metachromatic character of these granules is best shown with the Haust and Landing method (Appendix 3) after fixation in tetrahydrofuran-acetone, which will show more histiocytes containing acid mucopolysaccharide than otherwise suspected. In Maroteaux-Lamy syndrome (MPS 6) all granulocytes show Alder granulation which is strongly metachromatic after staining with Haust and Landing's method. In addition to the Alder granulation, occasional large intensely basophilic and metachromatic cells can be found. Too few studies on the variation between the marrow changes in the various types of mucopolysaccharidoses have been made for diagnostic points to have emerged but it would be expected that all types would show histiocytes containing an acid mucopolysaccharide, and this is best shown with Haust and Landing's method (1961). The deficiency of β-glucuronidase in MPS 7 can be shown using naphthol AS-BI-β-glucuronide as substrate (Peterson et al 1981). Histochemical methods are not available for the detection of the defects in the other mucopolysaccharidoses, apart from the β-galactosidase deficiency in Morquio type B (MPS 4B).

Sialidosis I (cherry red-spot myoclonus syndrome)

These patients show no hepatosplenomegaly, but storage cells are found in the bone marrow, albeit in small numbers. In Giemsa preparations the foamy storage cells have large vacuoles and the contents stain a blue-grey colour. The stored substance is strongly PAS-positive and negative with Sudan black. The periphery of the vacuoles is strongly positive for acid phosphatase activity. The cells are probably derived from plasma cells since the stored material (oligosaccharides with sialyl terminals) is very similar to those bodies erroneously attributed to Russell (Lendrum 1981).

Mannosidosis

The bone marrow in patients with mannosidosis contains numerous large foamy histiocytes with particularly well-defined vacuoles. Plasma cells also show marked, very well-defined vacuoles, the vacuoles being separated from each other by a rim of cytoplasm. The stored material is extremely water-soluble but with adequate protective measures (e.g. celloidinization) it should be PAS-positive. The patients present similarly to those with a mucopolysaccharidosis but a negative reaction with the toluidine blue method of Haust and Landing will distinguish mannosidosis from mucopolysaccharidosis. The histochemical method for demonstration of α mannosidase activity is not sufficiently sensitive or of sufficient selectivity to distinguish between the various isoenzymes of α-mannosidase.

The sea-blue histiocyte syndrome

Considerable confusion has been caused by the use of this term since there is no single cause for the occurrence of histiocytes containing deep-blue granules of lipofuscin. Profusion of these cells with their cytoplasm packed with granules suggests adult Niemann-Pick disease, while lesser numbers of this type of cell are found in lecithin-cholesterol acyltransferase (LCAT) deficiency. Smaller numbers, with fewer granules or diffusely blue granules, may be seen in thalassaemia, Fabry's disease, juvenile Batten's disease, chronic myeloid leukaemia or may represent busy histiocytes responding to an increased cell turnover. A careful search in most marrow samples will reveal an occasional histiocyte with some blue cytoplasmic granules, and these are of no real significance. Since there is no specificity, the term sea-blue histiocyte should not be used.

Acquired storage cells

In many situations where there is an excess of a particular substance, bone marrow aspirates will show loaded histiocytes. All forms of hyperlipidaemia, and Tangier disease, show large lipid-laden histiocytes staining positively with oil red O or Sudan black. Foamy histiocytes may be induced in cytotoxic therapy or in treatment with drugs designed to decrease lipid biosynthesis.

Foamy cells showing haemophagocytosis are a feature of familial lymphohistiocytosis and of the closely related viral associated haemophagocytic syndrome (see Landing 1987 for a review), although the number of cells in the marrow may be small. In acquired immunodeficiency states, opportunistic infections are common and sometimes large foam cells with granular cytoplasm (containing atypical mycobacteria) may be seen.

Langerhans cell histiocytosis (formerly histiocytosis X)

Large foamy histiocytes which contain no demonstrable substance in the vacoules may occur late in Langerhans cell histiocytosis and at that time they may be abundant and give the impression of a storage disease. The strong acid phosphatase reaction does not delineate the vacuoles but has a punctate localization in the cytoplasm which helps to distinguish these cells from those of the storage disorders (see also p. 307. More typical in the marrow is the presence of the rounded histiocytes, characteristically found in the skin lesions. These abnormal cells cannot be detected reliably in marrow films by routine Giemsa staining, but show very strong discrete patches of α-mannosidase activity, and can be readily distinguished from the scattered normal marrow cells displaying activity which can best be described as a pale blush.

SPLEEN AND LYMPH NODES IN STORAGE DISORDERS

Involvement of cells of the MP system results in evidence of the storage disorder in lymph nodes, spleen and other lymphoreticular structures. The morphological and staining characteristics of the storage cell are essentially the same whether they occur in the bone marrow, liver, spleen or lymph nodes. In some disorders, however (e.g. *mannosidosis, mucopolysaccharidosis type 4*), the evidence of storage may be minimal and missed in routine sections. If tonsils and adenoids are removed from patients with *mucopolysaccharidosis type 4 (Morquio)* not only is routine histology unhelpful but cryostat sections also do not show convincing mucopolysaccharide deposition. Thin (1 μm) sections of resin-embedded tissue may reveal a few foamy cells but electron microscopy is necessary for the detection of the storage cells and the membrane-bound empty vacuoles they contain. In *Hunter's disease (MPS type 2)* the adenoids show numerous foamy storage cells in the subepithelial areas and evidence of endothelial cell involvement is present throughout the tissue (Fig. 20.5).

In *Gaucher's disease* there is massive infiltration of storage cells in the spleen, which can weigh up to 3 kg in a child of under 10 years. The Gaucher cells, arising from the inability to degrade glucocerebroside derived from effete red and white blood cells, distort the architecture and the Malpighian follicles are not usually seen on the cut surface. The Gaucher cells are arranged characteristically in 'nests', the centres of which often appear empty and necrotic (Fig. 20.6).

Sphingomyelin and cholesterol accumulate in foamy storage cells in the spleen in a variety of conditions ranging non-specifically from *idiopathic thrombocytopenic purpura* (ITP) (Hill et al 1973) to specific deposition in *Niemann-Pick disease*. Some patients with *Niemann-Pick disease type C* also accumulate sphingomyelin and cholesterol in numerous foamy cells in the spleen but although storage cells are present in the liver, the sphingomyelin content of that organ is not increased (Neville et al 1973). Thus spleen should not be the only tissue examined in the diagnosis of a visceral storage disorder.

In ITP the lipid-laden foamy cells are scattered and should not give rise to confusion with a genuine storage condition.

The spleen in *Niemann-Pick disease* (type A & B) is filled with foamy cells and there is loss of

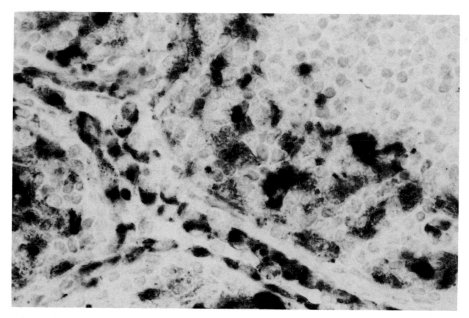

Fig. 20.5   Hunter's syndrome (MPS type 2); cryostat section of adenoid stained to show acid phosphatase activity (Gomori). Endothelial cells, normally negative, are positive and numerous large foamy cells are readily identified. A light nuclear counterstain has been added.   ×600

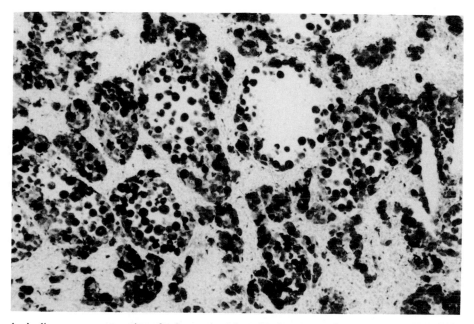

Fig. 20.6   Gaucher's disease; cryostat section of spleen stained for acid phosphatase activity (Gomori). Nests of strongly positive Gaucher cells are present, some of which have necrotic centres. A light nuclear counterstain has been added.   ×90

normal architecture. In contrast with conditions simulating Niemann-Pick disease, usually only those with sphingomyelinase deficiency show additional storage in the sinusoidal lining cells. This deposition which can be shown in frozen sections by Sudan black and by its autofluorescence has very similar properties to that of the 'sea-blue histiocytes' found in the marrow and spleen of patients with adult Niemann-Pick disease. The Niemann-Pick cells filling the spleen, lymph nodes, thymus and liver stain with Sudan black (usually giving red birefringence in polarized light) and with ferric haematoxylin after alkaline hydrolysis, and show variable PAS-positivity associated with a variable ganglioside content. The associated cholesterol can be demonstrated with the Schultz or PAN methods.

Although the storage of sphingomyelin and cholesterol in foam cells in the spleen of some patients with Niemann-Pick disease type C resembles that of types A & B the lack of involvement of the sinusoidal lining cells may serve to differentiate the two conditions. Those patients with type C who do not accumulate sphingomyelin in the spleen show massive infiltration of foamy cells which in frozen sections do not stain with Sudan black. The storage cells are PAS-positive (particularly in protected sections) and show rose-pink metachromasia with Feyrter's thionin. Their acid phosphatase activity is strong.

The precautions to prevent cystine from being removed from spleen and lymph nodes in *cystinosis* are the same as those described for liver and bone marrow, and cryostat sections are essential for the diagnosis and demonstration of cystine (Patrick & Lake 1968). In both spleen and lymph node the concentration of cystine is high enough for the Woollaston test to be positive (Fig. 20.7). The change of crystal shape in converting the free amino acid to the hydrochloride is characteristic of cystine.

OTHER CONDITIONS

Reactive lymph nodes and those involved in lymphoma and other histiocytic disorders may be difficult to interpret in routine sections and thin (1μm) sections of resin-embedded tissue may be preferred. However, cryostat or routine sections are helpful in the determination of the nature of the infiltrating cells by application of appropriate markers (see Ch. 19).

Langerhans cell histiocytosis (Histiocytosis X, eosinophilic granuloma, Letterer-Siwe, Hand-Schüller-Christian disease)

Langerhans cell histiocytosis is considered here (also in Ch. 11) because of the bone marrow involvement and because it has been variously

A   B

Fig. 20.7 Cystinosis; cryostat section of spleen. **A** The section has been stained with alcoholic basic fuchsin and photographed in partially polarized light. The characteristic crystal shapes are readily identified. ×825

A serial section (**B**) has been treated with concentrated hydrochloric acid by allowing the acid to be drawn, by capillary action under a coverslip placed over the air-dried section, and photographed in fully polarized light. The change of crystal shape from rectangular and hexagonal to fan-shaped needles is specific to cystine and constitutes the Woollaston test. ×170

described as a storage disorder or as a tumour. It is now known that the condition involves the antigen-presenting Langerhans cell, but the pathogenesis remains obscure. The lesions in this spectrum of disorders have been classified as 'benign' or 'malignant' (Bokkerink & deVaan 1980) depending on the degree of infiltration and types of cell present. However, more recent reviews (Jaffe 1987) show that this is not a valid basis for predicting prognosis.

In the skin the infiltrating histiocytes appear in the upper dermis. Among the rounded histiocytes occasional irregular-shaped stellate cells may be found. In lymph nodes both rounded and stellate cells are present. These cells and the multinucleate giant cells have slightly different staining characteristics which can be determined by the use of cryostat or routine sections (Elleder 1982, Thomas et al 1982, Jaffe 1987) (Fig. 20.8). The staining properties are summarized in Table 20.2. The

Table 20.2   Staining reactions of the histiocytes in Langerhans cell histiocytosis in comparison with Langerhans cells

Cell	Reaction: Acid phosphatase	Non-specific esterase	α-mannosidase	ATPase	Naphthyl-amidase	Ia-like antigen	HTA-1 antigen (OKT 6)	Anti-S100 protein	Peanut agglutinin
Large rounded histiocytes	+	+	+++	+/++	+	+	+	+	++
Irregular stellate histiocytes	++	++	−	+	+	+	+		
Giant multinucleate cells	+/+++	+++	−,+		±	±	−		+
Langerhans cells	±	±	+++	++	+	+	+	+	−,±

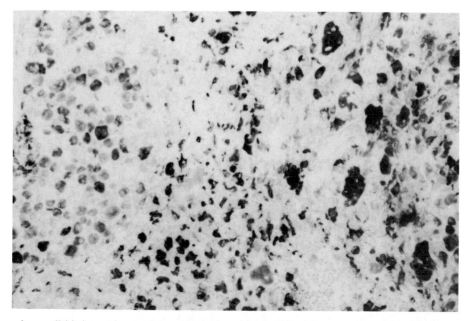

Fig. 20.8   Langerhans cell histiocytosis; cryostat section of lymph node stained for 'non-specific' esterase activity. Groups of rounded histiocytes (left), angular stellate histiocytes (centre) and large multinucleate histiocytes (right), with varying intensities of reaction are evident. ×140

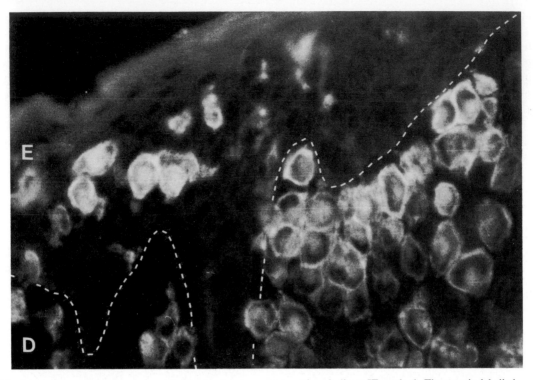

Fig. 20.9   Langerhans cell histiocytosis; skin biopsy, 6 μm cryostat section, indirect IF method. Fluorescein-labelled monoclonal antibody to human thymocyte antigen (HTA-1) reacts with large round or ovoid cells infiltrating the dermis (D) and epidermis (E). In normal skin, epidermal Langerhans cells have the HTA-1$^+$, HLA-DR$^+$ phenotype but morphologically similar cells in the normal dermis are HTA-1$^-$. (Photography by courtesy of Dr J. A. Thomas and Dr G. Janossy)

large rounded histiocytes have properties and ultrastructural features (Birbeck granules) which relate them to the Langerhans cell normally found in the epidermis. α-mannosidase activity, demonstrated by the semipermeable membrane technique (Elleder et al 1977), is present in both Langerhans cells of the epidermis and in the large rounded histiocytes, and this seems an effective marker together with the antibody OKT 6 for confirmation of the diagnosis in cryostat sections (Figs. 20.9 and 20.10). Peanut agglutinin and anti-S100 protein are useful markers in routine sections. Peanut agglutinin detects the abnormal cells in the dermal infiltrate in which dense surface and discrete paranuclear staining can be found (see Ch. 11). Langerhans cells in the epidermis are only poorly recognized. Histiocytes in general only display weak cytoplasmic staining with peanut agglutinin. The giant cells contain only residual

amounts of demonstrable lysozyme which appears to be derived from ingested granulocytes or histiocytes.

The large lipid-laden histiocytes sometimes seen contain cholesteryl esters, and will stain positively with oil red O and with the Schultz or PAN methods for cholesterol.

In the congenital self-healing form of Langerhans cell histiocytosis (Hashimoto and Pritzker type) the abnormal cells tend to involve the deeper dermis and are present in defined nodules. Their staining characteristics are indistinguishable from those of classical Langerhans cell histiocytosis. Juvenile xanthogranuloma occurs in patients with Langerhans cell histiocytosis, and any lesion of juvenile xanthogranuloma should be carefully studied using the markers outlined above to detect the presence of Langerhans cell histiocytosis. The changes, however, may be patchy.

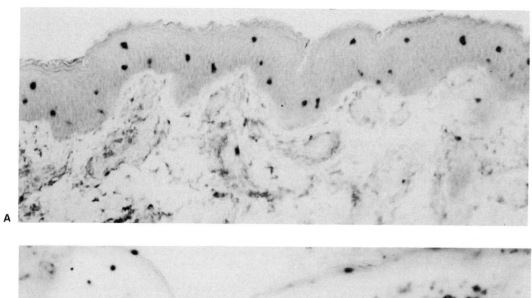

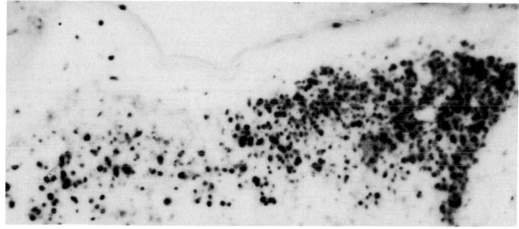

Fig. 20.10 A Skin biopsy. Cryostat section stained for α-mannosidase activity using the semipermeable membrane technique. Langerhans cells in the epidermis are strongly stained. Mast cells, around the vessels in the upper dermis show weaker diffuse activity in this case of mastocytosis. A light nuclear counterstain has been added. **B** Langerhans cell histiocytosis. Cryostat sections of a skin biopsy stained for α-mannosidase activity using the semipermeable membrane technique. The mass of infiltrating histiocytes just under the epidermis show very strong activity. Langerhans cells are markedly reduced in number in this lesion. ×180

REFERENCES

Adam G, Brereton R J, Agrawal M, Lake B D 1988 Biliary atresia and meconium ileus associated with Niemann-Pick disease. Journal of Pediatric Gastroenterology and Nutrition 7: 128–131

Bokkerink J P M, deVaan G A M 1980 Histocytosis X. European Journal of Pediatrics 135: 129–146

Crocker A C, Farber S 1958 Niemann-Pick disease: a review of 18 patients. Medicine 37: 1–95

Elleder M, Povysil C, Rozkovcová J, Cihula J 1977 Alpha-D-mannosidase activity in histiocytosis X. Virchows Archiv B. Cell Path 26: 139

Elleder M 1982 Enzyme histochemistry in histiocytosis X. Proceedings of the Royal Microscopical Society 17: 132

Hansen H G 1972 Hematologic studies in mucopolysaccharidoses and mucolipidoses. Birth Defects 8: 15–127

Haust M D, Landing B H 1961 Histochemical studies in Hurler's disease. A new method for localization of acid mucopolysaccharide, and an analysis of lead acetate 'fixation'. Journal of Histochemistry and Cytochemistry 9: 79–86

Hill J M, Speer R J, Gedikoglu A 1963 Secondary lipidosis of the spleen associated with thrombocytopenia and other blood dyscrasias treated with steroids. American Journal of Clinical Pathology 39: 607–615

Jaffe R 1987 Pathology of histiocytosis X. Perspectives in Pediatric Pathology 9: 4–47

Kitahara M, Eyre H J, Simonian Y, Atkin C I, Hasstedt S S J 1981 Hereditary myeloperoxidase deficiency. Blood 57: 888–893

Lake B D 1971 Wolman's disease. Histochemical detection of the enzyme deficiency in blood films. Journal of Clinical Pathology 24: 617–620

Lake B D 1974 An improved method for the detection of β-galactosidase activity, and its application to G_{M1}-gangliosidosis and mucopolysaccharidosis. Histochemical Journal 6: 211–218

Lake B D 1989 Metabolic disorders. General considerations. In: Berry C L (ed) Paediatric pathology 2nd edn. Springer-Verlag, Berlin, ch 14

Lake B D 1981 Blood and bone marrow biopsy as an aid to diagnosis in the cerebral lipidoses. In: Rose F C (ed) Metabolic disorders of the nervous system. Pitman, London, ch 13

Lake B D 1984 Lysosomal enzyme deficiencies. In: Adam J H, Corsellis JAN, Duchen L W (eds) Greenfield's neuropathology, 4th edn. Edward Arnold, London, p: 491–572

Lake B D, Milla P J, Taylor D S I, Young E P 1982 A mild variant of mucolipidosis type 4 (ML4). In: Berman E, Merin S, Maumenee I (eds) Genetics in ophthalmology. A R Liss, New York

Landing B H 1987 Lymphohistiocytosis in childhood. Pathologic comparison with fetal Letterer-Siwe disease (disseminated visceral histiocytosis X). Perspectives in Pediatric Pathology 9: 48–74

Lee R E 1981 The high incidence of malignant tumours in adults with Gaucher's disease. Laboratory Investigation 44: 37A (abstract)

Lendrum A C 1981 Misapplication of Russell's name (letter). Journal of Clinical Pathology 34: 689

Long R G, Lake B D, Pettit J E, Scheuer P J, Sherlock S 1977 Adult Niemann-Pick disease. Its relationship to the syndrome of the sea-blue histiocyte. American Journal of Medicine 62: 627–635

Lüllmann H, Lüllmann-Rauch R, Wassermann O 1973 Drug-induced phospholipidosis. German Medical Monthly 3: 128–135 (Translated from Deutsche Medizinische Wochenschrift 98: 1616–1623)

Neville B G R, Lake B D, Stephens R, Sanders M 1973 A neurovisceral storage disease with vertical supranuclear ophthalmoplegia and its relationship with Niemann-Pick's disease. A report of nine patients. Brain 96: 97–120

Patrick A D, Lake B D 1968 Cystinosis: electron microscopic evidence of lysosomal storage of cystine in lymph nodes. Journal of Clinical Pathology 21: 571–575

Peterson L, Nelson A, Parkin J 1981 Mucopolysaccharidosis type VII. A morphologic, cytochemical and ultrastructural study of the blood and bone marrow. Laboratory Investigation 44: 5p (abstract)

Thomas J A, Janossy G, Chilosi M, Pritchard J, Pincott J R 1982 Combined immunological and histochemical analysis of skin and lymph node lesions in histiocytosis X. Journal of Clinical Pathology 35: 327–337

Vanier M T, Wenger D A, Comley M E, Rousson R, Brady R O, Pentchev P G 1988 Niemann-Pick disease group C: clinical variability and diagnosis based on defective cholesterol esterification. A collaborative study on 70 patients. Clinical Genetics 3: 333–348

21. Soft tissue tumours

C. E. H. du Boulay

INTRODUCTION

Soft tissue (mesenchymal) neoplasms are rare, accounting for less than 1% of human malignancy, with the result that sarcomas are probably the least understood and most inadequately treated group of malignancies.

Soft tissue tumours appear in different guises and can mimic both reactive soft tissue lesions and epithelial neoplasms. It is essential that these lesions are accurately diagnosed and classified so that their biological behaviour can be predicted and treatment protocols devised. Histochemical and immunohistochemical techniques provide the pathologist with powerful methods for doing this. Many sarcomas are poorly differentiated and on routine H & E staining, give no clue as to their histogenesis or direction of differentiation. Histochemical stains are useful for elucidating broad tumour groups, but it is the application of panels of antisera, using the immunoperoxidase technique on paraffin embedded material, which has enhanced the diagnosis of sarcomas in recent years (Brooks 1982, du Boulay 1985, Roholl et al 1985).

HISTOCHEMISTRY

Mucins. Stains for mucopolysaccharides (glycosaminoglycans) are useful for distinguishing cartilaginous myxoid neoplasms from other myxoid sarcomas. Chondroid tumours tend to stain positively for Alcian blue at pH 2.5 and this staining is resistant to hyaluronidase (Mackenzie 1981).

Glycogen containing tumours. Many undifferentiated tumours of the malignant round cell type contain glycogen and therefore this is not a helpful discriminator. Intracellular glycogen is common in rhabdomyosarcomas and, although rare, may occur in neuroblastomas (Hashimoto et al 1983). Ewing's tumour is classically described as containing intracellular glycogen. Recently there has been debate about the origin of this tumour and its relationship to primitive neuroectodermal neoplasms (PNET) (Askin et al 1979, Navas-Palacios et al 1984). The coexistence of glycogen, neural markers (see Fig. 21.11) and dense core granules in PNET has been described (Linnoila et al 1986), dense core granules not occurring in Ewing's tumour.

Fat stains. These are of limited value. Although many well differentiated fatty tumours contain abundant lipid, others do not. In addition, other sarcomas, particularly malignant fibrous histiocytoma, also contain fat.

IMMUNOHISTOCHEMISTRY

One of the most important principles of diagnostic immunohistochemistry is the use of a panel of antisera for each case. Markers used in isolation are meaningless and this is particularly true when interpreting immunostaining in sarcomas. Many markers complement each other and it is often only by building up a profile of positive staining patterns that the diagnosis is made. A true positive staining pattern can be checked by looking for inbuilt control tissues such as blood vessels, nerves and muscle. Another important principle is that absence of staining does not necessarily indicate a negative diagnosis. Morphological criteria should always be paramount in the diagnosis of sarcomas and should override an apparently aberrant immunoperoxidase staining profile. Table 21.1 shows the most useful commercially available antisera in the diagnosis of mesenchymal tumours.

UNDIFFERENTIATED TUMOURS

A frequent diagnostic problem is that of the undifferentiated tumour which, on H & E preparations, gives no clue as to its origin. There are three main histological patterns which can be identified: the spindle cell tumour, the round cell tumour and the pleomorphic tumour. Using a simple panel of antisera initially, it is frequently possible to divide these into melanoma, sarcoma, carcinoma or lymphoma (Tables 21.2 and 21.3).

Cytoskeletal proteins such as cytokeratin and vimentin are intermediate-sized filaments which are present in many cell types and show some specificity for cell differentiation (Schlegel et al 1980, Denk et al 1983). Cytokeratins are seen in cells of epithelial origin, or in mesenchymal cells showing epithelial differentiation. Thus, in synovial sarcoma, mesothelioma and some types of lung carcinomas there is coexpression of vimentin and cytokeratin (Miettinen et al 1982a, Gatter et al 1986, Upton et al 1986). Despite this, it is still possible to divide tumours broadly into sarcoma or carcinoma, using sensible morphological criteria along with the immunoperoxidase panel.

Table 21.1 Immunohistochemical markers which are useful in the diagnosis of soft tissue tumours

Intermediate filaments

Prekeratin	(CK)(CAM 5.2)	Epithelial/mesenchymal cells
Vimentin	(Vim)	Mesenchymal/epithelial cells
Desmin	(Des)	Skeletal and smooth muscle
Neurofilament protein (NFP)		Neural cells/axons

Muscle markers

Myoglobin	(Myog)	Skeletal muscle
Myosin	(Myos)	Skeletal and smooth muscle
Actin	(Act)	Myoepithelial cells & pericytes
Desmin		Skeletal and smooth muscle

Reticuloendothelial markers

Leucocyte common antigen (LCA)	Leucocytes
α-1 anti-trypsin (A1AT)	Histiocytes and primitive mesenchyme
α-1 anti-chymotrypsin (A1ACT)	Melanocytes (rarely)
Lysozyme (Lys)	Reactive histiocytes, leucocytes

Endothelial markers

Factor VIII rag (F8)	Vascular endothelium
Ulex europaeus agglutinin	Vascular endothelium and lymphatic endothelium

Neural markers

S-100 protein	Neural cells, melanocytes, Langerhans cells, chondrocytes etc.
PGP 9.5	Neural/neuroendocrine cells
Myelin basic protein (MBP)	Schwann cells
Neuron specific enolase (NSE)	Neural/neuroendocrine cells

Table 21.2 Differential diagnosis of undifferentiated round cell tumours

	LCA	PGP	Vim	Des/Myog	
Ewing's sarcoma	−	−	+	−	−
Neuroepithelioma/blastoma	−	+	−	−	−
Rhabdomyosarcoma	−	−	+	+	+
Lymphoma	+	−	+	−	−

MESENCHYMAL TUMOURS

Having decided that an undifferentiated tumour is most likely to be of mesenchymal origin and not a lymphoma or carcinoma, it is then possible to determine more accurately its histological subtype using both morphology and immunostaining (Table 21.4).

1. Fibrohistiocytic tumours

The fibrous histiocytomas form the largest group of soft tissue tumours. They arise from primitive mesenchymal cells, which can differentiate towards collagen-producing fibroblasts and/or histiocytic cells (Lawson et al 1987). These can be identified by their expression of the 'histiocyte' markers, alpha-l-antitrypsin and alpha-l-antichymotrypsin (du Boulay 1982, Kindblom et al 1982). These markers are not exclusive to fibrous histiocytomas and can be seen in a variety of soft tissue tumours, carcinomas and melanoma (Leader et al 1986); however, when used as part of a panel, they remain diagnostically useful (Fig. 21.1). The presence of alpha-l-antitrypsin in a pleomorphic sarcoma may indicate

Table 21.3 Immunohistochemical staining of spindle cell tumours

	CK	A1AT	S100/PGP		Vim	Des
Melanoma	−	+	+	+	+	−
Squamous carcinoma	+	−	−	−	−	−
Fibrous histiocytoma	−	+	+/−	−	+	−
Smooth muscle tumour	+/−	−	−	−	+	+

Table 21.4 Immunoperoxidase staining profiles of soft tissue tumours

	CK	Vim	F8	Des	Myog	A1AT	S100	PGP
Angiosarcoma	−	+	+	−	−	−	−	−
Haemangiopericytoma	−	+	−	−	−	−	−	−
Kaposi's sarcoma	−	+	+/−	−	−	−	−	
Dermatofibrosarcoma	−	+	−	−	−	+/−	−	−
Fibrous histiocytoma	−	+	−	−	−	+	+/−	−
Fibrosarcoma	−	+	−	−	−	−	−	−
Synovial sarcoma	+	+	−	−	−	−	−	−
Epithelioid sarcoma	+	+	−	−	−	−	−	−
Leiomyosarcoma	+/−	+	−	+	−	−	−	−
Rhabdomyosarcoma	−	+	−	+	+	−	−	−
Liposarcoma (pleomorphic)	−	+	−	−	−	+/−	+/−	−
Liposarcoma	−	+	−	−	−	−	−	−
Neurofibroma	−	+	−	−	−	−	+	+
Schwannoma	−	+	−	−	−	−	+	+
Malignant nerve sheath tumour	−	+	−	−	−	−	+/−	+
Granular cell tumour	−	+	−	−	−	−	+	+
Alveolar soft part sarcoma	−	+	−	+	−	−	−	−
Clear cell sarcoma	−	+	−	−	−	−	+	−
Ewing's sarcoma	−	+	−	−	−	−	−	−
Chondrosarcoma	−	+	−	−	−	−	−	−
Chordoma	+	+	−	−	−	−	+	−

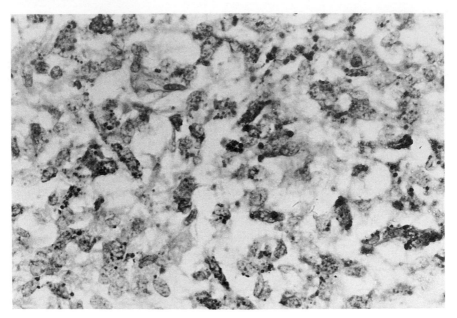

Fig. 21.1   Malignant fibrous histiocytoma shows strong cytoplasmic staining for alpha-1-antitrypsin. The staining pattern is intracytoplasmic and granular.

dedifferentiation of the tumour to a more primitive mesenchymal form. It is conceivable that this could be prognostically important and is an area which needs to be explored.

2. Synovial tumours

Synovial sarcoma is a rare tumour which presents in many morphological forms. Classically it shows a biphasic pattern of spindle cells and epithelial lined clefts, which in many cases cannot be seen on routine preparations. The tumour shows characteristic immunostaining with dual expression of cytokeratin and vimentin, the epithelial component often being highlighted (Fig. 21.2) and the biphasic nature is enhanced.

Epithelioid sarcoma is histogenetically related to synovial sarcoma and also shows cytokeratin and vimentin expression (Mills et al 1981, Miettinen et al 1982c).

3. Muscle tumours

Smooth muscle tumours

Histologically, both benign and malignant smooth muscle tumours may be confused with neural, fibrous-histiocytic or fibrous tumours. Most smooth muscle tumours show evidence of their derivation by staining positively for desmin, actin or myosin (du Boulay 1985, Saul et al 1987) (Fig. 21.3). Cytokeratin staining has been described in leiomyomas of the gut, although it is thought that this is due to crossreaction rather than the presence of intracellular cytokeratins (Brown et al 1987, Norton et al 1987). It should be remembered that malignant smooth muscle tumours may not express desmin and that immunostaining for desmin is often capricious.

Skeletal muscle tumours

Tumours of skeletal muscle origin, or tumours showing skeletal muscle differentiation can be recognized by using the myogenous markers, desmin and myoglobin (Mukai et al 1979, Eusebi et al 1986). The commonest diagnostic problem occurs in the childhood round cell tumours, which comprise alveolar or embryonal rhabdomyosarcoma, Ewing's sarcoma and neuroblastoma. All of these tumours stain positively for vimentin, but using muscle markers such as desmin and myoglobin together with neural markers S-100 and PGP 9.5 or NSE, rhabdomyosarcoma of childhood

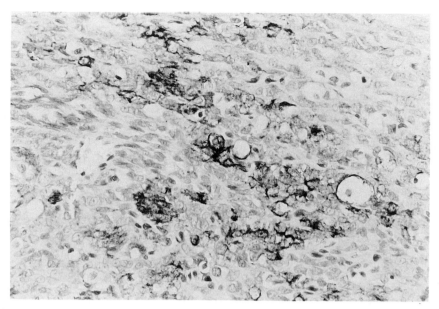

Fig. 21.2 The biphasic pattern of synovial sarcoma is enhanced when stained for cytokeratin (using antibody CAM 5.2).

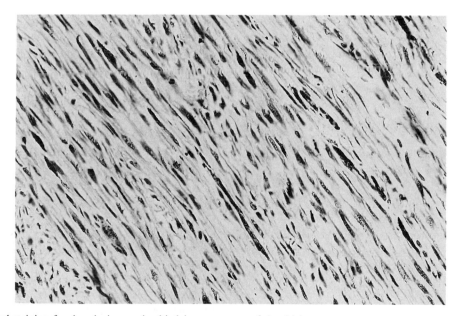

Fig. 21.3 Focal staining for desmin is seen in this leiomyosarcoma of the thigh.

can be separated from the neuroectodermal tumours (Fig. 21.4).

The interpretation of immunostaining for myoglobin should be carried out with care. Myoglobin can be taken up non-specifically by cells at the edge of a tumour and the unwary can mistake trapped myoblasts for tumour cells (Eusebi et al 1984) (Fig. 21.5). Desmin is more widely used as a muscle marker than myoglobin and is seen in a higher proportion of primitive rhabdomyosarcomas (Miettinen 1982b, Miettinen & Rapola 1989). These markers are also useful in

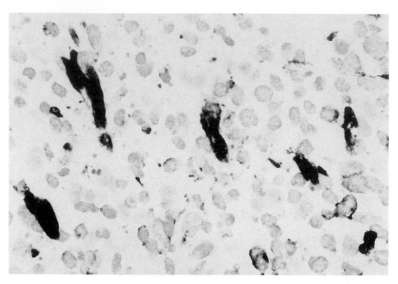

Fig. 21.4 Rhabdomyosarcoma. Strong desmin positivity is found in the more differentiated cells, but most cells show some discrete staining. ABC peroxidase and haematoxylin.

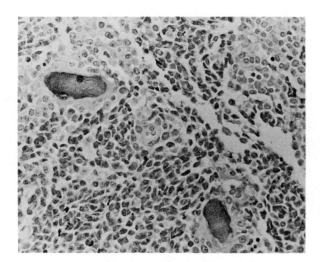

Fig. 21.5 Alveolar rhabdomyosarcoma stained for myoglobin. Note the trapped muscle fibres at the edge of the lesion. These should not be interpreted as malignant rhabdomyoblasts.

the identification of myogenous elements in Mullerian tumours, Wilms tumours and malignant nerve sheath (Triton) tumours.

4. Neural tumours

The most useful markers of peripheral nerve tumours S-100 protein, neuron specific enolase and PGP 9.5 (Stefansson et al 1982, Rode et al 1985) are nearly always positive in neural tumours and melanoma (Fig. 21.6). Schwannomas and neurofibromas stain strongly for both S-100 and PGP 9.5, but malignant nerve sheath tumours may lose their expression of both these neural markers. For further details see Chapter 7. The panel of markers should include the neural markers as well as desmin and cytokeratin to identify epithelial or muscle differentiation within nerve sheath lesions (Fig. 21.7).

5. Vascular tumours

The diagnosis of benign vascular tumours is usually straightforward, unlike malignant and poorly differentiated vasoformative tumours, which pose problems. Factor VIII-related antigen and a lectin derived from *Ulex europaeus* are the two most useful markers of vascular endothelium (Mukai et al 1980, Burgdorf et al 1981). Factor VIII-related antigen is not expressed in lymphatic endothelium whereas *Ulex europaeus* agglutin detects both lymphatic and vascular epithelium. Most angiosarcomas and haemangioendotheliomas show staining with both these antisera, but it is frequently focal and difficult to find, especially in anaplastic lesions (Fig. 21.8). To date the evidence for positive endothelial markers in Kaposi's sarcoma is

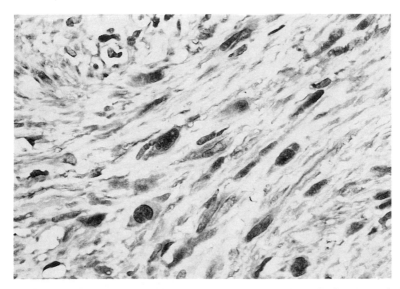

Fig. 21.6 An area of a benign Schwannoma showing loose cellularity. Positive staining for S-100 protein is seen in some nuclei and cellular processes.

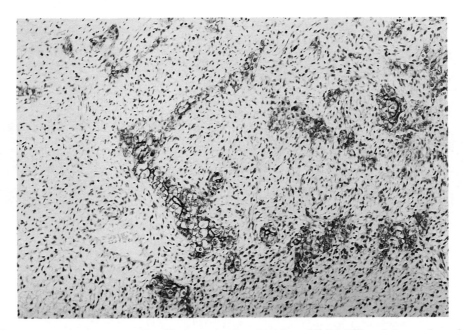

Fig. 21.7 Malignant nerve sheath tumour of mediastinum showing glandular epithelial differentiation as shown by cytokeratin (CAM 5.2) staining.

unconvincing (Guarda et al 1981, Leader et al 1986). The glomus tumour and haemangiopericytoma are tumours derived from pericytic cells and are negative for endothelial markers, but positive for the contractile marker actin (Hultberg et al 1988) (Fig. 21.9).

The application of factor VIII is a useful test in cases of epithelioid haemangioendothelioma, where positivity occurs focally around the intracellular lumina of newly formed vascular elements, helping to differentiate the tumour from metastatic signet ring carcinoma (Weiss & Enzinger 1982).

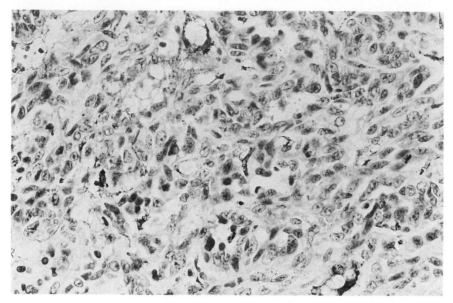

Fig. 21.8   Angiosarcoma shows strong staining of malignant endothelial cells for factor VIII-related antigen.

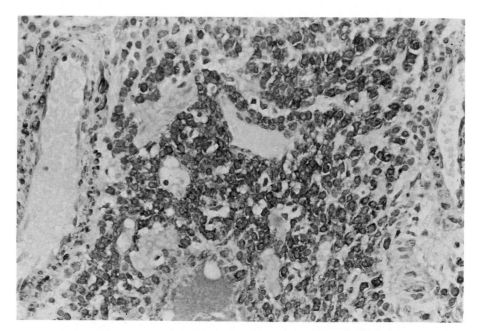

Fig. 21.9   Glomus tumour shows staining of pericytic cells for actin.

6. Tumours of fat

The diagnosis of liposarcoma still rests on the morphological identification of lipoblasts, there still being no suitable immunohistochemical marker. Some liposarcomas stain positively for S-100 protein, which can also be seen in the nuclei of normal fat cells, but this does not usually lead to diagnostic confusion when lipoblasts are present. Pleomorphic liposarcomas may stain positively for α-1-antitrypsin and there is some debate as to whether this staining represents dedifferentiation of such lesions to a more primitive, fibrohistiocytic phenotype.

7. Other tumours

Granular cell tumour

Originally known as granular myoblastoma, this tumour has distinctive morphological appearances, but its histogenesis has caused controversy. The cells contain PAS-positive granules and are arranged in sheets and clusters, with a variable amount of connective tissue. Immunohistochemical studies have shown that these tumours stain strongly for S-100 protein (Stefansson & Wollman 1982), and do not express myoglobin or desmin (Fig. 21.10). It is now thought that this lesion is of Schwannian rather than muscle origin (Chimelli et al 1984).

Extraskeletal Ewing's sarcoma

This is a primitive mesenchymal tumour composed of small round cells which appear to show no differentiation. They usually contain glycogen. Other soft tissue neoplasms which may mimic the appearance of Ewing's sarcoma are the Askin tumour, rhabdomyosarcoma, peripheral neuroblastoma and small cell osteosarcoma (Martin et al 1982). Ewing's tumour consistently stains for vimentin, but other markers show variable expression (Fig. 21.11 and see p. 321).

Extraskeletal myxoid chondrosarcoma

This tumour consists of small round cells arranged in strands in a myxoid matrix, which is Alcian blue positive, and resistant to hyaluronidase. It may resemble chordoma. Immunoperoxidase staining shows faint positive staining for lysozyme but none for S-100 protein. In contrast, chordomas are strongly positive for S-100 protein (Salisbury & Isaacson 1986) and have recently been shown to express cytokeratin (Nakamura et al 1983, Abenoza & Sibley 1986, Rutherfoord & Davies 1987). In addition to the markers mentioned above, chordomas contain characteristic physaliferous cells which are Alcian blue and PAS positive.

Alveolar soft part sarcoma

Many of these tumours contain distinctive PAS positive, diastase-resistant crystalline bodies which are unique to this neoplasm. For many years the histogenesis of this unusual tumour has been

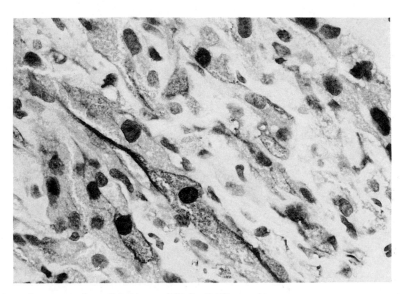

Fig. 21.10 In granular cell tumours the cells stain patchily for S-100 protein.

debated. Early electron microscopic findings suggested a myogenous derivation (Fisher & Reidbord 1971) and this has recently been supported by immunohistochemical studies which show positive desmin and actin staining (Mukai et al 1986).

Clear cell sarcoma of tendon sheath

There has been much debate on the histogenesis of clear cell sarcoma, which hitherto has been considered by some authors as a variant of synovial sarcoma. However, the observation that two-thirds of the cases contain either melanin or stain strongly for S-100 protein has led to the suggestion that this lesion is best thought of as a soft tissue melanoma (Chung & Enzinger 1983).

Osteogenic sarcoma

Alkaline phosphatase activity is strong in osteoblasts. Its presence in osteogenic sarcoma (Fig. 21.12) may be helpful in biopsy material where osteoid tissue is not readily identified (Sanerkin 1980, Pringle 1987). Alkaline phophatase activity must be demonstrated in cryostat or frozen sections since the bone isoenzyme is destroyed by normal processing. Activity is not found in malignant tumours of fibroblastic origin, and in chondrosarcomas alkaline phosphatase activity is either absent or weak. It is, however, present in aneurysmal bone cyst and differential diagnosis from telangiectatic osteogenic sarcoma may be difficult. Alkaline phosphatase is also present in endothelial cells and adipocytes and therefore may also be found in angiosarcomas and liposarcomas. These entities, however, are not likely to be confused with osteogenic sarcoma in any of its guises.

Oppedal et al (1989) have observed immunoreactivity of the antibody UJ13A in osteoblasts and in osteogenic sarcoma. This antibody has been used as a marker for neuroblastoma and its application in the examination of bone marrow for metastatic neuroblastoma must therefore be made with great care.

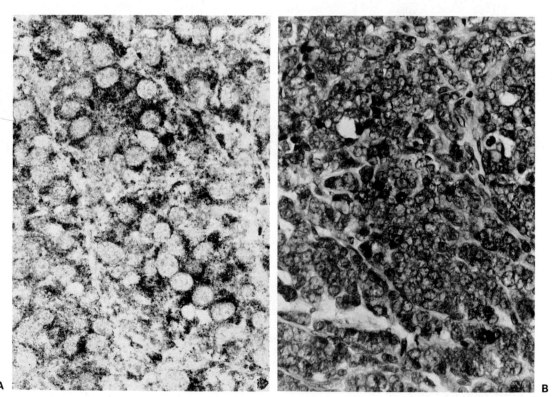

Fig. 21.11   A skin tumour (thoracopulmonary tumour of the chest wall) staining strongly for (**A**) PGP 9.5 and (**B**) NSE.

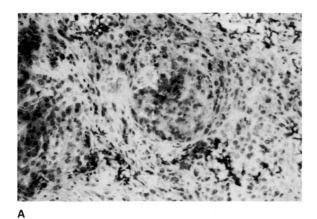

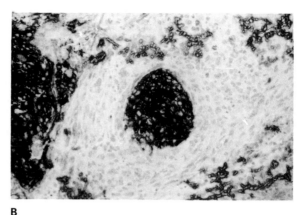

A   **B**

Fig. 21.12 Osteogenic sarcoma. **A** On H & E, nests of malignant cells are surrounded by a stroma without osteoid formation. In the serial (**B**) the osteoblastic nature of the malignant cells is demonstrated by strong alkaline phosphatase activity.

REFERENCES

Abenoza P, Sibley R K 1986 Chordoma. An immunohistochemical study. Human Pathology 17: 744–747

Askin F B, Rosai J, Sibley R K et al 1979 Malignant small cell tumor of thoracopulmonary region in childhood. Cancer 43: 2438–2451

Brooks J J 1982 Immunohistochemistry of soft tissue tumors. Progress and prospects. Human Pathology 13: 969–974

Brown D C, Theaker J M, Banks P M, Gatter K C, Mason D Y 1987 Cytokeratin expression in smooth muscle and smooth muscle tumours. Histopathology 11: 477–486

Burgdorf W H C, Mukai K, Rosai J 1981 Immunohistochemical identification of factor VIII rag in endothelial cells of cutaneous lesions of alleged vascular nature. American Journal of Clinical Pathology 75: 167–171

Chimelli L, Symon L, Scaravilli F 1984 Granular cell tumour of the fifth cranial nerve. Further evidence for Schwann cell origin. Journal of Neuropathology and Experimental Neurology 43: 634–642

Chung E B, Enzinger F M 1983 Malignant melanoma of soft parts. A reassessment of clear cell sarcoma. American Journal of Surgical Pathology 7: 405–413

Clark H B 1984 Immunohistochemistry of nervous system antigens: Diagnostic applications in surgical neuropathology. Seminars in Diagnostic Pathology 1: 309–317

Denk H, Krepler R, Artlieb U et al 1983 Proteins of intermediate filaments. An immunohistochemical and biochemical approach to the classification of soft tissue tumors. American Journal of Pathology 110: 193–208

du Boulay C E H 1982 Demonstration of alpha-1-antitrypsin and alpha-1-antichymotrypsin in fibrous histiocytomas using the immunoperoxidase technique. American Journal of Surgical Pathology 6: 559–564

du Boulay C E H, 1985 Immunohistochemistry of soft tissue tumours: A review. Journal of Pathology 146: 77–94

Eusebi V, Bondi A, Rosai J 1984 Immunohistochemical localization of myoglobin in non muscular cells. American Journal of Surgical Pathology 8: 51–55

Eusebi V, Ceccarelli C, Gorza L et al 1986 Immunocytochemistry of rhabdomyosarcoma. American Journal of Surgical Pathology 10: 293–299

Fisher C 1986 Synovial sarcoma. Ultrastructural and immunohistochemical features of epithelial differentiation in monophasic and biphasic tumours. Human Pathology 17: 996–1008

Fisher E R, Reidbord H 1971 Electron microscopic evidence suggesting the myogenous derivation of the so called alveolar soft part sarcoma. Cancer 27: 150–159

Gatter K C, Dunnill M S, Van Muijen G, Mason D 1986 Human lung tumours may coexpress different classes of intermediate filaments. Journal of Clinical Pathology 39: 950–954

Guarda L C, Silva E G, Ordonez N G, Smith L N 1981 Factor VIII in Kaposl's sarcoma. American Journal of Pathology 76: 197–200

Hashimoto H, Enjoji M, Nakajima T et al 1983 Malignant neuroepithelioma (peripheral neuroblastoma). American Journal of Surgical Pathology 7: 309–318

Hultberg B M, Daugaard S, Johansen H F, Mouridsen H T & Hou-Jensen K 1988 Malignant haemangiopericytomas and haemangioendotheliomas: an immunohistochemical study. Histopathology 12: 405–414

Kindblom L G, Jacobsen G K, Jacobsen M 1982 Immunohistochemical investigations of tumors of supposed fibroblastic-histiocytic origin. Human Pathology 13: 834–840

Lawson C W, Fisher C, Gatter K C 1987 An immunohistochemical study of differentiation in malignant fibrous histiocytoma. Histopathology 11: 375–383

Linnoila R I, Tsokos M, Triche T J et al 1986 Evidence for neural origin and PAS positive variants of the malignant small cell tumor of thoracopulmonary region (Askin tumor). American Journal of Surgical Pathology 10: 124–133

Leader M, Collins M, Patel J, Henry K 1986 Staining for factor VIII-related antigen and *Ulex europaeus* agglutinin I (UEA-I) in 230 tumours. An assessment of their specificity for angiosarcoma and Kaposi's sarcoma. Histopathology 10: 1153–1162

Leader M, Patel J, Collins M et al 1987
Anti-α-1-antichymotrypsin staining of 194 sarcomas, 38
carcinomas and 17 malignant melanomas. American
Journal of Surgical Pathology 11: 133–139

Mackenzie D H 1981 The myxoid tumours of somatic soft
tissues. American Journal of Surgical Pathology 5: 443–457

Martin S E, Dwyer A, Kissane J M, Costa J 1982 Small-cell
osteosarcoma. Cancer 50: 990–996

Miettinen M, Rapola J 1989. Immunohistochemical
spectrum of rhabdomyosarcoma and
rhabdomyosarcoma-like tumours. The American Journal
of Surgical Pathology 13: 120–132

Miettinen M, Lehto V-P, Badley R, Virtanen I 1982a
Expression of intermediate filaments in soft tissue
sarcomas. International Journal of Cancer 30: 541–546

Miettinen M, Lehto V-P, Badley R A, Virtanen I 1982b
Alveolar rhabdomyosarcoma. Demonstration of the muscle
type of intermediate filament protein desmin as a
diagnostic aid. American Journal of Pathology
108: 246–251

Miettinen M, Lehto V-P, Vartio T, Virtanen I 1982c
Epithelioid sarcoma. Archives of Pathology and
Laboratory Medicine 106: 620–623

Mills S E, Fechner R E, Burns D E et al 1981 Intermediate
filaments in eosinophilic cells of epithelioid sarcoma.
American Journal of Surgical Pathology 5: 195–202

Mukai K, Rosai J, Hallaway B E 1979 Localization of
myoglobin in normal and neoplastic human skeletal
muscle cells using an immunoperoxidase method.
American Journal of Surgical Pathology 3: 373–376

Mukai K, Rosai J, Burgdorf W H C 1980 Localization of
factor VIII rag in vascular cells using an
immunoperoxidase technique. American Journal of
Surgical Pathology 4: 273–276

Mukai M, Torikata C, Iri H et al 1986 Histogenesis of
alveolar soft part sarcoma. American Journal of Surgical
Pathology 10: 212–218

Nakamura Y, Becker L E, Marks A 1983 S100 protein in
human chordoma and rabbit notochord. Archives of
Pathology and Laboratory Medicine 107: 118–120

Navas-Palacios J J, Aparicio-Duque R E, Valdes M D 1984
On the histogenesis of Ewing's sarcoma. An
ultrastructural, immunohistochemical and cytochemical
study. Cancer 53: 1882–1901

Norton A J, Thomas J A, Isaacson P G 1987
Cytokeratin-specific monoclonal antibodies are reactive
with tumours of smooth muscle derivation. An
immunocytochemical and biochemical study using
antibodies to intermediate filament cytoskeletal proteins.
Histopathology 11: 487–499

Oppedal B R, Kemshead J T, Brandtzaeg P 1989 Bone
marrow examination in neuroblastoma patients. Human
Pathology in press.

Pringle J A S 1987 Pathology of bone tumours. Balliere's
Clinical Oncology 1: 21–63

Rode J, Dhillon A P, Doran J F, Jackson P J, Thompson
R J 1985 PGP 9.5, a new marker for human
neuroendocrine tumours. Histopathology 9: 147–158

Roholl P J M, De Jong A S H, Ramaekers F C S 1985
Application of markers in the diagnosis of soft tissue
tumours. Histopathology 9: 1019–1035

Rutherfoord G S, Davies A G 1987 Chordomas —
ultrastructure and immunohistochemistry: a report based
on the examination of six cases. Histopathology
11: 775–787

Salisbury J R, Isaacson P G 1986 Distinguishing chordoma
from chondrosarcoma by immunohistochemical
techniques. Journal of Pathology 148: 251–252

Sanerkin N G 1980 Definition of osteosarcoma,
chondrosarcoma and fibrosarcoma of bone. Cancer
46: 178–185

Saul S H, Rast M L, Brooks J J 1987 The
immunohistochemistry of gastrointestinal stromal tumours.
American Journal of Surgical Pathology 11: 464–473

Schlegel R, Banks-Schlegel S, McLeod J A, Pinkus G 1980
Immunoperoxidase localization of keratin in human
neoplasms. American Journal of Pathology 101: 41–49

Stefansson K, Wollmann R L 1982 S-100 protein in
granular cell tumours (granular cell myoblastomas).
Cancer 49: 1834–1838

Upton M P, Hirohashi S, Tome Y et al 1986 Expression of
vimentin in surgically resected adenocarcinomas and large
cell carcinomas of lung. American Journal of Surgical
Pathology 10: 560–567

Weiss S W, Enzinger F M 1982. Epithelioid
haemangioendothelioma. A vascular tumour often
mistaken for a carcinoma. Cancer 50: 970–981

22. Mesothelioma

M. Isabel Filipe and R. Poston

The differential diagnosis between mesothelioma (malignant) and adenocarcinoma, either primary in the lung or metastatic, is sometimes difficult and a definite diagnosis is not always possible on biopsy material using routine histological methods. Benign proliferative mesothelial lesions may also mimic malignancy.

Hyperplasia vs mesothelioma (malignant)

Mesothelioma cells have a great capacity to undergo reactive hyperplastic changes forming solid nests, papillary or tubular structures and may even mimic invasion. The differential diagnosis between benign proliferative mesothelial lesions and mesothelioma can be difficult. The clinical history and radiological findings are important and certain histological features, such as marked nuclear atypia, high nuclear/cytoplasmic ratio and fresh necrosis, favour malignancy.

Mucin histochemistry and/or immunocytochemistry are of no discriminatory value.

Mesothelioma vs carcinoma (primary or secondary)

Mesothelial cells can express a range of possible differentiation from cuboidal epithelial to small fibroblast-like, and this range is reflected in the diversity of patterns seen in mesotheliomas (epithelial, sarcomatoid, mixed and anaplastic types may be seen). In the well differentiated papillary-tubular tumours, intracellular and extracellular mucosubstances can be present.

It is important to bear in mind this unusual range of cellular differentiation in differential diag-

noses and in considering the use of histochemistry and immunocytochemical markers.

Mucin histochemistry

Mucosubstances containing hyaluronic acid have been demonstrated in mesotheliomas and in pleural and peritoneal effusions from patients with mesothelioma, but are absent from adenocarcinomas. The latter secrete glycoproteins which do not contain uronic acid. PAS, with or without diastase digestion, gives a negative reaction with hyaluronic acid, while glycoproteins in the majority of adenocarcinomas are positive after digestion. On the other hand, Alcian-blue pH2.5 stains both hyaluronic acid and glycoproteins with acidic groups. Pretreatment with hyaluronidase reduces or abolishes the Alcian-blue staining of hyaluronic acid, but has no effect on the glycoproteins. False negative results can occur in mesotheliomas with a low content of hyaluronic acid or where it has been extracted during fixation. However, a positive result could be helpful in a proportion of well differentiated tumours, but it is not a specific finding.

The accuracy of diagnosis can be improved by the use of immunocytochemical techniques described below.

Immunocytochemistry

Mesothelial cells express cytokeratins of both high (squamous epithelium-related) and lower (widespread distribution in epithelium) molecular forms (Blobel et al 1985). Only a few monoclonal antibodies exist which can selectively detect the high-molecular weight squamous-related cytokeratin types in paraffin sections. The antibody LP34 (ICRF Laboratories, London) reacts with both the low-molecular weight cytokeratin 18 and the high-molecular weight cytokeratins. Because immunoreactivity of LP34 towards cytokeratin 18 is differentially lost in paraffin sections (B. Lane, personal communication), this antibody gives strong staining of squamous epithelia and of epithelial-like mesothelial cells, with lesser or no reactivity in other epithelial cell types.

Mesothelial cells also bind CAM 5.2 (Fig. 22.1) thus strong reaction with LP34 and CAM 5.2 seems to be a feature of most epithelial-like mesothelial cells, whether benign or malignant. CAM 5.2 binding in connective tissue-like components of sarcomatoid mesotheliomas can be useful in the distinction from other spindle cell tumours, although some care is needed, as muscle-derived sarcomas can be positive (Montag et al 1988, Epstein & Budin 1986).

Mesothelioma tumour cells may exhibit vimentin positivity. Dual expression of vimentin and cytokeratins is not uncommon in carcinoma and mesothelioma, but in mesothelioma vimentin expression is usually more marked than that found in carcinomas. Mesotheliomas show also a positive reaction for EMA and S-100 protein (Rosai 1989). Mesotheliomas display other marked differences from carcinomas. They lack the expression of several carcinoma-related molecules. In our series (unpublished), none of 32 mesotheliomas gave any reaction with CEA, Ca19-9, B72.3 or Leu Ml, confirming previous reports (Ghosh et al 1987, Atkinson et al 1982, Lee et al 1986, Szpak et al 1986, Sheibani et al 1986, Poston & Sidhu 1986). All these antibodies share the property of failing to stain mesotheliomas, and giving positive results in a proportion of, but not all, adenocarcinomas and sometimes in other varieties of carcinoma. The reactions of adenocarcinomas are certainly capricious, and they may stain with one or other of these antibodies. A negative result with any one of

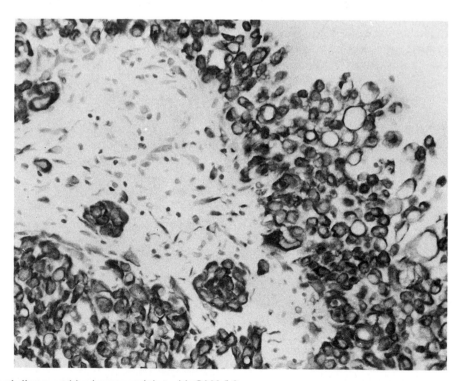

Fig. 22.1 Mesothelioma: positive immunostaining with CAM 5.2.

Table 22.1 Histochemical and immunocytochemical differential diagnosis of mesothelioma and adenocarcinoma

	Mesothelioma	Adenocarcinoma
Histochemical test		
D-PAS	−	+
AB pH2.5	+	+
Hyal/AB pH2.5	−	+
Immunocytochemistry		
CAM 5.2 (low mol. wt cytokeratin)	+ +	+ +
LP34 (high mol. wt cytokeratin)	+ +	+
EMA	+	+
Vimentin	+ +	+
S-100	+	+
CEA	−	+
Leu-M1	−	+
Cal9.9	−	+
B72.3	−	+

+ + Frequently present, + variable, positive or negative, − absent.

these could lead to misdiagnosis as mesothelioma. Thus the need for a panel of antibodies is reinforced (see Table 22.1).

A few other carcinoma markers have been used in a similar manner to those described above, but with less clear-cut results (Kondi-Paphitis & Addis 1986). Lectins do not seem to hold discriminatory ability (Rosai 1989).

The use of negative results with carcinoma markers to diagnose mesotheliomas is worrying, as immunohistochemistry can easily generate false negative results. Any marker that produces posi-tive specific reactions in mesotheliomas, which could discriminate between these tumours and carcinomas would clearly be of great value. Polyclonal antibodies have been raised to a celomic molecule of 200 kD molecular weight that stain effectively in paraffin sections (Donna et al 1986). Clearly these have a great potential. Recently, two monoclonal antibodies have been raised that react with specific surface membrane determinants of mesothelial cells (Stahel et al 1988).

These approaches seem to offer the best hope for improved mesothelioma diagnosis.

REFERENCES

Atkinson B F, Ernst C S, Herlyn M, Steplewski Z, Sears H F, Koprowski H 1982 Gastrointestinal cancer-associated antigen in immunoperoxidase assay. Cancer Research 42: 4820–4823

Blobel G A, Moll R, Franke W W, Kayser K W, Gould V E 1985 The intermediate filament cytoskeleton of malignant mesotheliomas and its diagnostic significance. American Journal of Pathology 121: 235–247

Donna A, Betta P G, Bellingeri D, Marchesini A 1986 New marker for mesothelioma: an immunoperoxidase study. Journal of Clinical Pathology 39: 361–368

Epstein J I, Budin R E 1986 Keratin and epithelial membrane antigen immunoreactivity in non-neoplastic fibrous pleural lesion. Implications for the diagnosis of desmoplastic mesothelioma. Human Pathology 17: 514–519

Ghosh A K, Gatter K C, Dunnill M S, Mason D Y 1987 Immunohistological staining of reactive mesothelium, mesothelioma and lung carcinoma with a panel of monoclonal antibodies. Journal of Clinical Pathology 40: 19–25

Kondi-Paphitis A, Addis B J 1986 Secretory component in pulmonary adenocarcinoma and mesothelioma. Histopathology 10: 1279–1287

Montag A G, Pinkus G S, Corson J M 1988 Keratin protein immunoreactivity of sarcomatoid and mixed type of diffuse malignant mesotheliomas: an immunoperoxidase study of 30 cases. Human Pathology 19: 336–342

Lee I, Radosevich J A, Cheijfec G et al 1986 Malignant mesotheliomas: Improved differential diagnosis from lung adenocarcinomas using monoclonal antibodies 44-3A6 and 624A12. American Journal of Pathology 123: 497–507

Poston R N, Sidhu Y S 1986 Diagnosing tumours on routine surgical sections by immunohistochemistry: use of cytokeratin, common leucocyte and other markers. Journal of Clinical Pathology 39: 514–523

Rosai J 1989 In: Ackerman's surgical pathology, vol 1. C V Mosby, New York, p 266

Stahel R A, O'Hara C J, Waibel R, Martin A 1988 Monoclonal antibodies against mesothelial membrane antigen discriminate between malignant mesothelioma and lung adenocarcinoma. International Journal of Cancer 15: 218–223

Sheibani K, Battifora H, Burke J S 1986 Antigenic phenotype of malignant mesotheliomas and pulmonary adenocarcinomas. An immunohistologic analysis demonstrating the value of the Leu M1 antigen. American Journal of Pathology 123: 212–219

Szpak C A, Johnston W W, Roggli V et al 1986. The diagnostic distinction between malignant mesothelioma of the pleura and adenocarcinoma of the lung as defined by a monoclonal antibody (B72.3). American Journal of Pathology 122: 252–260

23. Histochemistry in the investigation of breast disorders

J. P. Sloane   M. G. Ormerod

INTRODUCTION

Histochemical techniques currently have very limited value in the diagnosis of breast disorders, which is still largely achieved with haematoxylin and eosin stained sections. This is not to say that profound diagnostic problems do not exist, but rather that currently available histochemical methods are not generally useful in solving them despite the publication in the last 5 years of over 400 immunohistological studies of breast. There are, however, a few areas of breast pathology where conventional histochemistry and immunohistology may provide the pathologist with valuable assistance and others where there may be significant developments in the near future.

The pathologist has two major functions in reporting breast specimens, namely to make accurate diagnoses and to provide prognostically significant supplementary information so that management of patients can be planned more effectively. This chapter evaluates the role of currently available histochemical techniques in undertaking these two major functions. We have focused on the staining of histological sections, but it should be noted that fine needle aspiration cytology (FNAC) is playing an increasing role in the investigation of patients with breast disorders (see Ch. 30). In centres where cytological expertise is high, FNAC has resulted in a fall in the number of biopsies and intraoperative frozen sections. Many of the methods discussed below may also be used on cytological preparations, but others, particularly those which improve interpretation of architectural changes, are obviously not applicable.

DIAGNOSIS

Distinguishing ductal from lobular carcinoma in situ

There are two main forms of in situ carcinoma of the breast: ductal carcinoma in situ (DCIS) and lobular carcinoma in situ (LCIS). Both are associated with a substantial risk of subsequent infiltrating carcinoma. In DCIS, however, the invasive carcinomas tend to occur at the same site whereas in LCIS, which is almost invariably multicentric, infiltrative neoplasia may supervene in any location in either breast. Furthermore, the time interval between diagnosis of in situ neoplasia and infiltrative disease is usually longer in LCIS. The two forms of in situ carcinoma can usually be distinguished with ease in H & E sections by cytological and architectural criteria, but DCIS has a variable morphology and occasionally may assume cytological and architectural characteristics similar to those of LCIS.

A characteristic feature of LCIS is the presence of intracytoplasmic mucin globules, which can be demonstrated in most cases but are usually absent from DCIS. Some of these globules are small and multiple whereas others are single and large enough to fill half the cell (Gad & Azzopardi 1975). Using the Alcian blue pH2.5 — periodic acid Schiff method, the globules usually exhibit a blue rim with a magenta or purple dot in the centre, producing a bull's eye appearance (Fig. 23.1). These vacuoles appear to correspond to microvillus-lined intracytoplasmic lumina, which are visible by electron microscopy and may also be seen on light microscopy using antibodies to epithelial membrane antigen (Fig. 23.2) (Sloane & Ormerod 1981).

Antibodies to the product of the *c-erb B-2* (*neu*) oncogene react in paraffin sections in about 40% of cases of DCIS, but not in LCIS (Gusterson et al 1988a, De Vijver et al 1988). It is unlikely, however, that these antibodies will have much value in distinguishing these two forms of in situ carcinoma of the breast as staining is restricted to the large cell (comedo) variant of DCIS, which is easy to

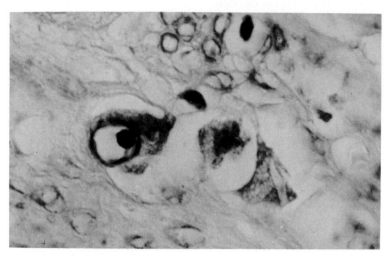

Fig. 23.1 Three metastatic breast carcinoma cells infiltrating the uterine cervix stained by the Alcian blue-periodic acid-Schiff method. One of the cells contains a large purple bull's eye globule which occupies about half the cell. ×1260

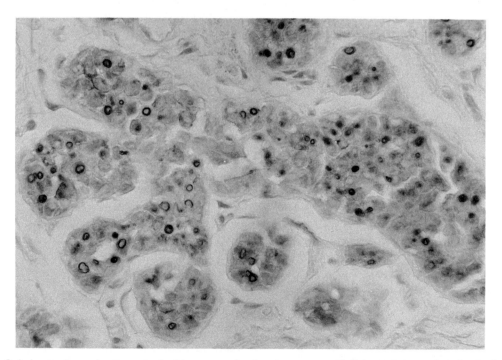

Fig. 23.2 Lobular carcinoma in situ stained with a conventional antiserum to epithelial membrane antigen by the indirect immunoperoxidase method. ×504

distinguish from LCIS by conventional histology. The small cell variants of DCIS which are sometimes confused with LCIS were uniformly negative in the study of De Vijver et al.

Distinguishing carcinoma in situ from hyperplasia

A major problem in diagnosing in situ carcinomas is distinguishing them from intralobular and intraductal hyperplasias. DCIS and LCIS form a continuous spectra with ductal hyperplasia (epitheliosis) and atypical lobular hyperplasia respectively. Unfortunately, no histochemical methods currently exist for determining whether breast proliferations are mono- or polyclonal, in contrast to those of lymphoid cells where immunoglobulin light chain staining and immunoglobulin and T cell receptor gene rearrangement studies have been effective in determining clonality (see Ch. 19).

In the study by Gusterson et al (1988b) immunostaining for the product of the *c-erb B-2* gene gave membrane positivity in 10/24 cases of pure DCIS but not in any of 19 cases of epitheliosis, including 13 cases of atypical epitheliosis. However, 6 of these cases showed weak cytoplasmic staining. Further studies to determine the role of these antibodies (and perhaps antibodies to other oncogene products) in providing a more objective means of distinguishing atypical hyperplasia from in situ carcinoma are clearly justified.

Assessing spread in DCIS

In evaluating mammary carcinoma in situ, it is important to determine if spread has occurred. In DCIS, spread of tumour cells may occur within the duct system or by breaching the duct wall to produce microinvasion. The extent of the former is important in determining the true extent of the tumour in order to assess the adequacy of excision. The latter is important as microinvasive tumours are capable of metastasizing. Neither of these considerations is important in LCIS as the lesion is almost invariably multicentric, making it pointless to assess excision margins, and microinvasive disease is rarely encountered. Histochemical stains

which can identify small numbers of tumour cells in otherwise non-neoplastic breast epithelium would help in detecting the extent of spread of DCIS. Antibodies to the *c-erb B-2* oncogene product may assist in detecting tumour cells in ducts and lobules which are only minimally involved by DCIS (Fig. 23.3) (Gusterson et al 1988a, 1988b).

A somewhat easier problem is the identification of cells of DCIS which have spread along the collecting ducts into the epidermis of the nipple to produce Paget's disease. The adenocarcinoma cells usually, but not always, exhibit obvious morphological difference from the epidermal cells. Figure 23.4A shows an example of the so-called Bowenoid form of Paget's disease in which the epidermis exhibits an atypical appearance and where the carcinoma cells blend with the keratinocytes. Histochemical stains that distinguish adenocarcinoma cells from keratinocytes and other epidermal cells are clearly of value in this situation. Methods which have been used successfully are conventional mucin stains (PAS after diastase) and the immunohistological detection of low molecular weight keratins not expressed by keratinocytes. The monoclonal antibody CAM5.2 is suitable for this purpose (Makin et al 1984). Polyclonal and monoclonal antibodies to epithelial membrane antigen give intense cytoplasmic and membrane staining for tumour cells, in contrast to the keratinocytes which exhibit weak membrane staining or complete negativity (Fig. 23.4B).

Microinvasion in DCIS may be impossible to identify confidently in H & E sections if there is extensive intraductal spread, with consequent enlargement and distortion of the duct system. Most studies on DCIS report examples of metastatic spread; the precise incidence at which this occurs is difficult to assess but appears to be about 1–2% (Ashikari et al 1971, Westbrook & Gallager 1975, Lagios et al 1982). Microinvasion may not be found in H & E sections in these cases, even after extensive sampling.

The entire duct system of the breast consists of a double layer of cells (inner epithelial and outer myoepithelial) surrounded by basement membrane, and the histochemical detection of myoepithelial cells and basement membrane

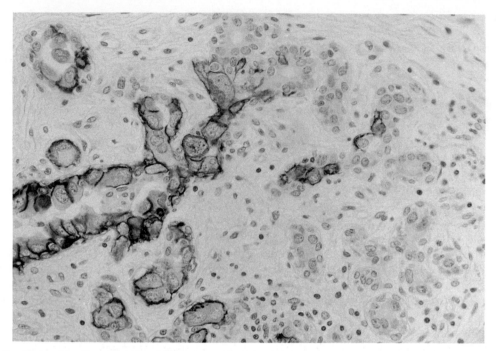

Fig. 23.3 Paraffin section of terminal duct lobular unit stained by the indrect immunoperoxidase method with monoclonal antibody to the product of the *c-erbB-2* oncogene. The terminal duct and acini on the left of the picture show infiltration by ductal carcinoma in situ whereas the acini on the right of the picture are largely uninvolved. ×300 (Photograph kindly supplied by Professor B. A. Gusterson)

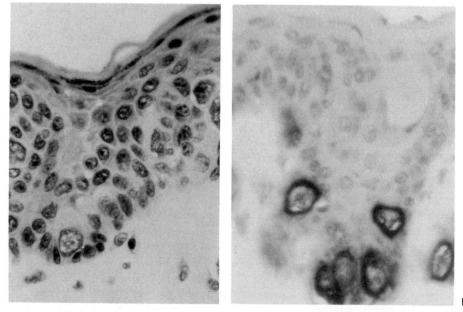

Fig. 23.4 Paraffin section of nipple showing Bowenoid form of Paget's disease stained by: **A** haematoxylin and eosin, **B** a conventional antiserum to epithelial membrane antigen using the indirect immunoperoxidase method. The malignant cells are clearly identified by the immunostaining. ×504

components may be used for identifying foci of invasion.

Basement membranes consist of an outer lamina densa, and an inner lamina lucida. The lamina lucida contains heparan sulphate proteoglycan, while both layers contain laminin and type IV collagen, which can be visualized with antibodies to these components (Birembaut et al 1985, Charpin et al 1986). Laminin and type IV collagen have a linear distribution around ducts, lobules and blood vessels.

Staining is generally weak in paraffin sections and it is usually necessary to pretreat sections with trypsin. Stronger and more consistent staining is obtained in cryostat sections, or from paraffin-embedded material fixed in methacarn (Gusterson et al 1982). However, recent antibodies against type IV collagen give strong consistent staining in paraffin sections. In the authors' experience, these immunohistological methods give clearer and more consistent visualization of the membrane than the traditional PAS or silver impregnation methods,

although the latter may be useful if appropriate immunostaining cannot be undertaken.

The basement membrane around DCIS is virtually intact, in contrast to infiltrating carcinomas where it is usually absent or present only in fragments around masses of infiltrating tumour cells, often where there is tubule formation. Some caution is required in interpreting the appearances around ducts as the membrane may show loss of definition, thickening, attenuation or small gaps. At the site of the gaps, the tumour cells may show no specific features or may protrude through the defect (Fig. 23.5). A similar phenomenon has been observed by electron microscopy and is strongly associated with intraductal malignancy in contrast to benign intraductal proliferations where the basement membrane is intact (Ozzello 1971). The prognostic significance of these small gaps in the basement membrane is not clearly understood. At present, it appears that microinvasion should only be considered when the basement membrane is either lost or grossly disrupted.

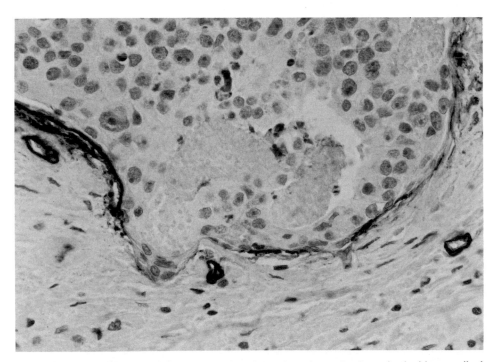

Fig. 23.5 Methacarn-fixed paraffin section of comedo variant of ductal carcinoma in situ stained with an antibody to type IV collagen by the indirect immunoalkaline phosphatase method. The basement membrane around the duct shows two defects through which small tongues of malignant cells protrude. There is also some fragmentation of the membrane at the top right of the picture. ×353
(Photograph kindly supplied by Professor B. A. Gusterson)

There are a number of methods for identifying myoepithelial cells. The traditional phosphotungstic acid haematoxylin (PTAH method) may give strong staining of myoepithelium, but this is usually achieved only when the cells exhibit a well-differentiated myoid appearance with long cell processes as seen in Figure 23.6. Myoepithelial cells of this appearance are usually encountered in involuting interlobular ducts. The more rounded cells of premenopausal ducts and lobules are visualized better by other techniques. Enzyme histochemical staining for alkaline phosphatase activity may give good visualization of myoepithelium in cryostat sections using any of the standard methods. The diazocoupling method using Naphtol AS: BI phosphate and brentamine fast Red gives the most pleasing results (Fig. 23.7). Capillary blood vessels are also strongly stained by this method.

Antibodies to the contractile proteins actin and myosin are currently the most effective means of demonstrating myoepithelial cells. Fixation and paraffin-embedding may destroy the epitopes recognized by some of the antibodies, but a number have been produced which react with paraffin sections (Bussolati et al 1980), especially if Carnoy's solution or methacarn is used as fixative (Gusterson et al 1982). Smooth muscle of vessel walls and stromal myofibroblasts also react. Antibodies to the common acute lymphoblastic leukaemia antigen (CALLA) (Gusterson et al 1986) and basal cell specific cytokeratin (Dairkee et al 1985) may also be used to demonstrate myeoepithelial cells.

In the majority of cases of DCIS, myoepithelial cells can be clearly identified at the periphery of the ducts (Fig. 23.8). In some areas, however, there is discontinuity of the myoepithelium, too extensive to be explained by attenuation of the myoepithelial cells (Gusterson et al 1982). This feature cannot be interpreted as evidence of early or incipient invasion. The major value of the myoepithelial cell staining is to establish that the structure in question is a dilated duct containing in situ tumour rather than a rounded mass of infiltrating carcinoma.

Distinguishing benign pseudoinfiltrative lesions from infiltrating carcinoma

There are two common benign breast lesions, sclerosing adenosis and the radial scar/complex sclerosing lesion, which appear to have no malignant potential, but exhibit infiltrative growth

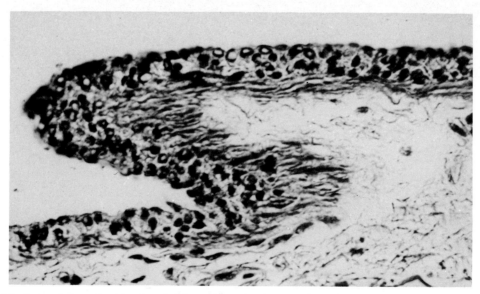

Fig. 23.6   Major breast ducts stained by the phosphotungstic acid haematoxylin method (PTAH) for myofibrils showing a well-developed myoepithelial layer. ×315

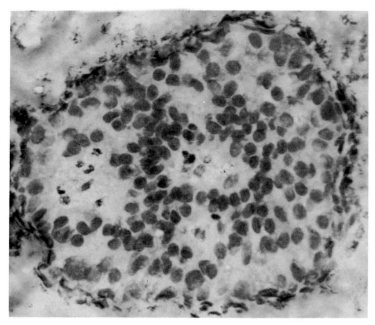

Fig. 23.7 Intraduct carcinoma stained for alkaline phosphatase by a diazo coupling method using Naphtol AS: BI phosphate and brentamine fast red R. The malignant cells are surrounded by a thin rim of myoepithelium. Haemalum counterstain. ×315

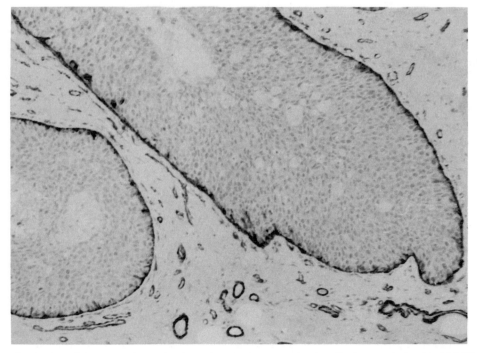

Fig. 23.8 Paraffin section of ductal carcinoma in situ stained with a monoclonal antibody to actin by the APAAP technique. Note the continuous rim of myoepithelial cells around the periphery. ×126

which may even involve perineural spaces and blood vessels (Taylor & Norris 1967, Davies 1973, Eusebi & Azzopardi 1976).

Sclerosing adenosis is a lobular proliferation in which enlarged lobules exhibit acini with distorted spiky infiltrative margins. The lesion may be confused with tubular carcinoma of the breast. In addition to the retention of lobular architecture and lack of cytological atypia, two major features of importance in distinguishing the lesion from carcinoma are the presence of myoepithelium and basement membrane around the infiltrative tubules, which may be demonstrated as described above (Figs 23.9, 23.10).

Radial scars are lesions usually less than 1 cm in size which exhibit a dense central zone of fibroelastotic tissue from which tubular structures radiate outwards. These tubules may show intraluminal hyperplasia. The complex sclerosing lesion is essentially similar in appearance, but is larger than 1 cm and exhibits a more complex pattern around the periphery, often with zones of sclerosing adenosis and papilloma formation. These lesions are particularly likely to be confused with tubular carcinoma in view of their infiltrative and more disorganized appearance, as well as the presence of elastosis. As in sclerosing adenosis, the infiltrating tubules are surrounded by myoepithelium and basement membrane which may be demonstrated with the appropriate antibodies. Elastin stains are of value in showing the distribution of elastic tissue, which is characteristically located in the centre of the lesion (Fig. 23.11), in contrast to tubular carcinomas where the distribution is more haphazard.

Determining the origin of infiltrating carcinoma in the breast

Most carcinomas encountered in the breast arise locally, but occasionally metastatic tumours from other sites are encountered. Pathologists consequently need to be wary of carcinomas in the breast with unusual appearances. Several

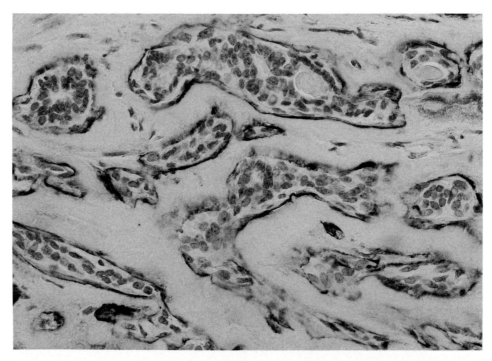

Fig. 23.9 Methacarn-fixed paraffin section of sclerosing adenosis stained with a conventional antiserum to type IV collagen by the indirect alkaline phosphatase method. Despite their infiltrative appearance, all the epithelial structures are surrounded by basement membrane. ×315
(Photograph kindly supplied by Professor B. A. Gusterson)

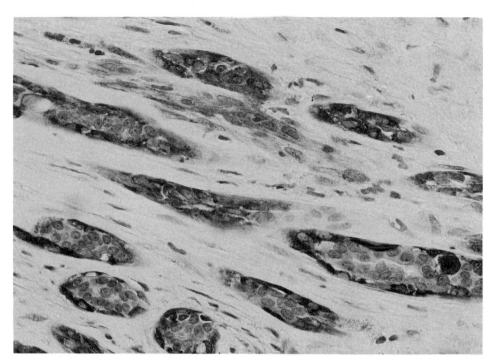

Fig. 23.10 Same case as Figure 23.9, stained with a conventional antiserum to myosin. The infiltrative epithelial structures are surrounded by myoepithelial cells. ×315
(Photograph kindly supplied by Professor B. A. Gusterson)

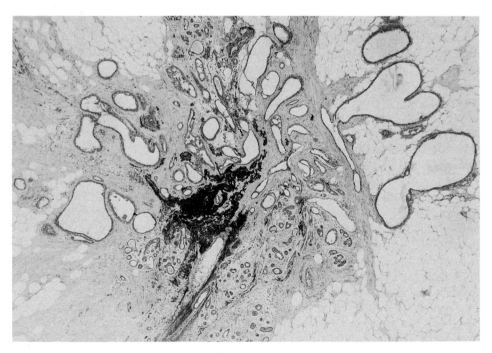

Fig. 23.11 A radial scar stained by the Weigert method for elastin which is characteristically located in the centre of the lesion. ×35

histochemical techniques can be of value in establishing that a tumour is of local origin.

The demonstration of in situ carcinoma as either DCIS or LCIS is important. Antibodies to myoepithelial cells and basement membrane components may assist in demonstrating DCIS as described earlier.

Epithelial membrane antigen (EMA) can be demonstrated to some degree in about 99% of carcinomas of the breast by immunostaining of paraffin sections. A rat monoclonal antibody, ICR.2 (Seralab) has been produced (Imrie et al, 1990) and gives staining closely similar to that of the original polyclonal antisera, as described by Sloane & Ormerod (1981). Failure to stain with ICR.2 strongly suggests that the carcinoma is not of breast origin, particularly if it is positive for cytokeratin (using CAM 5.2) and/or carcinoembryonic antigen (CEA).

Breast-specific immunohistological markers would be of great value in establishing that a breast carcinoma had arisen locally. The only antibodies with any real specificity for breast cells are those against the milk proteins, alpha-lactalbumin and casein.

Although some workers (e.g. Walker 1979) found that antibodies to alpha-lactalbumin stained breast carcinomas, we have detected the protein only in sections of pregnant or lactating breast, or in small lactational lesions occurring in non-pregnant patients. Breast carcinomas were negative even when they arose in lactating breasts (Bailey et al 1982). In experiments using alpha-lactalbumin cDNA probes, Hall et al (1981) showed that, although transcripts were present in normal mammary tissue in pregnancy and lactation, they could not be detected in human breast tumour tissue. They did, however, demonstrate in some breast tumours a peptide sharing antigenic determinants with human alpha-lactalbumin. Thus the balance of evidence at present suggests that most human breast carcinomas do not produce alpha-lactalbumin, but may secrete a similar molecule with which many antisera crossreact. An exception is the juvenile (secretory) carcinoma, but this is a rare tumour (see later).

Antisera raised by injecting rabbits with casein may react with breast carcinomas as well as other non-mammary tissues (Pich et al 1976, Fortt et al 1979). The tissue distribution of the antigen is very similar to that reported for epithelial membrane antigen (compare description of casein in Pich et al 1976 with that of EMA in Sloane & Ormerod 1981). Using a radioimmunoassay for EMA, we have shown that casein preparations can contain small amounts of EMA as an impurity (Ormerod et al 1982). It is likely that many anti-casein sera contain antibodies to EMA and this gives rise to the observed immunohistological staining patterns.

Conventional elastin stains may be of some value in deciding that a carcinoma is of breast origin. In the normal breast there is a cuff of fibrillary elastic tissue around the interlobular ducts which is absent from the terminal ducts and lobules. Larger ducts have more elastin than smaller ones and the material tends to disappear on involution. Increased amounts of elastic tissue (elastosis) have been reported in up to 86% of breast carcinomas (Azzopardi & Laurini 1974) around ducts and veins and occasionally diffusely in the stroma. It is much less prominent in medullary and colloid carcinomas. Elastosis is not specific for carcinoma and may also occur in association with benign lesions, particularly radial scars (see earlier). However, elastosis is very rarely associated with carcinoma metastatic to the breast and elastin stains may thus be of value in determining whether a breast carcinoma is of local origin.

Immunohistological stains for oestrogen receptor protein may have some role in identifying primary breast carcinomas, but, to the authors' knowledge, have not yet been fully investigated in this context.

Subtyping invasive breast carcinoma

There are many different types of breast carcinoma, but most (about 80%) are of the so-called infiltrative ductal variety. The remainder consist of a number of distinctive types, some of which are associated with a good prognosis. The identification of these subtypes is usually straightforward on H & E sections, but histochemical stains may sometimes assist in classifying them correctly.

Infiltrative lobular carcinomas consist of cells cytologically identical to those of the in situ lobular variety; they tend to infiltrate in a dissociated cell pattern with individual cells, Indian files and characteristic targetoid arrangements around ducts. Occasionally the pattern of growth may be more cohesive and the distinction from infiltrative ductal carcinoma may then be difficult. The demonstration of intracytoplasmic globules similar to those seen in LCIS (see earlier) may be helpful.

In colloid carcinomas which are associated with a better prognosis than infiltrative ductal tumours, there are islands of relatively uniform tumour cells in lakes of mucin which are hyaluronidase resistant and sialidase labile. In the rarer adenoid cystic carcinoma, there is a biphasic appearance on Alcian blue PAS staining with Alcian blue positive stromal 'cysts' and PAS positive ductal structures; this helps to distinguish it from morphologically similar carcinomas with a basaloid appearance and from cribriform intraduct carcinomas (Anthony & James 1975).

Histiocytoid carcinomas usually secrete hyaluronidase resistant, Alcian blue positive mucin, which can be used to distinguish them from granular cell myoblastoma and the various histiocytic lesions with which they may be confused (Hood et al 1973). Lipid-rich carcinomas, on the other hand, do not contain mucin but stain strongly for neutral lipid (Aboumrad et al 1963, Ramos & Taylor 1974). The rare glycogen-rich clear cell carcinoma (Azzopardi 1979, Hull et al 1981) may be identified by PAS staining before and after diastase digestion.

Some breast carcinomas exhibit argyrophilia. In many cases, double membrane bound neurosecretory granules have been demonstrated by electron microscopy (Fig. 23.12). Argyrophilic carcinomas exhibit diverse appearances. A significant proportion of colloid carcinomas are argyrophilic and show dense core granules on electron microscopy (Capella et al 1980). Partanen & Syrjanen (1981) and Azzopardi et al (1982) also reported a low incidence (under 5%) in breast carcinomas in general. Finally, a proportion of argyrophilic carcinomas exhibit morphological features in keeping with endocrine tumours of the carcinoid type. The term 'argyrophilic carcinoma' usually refers to tumours of this type. Morphologically, these tumours are fairly well circumscribed and vary in size from 1 to 5 cm. Many have occurred in the subareolar region. On histological examination they are cellular, cohesive tumours composed of uniform cells with round to ovoid anucleolate nuclei and abundant granular or clear cytoplasm. The growth pattern is characteristic of

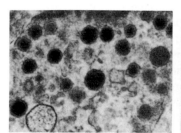
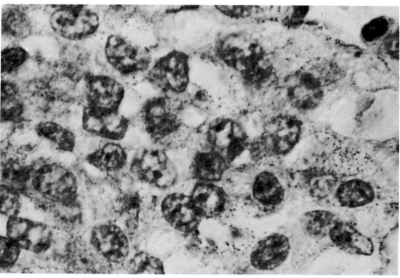

Fig. 23.12 Breast carcinoma stained by the Grimelius method for neurosecretory granules. ×1260
Inset is part of an electron micrograph of the same tumour showing that the granules are characteristic double membrane bound dense core vesicles. ×35 100

endocrine tumours, with ribbony arrangements, solid nests and perivascular pseudorosettes. The stroma is usually scanty and elastosis is not usually marked. Argyrophilic carcinomas occur in older people and may be associated with a better prognosis than ordinary infiltrative ductal carcinomas, but follow up has so far been rather short (Bussolati et al 1987).

Extensive immunohistological studies of argyrophilic carcinomas have been undertaken (Bussolati et al 1985, 1987). All were positive with antibodies to neurone specific enolase, but this does not necessarily indicate endocrine differentiation as normal myoepithelial cells may be positive. Immunostaining for prealbumin (Bussolati et al 1984) and PGP 9.5 (Rode et al 1985) was negative. About half the tumours give positive immunostaining with monoclonal antibodies to chromogranin, however (Bussolati et al 1985). These antibodies were found not to stain the benign or malignant breast lesions, but rare argyrophilic chromogranin-positive cells were seen in 7/66 blocks of normal breast tissue where they were interspersed between the epithelial and myoepithelial layers. These normal cells showed negative reactions with antibodies to calcitonin, somatostatin, bombesin, ACTH, insulin, glucagon, gastrin, VIP, GIP and cerulein (Bussolati et al 1985). The functional significance of these argyrophil cells is not yet understood.

Argentaffin positivity of breast carcinomas due to colonization by melanocytes has also been reported (Azzopardi & Eusebi 1977). It occurs in tumours which locally infiltrate the skin and appears to be a relatively common phenomenon in the superficial portions of such tumours. As the pigment may be taken up by the carcinoma cells themselves, there may be confusion with malignant melanoma. The mixture of dendritic melanocytes with tumour cells and the restriction of pigmentation to the zones of cutaneous involvement should enable the distinction to be made. Immunohistological staining with antibodies to epithelial membrane antigen and keratins should also be helpful as melanomas are negative.

Antibodies to alpha-lactalbumin generally show no staining of breast carcinomas (see earlier). We have found fairly consistent positivity, however, in tumours of the juvenile (secretory) type

(unpublished observations). These tumours are associated with a good prognosis and it is therefore important to distinguish them from other breast carcinomas. Their appearance is fairly distinctive, but immunostaining for alpha-lactalbumin may help confirm the diagnosis and also to exclude the possibility of an unusual metastatic tumour.

Diagnosing non-epithelial invasive malignancy

The two types of tumour most frequently mistaken for carcinoma (especially on frozen section) are malignant lymphomas and spindle cell sarcomas. The former may be mistaken for high-grade carcinomas lacking tubular differentiation, usually of the infiltrative ductal or medullary types. The solid variant of infiltrative lobular carcinoma may also cause problems. Conventional histochemical stains for mucin and elastin may be of value, as intracellular mucin and stromal elastosis are not encountered in lymphomas. The most effective way of making the distinction, however, is using immunohistological techniques with antibodies to leucocyte common antigen (Pizzolo et al 1980, Warnke et al 1983) and a panel of epithelial markers. Antibodies to EMA almost invariably give positivity on breast carcinomas, as mentioned earlier, but they also react with a small proportion of malignant lymphomas (Delsol et al 1984, Gatter et al 1985). It is therefore advisable to use a panel of epithelial markers combining EMA with antibodies to low molecular weight keratins (e.g. CAM 5.2) and CEA.

A minority of breast carcinomas exhibit spindle cell morphology and consequently may be misdiagnosed as spindle cell sarcomas. The epithelial marker panel described above may be useful for revealing epithelial differentiation. Antibodies reacting with higher molecular weight keratins may also be useful as spindle cell differentiation is not infrequently associated with squamous carcinomas of the breast. At the time of writing there are reliable markers only for specific mesenchymal cells, e.g. desmin and myoglobin for muscle, S100 protein for nerve sheath, and none for fibrosarcomas or undifferentiated sarcomas which form the majority of soft tissue tumours likely to be confused with spindle cell carcinoma in the breast (see Ch. 21). Mesenchymal tumours are thus frequently diagnosed by exclusion with immunohistological methods.

PROGNOSIS

Oestrogen receptors

About two-thirds of mammary carcinomas express oestrogen receptors and about half of these tumours respond to endocrine therapy. Oestrogen receptor positivity is also associated with an improved prognosis. While this measurement is usually made biochemically, oestrogen receptor can be visualized immunohistologically using antibodies to the oestrogen receptor protein (King & Greene 1984). An immunohistological assay has the advantage of overcoming sampling problems, ensuring that tumour tissue is evaluated and that the results are not influenced by the amount of stroma. In a biochemical assay, an average value is obtained of the whole tissue, which is sampled by macroscopic examination alone. If the tumour cells are in a minority, a falsely low reading may result. Furthermore, immunohistological examination can be made on small lesions without compromising the histological diagnosis. Needle biopsies and fine needle aspiration samples may also be investigated. These advantages are important in breast cancer screening where many of the detected lesions are small.

Most work has been undertaken with the monoclonal antibody H222 (Abbott Laboratories). At the time of writing, the high cost of this reagent has limited research in this area.

Immunostaining is usually undertaken on cryostat sections fixed in buffered formalin or paraformaldehyde (Charpin et al 1988, Reiner et al 1988). Semiquantitative estimates of the degree of positivity are often attempted by scoring the percentage of positive cells and intensity of staining. Immunostaining for oestrogen receptors using antibody H222 has shown a high level of concordance with results obtained by the conventional dextran charcoal assay (Charpin et al 1988) and has been found to be predictive of response to endocrine therapy (McClelland et al 1986).

Immunostaining varies with tumour type. Reiner et al (1988) found positivity in 67% of infiltrative ductal carcinomas, 84% of infiltrative

lobular carcinomas and about 50% of ductal carcinomas in situ. Carcinomas of the colloid, tubular and papillary types were almost invariably positive, whereas medullary carcinomas were usually negative. In infiltrative ductal carcinoma, positivity was related to low tumour grade and small size. In all cases staining is restricted to nuclei (Fig. 23.13); immunoelectron microscope studies undertaken by Charpin et al (1988) showed diffuse nuclear positivity with negativity of the nucleolus. Different fields and levels of the same blocks tended to show a similar distribution of staining, but different areas of large tumours sometimes showed significant variations.

In normal tissues and benign lesions, patchily positive cells are seen in normal lobules and ducts as well as in adenosis, epitheliosis and fibroadenomas. Weaker staining is seen in cystic disease (Charpin et al 1988). No significant differences have been observed in pre- or postmenopausal breasts.

There have also been reports of oestrogen receptor staining in paraffin sections, usually after treating the sections with DNase with or without trypsinization (Shintaku & Said 1987, Hiort et al 1988, Andersen et al 1988). A high level of concordance has been reported with the standard dextran charcoal assay and the immunohistological technique on paraffin sections (Shintaku & Said 1987, Andersen et al 1988). Although positivity can be obtained in paraffin blocks stored for long periods, fixation appears to be critical and routine handling is often suboptimal (De Rosa et al 1987). Fixation in formol saline for 4–12 hours has been found to give good results (Shintaku & Said 1987, De Rosa et al 1987) and enzyme pretreatment may improve staining (Andersen et al 1988). It is clear that laboratories wishing to undertake immunostaining for oestrogen receptors on formalin-fixed, paraffin-embedded sections should evaluate the methodology very carefully before undertaking a clinical service.

The presence of oestrogen-dependent progesterone receptors is also correlated with a response to endocrine therapy. The successful staining of these receptors in routinely-fixed, paraffin-embedded tumours, using the Abbott Laboratories monoclonal antibody KD68, has been reported (Muller-Holzner et al 1989).

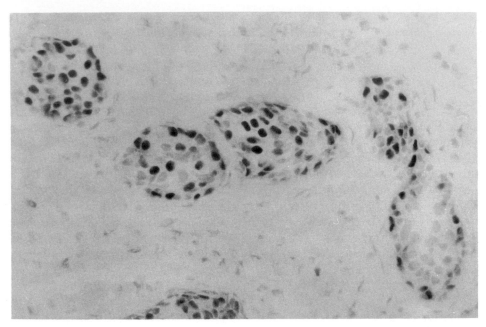

Fig. 23.13 Frozen section of lobular carcinoma in situ stained for oestrogen receptor using antibody H222 (Abbott). A high proportion of cells show nuclear staining. ×504

Monoclonal antibodies have also been obtained to an oestrogen receptor related protein (ER-D5; King et al 1985) and an oestrogen regulated protein (P24; Adams et al 1983). It has been claimed that there is a good correlation between the presence of oestrogen receptors and the expression of D5 (Cano et al 1986) and P24 (Adams et al 1985) but this has not been confirmed by other workers (Giri et al 1987, Horne et al 1988).

Epidermal growth factor receptor

Nicholson et al (1988) have reported a strong inverse correlation between receptors for epidermal growth factor (EGFr) (as measured by binding of EGF to membranes) and oestrogen receptors as well as a positive correlation with early recurrence of carcinoma and death. In a study of EGFr status in various histological subtypes of breast cancer, Sainsbury et al (1988) found that 34% of infiltrating ductal and 21% of infiltrating lobular carcinomas were positive. All three tubular carcinomas were negative as well as two out of three mucoid carcinomas. EGFr can also be

demonstrated immunohistochemically and, for the reasons given above in respect of oestrogen receptors, this method may offer some advantages over the biochemical assay. Using cryostat sections fixed in acetone, Horne et al (1988) found a significant correlation between immunostaining for EGFr and the biochemical assay.

Oncogene products

Receptors for various growth factors and the predicted products of some of the oncogenes are closely related. This holds for the oncogene, *c-erbB-2 (HER2/neu)* and EGFr. In many tumours, the *c-erbB-2* gene is amplified and there is a good correlation in breast tumours between gene amplification and immunohistochemical reaction on paraffin sections with an antibody raised against a predicted peptide sequence of the gene product (Gusterson et al 1988c). Slamon et al (1987) have reported that amplification of the oncogene has prognostic significance, but Barnes et al (1988) were unable to demonstrate a significant association between positive immunostaining and clinical outcome although there was a tendency for patients with stained tumours to have a worse prognosis.

Other oncogenes, such as c-*myc*, are amplified in a proportion of tumours (Escot et al 1986) and the gene may be transcribed (Biunno et al 1988). The significance of these findings is as yet unclear.

Epithelial membrane antigen and related antigens

It has been claimed that the extent of expression of some of the epitopes on EMA (as revealed by staining with certain monoclonal antibodies) has prognostic significance (Ellis et al 1985, Wilkinson et al 1984), although this has not been confirmed by later studies (Berry et al 1985, Angus et al 1986). It is possible that these differences might be related to different methodologies. Also, some variations in expression of EMA may reflect the degree of differentiation, which has long been recognized as a prognostic factor in mammary tumours. The link between a good prognosis and EMA staining is too weak to justify using the immunohistochemical stain for this purpose.

Lectins

Recent studies have shown a relationship between affinity of breast carcinomas for certain lectins, and prognosis. Fenlon et al (1987) reported that the proportion of cells staining with a lectin from *Ulex europeus*, related to earlier local recurrence and shorter survival. Staining with a lectin from the snail *Helix pomatia* related to lymph node stage, time to local or regional recurrence and survival. Leathem & Brooks (1987) found that patients whose tumours stained with the *Helix pomatia* lectin exhibited a shorter time to first recurrence of tumour and shorter survival time. These relationships held only for premenopausal patients in the latter study, however. These findings are currently undergoing further evaluation.

Markers of proliferation

The fraction of cells in the S phase of the cell cycle (SPF) and the thymidine labelling index (TLI) in-dicate the proportion of carcinoma cells synthesizing DNA and consequently reflect the rate of proliferation. TLI and SPF relate to nuclear anaplasia, correlate inversely with oestrogen and progesterone receptor content, have a weak positive correlation with tumour size, but have no relationship to axillary lymph nodal status. High TLI relates to aneuploidy and is associated with a high risk of early relapse after primary therapy (for review, see Meyer et al 1984). Neither SPF nor TLI can be applied to routine diagnostic histopathology; the former is measured by flow cytometry (see Hedley et al 1987) and the latter by autoradiography after culturing tumour in medium containing radiolabelled thymidine.

Antibodies which react with 5'-bromo-deoxyuridine (BrdU), an analogue of thymidine, have been produced (see references in Gray & Mayall 1985). In combination with an immunohistochemical technique, this permits the use of a non-radioactive precursor for the measurement of TLI (e.g. see Nagashima et al 1988). However, either the fresh tumour must be incubated in medium containing BrdU (see Appendix 2) or BrdU must be injected into the patient prior to surgery.

A monoclonal antibody, Ki-67 (Gerdes et al 1983) reacts in cryostat sections with an antigen expressed in the nuclei of cells in all phases of the cell cycle except G_0 (Gerdes et al 1984). The precise relationship to cell cycle is not entirely clear at present, but Ki-67 positivity is undoubtedly related in some way to cell proliferation (Walker & Camplejohn 1988). Barnard et al (1987) recently found that Ki-67 score correlated strongly with mitotic index and weakly with tumour grade, necrosis and cellular reaction. No relationship with tumour size, lymph nodal status or oestrogen receptor levels was found. Further work is needed to evaluate the prognostic value of this antibody.

Demonstration of vascular invasion

Vascular invasion may be an important prognostic factor (Bettelheim & Neville 1981) and immunostaining for factor VIII related antigen or collagen type IV can assist in the identification of small blood and lymphatic vessels containing tumour

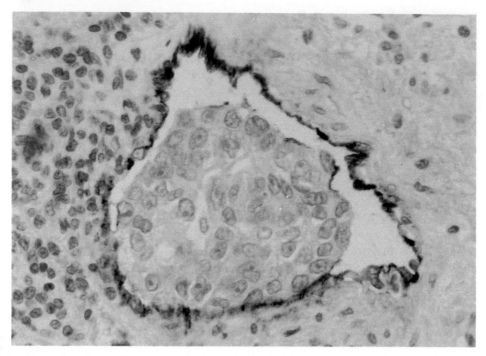

Fig. 23.14 Paraffin section of a small mammary blood vessel stained with a conventional antiserum to factor VIII related antigen after trypsinization. Note the strong endothelial cell staining around the small clump of carcinoma cells in the lumen. Indirect inmmunoperoxidase. ×504

(Fig. 23.14) (Bettelheim et al 1984). Staining for factor VIII related antigen can be undertaken after trypsinization on paraffin sections, but stronger staining is usually obtained in cryostat sections or paraffin-embedded tissue fixed in methacarn.

Staging of breast cancer

Approximately 99% of breast carcinomas express EMA. Although EMA is an antigen found on epithelial surfaces in a wide range of normal tissues, it has a useful role in visualizing small numbers of infiltrating mammary carcinoma cells in non-epithelial environments. EMA is frequently found in the cytoplasm and around the whole plasma membrane in infiltrating mammary carcinoma cells, in contrast to normal breast cells where expression is limited to the luminal surface.

Carcinoma cells may be detected using EMA antibodies in sections of biopsies taken from patients for staging purposes where conventional stains give equivocal results (Sloane et al 1983) (Fig. 23.15).

Antibodies to EMA have been studied extensively in detecting single carcinoma cells in bone marrow. Anti-keratin antibodies have also been used for this purpose (Schlimok et al 1987). If aspirates of bone marrow are taken at the time of an operation to remove a primary breast tumour, single metastatic tumour cells can be found in samples from approximately 27% of the patients (Redding et al 1983). On first consideration, it might be thought that identifying tumour cells at distant sites at presentation would indicate a poor prognosis. A careful study of over 300 patients showed that this was only a weak prognostic factor, although it is stronger if the numbers of cells detected are taken into account and within certain subsets of patients (Coombes et al 1986). At present, the work involved in preparing smears, staining for EMA and screening is not justified as a routine procedure.

The presence of tumour in axillary lymph nodes is important in tumour staging. In our study of 545 nodes from 39 patients, immunostaining for EMA revealed a greater extent of infiltration in

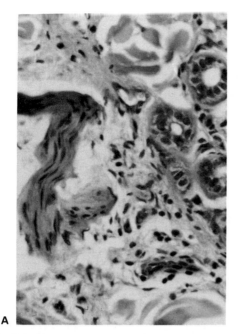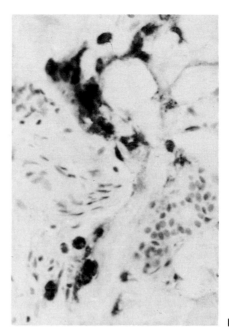

A   B

Fig. 23.15 Skin biopsy taken from the vicinity of a mastectomy scar of a patient with breast carcinoma (**A**). H & E stain showing dermal infiltration between a sweat gland and a nerve by cells with small, densely staining nuclei and copious cytoplasm. **B** Indirect immunoperoxidase method for epithelial membrane antigen showing the epithelial nature of the infiltrating cells. ×315

those nodes already known to contain tumour but the number of positive nodes was not increased (Sloane et al 1980). In contrast, when Wells et al (1984) made a similar study of nodes from 45 patients reported as node negative by conventional staining, they found that 7 cases were positive. The apparent discrepancy between these studies might lie in the sensitivity of the methods used or in a difference in the original reporting. With experience of examining micrometastases visualized

by immunohistochemical stains, the pathologist can revise the criteria for identifying small deposits and may consequently improve his interpretation of conventionally stained sections. However, it is often quicker and easier to pick up small tumour deposits using immunohistochemical stains. Whether the scoring of extra small deposits has any prognostic significance is open to doubt (Wilkinson et al 1982), but worthy of further investigation.

REFERENCES

Aboumrad M H, Horn R C, Fine G 1963 Lipid-secreting mammary carcinoma: report of a case associated with Paget's disease of the nipple. Cancer 16: 521–525

Adams D J, Haji H, Edwards D P, Bjercke J, McGuire L 1983 Detection of a M 24 000 estrogen-regulated protein in human breast cancer by monoclonal antibodies. Cancer Research 43: 4297–4301

Adams D J, McGuire W L 1985 Quantitative enzyme-linked immunosorbent assay for the estrogen-regulated M 24 000 protein in human breast tumors: correlation with estrogen and progesterone receptors. Cancer Research 45: 2445–2449

Andersen J, Orntoft T F, Poulsen H S 1988

Immunohistochemical demonstration of estrogen receptors (ER) in formalin-fixed, paraffin-embedded human breast cancer tissue by use of a monoclonal antibody to ER. Journal of Histochemistry and Cytochemistry 36: 1553–1560

Angus B, Napier J, Purvis J et al 1986 Survival in breast cancer related to tumour oestrogen receptor status and immunohistochemical staining for NCRC 11. Journal of Pathology 149: 301–306

Anthony P P, James P D 1975 Adenoid cystic carcinoma of the breast: prevalence, diagnostic criteria, and histogenesis. Journal of Clinical Pathology 28: 647–655

Ashikari R, Hajdu S I, Robbins G F 1971 Intraductal carcinoma of the breast (1960–1969). Cancer 28: 1182–1187

Azzopardi J G 1979 Problems in breast pathology W B Saunders, Philadelphia

Azzopardi J G, Laurini R N 1974 Elastosis in breast cancer. Cancer 33: 174–183

Azzopardi J G, Eusebi V 1977 Melanocyte colonization and pigmentation of breast carcinoma. Histopathology 1: 21–30

Azzopardi J G, Muretto P, Goddeeris P, Eusebi V, Lauweryns J M 1982 'Carcinoid' tumours of the breast: the morphological spectrum of argyrophil carcinomas. Histopathology 6: 549–569

Bailey A J, Sloane J P, Trickey B A, Ormerod M G 1982 An immunocytochemical study of α-lactalbumin in human breast tissue. Journal of Pathology 137: 13–23

Barnard N J, Hall P A, Lemoine N R, Kadar N 1987 Proliferative index in breast carcinoma determined in situ by Ki67 immunostaining and its relationship to clinical and pathological variables. Journal of Pathology 152: 287–295

Barnes D M, Lammie G A, Millis R R, Gullick W L, Allen D S, Altman D G 1988 An immunohistochemical evaluation of c-erbB-2 expression in human breast carcinoma. British Journal of Cancer 58: 448–452

Berry N, Jones D B, Smallwood J, Taylor I, Kirkham N, Taylor-Papadimitriou J 1985 The prognostic value of the monoclonal antibodies HMFG1 and HMFG2 in breast cancer. British Journal of Cancer 51: 179–186

Bettelheim R, Neville A M 1981 Lymphatic and vascular channel involvement within infiltrative breast carcinomas as a guide to prognosis at the time of primary surgical treatment. Lancet ii: 631

Bettelheim R, Mitchell D, Gusterson B A 1984 Immunocytochemistry in the identification of vascular invasion in breast cancer. Journal of Clinical Pathology 37: 364–366

Birembaut P, Caron Y, Adnet J-J 1985 Usefulness of basement membrane markers in tumoural pathology. Journal of Pathology 145: 283–296

Biunno I, Pozzi M R, Pierotti M A, Pilotti S, Cattoretti G, Della Porta G 1988 Structure and expression of oncogenes in surgical specimens of human breast carcinomas. British Journal of Cancer 57: 464–468

Bussolati G, Papotti M, Sapino A 1984 Binding of antibodies against human prealbumin to intestinal and bronchial carcinoids and to pancreatic endocrine tumours. Virchows Archive (Cell Pathol) 45: 15–22

Bussolati G, Alfani V, Weber K, Osborn M 1980 Immunocytochemical detection of actin on fixed and embedded tissues: its potential use in routine pathology. Journal of Histochemistry and Cytochemistry 28: 169–173

Bussolati G, Gugliotta P, Sapino A, Eusebi V, Lloyd R V 1985 Chromogranin-reactive endocrine cells in argyrophilic carcinomas ('carcinoids') and normal tissue of the breast. American Journal of Pathology 120: 186–192

Bussolati G, Papotti M, Sapino A, Gugliotta P, Ghiringhello B, Azzopardi J G 1987 Endocrine markers in argyrophilic carcinomas of the breast. American Journal of Surgical Pathology 11: 248–256

Cano A, Coffer A I, Adatia R, Millis R R, Rubens R D, King R B J 1986 Histochemical studies with an estrogen receptor related protein in human breast tumors. Cancer Research 46: 6475–6480

Capella C, Eusebi V, Mann B, Azzopardi J G 1980 Endocrine differentiation in mucoid carcinoma of the breast. Histopathology 4: 613–630

Charpin C, Lisitzky J C, Jacquemier J et al 1986 Immunohistochemical detection of laminin in 98 human breast carcinomas: a light and electron microscopic study. Human Pathology 17: 355–365

Charpin C, Martin P M, De Victor B et al 1988 Multiparametric study (SAMBA 200) of estrogen receptor immunocytochemical assay in 400 human breast carcinomas: analysis of estrogen receptor distribution heterogeneity in tissues and correlations with dextran coated charcoal assays and morphological data. Cancer Research 48: 1578–1586

Coombes R C, Berger U, Mansi J et al 1986 Prognostic significance of micrometastases in bone marrow in patients with primary breast cancer. NCI Monograph 1: 51–53

Dairkee S H, Blayney C, Smith H S, Hackett A J 1985 Monoclonal antibody that defines human myoepithelium. Proceedings of the National Academy of Sciences USA 82: 7409–7413

Davies J D 1973 Neural invasion in benign mammary dysplasia. Journal of Pathology 109: 225–231

Delsol G, Stein H, Pulford K A F et al 1984 Human lymphoid cells express epithelial membrane antigen: implications for diagnosis of human neoplasms. Lancet ii: 1124–1128

De Rosa C M, Ozzello L, Greene G L, Habif D V 1987 Immunostaining of estrogen receptor in paraffin sections of breast carcinomas using monoclonal antibody D75P3: effects of fixation. American Journal of Surgical Pathology 11: 943–950

De Vijver M J, Peterse J L, Mooi W J et al 1988 Neu-protein overexpression in breast cancer: association with comedo-type ductal carcinoma in situ and limited prognostic value in stage II breast cancer. The New England Journal of Medicine 319: 1239–1245

Ellis I O, Hinton C P, MacNay J et al 1985 Immunocytochemical staining of breast carcinoma with the monoclonal antibody NCRC 11: a new prognostic indicator. British Medical Journal 290: 881–883

Escot C, Theillet C, Lidereau R et al 1986 Genetic alteration of the c-myc protooncogene (MYC) in human primary breast carcinomas. Proceedings of the National Academy of Sciences USA 83: 4834–4838

Eusebi V, Azzopardi J G 1976 Vascular infiltration in benign breast disease. Journal of Pathology 118: 9–25

Fenlon S, Ellis I O, Bell J, Todd J H, Elston C W, Blamey R W 1987 Helix pomatia and Ulex europeus lectin binding in human breast carcinoma. Journal of Pathology 152: 169–176

Fortt R W, Gibbs A R, Williams D, Hansen J, Williams I 1979 The identification of 'casein' in human breast cancer. Histopathology 3: 395–406

Gad A, Azzopardi J G 1975 Lobular carcinoma of the breast: a special variant of mucin-secreting carcinoma. Journal of Clinical Pathology 28: 711–716

Gatter K C, Heryet A, Alcock C, Mason D Y 1985 Clinical importance of analysing malignant tumours of uncertain origin with immunohistological techniques. Lancet i: 1302–1305

Gerdes J, Schwab U, Lemke H, Stein H 1983 Production of a mouse monoclonal antibody reactive with a human nuclear antigen associated with cell proliferation. International Journal of Cancer 31: 13–20

Gerdes J, Lemke H, Baisch H, Wacker H H, Schwab U, Stein H 1984 Cell cycle analysis of a cell proliferation associated human nuclear antigen defined by the monoclonal antibody Ki67. Journal of Immunology 133: 1710–1716

Giri D D, Dangerfield V J M, Lonsdale R, Rogers K, Underwood J C E 1987 Immunohistology of oestrogen receptor and D5 antigen in breast cancer: correlation with oestrogen receptor content of adjacent cryostat sections assayed by radioligand binding and enzyme immunoassay. Journal of Clinical Pathology 40: 734–740

Gray J W, Mayall B H (eds) 1985 Monoclonal antibodies against bromodeoxyuridine. Alan R Liss, New York

Gusterson B A, Monaghan P, Mahendran R, Ellis J, O'Hare K J 1986 Identification of myoepithelial cells in human and rat breast by anti-common acute lymphoblastic leukemia antigen antibody A12. Journal of National Cancer Institute 77: 343–349

Gusterson B A, Warburton M J, Mitchell D, Ellison M, Neville A M, Rudland P S 1982 Distribution of myoepithelial cells and basement membrane proteins in the normal breast and in benign and malignant breast diseases. Cancer Research 42: 4763–4670

Gusterson B A, Machin L G, Gullick W J et al 1988a Immunohistochemical distribution of c-erbB-2 in infiltrating and in situ breast cancer. International Journal of Cancer 42: 842–845

Gusterson B A, Machin L G, Gullick W J et al 1988b c-erbB-2 expression in benign and malignant breast disease. British Journal of Cancer 58: 453–457

Gusterson B A, Gullick W J, Venter D J et al 1988c Immunohistochemical localization of c-erbB-2 in human breast carcinomas. Molecular and Cellular Probes 2: 383–391

Hall L, Craig R K, Davies M S, Ralphs D N L, Campbell P N 1981 Lactalbumin is not a marker of human hormone-dependent breast cancer. Nature 290: 602–604

Hedley D W, Rugg C A, Gelber R D 1987 Association of DNA index and S-phase fraction with prognosis of nodes positive early breast cancer. Cancer Research 47: 4729–4735

Hiort O, Kwan P W L, DeLellis R A 1988 Immunohistochemistry of estrogen receptor protein in paraffin sections: effects of enzymatic pretreatment and cobalt chloride intensification. American Journal of Pathology 90: 559–563

Hood C I, Font R L, Zimmerman L E 1973 Metastatic mammary carcinoma in the eyelid with histiocytoid appearance. Cancer 31: 793–800

Horne G M, Angus B, Wright C et al 1988 Relationships between oestrogen receptor, epidermal growth factor receptor, ER-D5, and P24 oestrogen regulated protein in human breast cancer. Journal of Pathology 155: 143–150

Hull M T, Priest J B, Broadie T A, Ransburg R C, McCarthy L J 1981 Glycogen-rich clear cell carcinoma of the breast: a light and electron microscopic study. Cancer 48: 2003–2009

Imrie S F, Sloane J P, Ormerod M G et al 1990 Detailed investigation of the diagnostic value in tumour pathology of ICR.2, a new monoclonal antibody to EMA. Histopathology In Press.

King W J, Greene G L 1984 Monoclonal antibodies localize oestrogen receptor in the nuclei of target cells. Nature 307: 745–747

King R J B, Coffer A I, Gilbert J et al 1985 Histochemical studies with a monoclonal antibody raised against a partially purified soluble estradiol receptor preparation from human myometrium. Cancer Research 45: 5728–5733

Lagios M D, Westdahl P R, Margolin F R, Rose M R 1982 Duct carcinoma in situ: relationship of extent of noninvasive disease to the frequency of occult invasion, multicentricity, lymph node metastases, and short-term treatment failures. Cancer 50: 1309–1314

Leathem A J, Brooks S A 1987 Predictive value of lectin binding on breast-cancer recurrence and survival. Lancet i: 1054–1056

Makin C A, Bobrow L G, Bodmer W F 1984 Monoclonal antibody to cytokeratin for use in routine histopathology. Journal of Clinical Pathology 37: 975–983

McClelland R A, Berger U, Miller L S, Powles T J, Jensen E V, Coombes R C 1986 Immunocytochemical assay for estrogen receptor: relationship to outcome of therapy in patients with advanced breast cancer. Cancer Research (Suppl) 46: 4241s–4243s

Meyer J S, McDivitt R W, Stone K R, Prey M U, Bauer W C 1984 Practical breast carcinoma cell kinetics: review and update. Breast Cancer Research and Treatment 4: 79–88

Muller-Holzner E, Zeimet A, Muller L C, Daxenbichler G, Dapunt O 1989 Monoclonal technique to aid decision on endocrine therapy in breast cancer. Lancet i: 1147–1148

Nagashima T, Hoshino T, Cho K G, Senegor M, Waldman F, Nomura K 1988 Comparison of bromodeoxyuridine labeling indices obtained from tissue sections and flow cytometry of brain tumors. Journal of Neurosurgery 68: 388–392

Nicholson S, Sainsbury J R C, Needham G K, Chambers P, Farndon J R, Harris A L 1988 Quantitative assays of epidermal growth factor receptor in human breast cancer: cut-off points of clinical relevance. International Journal of Cancer 42: 36–41

Ormerod M G, Bussolati G, Sloane J P, Steele K, Gugliotta P 1982 Similarities of antisera to casein and epithelial membrane antigen. Virchows Archiv (Pathol Anat) 397: 327–333

Ozzello L 1971 Ultrastructure of the human mammary gland. Pathology Annual 6: 1–59

Partanen S, Syrjanen K 1981 Argyrophilic cells in carcinoma of the female breast. Virchows Archiv (Pathol Anat) 391: 45–51

Pich A, Bussolati G, Carbonara A 1976 Immunocytochemical detection of casein and casein-like proteins in human tissues. Journal of Histochemistry and Cytochemistry 24: 940–947

Pizzolo G, Sloane J, Beverley P et al 1980 Differential diagnosis of malignant lymphoma and nonlymphoid tumors using monoclonal anti-leucocyte antibody. Cancer 46: 2640–2647

Ramos C V, Taylor H B 1974 Lipid-rich carcinoma of the breast: a clinicopathologic analysis of 13 examples. Cancer 33: 812–819

Redding W H, Monaghan P, Imrie S F et al 1983 Detection of micrometastases in patients with primary breast cancer. Lancet ii: 1271

Reiner A, Reiner G, Spona J, Schemper M, Holzner J H 1988 Histopathologic characterization of human breast cancer in correlation with estrogen receptor status: a comparison of immunocytochemical and biochemical analysis. Cancer 61: 1149–1154

Rode J, Dhillon A P, Doran J F, Jackson P, Thompson R J 1985 PGP 9.5, a new marker for human neuroendocrine tumours. Histopathology 9: 147–158

Sainsbury J R C, Nicholson S, Angus B, Farndon J R, Malcolm A J, Harris A L 1988 Epidermal growth factor receptor status of histological sub-types of breast cancer. British Journal of Cancer 58: 458–460

Schlimok G, Funke I, Holzmann B et al 1987 Micrometastatic cancer cells in bone marrow: in vitro detection with anti-cytokeratin and in vivo labeling with anti-17-1A monoclonal antibodies. Proceedings of the National Academy of Sciences USA 84: 8672–8676

Shintaku I P, Said J W 1987 Detection of estrogen receptors with monoclonal antibodies in routinely processed formalin-fixed paraffin sections of breast carcinoma: use of DNase pretreatment to enhance sensitivity of the reaction. American Journal of Clinical Pathology 87: 161–167

Slamon D J, Clark G M, Wong S G, Levin W J, Ullrich A, McGuire W L 1987 Human breast cancer: correlation of relapse and survival with amplification of the HER-2/neu oncogene. Science 235: 177–182

Sloane J P, Ormerod M G 1981 Distribution of epithelial membrane antigen in normal and neoplastic tissues and its value in diagnostic tumor pathology. Cancer 47: 1786–1795

Sloane J P, Hughes F, Ormerod M G 1983 An assessment of the value of epithelial membrane antigen and other epithelial markers in solving diagnostic problems in tumour histopathology. Histochemical Journal 15: 645–654

Sloane J P, Ormerod M G, Imrie S F, Coombes R C 1980 The use of antisera to epithelial membrane antigen in detecting micrometastases in histological sections. British Journal of Cancer 42: 393–398

Taylor H B, Norris H J 1967 Epithelial invasion of nerves in benign diseases of the breast. Cancer 20: 2245–2249

Walker R A 1979 The demonstration of α-lactalbumin in human breast carcinomas. Journal of Pathology 129: 37–42

Walker R A, Camplejohn R S 1988 Comparison of monoclonal antibody Ki-67 reactivity with grade and DNA flow cytometry of breast carcinomas. British Journal of Cancer 57: 281–283

Warnke R A, Gatter K C, Falini B et al 1983 Diagnosis of human lymphoma with monoclonal antileukocyte antibodies. New England Journal of Medicine 309: 1275–1281

Wells C A, Heryet A, Brochier J, Gatter K C, Mason D Y 1984 The immunocytochemical detection of axillary micrometastases in breast cancer. British Journal of Cancer 50: 193–197

Westbrook K C, Gallager H S 1975 Intraductal carcinoma of the breast: a comparative study. American Journal of Surgery 130: 667–670

Wilkinson E J, Hause L L, Hoffman R G et al 1982 Occult axillary lymph node metastases in invasive breast carcinoma: characteristics of the primary tumor and significance of the metastases. Pathology Annual 17: 67–91

Wilkinson M J, Howell A, Harris M, Taylor-Papadimitriou J, Swindell R, Sellwood R A 1984 The prognostic significance of two epithelial membrane antigens expressed by human mammary carcinomas. International Journal of Cancer 33: 299–304

24. Immunohistological approach to the diagnosis of ovarian and testicular neoplasms

A. E. Sherrod   C. R. Taylor   N. E. Warner

Willis, in *Pathology of Tumours* (Willis 1948) prefaces the chapter pertaining to ovarian neoplasms with the following statement:

Several factors have combined to cause confusion in their [ovarian tumours] classification and nomenclature, namely (1) uncertainties as to the histogenetic relationships of some of the tissues of the normal ovary, (2) uncertainties as to the derivation of particular tumours from particular ovarian tissues, (3) practical difficulties in some cases of distinguishing growths of ovarian origin from those of parovarian or tubal origin, and (4) the frequent mistaking of secondary for primary growths in the ovaries.

In his chapter on tumours of the testis, Willis encounters a similar veil of confusion. Critical review of the literature reveals that with the passing of time one particular concept may attain popularity and become so widely accepted that it is mistaken for fact. Within the study of testicular and ovarian neoplasms, difficulties of classification are compounded by the lack of a firm scientific basis for the recognition and distinction of individual normal cells, and their neoplastic counterparts. It is often forgotten that normally proliferating cells of different cell lines may appear remarkably similar; tumours composed of a high proportion of proliferating cells may appear similar on this basis alone. Often pathologists distinguish such tumours, not by the majority components — for these are quite similar, but by a minority population displaying some degree of differentiation along one or another cell line. In the final analysis, such distinctions, based as they are on subtle subjective morphological judgments, may be very difficult to make.

CLASSIFICATION OF OVARIAN AND TESTICULAR TUMOURS

The basic classifications of ovarian and testicular neoplasms used within this chapter have been adopted on the basis of convenience and popularity. The adoption of these classifications is not intended to convey support for their scientific rectitude, or any preference for these schemata over those championed by other investigators. The classification of testicular neoplasms (Table 24.1) is based on that used by Mostofi & Price in the Armed Forces Institute of Pathology (AFIP) Atlas (Mostofi & Price 1973). This in turn was founded upon the conceptual view advanced by Friedman

Table 24.1 Classification of testicular neoplasms (modified from Mostofi & Price 1973)

1. Germ cell tumours:
   a. Single histological type:
     (i) Seminoma
     (ii) Embryonal carcinoma
     (iii) Choriocarcinoma
     (iv) Teratoma
   b. More than one histological type:
       Any combinations of type within (a)

2. Gonadal stromal tumours:
   a. Undifferentiated
   b. Leydig cell
   c. Sertoli (granulosa-theca) cell
   d. Combinations of these

3. Mixed germ cell and gonadal stromal tumours

4. Metastatic and other

Note: The British Testicular Tumour Panel (BTTP) classifies germ cell tumours as follows: 1. Seminoma, 2. malignant teratoma — a) differentiated, b) intermediate, c) undifferentiated, d) trophoblastic, and recognizes overlap between these groups.

Table 24.2 Classification of ovarian neoplasms (modified from WHO, Serov et al 1973) (bracket entries are relatively rare; they are included to maintain the parallel with Table 24.1)

1. Germ cell tumours:
   a. Dysgerminoma (seminoma-like)
   b. (Embryonal carcinoma)
   c. Choriocarcinoma
   d. Teratoma
   e. Endodermal sinus (yolk sac) tumour
   f. (Mixtures of the above)

2. Gonadal stromal tumours:
   a. (Undifferentiated)
   b. Sertoli-Leydig cell (androblatoma)
   c. Granulosa-theca cell

3. Mixed germ cell and gonadal stromal tumours

4. Tumours of ovarian surface epithelium:
   a. Serous cystadenoma and carcinoma
   b. Mucinous cystadenoma and carcinoma
   c. Endometrioid carcinoma
   d. Undifferentiated carcinoma

5. Metastatic and other

and Moore that embryonal carcinomas, teratomas and choriocarcinomas all originate from the pluripotent germ cells. Approximately 93% of the AFIP cases can then be considered to be of germ cell origin.

The simplistic classification given for ovarian neoplasms (Table 24.2) is based upon the 1973 World Health Organization recommendations (Sero et al 1973), somewhat condensed for the purposes of this discussion.

Clearly many parallels exist between tumours of the ovary and tumours of the testis, as might be expected considering that the cellular composition of these two organs is identical in embryological terms and differs only to the extent that the cells are subject to differing hormonal environments during embryological differentiation and subsequent maturation. It will be shown that immunohistological techniques have given new insight into the origins and interrelations of these tumours.

THE ADVENT OF IMMUNOHISTOLOGICAL METHODS

The histopathologic diagnosis of neoplasia is at times hedged with uncertainty, owing in part to conceptual difficulties regarding the nature and origin of the neoplastic cells and also to a continuing lack of alternative methods of cell identification with which to validate the morphologic criteria employed. (Taylor 1978)

Histochemical methods have proved extremely valuable in some areas of pathology, but have had limited application in the realms of testicular and ovarian neoplasms, for the normal cells of the ovary and testis do not show uniquely characteristic histochemical patterns.

The development of immunohistological methods, depending upon the specificity of an antibody marker for a particular cell or tissue antigen, offers the prospect of developing an alternative form of 'special stain' for use in routinely processed tissues. The limitations of the methods, the practical aspects and controls necessary for successful interpretation are outlined in Chapter 3. Table 24.3 lists commercially available immunostaining kits relevant to the study of ovarian and testicular pathologies.

IMMUNOSTAINING OF OVARIAN AND TESTICULAR TUMOURS

Table 24.4 lists the principal types of testicular and ovarian neoplasms and the various immunostains that may be employed to assist in their discrimination from one another. It must be emphasized that, in devising immunohistological staining methods, the pathologist is not casting aside morphological criteria, but rather is taking advantage of the remarkable specific staining offered by immunological techniques in order to

Table 24.3 Commercially available immunostaining kits for ovarian and testicular pathology

Primary tumours	Human chorionic gonadotrophin
	Pregnancy-specific beta-l-glycoprotein
	Human placental lactogen
	Oestradiol
	Testosterone
	Alpha fetoprotein
Metastases	Carcinoembryonic antigen
	Prostate-specific acid phosphatase

validate, refine and redefine, if necessary, the morphological criteria upon which diagnosis is traditionally based.

OBSERVED PATTERNS OF STAINING — EPITHELIAL OVARIAN TUMOURS

In recent years a variety of antibodies have been described which identify antigens associated with ovarian epithelial tumours. Most of these antibodies are not commercially available, having been developed and characterized in research laboratories. One antibody, however, which is more widely available recognizes the CA125 antigen, a proposed glycoprotein derived from a serous ovarian carcinoma cell line. Initially there were two disadvantages to using the CA125 antibody. The first, which still remains a limiting factor, is the antibody's restriction to identification of predominantly serous tumours (most mucinous tumours are negative). The second disadvantage was related to the requirement of frozen tissue for immunohistological purposes. This problem has reportedly been resolved by Shishi and colleagues (1986) who propose the use of pronase on paraffin sections to expose the presence of this antigen.

A much more sensitive but non-specific marker of both benign and malignant surface epithelial ovarian neoplasms is cytokeratin, an intermediate filament which composes the cytoskeleton of cells of epithelial origin (see Ch. 4). Cytokeratin is but one in a group of intermediate filaments, the others being vimentin, desmin, neurofilament and glial fibrillary acidic protein (GFAP). These filaments compose the major framework or cytoskeleton of most cell types and are also considered cell type specific. Carcinomas are characterized by cytokeratins, sarcomas of muscle cells by desmin, non-muscle sarcomas by vimentin and gliomas by GFAP. Unlike the other intermediate filaments, cytokeratin can be subdivided into different types, 19 in all, allowing further classification of tumours according to cytokeratin profiles (Osborn et al 1983, Moll et al 1982) (see also Ch. 4). As one would predict, the epithelial ovarian tumours are almost universally negative for vimentin and the other intermediate filaments. As discussed below, antibodies to cytokeratin and vimentin are used primarily to characterize undifferentiated tumours.

OBSERVED PATTERNS OF STAINING — GERM CELL TUMOURS

The seminoma/dysgerminoma group

This represents the most commonly encountered malignant tumour of germ cell origin. So-called 'pure' examples of this group of tumours do not show staining either for HCG or for alpha fetoprotein (AFP), either in the majority of the tumour cells or in the associated tumour giant cells that may be present (Kurman & Scardino 1981). However, some cases contain large multinucleated giant cells that do give positive immunostaining for HCG, but not for alpha fetoprotein; these cells are thought to represent isolated examples of syncytiotrophoblastic differentiation within a tumour that otherwise is pure seminoma. Kurman points out that prior to immunohistological staining these cells had often been termed pseudo-trophoblast, and their relationship to functional true syncytiotrophoblastic elements had not been appreciated. Seminomas may also reveal a minority of cells giving low intensity staining reactions with anti-oestradiol and anti-testosterone antisera, most likely representing the presence of the steroid hormone bound in vivo by active receptors or binding globulins. Finally, although there is some disagreement among investigators, most reports claim either an absence of inter-mediate filaments, particularly cytokeratin, or at least a significant reduction in the expression of these antigens (Denk et al 1987).

Embryonal carcinomas

These frequently contain syncytiotrophoblastic elements, emphasizing the close biological relationship of these two germ cell-derived neoplasms. The embryonal carcinoma cells per se typically do not stain for HCG. However, syncytiotrophoblastic elements commonly present within these tumours do show a strong positive reaction. On the other hand, alpha fetoprotein (AFP) staining is never observed within the syn-cytiotrophoblastic elements, but may often be seen within an apparently pure embryonal carcinoma, that seems to have functional differentiation even in the absence of histological yolk sac differentiation (Kurman & Scardino 1981). Both

Table 24.4   Anticipated patterns of staining in gonadal stromal tumours

Tumour category	HCG	SP1	AFP	Oestradiol	Testosterone	Progesterone	CEA	Breast antigens[1]	CA125
Seminoma	(+)[2]			(+)[3]	(+)[3]				
Embryonal carcinoma	++	+	+						
Choriocarcinoma	+++	++							
Teratoma									
With yolk sac differentiation			++						
With trophoblastic differentiation	+++	++							
With other secretory differentiation[4]							(+)[4]	(+)	
Endodermal sinus (yolk sac)	+	+	+++						
Leydig cell				+	+++	+			
Sertoli				+[3]	+	+			
Androblastoma (Sertoli-Leydig)				+[3]	+	+			
Granulosa-theca									
Granulosa				++	+	+			
Luteinized theca				++	+	++			
Undifferentiated gonadal stromal				++	++	−			
Mucinous cystadenocarcinoma							+		(+)[2]
Serous cystadenocarcinoma									++
Metastic and direct spread									
Colon (GI tract)							+++		
Breast	+[5]	+[5]		+[2,6]		+[3]		++	
Melanoma									
Lymphoma									
Other	+[5]	+[5]	+[7]		(+)[6]				

[1] Breast-associated antigens include 'T antigen', mammary epithelial membrane antigen and a variety of other antigens identifiable by use of monoclonal antibodies.

[2] Semiquantitative scoring, + to +++; not every case, nor every cell of positive cases, will be positive. Bracketed entries may, or may not, show positivity.

[3] Positivity of some cell elements, such as Sertoli cells or even breast epithelial cells, for oestradiol or other steroid hormones, may reflect receptor binding of the hormone rather than intrinsic production.

[4] Rare teratomas containing 'thyroid' elements will show immunostaining for thyroglobulin; teratomas with 'gastrointestinal' type epithelium may give a reaction for CEA; additional elements may show other staining patterns according to product.

Table 24.4 Cont'd

Tumour category	Cytokeratin	Vimentin	Neurofilament	Desmin/muscle specific actin	GFAP	Leucocyte common antigen	Other
Seminoma	$(+)^2$	$(+)^2$					
Embryonal carcinoma	+++						
Choriocarcinoma	++						
Teratoma	$+^4$	$+^4$	$+^4$	$+^4$	$+^4$		
With yolk sac differentiation	++						
With trophoblastic differentiation	++						
With other secretory differentiation[4]	+++						$+^4$
Endodermal sinus (yolk sac)	++						
Leydig cell		+					
Sertoli							
Androblastoma (Sertoli-Leydig)							
Granulosa-theca							
Granulosa		++					
Luteinized-theca		++					
Undifferentiated gonadal stromal		++					
Mucinous cystadenocarcinoma	++						
Serous cystadenocarcinoma	+++						
Metastatic and direct spread							
Colon (GI tract)	+++						
Breast	+++						
Melanoma		+++					
Lymphoma						+++	
Other		$(+)^2$					$(+)^8$

[5] Non-trophoblastic tumours sometimes contain HCG or SP1, e.g. carcinoma of breast, for reasons unknown.

[6] Non-ovarian/testicular tumours may show positivity, presumably also due to 'receptor binding of steroid hormone'; breast particularly is often positive for oestradiol (Taylor et al 1981); prostate is sometimes positive for testosterone.

[7] Primary hepatocellular carcinomas typically stain strongly for AFP.

[8] Hormonally active tumours of many types may produce secondary deposits in the ovary, and rarely the testis; immunostains are available against many of the hormones and for many other cell products (Taylor & Kledzik 1981).

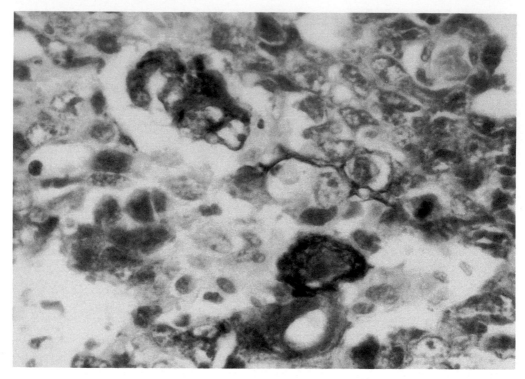

Fig. 24.1   Embryonal carcinoma metastatic to retroperitoneal lymph node of a young man, immunostained for cytokeratin by the ABC method. Grey-black positive staining is present in the cytoplasm of scattered tumour cells. Paraffin section, counterstain haematoxylin. ×400

components, syncytiotrophoblastic elements and embryonal carcinoma (Fig. 24.1), have been shown to express cytokeratin (Denk et al 1987).

Choriocarcinomas

Choriocarcinomas, representing a malignant counterpart of the trophoblast, typically produce large amounts of HCG and stain strongly with anti-HCG sera (Fig. 24.2). The normal trophoblast also produces beta-1-pregnancy-specific glycoprotein (SP1); choriocarcinoma typically mimics this production. It should be emphasized that, while the pattern of staining of choriocarcinomas for both HCG and SP1 is very striking and very characteristic, the mere presence of staining with either of these antisera does not of itself define the tumour as of trophoblastic origin. For example, as many as 50% of cases of breast carcinoma have been observed to stain for HCG or for SP1 (Horne et al 1977), or for both, and HCG staining has been reported less often in

other tumours (Taylor & Kledzik 1981). Choriocarcinomas will label for cytokeratin but lack immunoreactivity for other intermediate filaments (Denk et al 1987).

Teratomas

These do not show positive staining for either HCG or AFP unless, within the teratoma, there is differentiation either towards trophoblastic elements or yolk sac elements. In such instances the differentiated elements show the pattern of staining typical of trophoblast or of yolk sac respectively. It should be remembered that teratomas may occasionally produce functional epithelial and glandular elements and that these may be stained for the appropriate cell product (e.g. thyroglobulin within 'teratomatous thyroid', or carcinoembryonic antigen within teratomatous gastrointestinal type epithelium). Furthermore, depending on the germ cell elements represented in the tumor there may be immuno-

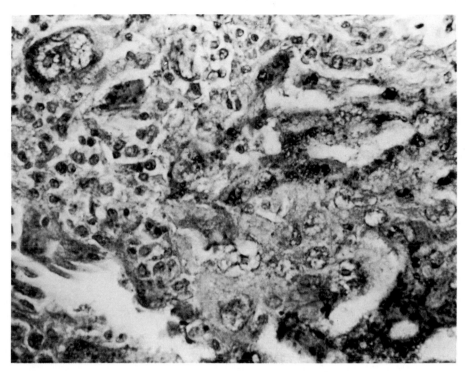

Fig. 24.2 Choriocarcinoma presenting in the lung of a young girl, immunostained for human chorionic gonadotrophin by the PAP method. Positive staining appears grey-black in the cytoplasm of tumour cells, including giant cells. Paraffin section, counterstain haematoxylin. ×500

reactivity identified for cytokeratin, vimentin, neurofilament, GFAP, and desmin for epithelial, mesenchymal, neural, astrocytic and muscle components respectively (Denk et al 1987).

Endodermal sinus (yolk sac) tumours

These tumours show a very typical pattern of staining regardless of the fine histological subtypes. It is important to recognize that positivity for AFP may be distributed throughout the tumour, or may occur only focally (Fig. 24.3). Typically, morphological elements in the tumour that resemble yolk sacs show intense positive staining. However, in other areas of the tumour, or in other tumours that show no evidence of morphological yolk sac differentiation, it often is possible to demonstrate variable amounts of staining for AFP. Histologically it has been shown that AFP staining correlates with intracellular and extracellular hyaline eosinophilic globules identifiable in routine H & E

sections. Many examples of tumours showing yolk sac differentiation will also stain for other liver cell products, such as transferrin and alpha-1-antitrypsin. Positive staining for AFP can be used to identify an endodermal sinus or yolk sac tumour, even when there is little evidence of morphological yolk sac differentiation. Yolk sac tumours will also give positive staining for cytokeratin while showing no immunoreactivity for other intermediate filaments (Eglen & Ulbright 1987).

OBSERVED PATTERNS OF STAINING — GONADAL STROMAL TUMOURS

Immunostaining for the different steroid hormones has shed light upon the histogenesis and the histological differentiation of hormonally active tumours within the ovary and testis. Prior to this, the nomenclature and classification of these tumours has largely been dependent upon morphological features coupled with clinical status and

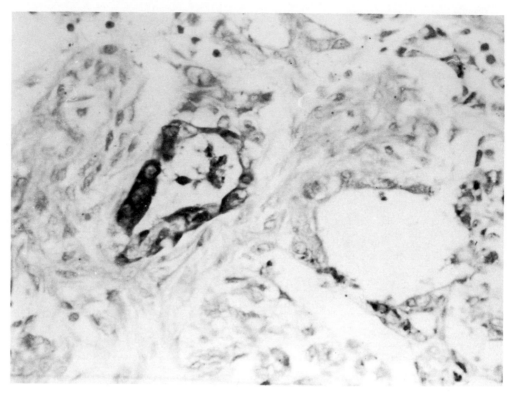

Fig. 24.3 Endodermal sinus (yolk sac) carcinoma immunostained for alpha-fetoprotein by the PAP method. Positive black staining is visible in a recognizable 'yolk sac' and in isolated scattered cells elsewhere in the tumour. Paraffin section, counterstain haematoxylin. ×720

evidence of endocrine manifestations (Kurman et al 1978, 1979, 1981, Taylor et al 1978). The fact that these neoplasms often present variable patterns of hormone production and that the tumours themselves may display diverse histological patterns has led to considerable difficulties in developing a consistent working hypothesis relating individual cell types to the production of particular hormones. The feasibility of preparing antisera against the different steroid hormones and of applying these antisera to the demonstration of these hormones in formalin paraffin wax sections promises to shed new light on this controversy.

Immunohistological techniques have proved remarkably successful, with some reservations, as discussed below. In general terms, staining for a particular steroid hormone within cell cytoplasm is interpreted as evidence of intracellular synthesis, although, of course, the possibility of binding to various intracellular binding proteins or receptors is a real one. It has, for example, been recognized

that Sertoli cells contain receptors for various androgens, while granulosa cells apparently contain oestrogen-binding proteins (Christensen et al 1977, Schreiber et al 1976). In addition, when using antisera to steroid hormones, one must be aware of the possibility of cross-reactivity between the different antisera. In practical terms, this can be controlled to some degree by assessing the specificity of the antisera, one against another, in radioimmunoassay systems. With regard to immunohistological studies, it is important also to include a biological control, as for example in the intense staining of normal interstitial Leydig cells in testis with anti-testosterone serum, and the lesser degree of staining or absence of staining with anti-oestradiol serum. Another potential problem is that in order to generate antisera against the steroid hormones, the hormones are first linked to some sort of carrier molecule. Bovine serum albumin (BSA) has often been used for this purpose; the resulting antiserum will thus contain antibodies against the steroid moiety, and also against the bovine serum albumin. These anti-albumin antibodies will show varying degrees of crossreactivity with human albumin and will cause immunohistological staining that may be confused with specific staining for anti-oestradiol. For these reasons, antisera that are to be used for immunohistological studies must, if coupled with BSA, be extensively absorbed with human albumin. Keyhole limpet haemocyanin as a carrier avoids the possibility of crossreacting antisera. Monoclonal antibody technology may eventually produce a specific monoclonal antibody having no crossreactivity with carrier molecules.

The intermediate filament expressed by these tumours is restricted to vimentin in contrast to ovarian epithelial tumours and many germ cell neoplasms (Miettinen et al 1985). The different patterns of staining anticipated among gonadal stromal tumours are illustrated in Table 24.4.

Leydig cell tumours and Sertoli-Leydig cell tumours (androblastomas)

Characteristically, these show intense staining for testosterone in the recognizable Leydig cell elements. Leydig cells may also show positive staining with antiserum to oestradiol and less often with antiserum to progesterone. Occasionally the recognizable Sertoli cells, forming more or less well-organized tubules, also show staining for testosterone and/or oestradiol.

Granulosa-theca cell tumours

Quite commonly these tumours are associated with clinical manifestations of hormone production. Classically, oestrogen synthesis has been considered to be confined to theca cells while the luteinized granulosa cells have been held responsible for progesterone production. These rigid definitions of cell type correlated to hormone production do not appear to hold true according to the evidence of immunoperoxidase staining. Granulosa cells have been observed to show positive staining, not only for oestradiol, but also for progesterone and even for testosterone. Similarly, luteinized theca cells contain progesterone, but also may show staining for oestradiol or testosterone.

These findings challenge the time-honoured concept that individual morphological cell types are responsible for, and restricted to, the production of single hormones; theca cells for oestrogen, luteinized granulosa cells for progesterone and Leydig cells for testosterone. Immunoperoxidase studies suggest that the neoplastic counterparts of these cells certainly do not recognize these restrictions; there is increasing evidence that normal cells also break the accepted rules (Kurman et al 1981). Clearly, immunoperoxidase studies, staining for these specific steroid hormones, may be expected to contribute to our understanding of the biology of these neoplasms. Such studies can be performed upon tissue biopsies in order to gain a more realistic assessment of the potential hormonal manifestations of individual tumours in individual patients, for it is clear that morphology alone is not a reliable indicator of the type of hormone produced by any individual morphologic type of neoplasm.

Undifferentiated gonadal stromal cell tumours

Morphologically these tumours consist of highly undifferentiated 'primitive' appearing cells, reflecting the high proportion of cells in the S-(synthesis) phase of cell cycle. These neoplasms bear a close resemblance to other undifferentiated neoplasms that may be encountered in ovary and testis and, in our experience, have sometimes been misdiagnosed as reticulum cell sarcoma (histiocytic lymphomas, or, in more modern terminology, immunoblastic sarcoma) of the testis. The staining of such tumours for steroid hormones, specifically the demonstration of testosterone and/or oestradiol in the cytoplasm of the neoplastic cells, provides support for the diagnosis of gonadal stromal neoplasm (Fig. 24.4) (Maurer et al 1980). Evaluation for intermediate filaments in these tumours can be particularly helpful as they are positive for vimentin and lack cytokeratin.

METASTATIC TUMOURS OF OVARY AND TESTIS

Metastatic neoplasms may present difficulties in differential diagnosis of ovarian tumours. This is less often a problem with reference to the testis,

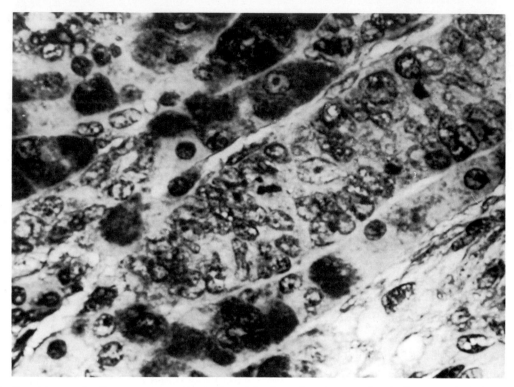

Fig. 24.4 Gonadal stromal cell tumour, presenting in the pelvis of an otherwise normal male, immunostained for oestradiol. Positive intense black staining is seen in cells having some resemblance to Leydig cells; Sertoli-like tubules showed scattered light staining. Antiserum to testosterone gave an even more intense staining pattern. Paraffin section, counterstain haematoxylin. ×720

(This case was reported Maurer et al 1980)

except perhaps for some of the malignant lymphomas.

Whereas a positive immunoperoxidase stain for HCG, AFP, or one of the steroid hormones may be of real value in establishing a diagnosis of primary neoplasm of the gonad, a negative stain is of much less value. Even in the presence of procedural controls and biological 'positive' and 'negative' controls, one can never entirely exclude a false-negative result due to technical reasons. In addition, tumours of a lineage with the potential for producing a particular cell product do not necessarily produce that product in amounts detectable by the techniques described. Thus a negative immunoperoxidase stain cannot be given too much weight in considering a differential diagnosis of primary versus metastatic tumour within the testis or ovary. At the present time, the range of antisera available, with specificity against antigens associated with other tumours that might

occur as metastases within the gonads, is quite limited. Nonetheless, new reagents are becoming available. Antisera against carcinoembryonic antigen can be useful in recognizing metastases from a colonic primary. However, primary cystadenocarcinomas of the ovary may also be carcinoembryonic antigen-positive, mucinous carcinomas more so than their serous counterparts (Heald et al 1979). Antisera against prostatic acid phosphatase or prostatic epithelial antigen may assist in the recognition of lymph nodal metastases as of prostatic origin versus a primary testicular neoplasm (Taylor & Kledzik 1981), and antisera against breast-associated antigens may be helpful in recognizing poorly differentiated breast carcinoma metastatic to the ovary (Imam & Tokes 1981). A battery of antibodies may prove helpful in determining the cell of origin in poorly differentiated tumours. While the demonstration of cytokeratin positivity may not differentiate between a primary or metastatic tumour, it will at least define the cell of origin as either epithelial or germ cell in nature. The absence of cytokeratin combined with positivity for vimentin, S-100 protein, and melanoma associated antigens will identify the tumour as metastatic melanoma (Gown et al 1986). Immunoreactivity for the common leucocyte antigen would obviously characterize a tumour as lymphoma.

CONCLUSION

It has been written that 'the importance of immunohistologic methods in surgical pathology can scarcely be overestimated' (Bosman & Kruseman 1979). Immunohistological techniques applicable to paraffin wax sections are only 12 years old, but already the variety of specific immunostains available is enormous, many of which can be applied quite routinely in diagnostic and investigative laboratories (Taylor & Kledzik 1981, Taylor 1986). It is tempting to speculate that this represents the beginning of a new era in diagnostic pathology in which the pathologist will learn to rely, not only upon traditional histological methods, but also upon specific immunostains for cell and tumour recognition. With regard to neoplasms of the testis and ovary, a beginning has been made, but it still remains only a beginning.

REFERENCES

Bosman F T, Kruseman A C N 1979 Clinical applications of the enzyme-labeled antibody method. Immunoperoxidase methods in diagnostic pathology. Journal of Histochemistry and Cytochemistry 27: 1140–1147

Christensen A K, Wisner J R, Orth J 1977 Preliminary observations on the localization of androgen and FSH receptors in the rat seminiferous tubule, studied by autoradiography at the light microscope level. In: Troen P, Nankin HR (eds) The testis in normal and infertile men. Raven Press, New York, p 153

Denk H, Moll R, Weybora W et al 1987 Intermediate filaments and desmosomal plaque proteins in testicular seminomas and non-seminomatous germ cell tumors as revealed by immunohistochemistry. Virchows Archives [A] 410: 295–307

Eglen D, Ulbright T 1987 The differential diagnosis of yolk sac tumor and seminoma. Usefulness of cytokeratin, alpha-fetoprotein, and alpha-1-antitrypsin immunoperoxidase reactions. American Journal of Clinical Pathology 88: 328–332

Heald J, Buckley C H, Fox H 1979 An immunohistochemical study of the distribution of CEA in epithelial tumours of the ovary. Journal of Clinical Pathology 32: 910–926

Horne C H W, Reid I N, Milne G D 1976 Prognostic significance of inappropriate production of pregnancy proteins by breast cancer. Lancet ii: 279–282

Gown A, Vogel A, Hoak D, Gough F, McNutt M 1986 Monoclonal antibodies specific for melanocytic tumors distinguish subpopulations of melanocytes. American Journal of Pathology 123: 195–203

Hsu S M, Raine L, Fanger H 1981 Use of avidin-biotin-peroxidase complex (ABC) in immunoperoxidase techniques: a comparison of ABC and unlabeled antibody (PAP) procedure. Journal of Histochemistry and Cytochemistry 29: 577–580

Imam A, Tokes Z A 1981 Immunoperoxidase localization of a glycoprotein on plasma membrane of secretory epithelium from human breast. Journal of Histochemistry and Cytochemistry 29: 581–584

Kurman R J, Scardino P T 1981 Alpha-fetoprotein and human chorionic gonadotropin in ovarian and testicular germ cell tumors. In: DeLellis R A (ed) Diagnostic immunohistochemistry. Masson Monographs in Diagnostic Pathology. Masson, New York, ch 17, p 277

Kurman R J, Goebelsmann U, Taylor C R 1979 An immunohistological study of steroid localization in granulosa-theca tumors of the ovary. Cancer 43: 2377–2384

Kurman R J, Goebelsmann U, Taylor C R 1981 Steroid hormones in functional ovarian tumours. In: DeLellis R A (ed) Diagnostic immunohistochemistry. Masson Monographs in Diagnostic Pathology. Masson, New York, ch 8, p 137

Kurman R J, Andrade D, Goebelsmann U, Taylor C R 1978 An immunohistological study of steroid localization in Sertoli-Leydig tumors of the ovary and testis. Cancer 42: 1772–1783

Maurer R, Taylor C R, Schmucki O, Hedinger C E 1980 Extratesticular gonadal stromal tumor in the pelvis: a case report with immunoperoxidase findings. Cancer 44: 985–990

Miettinen M, Virtanen I, Talerman A 1985 Intermediate filament proteins in human testis and testicular germ-cell tumors. American Journal of Pathology 120: 402–410

Moll R, Franke W W, Schiller D L, Geiger B, Reinhard K 1982 The catalog of human cytokeratins: patterns of expression in normal epithelia, tumors and cultured cells. Cell 31: 11–24

Mostofi F K, Price E B Jr 1973 Tumors of the male genital system. Atlas of tumor pathology, fascicle 7, series 2. Armed Forces Institute of Pathology, Washington D C

Osborn M, Weber K 1983 Biology of disease. Tumor diagnosis by intermediate filament typing: a novel tool for surgical pathology. Laboratory Investigation 48: 372–394

Schreiber J R, Reid R, Ross G T 1976 A receptor like testosterone binding protein in ovaries from immature female rats. Endocrinology 98: 1206–1213

Serov S F et al 1973 International histological classification of tumors (benign). Histologic typing of ovarian tumors. World Health Organization, Geneva

Shishi J, Ghazizadeh M, Oguro T, Aihara K, Araki T 1986 Immunohistochemical localization of CA125 antigen in formalin-fixed paraffin sections of ovarian tumors with the use of pronase. American Journal of Clinical Pathology 85: 595–598

Taylor C R 1978 Classification of lymphomas: 'new thinking' on old thought. Archives of Pathology and Laboratory Medicine 102: 549–554

Taylor C R 1986 Immunomicroscopy: a diagnostic tool for the surgical pathologist. Saunders, London

Taylor C R, Kledzik G 1981 Immunohistologic techniques in surgical pathology — a spectrum of 'new' special stains.

Human Pathology 12: 590–596

Taylor C R, Kurman R J, Warner N E 1978 The potential value of immunohistological techniques in the classification of ovarian and testicular tumors. Human Pathology 9: 417–427

Taylor C R, Cooper C L, Kurman R J, Goebelsmann U, Markland F S Jr 1981 Detection of estrogen receptor in breast and endometrial carcinoma by the immunoperoxidase technique. Cancer 47: 2634–2640

Willis R A 1948 Pathology of tumours. Butterworths, London

25. Prostate

A. C. Jöbsis

INTRODUCTION

There are two major diagnostic problems in tumour pathology of the prostate in which histochemistry can provide helpful discriminative data: (1) Is the prostatic epithelial proliferation malignant? and (2) Could the carcinoma encountered be of prostatic origin?

Problem (2) can be resolved by the demonstration of the isoenzyme prostatic acid phosphatase (PRAP) and prostate specific antigen (PSA). Problem (1) may be resolved by mucin histochemistry.

At present, no reliable methods are available for predicting malignant transformation or the biological behaviour of prostatic cancer — for instance, with regard to hormonal treatment. The latter prognostic issue has been extensively studied using different markers, but so far no solid evidence has emerged that would be applicable in a routine setting.

MUCIN HISTOCHEMISTRY

With regard to the differential diagnosis between adenocarcinoma and epithelial hyperplasia, no reliable data can be obtained by the practising pathologist using investigations based on tumour-associated antigens or loss of blood group antigens. Although the cytophotometric determination of DNA, or the investigation of nucleolar organizer region-associated proteins (Smith & Crocker 1988) may offer a solution for this differential diagnostic problem, as they do in some other organs, these techniques will not be discussed here since their results in the field of prostatic lesions are still too preliminary for a routine setting. The prostate is one of the few organs in which mucin histochemistry may yield discriminative results in the investigation of glandular-like proliferations suspected of malignancy. The potential value of some of the mucin stains as a diagnostic tool in prostatic pathology was described in the 1960s by, among others, Hukill & Vidone (1967).

Acid mucins, which are not generally present in non-neoplastic prostatic glands, can be demonstrated in about 50–70% of well differentiated prostatic adenocarcinomas. These mucins are rarely seen in less well differentiated adenocarcinomas and not seen in undifferentiated prostatic carcinomas. A high content encountered in well-differentiated prostatic carcinomas (small gland type) allows differentiation between malignant and benign in about two out of three questionable cases. The neutral mucins are of little or no discriminative value in the prostate.

The acid mucins can be classified histochemically as: sulphate-containing mucins (sulphomucins), and sialic acid-containing mucins (sialomucins). The most suitable technique for routine use is the high-iron diamine method (HID) combined with Alcian blue (AB) staining (see Appendix 3), which demonstrates sulphomucins (brown-black) and sialomucins (blue) in one section of routinely fixed paraffin wax-embedded prostatic tissue (Fig. 25.1). A promising application is the use of lectins. For instance, Söderström (1987) reported a better discrimination between benign and malignant prostatic epithelial proliferations by means of soya bean lectin than with the HID/AB stain.

Sulphomucins and sialomucins, either separately or together, are more frequently seen in the small gland type of well differentiated prostatic

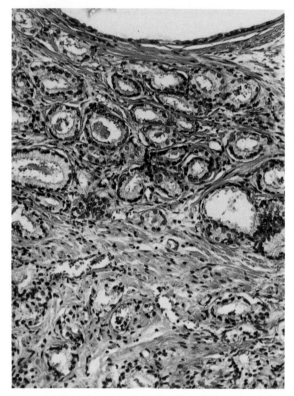

Fig. 25.1 HID/AB staining for acid mucins which are best discernible in the lumen of the gland-like malignant proliferations. Nuclei counterstained. ×126

carcinomas than in the large gland or the cribriforme type. The mucins are usually located in the lumen or on the luminal surface and are only rarely detectable in the cytoplasm (Fig. 25.2). Colloid prostatic carcinomas stain strongly positive. Sporadically, a positive HID/AB reaction is found in a non-neoplastic gland. These results, which are in keeping with those reported by others employing the same or analogous methods (Hukill & Vidone 1967, Dollbert 1980) may therefore offer some help in doubtful cases, but cannot in themselves be decisive.

The presence of acid mucins, which is neither of prognostic value within the group of well-differentiated adenocarcinomas nor of discriminative value with regard to bladder carcinomas (Hukill & Vidone 1967), may increase diagnostic accuracy, especially in cases with crush artefacts or tiny fragments (needle biopsy).

ENZYME HISTOCHEMISTRY

The question: 'Does this carcinoma originate in the prostate?' can be answered by demonstration of prostatic acid phosphatase activity. This iso-enzyme of acid phosphatase is almost exclusively associated with prostatic epithelial cells. Excep-

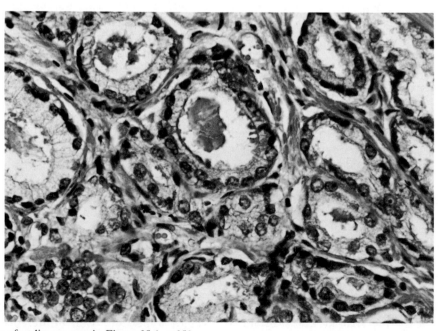

Fig. 25.2 Details of malignant part in Figure 25.1. ×350

tions, which include the B-cells in the islets of Langerhans, some insulinomas, some carcinoids and adenoma of the middle ear do not essentially diminish the differential diagnostic value of this histochemical feature.

PRAP is distinguishable from other acid phosphatase isoenzymes in epithelial cells by its relatively high resistance to formaldehyde (Pearse 1968) and by its almost absolute substrate specificity for phosphorylcholine (Serrano et al 1976). The sensitivity of PRAP for L(+) tartrate is relatively high but does not offer a reliable discriminative histochemical feature. These biochemical aspects can be demonstrated in cryostat sections of fresh (Figs 25.3–25.6) or formaldehyde-fixed tissues. The most reliable enzyme histochemical method uses phosphorylcholine as substrate (Appendix 5). Positive staining in serial cryostat sections, one with no fixation and the other after formalin fixation indicates the prostatic nature of the acid phosphatase activity. Although PRAP is relatively insensitive to formalin, the acid phosphatases of epithelial cells of other organs may display, after fixation, some activity which may be demonstrated using substrates such as β-glycerophosphate and α-naphthylphosphate (Pearse 1968), which may cause some problems of interpretation (Figs 25.3 and 25.4). A section of non-prostatic tissue should therefore be incorporated as a control. If only formalin-fixed tissue is available, PRAP activity may be demonstrated in frozen sections, preferably with phosphorylcholine as substrate (Fig. 25.5). A

negative result under these circumstances does not exclude the prostatic origin of the carcinoma, since some reduction in PRAP activity does occur in formalin-fixed tissue at room temperature. The critical time period depends on the size of the

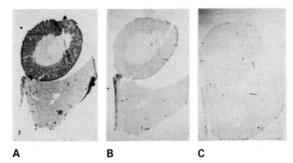

Fig. 25.4 Acid phosphatase activity in three consecutive cryostat sections of a composite block containing unfixed liver and kidney. The incubation circumstances were identical with those for the sections **A**, **B** and **C** shown in Figure 25.3. There is a significant degree of inhibition by means of formalin, so these organs contain little or no PRAP isoenzyme. ×1

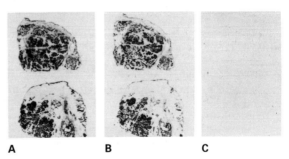

Fig. 25.5 Acid phosphatase activity in three consecutive cryostat sections of the same unfixed prostate as in Figure 25.3 with phosphorylcholine as substrate. The incubation circumstances were otherwise identical with those for the sections **A**, **B** and **C** shown in Figure 25.3. There is no inhibition by means of formalin, so the staining is due to the activity of the PRAP isoenzyme. ×1

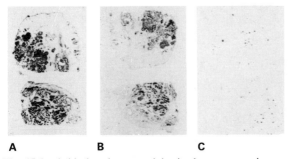

A   **B**   **C**

Fig. 25.3 Acid phosphatase activity in three consecutive cryostat sections of unfixed prostate with beta-glycerophosphate as substrate. In **A** the formalin prefixation was omitted, whereas in **B** the intensity of staining is reduced slightly due to this prefixation. In **C** no staining has developed due to the inhibition of the enzyme activity by 10 mmol/l L(+) tartrate. No counterstain. ×1

Fig. 25.6 No staining of acid phosphatase activity with phosphorylcholine as substrate in this cryostat section of unfixed liver and kidney. The incubation circumstances were identical with those for the section shown in Figure 25.3A. ×1

tissue and, of course, the amount of PRAP in the carcinoma cells.

The enzyme histochemical methods discussed, which can also be applied to cytological smears (imprint or aspiration) (Fig. 25.7), stain the cytoplasm of carcinoma cells less intensely than that of benign epithelial cells. There is no direct relationship between the degree of differentiation and the staining intensity. Papillary adenocarcinomas usually show a weak apical reaction. The staining intensity of carcinomas varies greatly, even in one section. This implies that a negative result in a small fragment of tumour cannot exclude a prostatic origin. Metastatic foci have essentially the same histochemical properties as the primary tumour. Androgen deprivation or oestrogen therapy may reduce the staining intensity. PRAP activity is irreversibly inhibited by alcohol and by most decalcifying solutions. PRAP activity cannot therefore be investigated in paraffin wax sections or in skeletal metastases. These limitations, however, can be circumvented by immunohistochemistry.

IMMUNOHISTOCHEMISTRY

Essentially similar results are obtained, both in sections and in smears, by the enzyme histochemical method using phosphorylcholine or by applying the specific antiserum against either human PRAP or human enzymatically inactive glycoprotein prostate specific antigen (PSA) (Jöbsis et al 1978, Wang et al 1979, Nadji 1980, Jöbsis et al 1981).

However, the great advantage of the immunological method — when applying a primary heteroantiserum — is that these two antigens stay largely unaltered by fixation, decalcification or embedding procedures. Exceptions to this are fixation in Zenker's solution, Susa fixation and the almost obsolete decalcification procedure with nitric acid. Another potential source of false-negative results is the small size of the biopsy, since the staining intensity may vary greatly within a tumour (Fig. 25.8). The staining intensity, which has no predictive value regarding response to hormonal treatment, may be so low that it is advisable to repeat the incubation procedure under different conditions. The same changes must also be made in both control incubation procedures. In repeating the procedure under carefully selected conditions some cases originally negative (small fragments of tumour or hormonally influenced tumours) have proved positive without false-positive results.

The most likely causes of false-positive results are non-specificity of the antiserum against human

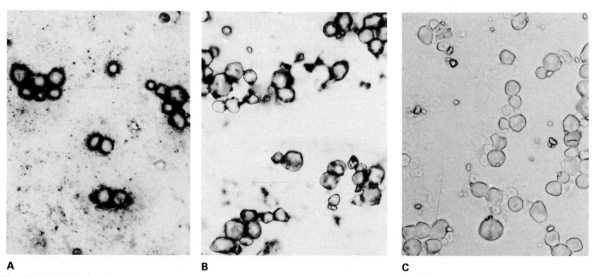

A   **B**   **C**

Fig. 25.7   Acid phosphatase activity in three imprint smears of unfixed prostatic tissue with beta-glycerophosphate as substrate. In **A** no formalin prefixation, in **B** formalin prefixation and in **C** the activity was inhibited by 10 mmol/l L(+) tartrate No counterstain.   ×280

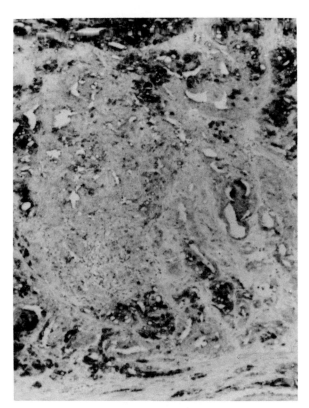

Fig. 25.8 Immunohistochemical demonstration of PRAP in paraffin wax section of prostatic carcinoma. The great variation of staining intensity (peroxidase technique) within one carcinoma is shown. Weak counterstain. ×81

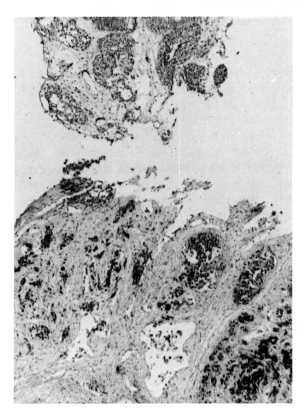

Fig. 25.9 Immunohistochemical demonstration of PRAP in paraffin wax section of a prostatic carcinoma. The primary (big fragment) shows a greater intensity of staining than the secondary (small fragment), which was biopsied after 6 months of hormonal treatment. Weak counterstain. ×50

PRAP or inappropriate handling of the antiserum (repeated freezing and thawing, for instance) and other technical shortcomings (Aumuller et al 1981, Jöbsis et al 1981). The problem of non-specificity can be tested in the case of PRAP either by the absorbance technique or by the mixed aggregation immunocytochemical technique (Jöbsis et al 1981).

The specificity and sensitivity of the immunohistochemical technique with the heteroantiserum against PRAP in paraffin wax sections were 96% and 98% respectively in a series of 200 proven, therapeutically uninfluenced prostatic carcinomas and 339 proven non-prostatic tumours (23 different types) and their normal counterparts (Jöbsis et al 1981). Metastases of prostatic carcinomas (Fig. 25.9) were positive, as were the primaries, whereas 50 bladder carcinomas were negative. Apart from the positive reaction in the islets of Langerhans (beta-cell pattern) and in several insulinomas (a finding which may offer the pathologist an unexpected differential diagnostic tool in that area!), the anti-PRAP serum detected the prostatic origin of the carcinomas. The specificity and sensitivity depended mainly on the quality of the primary antiserum. Less important aspects were the degree of differentiation of the tumour and some therapeutic measures (Grignon & Troster 1985, Keillor & Aterman 1987).

The immunohistochemical results for PSA indicate in general a somewhat higher degree of specificty and a lower degree of sensitivity (Feiner & Gonzales 1986, Keillor & Aterman 1987). Although PSA-expression is presumed to be under androgen control, immunohistochemical data do not as yet provide information concerning a possible favourable effect of hormonal intervention. The same restrictions as described for the interpretation of immunohistochemical results of PRAP can be made for PSA.

It is advisable to apply both heteroantisera — one against PRAP and the other against PSA — as a complementary system, especially in cases of poorly differentiated carcinomas, therapeutically influenced carcinomas and in tiny fragments of carcinoma. Experience has taught us that only one of the two antigens may be demonstrable.

A comparison of heteroantisera and a monoclonal antibody raised against malignant prostatic tissue showed that the monoclonal antibody was less sensitive (Schevchuk et al, 1983). In a routine setting monoclonal antibodies appear less useful than polyclonal antibodies in the diagnosis of prostatic carcinoma (Gallee et al 1986).

The presence of immunoreactive PSA and PRAP in sites other than prostate and metastases of prostatic carcinoma has been mentioned. Nowels et al (1988) have reported on PSA and PRAP in cystitis cystica and cystitis glandularis (positive in 14 out of 38 cases) and in 4 out of 12 postmortem bladder samples from non-urinary tract associated deaths. They found that immunoreactivity was present in both male and female cases and speculated on the potential for bladder epithelium to differentiate to prostatic type.

REFERENCES

Aumuller G, Pohl C, van Etten R L, Seits J 1981 Immunohistochemistry of acid phosphatase in the human prostate: normal and pathologic. Virchows Archiv B 35: 249

Dollbert L 1980 Early prostatic carcinoma. Human Pathology 11: 688

Feiner H D, Gonzalez R 1986 Carcinoma of the prostate with atypical immunohistological features. American Journal of Surgical Pathology 10: 765

Gallee M P W, van Vroonhoven C C J, van der Korput A G M et al 1986 Characterization of monoclonal antibodies raised against the prostatic cancer cell line PC-82. The Prostate 9: 33

Grignon D, Troster M 1985 Changes in immunohistochemical staining in prostatic adenocarcinoma following diethylstilbestrol therapy. The Prostate 7: 195

Hukill P B, Vidone R A 1967 Histochemistry of mucins and other polysaccharides in tumours. Laboratory Investigation 16: 395

Jöbsis A C, de Vries G P, Anholt R R H, Sanders G T B 1978 Demonstration of the prostatic origin of metastases. Cancer 41: 1788

Jöbsis A C, de Vries G P, Meijer A E F H, Ploem J S 1981 Prostatic acid phosphatase immunologically detected: possibilities and limitations with regard to tumour histochemistry. Histochemical Journal 13: 961

Keiller J S, Aterman K 1987 The response of poorly differentiated prostatic tumors to staining for prostate specific antigen and prostatic acid phosphatase: A comparative study. Journal of Urology 137: 894

Nadji M 1980 The potential value of immunoperoxidase techniques in diagnostic cytology. Acta Cytologica 24: 442

Nowels K, Kent E, Rinsho K, Oyasu R 1988 Prostate specific antigen and acid phosphatase-reactive cells in cystitis cystica and glandularis. Archives of Pathology and Laboratory Medicine 112: 734-737

Pearse A G E 1968 Acid phosphatases. Histochemistry, 3rd edn. Churchill, London, ch 16, p 547

Serrano J A, Shannon W A Jr, Sternberger N J, Wasserkrug H L, Seligman A M 1976 The cytochemical demonstration of prostatic acid phosphatase using a new substrate, phosphorylcholine. Journal of Histochemistry and Cytochemistry 24: 1046

Shevchuk M M, Romas N A, Ng P Y, Tannenbaum M, Olsson C A 1983 Acid phosphatase localization in prostatic carcinoma. Cancer 52: 1642-1646

Smith R, Crocker J 1988 Evaluation of nucleolar organizer region-associated proteins in breast malignancy. Histopathology 12: 113

Söderström K O 1987 Lectin binding to prostatic adenocarcinoma. Cancer 60: 1823

Wang M C, Valenzuela L A, Murphy G P, Chu T M 1979 Purification of a human prostate specific antigen. Investigative Urology 17: 159

26. Thyroid

M. C. Sambade   M. Sobrinho-Simões

INTRODUCTION

Thyroid oncology is based upon the concept that there are two different lineages of epithelial cells within the normal gland: follicular cells and parafollicular or C cells. According to most authors these lineages represent two histogenetically distinct cell populations and give rise to two clearly different types of thyroid malignancies: follicular or papillary carcinomas and medullary carcinomas (also termed C cell carcinomas) (Hedinger & Sobin 1974). Although at present this concept is debatable (Sobrinho-Simões et al 1985b), it is a convenient basis for classification.

The most important contributions of histochemistry, immunocytochemistry and lectin histochemistry in the diagnosis of thyroid tumours are reviewed here using a problem-oriented approach.

NORMAL THYROID AND HYPERPLASTIC LESIONS

Follicular cells may be distinguished from parafollicular cells by *histochemical* silver impregnation techniques which disclose the presence of neurosecretory granules in the latter (Solcia et al 1986). Care should be exercised, however, when using these techniques since the dense lysosomal granules of the follicular cells can also give rise to a granular staining pattern which can be (mis)-interpreted as a positive reaction. These two types of dense granules can be distinguished by their different immunoreactivity for chromogranin, a neuroendocrine marker that does not recognize non-neuroendocrine argyrophil granules (Wilson & Lloyd 1984, Solcia et al 1986).

Immunocytochemistry plays a major role in the identification of thyroid cell types as far as 'product' characterization is concerned. Follicular cells display immunoreactivity for thyroglobulin (Fig. 26.1), regardless of the cytoplasmic appearance (clear, amphophilic or eosinophilic) (Civantos et al 1984, Nesland et al 1985), whereas parafollicular cells stain usually for calcitonin and calcitonin-gene-related-peptide (CGRP) (Fig. 26.1) (Albores-Saavedra et al 1985). A 'non-functional' identification of normal, hyperplastic and neoplastic parafollicular cells is also usually possible using antisera for carcinoembryonic antigen (CEA) (Fig. 26.2) (DeLellis et al 1978, Kodama et al 1980, Mendelsohn et al 1984).

In our experience, the use of neuron specific enolase (NSE) to differentiate follicular and parafollicular cells does not give consistent results, despite claims of it being a useful marker of the diffuse neuroendocrine system (Carlei & Polak 1984). The same lack of specificity holds true for neoplastic conditions, i.e. immunoreactivity for NSE is frequently found in many follicular cell 'derived' tumours of the thyroid (Fig. 26.2) unlike previous findings (Lloyd et al 1983).

Martin-Lacave et al (1988) claim that lectin from *Ulex europaeus I* may in some cases discriminate follicular cells from C cells. In our hands, this lectin clearly identified the endothelial cells, especially in frozen material and protease treated paraffin sections, but did not distinguish normal and/or neoplastic C cells from follicular cells (Sobrinho-Simões & Damjanov 1986, Sobrinho-Simões et al 1987).

For the identification of other types of cells present in the thyroid gland, namely lymphoid and antigen-presenting cells, and the demonstration of

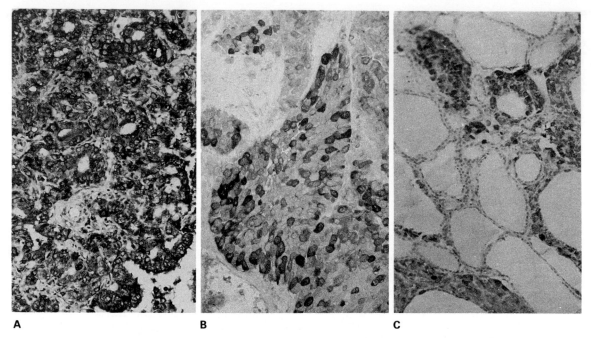

A   **B**   **C**

Fig. 26.1 Neoplastic cells of thyroid tumours usually stain either for thyroglobulin or for calcitonin and/or CGRP.
A Follicular variant of papillary carcinoma showing intense immunoreactivity for thyroglobulin. **B** Some of the neoplastic cells of this MCT display immunoreactivity for CGRP. **C** At the periphery of the same MCT depicted in Figure 26.2**B** one can see isolated nests of calcitonin positive cells which should not be interpreted in this context as a definite sign of C-cell hyperplasia. Avidin-biotin-peroxidase. **A** ×187, **B** ×375, **C** ×187

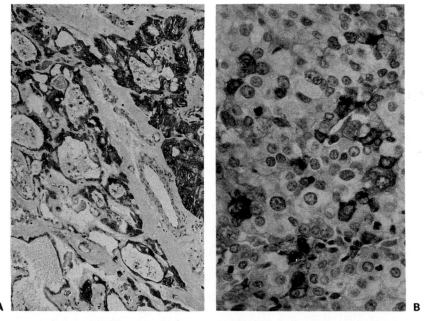

A   **B**

Fig. 26.2 A   The lack of specificity of NSE immunoreactivity is documented by the intense staining of this thyroglobulin-positive 'hyalinizing trabecular adenoma' **B**. CEA immunoreactivity is not restricted to MCT; some CEA positive cells are observed in this poorly differentiated thyroglobulin-positive carcinoma. Aviden-biotin-peroxidase. **A** ×187, **B** ×625

changes in the HLA-DR phenotype of follicular cells, the use of both B and T cell markers and tissue typing (see Ch. 19) can provide the basis for considerable insight into some hyperplastic lesions such as Graves' disease and Hashimoto thyroiditis (Hanafusa et al 1983, Aichinger et al 1985).

THYROID TUMOURS

Is conventional histochemistry still useful in the diagnosis of thyroid tumours?

The PAS reaction is insufficiently selective to differentiate thyroglobulin from any of the substances that may mimic thyroid colloid. Also, the widespread use of immunocytochemistry has largely replaced the interest of silver impregnation techniques and amyloid stains in the diagnosis of medullary carcinoma of the thyroid (MCT).

Some emphasis has been put on the occasional identification of 'mucins' in thyroid tumours (Golough et al 1985, Rigaud et al 1985). It is now recognized that the presence of alcianophilic 'sialomucins' is probably meaningless since the

staining is due to carbohydrate components or breakdown products of thyroglobulin and colloid (Gherardi 1987, Rigaud & Bogomoletz 1987). The same does not apply, however, to the demonstration of sulphomucins by the high iron diamine-Alcian blue method, since normal thyroglobulin does not contain sulphated groups (Gherardi 1987). In our experience the presence of sulphomucin is restricted to rare variants of thyroid tumours (mucin-producing MCT, mucinous carcinoma and mucoepidermoid carcinoma) (Fig. 26.3) and the finding of sulphomucin can help in establishing a diagnosis if the setting is appropriate for a primary thyroid tumour. However, whenever there is a sulphomucin positive mucinous tumour in the thyroid the possibility of a metastasis should always be the first consideration.

Immunocytochemistry in thyroid oncology

Is it a primary thyroid tumour?

The answer to this question is obtained in most instances by the search of immunoreactivity for

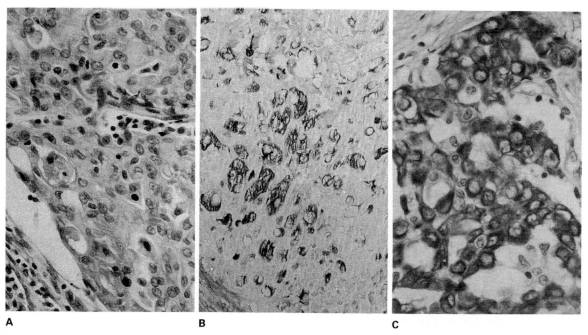

A **B** **C**

Fig. 26.3 Mucoepidermoid carcinoma of the thyroid displays typically a solid pattern of growth **A** with eosinophilic hyaline bodies and clear vacuoles. **B** where sulphomucins can be demonstrated. **C** Immunoreactivity for epidermal prekeratin C is a constant though not pathognomonic feature of this type of tumour. **A** H & E ×625, **B** high-iron diamine-Alcian blue ×187, **C** avidin-biotin-peroxidase ×625

follicular and parafollicular cell markers (thyro-globulin and calcitonin or CGRP, respectively).

Immunoreactivity for CEA is also useful for identifying MCT, though it does not exclude by itself a metastatic carcinoma nor even a follicular cell 'derived' carcinoma (Sobrinho-Simões et al 1985a).

The absence of all the follicular and para-follicular cell markers in a clearly epithelial tumour raises the problem of differential diagnosis between anaplastic carcinoma of the thyroid and metastasis from a carcinoma originating elsewhere. As in other fields of oncology this problem should be dealt with mainly using a clinical approach.

The diagnosis of calcitonin and CGRP-negative MCT should only be considered if one is deal-ing with a definite neuroendocrine tumour (chromogranin- and/or synaptophysin-positive) in a patient with a familial history of MEN II and no signs of primary neuroendocrine carcinoma elsewhere.

Is it a non-epithelial tumour?

The possibility of a non-epithelial tumour arises whenever one is faced with a small cell-, a large cell- or a spindle cell tumour in the thyroid gland.

The differential diagnosis of 'small cell' tumours is between MCT and lymphoma, which can be easily solved by the use of antibodies to calcitonin/CGRP and leucocyte common antigen (LCA). A few thyroglobulin-positive small cell tumours have also been reported as poorly dif-ferentiated carcinomas (Carcangiu et al 1984).

A large cell tumour of the thyroid can be an anaplastic carcinoma or a lymphoma of high grade of malignancy. The demonstration of LCA positivity remains the most reliable method to achieve this differential diagnosis, since many anaplastic carcinomas of the thyroid fail to react with anti-cytokeratin antibodies (Carcangiu et al 1985, Schröder et al 1986), and the epithelial membrane antigen (EMA) can no longer be ac-cepted as an epithelial marker (Thomas & Battifora 1987).

Finally, spindle cell malignant tumours of the thyroid can be MCT, anaplastic carcinomas or sar-comas. A straightforward diagnosis should only be made in these cases if the neoplastic cells show im-munoreactivity for calcitonin, cytokeratins or specific mesenchymal markers respectively. There is, however, a fairly large number of spindle cell tumours of the thyroid which fail to react with any of the aforementioned antisera; these cases are in-cluded, for practical reasons, in the group of anaplastic carcinomas (Rosai et al 1985).

Is it a medullary carcinoma with an unusual pattern?

MCT with giant cells, with follicular structures or with a papillary pattern of growth will be recog-nized by their reactivity to antisera for calcitonin and CGRP (Albores-Saavedra et al 1985). The biological and prognostic implications of these fea-tures, however, remain to be clarified (Sambade et al 1988).

Mixed medullary and follicular carcinomas: do they exist?

There are rare cases of thyroid carcinoma showing both the morphological features of MCT with immunoreactivity for calcitonin, and the mor-phological features of follicular carcinoma with immunoreactivity for thyroglobulin. These tumours are classified as mixed medullary-follicular carcinomas in the 2nd edition of the World Health Organization's booklet on thyroid tumours (Hedinger & Sobin 1988).

Apart from these mixed carcinomas there are thyroglobulin-positive tumours coexpressing neuroendocrine markers within the groups of poorly differentiated carcinomas (Ljungberg et al 1984), and mucoepidermoid carcinomas (Franssila et al 1984, Sambade et al 1987) as well as a variable percentage of MCT in which immuno-reactivity for thyroglobulin has been found (Uribe at al 1987, Holm et al 1987).

These findings, together with the demonstration of thyroglobulin and calcitonin in the same neoplastic cells using double immunocytochemical methods (Holm et al 1986), challenge the concept of dual histogenesis of thyroid tumours.

Is the tumour benign or malignant?

If one excludes the presence of calcitonin, which implies, from a practical standpoint, the diagnosis of medullary carcinoma even if the tumour is encapsulated, there is at the moment no reliable immunocytochemical basis for diagnosing malignancy.

Miettinen et al (1984) claimed that papillary carcinomas express epidermal prekeratins, whereas only (low-molecular-weight) cytokeratins are expressed in follicular carcinomas and follicular adenomas. The authors suggested that this might be used to distinguish follicular tumours from the encapsulated follicular variant of papillary carcinoma. Unfortunately, the study of a large series of tumours (Schröder et al 1986, Buley et al 1987), as well as our own results, do not provide enough evidence to consider the above mentioned intermediate filament expression as a reliable diagnostic clue (Fig. 26.3).

The putative usefulness of the immunoreactivity for the iron-binding-proteins lactoferrin and transferrin in the diagnosis of malignancy in thyroid lesions (Barresi & Tuccari 1987) needs confirmation.

The evaluation of the expression of metallothionein (Nartey et al 1987) and ras-oncogene p21 (Johnson et al 1987) allows the identification of the proliferative status, but does not provide evidence for the benign or malignant nature of the lesion.

Are there any prognostic indicators?

The contribution of immunocytochemistry to prognosis is rather limited since the outcome of patients with thyroid cancer can be generally predicted using sex and age of patients, histopathology and staging.

Neuroendocrine features in poorly differentiated thyroglobulin positive carcinomas seem to carry a poor prognosis (Ljungberg et al 1984). On the other hand, MCT with thyroglobulin immunoreactivity seem to behave less aggressively than common MCT (Holm et al 1987). However, the small size of the published series does not allow any definite conclusions on the issue.

The prognostic significance of the inverse relationship of tissue calcitonin and CEA in MCT remains also to be proven (Mendelsohn et al 1984).

It has been recently shown that the expression of Leu-M1 by the neoplastic cells of papillary carcinomas, and the density of Langerhans cells infiltrating these tumours, may add some prognostic information (Schröder et al 1987, Schröder et al 1988); larger series appear nevertheless to be necessary to validate these data.

What role is played by lectin histochemistry?

The comparison of the lectin binding properties of thyroid tumours with those of normal or goitrous glands discloses the existence of relatively constant differences between them (Sobrinho-Simões & Damjanov 1986). This is particularly obvious when one is dealing with papillary carcinomas or overtly invasive follicular carcinomas which show focal reactivity with some lectins that are non-reactive with normal follicular cells (*Solanum tuberosum*, soybean, *Ulex europaeus I* and *Dolichos biflorus*) (Figs 26.4 and 26.5).

Despite these lectin-defined differences in the composition of glycoconjugates of benign and malignant thyroid cells, lectin histochemistry is of no help in assessing whether a certain follicular tumour is malignant. Well differentiated (minimally invasive) follicular carcinomas show a lectin-binding pattern essentially identical to that of the normal thyroid gland and of benign adenomatous lesions (Sobrinho-Simões & Damjanov 1986).

In our hands the demonstration of endothelial cells by the lectin of *Ulex europaeus I* is not helpful in separating a benign from a malignant encapsulated follicular tumour. Such distinction has to be based, in our opinion, upon the presence of capsular perforation and/or invasion of sufficiently large capsular veins to be unequivocally identifiable in H & E sections.

Lectin histochemistry has also failed to show consistent differences in the glycoconjugate composition of poorly differentiated thyroid carcinomas (Fig. 26.6) and MCT (Sobrinho-Simões et al 1987) when compared to those of follicular and papillary carcinomas. Thus, despite the enormous potential of the method, it has no place in the diagnostic histopathology of the thyroid (Damjanov 1987).

CONCLUSIONS

Immunocytochemistry plays a major role in the diagnosis of thyroid tumours and is crucial for the

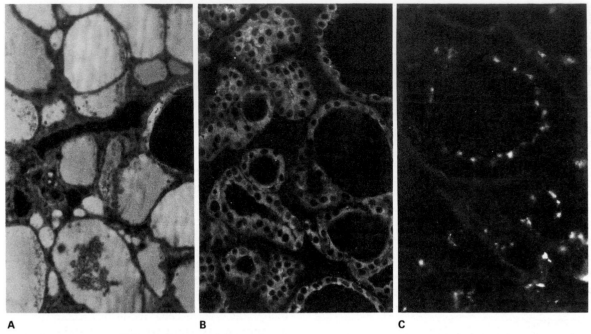

A **B** **C**

Fig. 26.4 'Normal' thyroid tissue reacted with fluorescein isothiocyanate-labelled lectins. Lectins from **A** wheat germ and **B** *Canavalia ensiformis* reacts with the cytoplasm of follicular cells and colloid. **C** Focal apical cell membrane reactivity is occasionally seen with the lectin from *Maclura pomifera* ×320

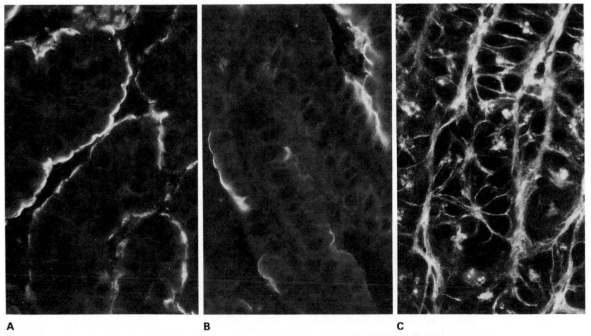

A **B** **C**

Fig. 26.5 Two papillary carcinomas and one overtly invasive follicular carcinoma reacted with fluorescein isothiocyanate-labelled lectins. The apical cell membrane reactivity of papillary carcinomas can be either **A** continuous (—lectin from *Solanum tuberosum*) or **B** focal (—lectin from *Dolichos biflorus*). Cells in the normal thyroid are unreactive with these lectins. **C** Both the cell membranes and the content of small lumina of this overtly invasive follicular carcinoma react intensely with the lectin from *Maclura pomifera*; this staining pattern is completely different from that occasionally found in normal thyroid. ×320

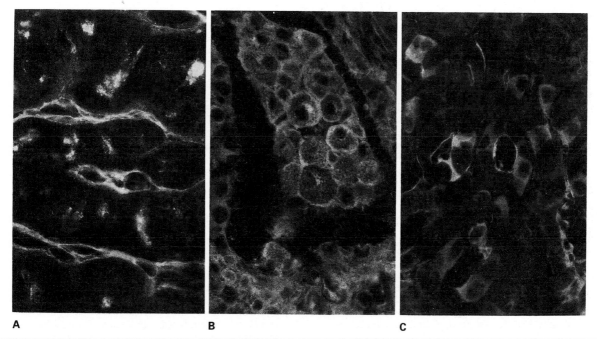

Fig. 26.6 Three poorly differentiated thyroid carcinomas reacted with fluorescein isothiocyanate-labelled lectins. **A** The lectin from wheat germ stains focally the cytoplasm of the neoplastic cells, the content of small lumina and the basal membrane. The cytoplasm of the neoplastic cells shows a granular or a diffuse pattern of reactivity with the lectins from **B** *Ricinus communis II* and **C** *Phaseolus vulgaris E4*, respectively. These lectins also stain the cells of normal thyroid and benign adenomatous lesions. ×320

understanding of the histopathogenesis of most thyroid lesions. Despite this, the contribution of immunocytochemistry and of the study of mucins and lectin histochemistry to the diagnosis of malignancy and the prediction of prognosis remains rather limited, at least for the time being.

REFERENCES

Aichinger G, Fill H, Wick G 1985 In situ immune complexes, lymphocyte subpopulations and HLA-DR positive epithelial cells in Hashimoto thyroiditis. Laboratory Investigation 52: 132–140

Albores-Saavedra J, LiVolsi V A, Williams E D 1985 Medullary carcinoma. Seminars in Diagnostic Pathology 2: 137–146

Barresi G, Tuccari G 1987 Iron-binding proteins in thyroid tumours. An immunocytochemical study. Pathology Research and Practice 182: 344–351

Buley I D, Gatter K C, Heryet A, Mason D Y 1987 Expression of intermediate filament proteins in normal and diseased thyroid glands. Journal of Clinical Pathology 40: 136–142

Carcangiu M L, Zampi G, Rosai J 1984 Poorly differentiated ('insular') thyroid carcinoma. A reinterpretation of Langhans' 'wuchernde struma'. The American Journal of Surgical Pathology 8: 655–668

Carcangiu M L, Steeper T, Zampi G, Rosai J 1985 Anaplastic thyroid carcinoma. A study of 70 cases. American Journal of Clinical Pathology 83: 135–158

Carlei F, Polak J M 1984 Antibodies to neuron-specific enolase for the delineation of the entire diffuse neuroendocrine system in health and disease. Seminars in Diagnostic Pathology 1: 59–70

Civantos F, Albores-Saavedra J, Nadji M, Morales A R 1984 Clear cell variant of thyroid carcinoma. The American Journal of Surgical Pathology 8: 187–192

Damjanov I 1987 Lectin cytochemistry and histochemistry. Laboratory Investigation 57: 5–20

DeLellis R A, Rule A H, Spiler I, Nathanson L, Tashjian A H Jr, Wolfe H J 1978 Calcitonin and carcinoembryonic antigen as tumor markers in medullary thyroid carcinoma. American Journal of Clinical Pathology 70: 587–594

Franssila K O, Harach H R, Wasenius V-M 1984 Mucoepidermoid carcinoma of the thyroid. Histopathology 8: 847–860

Gherardi G 1987 Signet ring cell 'mucinous' thyroid adenoma: a follicle cell tumour with abnormal accumulation of thyroglobulin and a peculiar histochemical profile. Histopathology 11: 317–326

Golouh R, Us-Krasovec M, Auersperg M, Jancar J, Bondi A, Eusebi V 1985 Amphicrine—composite calcitonin and mucin producing—carcinoma of the thyroid. Ultrastructural Pathology 8: 197–206

Hanafusa T, Chiovato L, Doniach D, Pujol-Borrell R, Russèl R C G, Botazzo G F 1983 Aberrant expression of HLA-DR antigen on thyrocytes in Graves' disease: relevance of autoimmunity. Lancet ii: 1111–1115

Hedinger Chr, Sobin L 1974 Histologic typing of thyroid tumours. In: International histological classification of tumours, vol ll. World Health Organization, Geneva

Hedinger Chr, Sobin L 1988 Histologic typing of thyroid tumours, 2nd edn. World Health Organization, Zurich (in press)

Holm R, Sobrinho-Simões M, Nesland J M, Johannessen J V 1986 Concurrent production of calcitonin and thyroglobulin by the same neoplastic cells. Ultrastructural Pathology 10: 241–248

Holm R, Sobrinho-Simões M, Nesland J M, Sambade C, Johannessen J V 1987 Medullary thyroid carcinoma with thyroglobulin immunoreactivity. A special entity? Laboratory Investigation 57: 258–268

Johnson T L, Lloyd R V, Thor A 1987 Expression of ras oncogene p2l antigen in normal and proliferative thyroid tissues. American Journal of Pathology 127: 60–65

Kodama T, Fujino M, Endo Y, Obara T, Fujimoto Y, Oda T, Wada T 1980 Identification of carcinoembryonic antigen in the C-cell of the normal thyroid. Cancer 45: 98–101

Ljungberg O, Bondeson L, Bondeson A-G 1984 Differentiated thyroid carcinoma, intermediate type: a new tumor entity with features of follicular and parafollicular cell carcinoma. Human Pathology 15: 218–228

Lloyd R V, Sisson J C, Marangos P J 1983 Calcitonin, carcinoembryonic antigen and neuron-specific enolase in medullary thyroid carcinoma: an immunohistochemical study. Cancer 51: 2234–2239

Martin-Lacave I, Gonzalez-Campora R, Moreno-Fernandez A, Sanchez-Gallego F, Montero C, Galera H 1988 Mucosubstances in medullary carcinoma of the thyroid. Histopathology 13: 55–66

Mendelsohn G, Wells S A, Baylin S B 1984 Relationship of tissue carcinoembryonic antigen and calcitonin to tumor virulence in medullary thyroid cracinoma. Cancer 54: 657–662

Miettinen M, Franssila K, Lehto V-P, Paasivuo R, Virtanen I 1984 Expression of intermediate filament proteins in thyroid gland and thyroid tumors. Laboratory Investigation 50: 262–270

Nartey N, Cherian M G, Banerjee D 1987 Immunohistochemical localization of metallothionein in human thyroid tumors. American Journal of Pathology 129: 177–182

Nesland J M, Sobrinho-Simões M A, Holm R, Sambade M C, Johannessen J V 1985 Hürthle-cell lesions of the thyroid: a combined study using transmission electron microscopy, scanning electron microscopy and immunocytochemistry. Ultrastructural Pathology 8: 269–290

Rigaud C, Bogomoletz W V 1987 'Mucin secreting' and 'mucinous' primary thyroid carcinomas: pitfalls in mucin

histochemistry applied to thyroid tumours. Journal of Clinical Pathology 40: 890–895

Rigaud C, Peltier F, Bogomoletz W V 1985 Mucin producing microfollicular adenoma of the thyroid. Journal of Clinical Pathology 38: 277–280

Rosai J, Saxén E A, Woolner L 1985 Undifferentiated and poorly differentiated carcinoma. Seminars in Diagnostic Pathology 2: 123–136

Sambade C, Sobrinho-Simões M, Franssila K 1987 Muco-epidermoid carcinoma of the thyroid revisited. Pathology Research and Practice 182: 552–553 (abstract)

Sambade C, Badaque-Faria A, Cardoso-Oliveira M, Sobrinho-Simões M, 1989 Follicular and papillary variants of medullary carcinoma of the thyroid. Pathology Research and Practice 184: 98–103

Schröder S, Dockhorn-Dworniczak B, Kastendieck H, Böcker W, Franke W W 1986 Intermediate-filament expression in thyroid gland carcinomas. Virchows Archiv [Pathol Anat] 409: 751–766

Schröder S, Schwarz W, Rehpenning W, Löning Th, Böcker W 1987 Prognostic significance of Leu-M1 immunostaining in papillary carcinomas of the thyroid gland. Virchows Archiv [Pathol Anat] 411: 435–439

Schröder S, Schwarz W, Rehpenning W, Lönning Th, Böcker W 1988 Dendritic/Langerhans cells and prognosis in patients with papillary thyroid carcinomas. Immunocytochemical study of 106 thyroid neoplasms correlated to follow-up data. American Journal of Clinical Pathology 89: 295–300

Sobrinho-Simões M, Damjanov I 1986 Lectin histochemistry of papillary and follicular carcinoma of the thyroid gland. Archives of Pathology & Laboratory Medicine 110: 722–729

Sobrinho-Simões M, Nesland J M, Johannessen J V 1985b Farewell to the dual histogenesis of thyroid tumors? Ultrastructural Pathology 8:iii–v (editorial)

Sobrinho-Simões M, Nesland J M, Damjanov I 1987 Lectin histochemistry of medullary carcinoma of the thyroid gland. Pathology Research and Practice 182: 558 (abstract)

Sobrinho-Simões M A, Nesland J M, Holm R, Sambade M C, Johannessen J V 1985a Hürthle cell and mitochondrion-rich papillary carcinomas of the thyroid gland: an ultrastructural and immunocytochemical study. Ultrastructural Pathology 8: 131–142

Solcia E, Capella G, Buffa R, Tenti P, Rindi G, Cornaggia M 1986 Antigenic markers of neuroendocrine tumors: their diagnostic and prognostic value. In: Fenoglio-Preiser C M, Weinstein R S, Kaufman N (eds) New concepts in neoplasia as applied to diagnostic pathology. Willimas & Wilkins, Baltimore

Thomas P, Battifora H 1987 Keratins versus epithelial membrane antigen in tumor diagnosis: an immunohistochemical comparision of five monoclonal antibodies. Human Pathology 18: 728–734

Uribe M, Fenoglio-Preiser C M, Grimes M, Feind C 1985 Medullary carcinoma of the thyroid gland. The American Journal of Surgical Pathology 9: 577–594

Wilson B S, Lloyd R V 1984 Detection of chromogranin in neuroendocrine cells with a monoclonal antibody. American Journal of Pathology 115: 458–468

27. Pancreas: tumours

R. N. Poston   M. Isabel Filipe

All parts of the pancreas share a common embryological origin from the fused dorsal and ventral pancreatic buds of the foregut. These include the cells of the islets of Langerhans, which develop as outgrowths from the primitive ducts. The common tumours, the ductular carcinomas and the islet cell tumours do, however, reflect the distinct division of the adult gland into exocrine and endocrine components.

The majority of pancreatic exocrine tumours are of ductal origin, of these the well differentiated adenocarcinoma is the most common. The benign ductal adenomas, serous and mucinous cystadenomas are rare. However, these lesions have malignant potential, and a differential diagnosis between a mucinous cystadenoma with atypical epithelium and a mucinous adenocarcinoma can be difficult or impossible in biopsy material. Furthermore, some adenocarcinomas are so well differentiated, uniform, and evoke such a marked desmoplastic reaction that they may be difficult to differentiate from the distorted normal ducts found in chronic pancreatitis. The absence of calcification and the more regular and concentric sclerosis around ducts and lobes are signs indicative of pancreatitis. On the other hand, the presence of ductal changes, such as ductal papillary hyperplasia, atypical epithelial proliferation or intraduct carcinoma may precede ductal carcinoma and will support a diagnosis of malignancy.

Only a small percentage of pancreatic carcinomas arise from acinar cells, and a few others, e.g. pancreatoblastoma, are of uncertain histogenesis.

Other tumours to be considered in the region are ampullary and periampullary lesions, the majority being well differentiated adenocarcinomas, and a few carcinoids and pseudotumours.

Histochemistry can assist with diagnosis in some major areas of pancreatic histopathology and will be considered with the following aims in mind:

1. The differential diagnosis of tumours of the pancreas.
2. The possibility of pancreatic origin of metastatic or undifferentiated tumour.
3. The differential diagnosis of pancreatic carcinoma from chronic pancreatitis.

Most well differentiated ductal carcinomas resemble the normal epithelium in secreting both neutral (PAS positive) and weakly sulphated acid mucins. PAS positive granules are also seen in the cytoplasm of acinar cell tumours and in pancreatoblastomas. Strong affinity for the *Ulex europeus* (UEA) and wheat germ (WGA) lectins have been described in pancreatic ductal carcinoma cells.

There have been many immunohistochemical studies on markers related to duct carcinomas, and a selected list is given in Table 27.1. These carcinomas show similarities to other adenocarcinomas in their pattern of marker expression, but also exhibit some more specific features. They contain cytokeratin 19, a low molecular weight cytokeratin present in many epithelia, and in addition express cytokeratin 7, a higher molecular weight molecule found in squamous cells (Osborn et al 1986). Such squamous type cytokeratins are also present in certain other ductular carcinomas, such as cholangiocarcinomas and those of the breast and sweat glands. Alpha-1-antitrypsin is also found in a similar range of tumours (Table 27.1).

Table 27.1 Antigens associated with carcinoma of the exocrine pancreas

Pancreatic oncofetal antigen +ve CaP, −ve normal pancreas	Gelder et al 1978
Polyclonal anti-mucin antibody rendered pancreas-specific by absorption with ovary	Gold et al 1983
Cytokeratin 7 +ve in Ca pancreas, bile duct −ve in Ca stomach, colon and liver	Osborn et al 1986
Alpha-1-antitrypsin/chymotrypsin: +ve Ca breast, bronchus and pancreas −ve Ca stomach and ovary	Permanetter & Wiesinger 1987
CEA in CaP, NCA in normal pancreas and CaP (CEA also +ve in many other Ca)	Batdge et al 1986, Tsutsumi et al 1984
CA19-9 increased in CaP, but +ve in normal pancreatic ducts	Makovitzky 1986
DU-PAN-2 present in normal pancreas and other tissues, increased in CaP	Metzger et al 1982, Haviland et al, 1987
B72.3 reactive with CaP and other Ca, −ve normal pancreas	Lyubsky et al 1988
YH206 reacts with Ca lung, stomach and CaP, little in normal tissues	Yachi et al 1986
Pancreatic cancer-associated antigen (PCAA) in 95% CaP, 35% other Ca, normal tissues −ve	Chu et al 1987
YPan2 +ve in 74% CaP, some in other Ca, little in normal pancreas	Yuan et al 1985
SSEA-1 +ve in 55% CaP −ve normal tissues	Satomura et al 1986

Ca — carcinoma, CaP — carcinoma of pancreatic ducts, CEA — carcinoembryonic antigen, NCA — non-specific crossreacting antigen.

Carcinoembryonic antigen (CEA) is found in most ductular carcinomas, but not usually in as large quantities as are found in tumours of the gut. This can result in negative reactions with some tumours, particularly in paraffin sections. In the past, considerable difficulties were caused by the overlap of multiple epitopes between CEA and the related molecule, non-specific crossreacting antigen (NCA). Recent work has demonstrated that new specific monoclonal antibodies against CEA will stain almost all ductular carcinomas in paraffin sections, but do not react with the normal gland (Batge et al 1986). By contrast, many antibodies reactive with NCA stain the normal ducts (Tsutsumi et al 1984).

The monoclonal antibody CA 19-9 (Koprowski et al 1979) reacts with an antigen termed GI tract carcinoma antigen (GICA), which has been identified as a sialated form of the Lewis a blood group, known as sialyl-lacto-N-fucopentaose II. It is present in many gastrointestinal carcinomas, but also in some adenocarcinomas from other sites, e.g. lung. It is frequently found in pancreatic duct carcinomas (80% positive — Atkinson et al 1982), but reactivity can be lost with dedifferentiation, or metastasis. The staining of duct carcinomas is understandable when compared to the normal gland, as the normal pancreatic duct is strongly immunoreactive. The 5% of the population who are negative for both the Lewis a and b antigens are also unable to express the CA 19-9 antigen. DU-PAN-2 is a monoclonal antibody against another carbohydrate determinant, which likewise stains pancreatic duct carcinomas, but is also expressed in other carcinomas and normal tissues.

There is an extensive literature on pancreas-related oncofetal antigens, some examples of which are included in Table 27.1. One of the first reported was the pancreas cancer-associated antigen (PCAA) of Chu et al (1977), a 1000 kD glycoprotein present in 95% of pancreatic carcinomas and 30% of other carcinomas. This antigen could be detected by both polyclonal and monoclonal antibodies. A point of considerable interest is that it was consistently absent in benign pancreatic lesions (Chu et al 1987). Two monoclonal antibodies against sialated carbohydrates, YPan2 and SSEA-1 (Table 27.1) have been reported, which show similar discrimination between neoplastic and benign pancreatic tissue (Yuan et al 1985, Satomura et al 1987).

The monoclonal antibody B72.3 recognizes a molecule, TAG72, which is strongly carcinoma-associated, and present in lesions from a wide range of organs. Recent work by Lyubsky et al (1988), has shown good results in differentiating pancreatic carcinoma from pancreatitis; 21/25 primary carcinomas and 16/16 metastases were positive, while pancreatitis showed at the most only traces of reactivity.

Useful reagents also exist for distinguishing carcinomas of pancreatic origin from those of other organs. The pancreas-specific antigen of Chu et al (1987) is normally present on acinar cells, is also expressed on pancreatic duct carcinoma, but is not found on any other cell type. Likewise, the antibodies produced Gold et al (1983) and Yuan et al (1985) (Table 27.1) have shown pancreas specificity.

A less common tumour of the exocrine pancreas is that derived from the acinar cells. A useful study by Morohoshi et al (1987) has shown that these tumours contain lipase, trypsinogen and chymotrypsinogen, a convincing confirmation of their origin. This was specially significant, as ductular carcinomas and solid/cystic tumours were negative (see below). Two cases of the rare pancreatoblastoma tumour gave the same results as the acinar tumors.

Another rare tumour is the solid and cystic tumour (also known as solid and papillary, or papillary and cystic). These are benign or low grade malignant tumours that occur in young women. Recently these have been intensively investigated by immunohistochemistry, but to somewhat confusing effect. The epithelial origin has been confirmed by the presence of cytokeratins, but in addition neuron specific enolase (NSE) has been found by most groups (Morohoshi et al 1987, Kamisawa et al 1987, Chott et al 1987). Sometimes this is only present in a minority of cells, when it may be accompanied by polypeptide hormone molecules (Meittinen et al 1987). However, the lack of specificity of many anti-NSE antisera makes dubious the conclusion that the cells are of neuroendocrine origin, especially when positive cells without endocrine granules are found, as has been reported. Some authors have favoured an acinar cell origin (Arai et al 1986, Leiber et al 1987), but the presence of vimentin and the scattered neuroendocrine cells suggest that they may arise from primitive pleuripotential cell elements (Meittinen et al 1987).

The islet cell tumours are an important group of pancreatic tumours that have been thoroughly analysed by immunohistochemistry. They arise from the normal cells of the islets, and may either secrete similar polypeptide hormones, or produce atypical, inappropriate peptides. The normal islet is composed of four principal cell types secreting specific hormones: A cells — glucagon, B cells — insulin, D cells — somatostatin, and PP cells — pancreatic polypeptide. These neoplasms are similar to other peripheral neuroendocrine tumours in that they express epithelial markers, such as low molecular weight cytokeratins, and thus are stained by the monoclonal antibody CAM5.2.

The immunostaining of peptide hormones has been a story of success, as almost all the peptides are immunogenic, have specific epitopes, and are molecules that survive in paraffin sections. The correlation of particular peptides with specific clinical syndromes does much for their interest and importance. Many excellent antipeptide polyclonal antibodies are available commercially, and their adequacy for diagnostic tasks probably explains the relative paucity of monoclonal antibodies.

Islet cell tumours react with neuroendocrine markers of wide distribution, of which NSE has been the most used. Leu-7 has also been studied, but the quality of staining with both is often poor. Recently PGP9.5, chromogranin A and synaptophysin have been introduced which offer superior specificity. The chromogranins are components of the characteristic cytoplasmic neuroendocrine granules and are probably the silver ligand in the Grimelius reaction (Solcia et al 1986). The overall sensitivity on islets cell tumours is not high, as 9/15 were stained with the monoclonal antibody LK2Hl0, reactive against chromogranin A. This, however, probably relates to the selective staining of only the A and PP cells in the normal islet, as this antibody is otherwise an excellent marker of neuroendocrine cells. Nevertheless, it is a screening marker of choice through its ability to react with non-hormone-secreting tumours (Sobol et al 1989). Synaptophysin reacts with a separate cytoplasmic component, the small synaptic vesicle, and may be a good agent for detecting islet cell tumours, as 14/15 were positive (Buffa et al 1988); however, immunoreactivity can be lessened in paraffin sections.

Insulinomas are the most common type of islet cell tumour, and may be associated with hypoglycaemic syndromes. Insulin can be demonstrated readily in the tumour, and this should be a standard diagnostic procedure. The vast majority of these tumours are small benign adenomas; about 10% are malignant, but no useful guide to malignant behaviour has been established from any aspect of the histology. Tumours producing somatostatin or glucagon are usually larger, and over half are malignant. Minor populations of cells producing other peptides can also be present, and indeed may exist in any of the islet tumours. Pancreatic polypeptide producing tumours are usually without hormonal effect, but can be malignant (Strodel et al 1984). By contrast, tumours synthesizing gastrin and vasoactive intestinal polypeptide cause characteristic clinical outcomes, namely the Zollinger-Ellison and watery diarrhoea syndromes, and each can be detected by appropriate antibodies. (For nesidiodysplasia, see Ch. 29.)

It can be said in conclusion that immunohistochemistry has made great strides in recent years in providing methods and reagents that can dissect pancreatic neoplasms biochemically with considerable accuracy. The newer and probably more specific monoclonal antibodies need further trials and general availability before their precise diagnostic utility can be established. In the meantime, CEA & CA 19-9 can be used for screening exocrine pancreatic neoplasms, while a widely reactive endocrine marker such as PGP9.5, chromogranin or synaptophysin can be used for endocrine tumours of the pancreas. Antibodies directed towards pancreatic acinar enzymes will give useful additional information.

REFERENCES

Arai T, Kino I, Nakamura S, Koda K 1986 Solid and cystic acinar tumours of the pancreas. A report of two cases with immunohistochemical and ultrastructural studies. Acta Pathologica Japan 36: 1887–1896

Atkinson B F, Ernst C S, Herlyn M et al 1982 Gastrointestinal cancer-associated antigen in immunoperoxidase essay. Cancer Research 42: 4820–4823

Batge B, Bosslet K, Sedlacek H H, Kern H F, Kloppel G 1986 Monoclonal antibodies against CEA related components discriminate between pancreatic duct type carcinomas and non neoplastic duct lesions as well as non duct type neoplasias. Virchows Archives [A] 408: 361–374

Buffa R, Rindi G, Sessa F et al 1988 Synaptophysin immunoreactivity and small clear vesicles in neuroendocrine cells and related tumours. Molecular and Cellular Probes 2: 367–381

Chott A, Kloppel C, Buxbaum P, Heitz P U 1987 Neuron specific enolase demonstration in the diagnosis of a solid cystic (papillary cystic) tumour of the pancreas. Virchows Archives [A] 410: 397–402

Chu T, Holyoke E, Douglass H 1977 Isolation of a glycoprotein antigen from ascites fluid of pancreatic cancer. Cancer Research 37: 1525–1529

Chu T M, Loor R M, Tan M H et al 1987 Tumour antigens in pancreatic cancer (abstract). Workshop — antigens of human pancreatic adenocarcinomas: their role in diagnosis and therapy. Organ Systems Coordinating and Duke University Cancer Centres, Rochville, Maryland USA

Ichihara T 1988 Immunohistochemical localization of CA19-9, and CEA in pancreatic carcinoma and associated diseases. Cancer 61: 324–337

Gelder F B, Reese C J, Moossa A R, Hall T, Hunter R 1978 Purification, partial characterisation and clinical evaluation of a pancreatic oncofetal antigen. Cancer Research 38: 313–324

Gold D V, Hollingworth P, Kremer T, Nelson D 1983 Identification of a human pancreatic duct tissue specific antigen. Cancer Research 43: 235–238

Kamisawa T, Fukayama M, Koike M, Tabata I, Okamoto A 1987 So-called 'papillary and cystic neoplasm of the pancreas'. An immunohistochemical and ultrastructural study. Acta Pathologica Japan 37: 785–794

Koprowski H, Steplewski Z, Mitchell K, Herlyn M, Fulner P 1979 Colorectal carcinoma antigens detected by hybridoma antibodies. Somatic Cell Genetics 5: 957–972

Lieber M R, Lack E E, Roberts J R et al 1987 Solid and papillary epithelial neoplasia of the pancreas. An ultrastructural and immunocytochemical study of six cases. American Journal of Surgical Pathology 11: 85–93

Lyubsky S, Malariaua J, Lozowski M et al 1988 A tumour associated antigen in carcinoma of the pancreas: defined by monoclonal antibody B72.3. American Journal of Clinical Pathology 89: 160–167

Makovitzky J 1986 The distribution and localization of the monoclonal antibody defined antigen 19–9 (CA19–9) in chronic pancreatitis and pancreatic carcinoma. An immunohistochemical study. Virchows Archives [B] 51: 535–544

Maruyama H, Mori T, Shimano T, Inaji H, Chu T M 1983 Differential distribution of the pancreatic cancer-associated antigen (SCAA) and pancreatic tissue antigen (PaA) in pancreatic and gastrointestinal cancer tissues. Annals of the New York Academy of Sciences 417: 240–250

Meittinen M, Partanen S, Fraki O, Kivilaakso E 1987 Papillary cystic tumour of the pancreas. An analysis of cellular differentiation by electron microscopy and immunohistochemistry. American Journal of Surgical Pathology 11: 855–865

Metzger R, Gaillard M, Levine S, Tuck F, Bossen E K, Borowitz M 1982 Antigens of human pancreatic adenocarcinoma. Cancer Research 42: 601–608

Morohoshi T, Kanda M, Horie A, Chott A, Dreyer T, Klöppel G, Heitz P V 1987 Immunocytochemical markers of uncommon pancreatic tumours. Acinar cell carcinoma, pancreatoblastoma and solid cystic tumour. Cancer 59: 739–747.

Osborn M, van-Lessen G, Weber K, Kloppel G, Altmannsberger M 1986 Differential diagnosis of gastrointestinal carcinomas by using monoclonal antibodies for individual keratin polypeptides. Laboratory Investigation 55: 497–504

Permanetter W, Wiesinger H 1987 Immunohistochemical study of lysozyme, alpha-1-antitrypsin, tissue polypeptide antigen, keratin & CEA in effusion sediments. Acta Cytologica 31: 104

Rindi G, Buffa R, Sessa F, Tortora O, Solcia E 1986 Chromogranin A, B and C in immunoreactivities of mammalian endocrine cells. Distribution, distinction from hormones-prohormones and relationship with the argyrophil component of secretory granules. Histochemistry 85: 19–28

Sack T L Kim Y S 1986 Pancreatic cancer-associated carbohydrate antigens. Frontiers of Gastrointestinal Research 12: 48–59

Satomura Y, Sawabu N, Takemori Y, Ohta H 1987 Expression of various sialylated carbohydrate antigens in human malignant and non-malignant pancreatic tissues (abstract). Workshop — Antigens of human pancreatic adenocarcinomas: their role in diagnosis and therapy. Organ System Coordinating and Duke University Cancer Centers, Rockville, Maryland, USA

Sobol R E, Memoli V, Deftos J 1989 Hormone-negative, chromogranin A-positive endocrine tumours. New England Journal of Medicine 320: 444–447

Solcia E, Capella C, Buffa R, Tenti P, Rindi G, Cornaggia M 1986 Antigenic markers of neuroendocrine tumours: their diagnostic and prognostic value. In: New concepts in neoplasia as applied to diagnostic pathology, IAP Monograph. Williams and Wilkins, Baltimore, p 242–261

Strodel W E, Vinik A I, Lloyd R V et al 1984 Pancreatic polypeptide producing tumours. Silent lesions of the pancreas. Archives of Surgery 119: 508–514

Tsutsumi Y, Nagura H, Watanabe K 1984 Immunohistochemical observations of carcinoembryonic antigen and CEA-related substances in normal and neoplastic pancreas. Pitfalls and caveats in CEA immunohistochemistry. American Journal of Clinical Pathology 82: 535–542

Yachi A, Imai K, Endo T, Hinoda Y 1986 Immunohistochemical analysis of human adenocarcinoma-associated antigen YH206 detected by a monoclonal antibody. Japanese Journal of Medicine 25: 127–134

Yuan S, Ho J, Yuan M, Kim Y S 1985 Human pancreatic cancer-associated antigen detected by murine monoclonal antibodies. Cancer Research 45: 6179–6187

28. The pituitary

S. E. Daniel   F. Scaravilli

INTRODUCTION

Traditional histochemical stains and, more recently, immunocytochemical methods have contributed greatly to the understanding of normal pituitary anatomy and function and to pituitary pathology. The aim of this chapter is to discuss the most useful application of these techniques in the examination of the anterior pituitary (adenohypophysis) and to tumours derived from this part of the gland — the pituitary adenomas. For diagnosis of pituitary specimens most laboratories practise a combined histochemical and immunocytochemical approach using paraffin sections of formalin fixed tissues. Histochemical methods characterize the different cell populations according to their content of complex carbohydrates or basic proteins, while immunocytochemical methods identify cell types on the basis of hormonal and non-hormonal markers. The results so obtained provide information concerning the origin, differentiation, structure and function of cells comprising neoplasms or other pituitary lesions; in certain instances ultrastructural examination may provide additional useful information.

STAINING PROPERTIES OF NORMAL ANTERIOR PITUITARY

The anterior pituitary, derived from pharyngeal ectoderm, produces at least six separate hormones: growth hormone (GH), prolactin (PRL), thyroid-stimulating hormone (TSH), adrenocorticotrophic hormone (ACTH) and the gonadotrophins, luteinizing hormone (LH) and follicle-stimulating hormone (FSH). These hormones are all polypeptides or proteins, and the gonadotrophins, TSH and large precursor form of the ACTH family of peptides — pro-opiomelanocortin (POMC), also contain a carbohydrate moiety. Modern staining techniques are aimed at identifying cell types according to hormone-production. Chronologically, however, well before the secretory function of the adenohypophysis was established, cytochemical stains were applied to differentiate cell populations (Rolleston 1936). The dyes used coloured the cytoplasmic granules of acidophils red and those of basophil blue, while the chromophobes were unstained. Later, Pearse (1949) introduced the periodic acid-Schiff histochemical method combined with Orange G (PAS/OG) for distinguishing various cell types and this stain still remains one of the most useful basic techniques applicable to pituitary histology. The PAS reaction detects the muco- and glycoprotein hormones (POMC, TSH, FSH, LH). Orange-G stains granules of cells containing basic non-glycosylated protein (GH, PRL) yellow-orange. The PAS-positive cells, termed mucoid cells, correspond to the basophils, while those containing granules stained with Orange G are acidophils.

Using immunocytochemical staining it is possible to localize production of a specific hormone at a cellular level and thus to define cell types according to their secretory function. Either the peroxidase-antiperoxidase (PAP) complex technique (Sternberger 1979), or the more sensitive avidin-biotin-peroxidase complex method (Hsu et al 1981) can be used with monoclonal antibodies to the six main types of hormone produced; both labelling techniques are applicable for electron immunocytochemistry (Childs & Unabia 1982, Varndell & Polak 1986) and furthermore, with double or multiple immunoenzyme stains, colocalization of hormones can be studied (Van Noorden et al 1985b, and Ch. 3).

The distribution of the various cell populations within the anterior pituitary is non-random. In horizontal section (Fig. 28.1) cells producing GH are the most numerous (50% of total cell population) and are concentrated in the lateral part of the gland, while PRL secreting cells (15–20% of total) occur in a posterolateral distribution; both cell types are acidophil (Fig. 28.2) when sufficiently well-granulated. Cells producing ACTH (15–20% of total) and TSH (5% of total) are distributed mainly in the median wedge (Fig. 28.1) and gonadotrophin secreting cells (10% of total) are scattered throughout the gland; these cell populations are predominantly basophil (Fig. 28.2). Chromophobe cells are found throughout the adenohypophysis and can often be shown by immunostaining to contain hormone, presumably insufficient in amount to cause positive histochemical staining and perhaps representing a highly secretory rather than a storage phase of a particular cell type.

Immunocytochemical methods may also be applied to the identification of either subunits or fragments of the glycoprotein hormones. Thus the closely related molecules of LH, FSH and TSH can be distinguished by their specific β subunits, whereas the α subunits are virtually identical. In corticotrophs the processing of POMC results in a variety of smaller peptides including ACTH, lipotrophin (LPH), melanotrophins (MSH), endorphins and corticotrophin-like intermediate lobe peptide (CLIP) (Van Noorden et al 1985a). The nature of the POMC fragments are both site- and species-dependent; in adult man there is no true intermediate lobe and only a small number of corticotrophs belong to the subset ('intermediate lobe-like') producing α-MSH and CLIP; these are distributed in the intermediate zone as well as more anteriorly and in the basophil invasion of the posterior lobe (Doniach 1985).

Recently, in addition to the classical hormones, other substances including neuron specific enolase, a variety of neuropeptides, intermediate filament proteins and S-100 protein have been identified in the anterior pituitary (Ordronneau & Petrusz 1986, McNicol 1987). Many of these non-

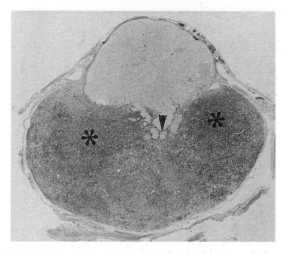

Fig. 28.1 Mid-horizontal section of normal human pituitary obtained at postmortem. The posterior lobe (above) is pale, as are colloid-filled cysts of the zona intermedia (arrowhead). Most of the acidophil cells are concentrated in the postero-lateral wings (asterisks) which appear darker than the median wedge. Mallory's acid fuchsin, aniline blue-Orange G. ×4

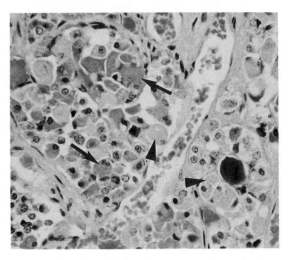

Fig. 28.2 Photomicrograph of a normal pituitary showing the cytochemical types of cells as revealed by the PAS/OG method. Basophil cells (arrows) which have an affinity for PAS appear darker than acidophil cells (arrowheads) stained by OG. ×300

hormonal immunocytochemical markers are being increasingly applied to diagnostic pathology and knowledge of their distribution in normal pituitary is essential. Cytokeratins are present in epithelial derived cells of the pituitary and thus localized to glandular cells of the adenohypophysis where there is a different subcellular distribution in the various secretory cell types (Höfler et al 1984). Immuno-staining with anti-S100 protein can be used to demonstrate an otherwise unidentifiable popu-lation of non-secretory folliculostellate cells which are widely distributed in the normal adenohypophysis. The function of these cells (Vila-Porcile & Olivier 1984) and whether or not they have a neoplastic counterpart remains uncertain.

The value of routine histological stains to anterior pituitary morphology deserves mention. We have found reticulin silver impregnation par-ticularly useful in distinguishing a normal architectural pattern with clusters of cells forming acini from a diffuse hyperplasia or pituitary adenoma in which discrete acinar structure is lack-ing (Fig. 28.3A, B).

THE PITUITARY ADENOMAS

Classification

Before the advent of immunocytochemistry, classification of pituitary adenomas was based on histological growth pattern (diffuse, papillary or sinusoidal) and on staining properties of the tumour cell cytoplasm (acidophil, basophil or chromophobe). With the establishment of im-munostaining it was soon recognized that neither histological pattern nor staining properties were criteria with any reliable clinical or functional sig-nificance. A large proportion of chromophobe adenomas, long considered to be non-secretory, were found to produce the full spectrum of pituitary hormones. Acidophil adenomas pre-viously associated only with acromegaly and GH production were recognized to secrete PRL or al-ternatively to be hormonally inactive. Similarly, it

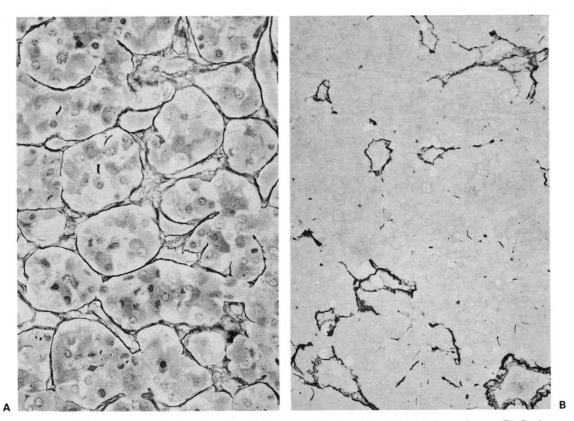

Fig. 28.3 Photomicrograph showing the reticulin pattern in the normal pituitary (**A**) and in a pituitary adenoma (**B**). In the normal, the acinar structure is outlined by a delicate reticulin network, which is only irregularly represented in adenomas. Gordon and Sweet's reticulin method. ×300

was found that while the majority of basophil adenomas occurred in patients with either Cushing's disease or Nelson's syndrome due to excess ACTH, they could also be non-secretory and clinically silent.

The now widespread application of commercially available antibodies to the various pituitary hormones has led to the development of a new and more extensive classification of the adenomas (Horvath & Kovacs 1986) in which at least 14 tumour types are described (Table 28.1). Identification of some of these adenomas requires a combined histological, ultrastructural and immunocytochemical approach correlated with biochemical assay for evidence of hormone excess. Already this more complex separation of the various pituitary adenomas is proving worthwhile and facilitating correlation of morphological features with endocrine activity, therapeutic response and prognosis. However, when the available information is incomplete a simplified form of the classification may be necessary.

Table 28.1 Prevalence of the various types of pituitary adenoma found in a series of 1200 surgically-removed pituitary tumours (modified with kind permission from Horvath & Kovacs 1986 and Churchill Livingstone)

Adenoma type	Prevalence (%)
Growth hormone cell	
Densely granulated	7.1
Sparsely granulated	7.9
Prolactin cell	
Densely granulated	0.5
Sparsely granulated	27.5
Mixed growth hormone cell-prolactin cell	5.2
Acidophil stem cell	2.5
Mammosomatotroph cell	1.4
Corticotroph cell	
Functional	8.1
Silent	5.7
Gonadotroph cell	4.5
Thyrotroph cell	0.6
Plurihormonal	4.3
Null cell	
Non-oncocytic	17.4
Oncocytic	7.3

Rapid diagnosis

Often the pathologist is required to assess pituitary tissue at the time of surgery. We find the smear technique with toluidine blue staining preferable to either touch preparation or frozen section; it is quick to perform (2 minutes) and provides excellent preservation of nuclear and cytological detail. Using this method, pituitary adenoma is usually easily recognizable from other pathology and, with experience, from normal pituitary tissue (Adams et al 1981). Histochemical stains can be applied to smear preparations of pituitary tissue (Landolt & Krayenbühl 1972); likewise, immunocytochemical techniques, which have been used on smears of general pathological (Springall 1986) and neuropathological (Budka et al 1985) material, may also be applied to pituitary tumours. However, both methods are time consuming and we believe that if the distinction between adenoma and normal gland has been made, identification of cell type can await paraffin sections.

Differentiating functional types of adenoma on paraffin sections

Only the light microscopic staining patterns of the pituitary adenomas will be discussed here. For a full classification ultrastructural features may be necessary and are described by Scheithauer (1984a, b) and Horvath & Kovacs (1986). The intensity of immunostaining obtained with either the avidin-biotin or PAP technique may vary considerably from case to case and also between cells of a single tumour; in many instances only a proportion of tumour cells appear to be positively stained. In general, the intensity of immunoreactivity for a hormone is roughly indicative of the cell granule content but provides no indication of rate of secretion or serum hormone levels which are better correlated with tumour size (Scheithauer et al 1986).

Prolactin cell adenoma

This is the commonest type of pituitary adenoma (Table 28.1). Sparsely granulated tumours are chromophobe or weakly acidophilic, while the much more rarely encountered densely granulated tumours are acidophil adenomas. These two sub-types are more easily distinguishable by immunostaining for prolactin, which in the poorly granular tumours, is concentrated in a paranuclear position (Fig. 28.4) corresponding to the Golgi region (Duello & Halmi 1980), while in the densely granular tumours prolactin is present diffusely throughout the cell cytoplasm.

Growth hormone cell adenomas

Subtypes of sparsely and densely granulated adenomas occur in this group with approximately equal frequency (Table 28.1). The majority are to some extent acidophil although the poorly granular variety may be chromophobe. Sparsely granulated GH cells often contain a characteristic juxtanuclear area, which is cytokeratin positive but unreactive for GH (Fig. 28.5) and is hyaline on haemato-xylineosin staining. This cytoplasmic feature corresponds ultrastructurally to a fibrous body (Horvath & Kovacs 1986) and may also occur in the sparsely granular GH secreting cells of mixed growth hormone-prolactin cell adenomas and in acidophil stem cell adenomas. Sparsely granulated GH adenomas are more often invasive tumours and larger than the densely granulated variety (Robert 1979).

Adenomas producing prolactin and growth hormone

Tumours immunoreactive for both PRL and GH (Fig. 28.6a, b) are common (Table 28.1) and include three types of adenoma, which are difficult to separate in the absence of either double immunostaining techniques or ultrastructural examination.

Growth hormone cell-prolactin cell adenomas are composed of two distinct, albeit related, cell lines (bihormonal, bicellular); the granularity of each of the cell types determines the staining properties and degree of positivity with immunostaining for GH and PRL.

Mammosomatotroph cell adenomas and acidophil stem cell adenomas are composed of a single cell type which is capable of producing both GH and PRL (bihormonal, monocellular). In the normal pituitary gland there are very few or no mammosomatotroph cells, but immature cells resembling the acidophil stem cell are identifiable and may be the progenitor cell of somatotrophs and lactotrophs. Mammosomatotroph adenomas are usually acidophilic while acidophil stem cell tumours may be chromophobe or only weakly acidophilic; the staining patterns reflect the predominant production of GH by mammosomatotroph tumours and PRL by acidophil stem cell adenomas. Results of GH and PRL

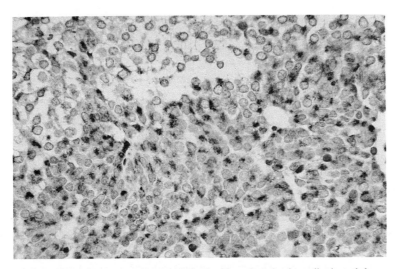

Fig. 28.4 Photomicrograph of a pituitary adenoma immunostained with anti-prolactin antibody and the peroxidase-antiperoxidase technique. Clumps of the reaction product, in close proximity to the nucleus and corresponding to the Golgi region, are characteristic of the poorly granular type of prolactin cell adenoma. ×300

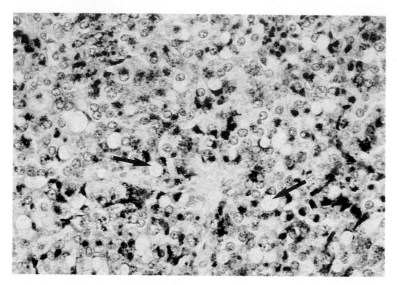

Fig. 28.5 Photomicrograph of a GH secreting adenoma immunostained with anti-growth hormone antibody using the avidin-biotin-peroxidase complex method. Most cells are sparsely granulated and some of them contain a round juxtanuclear area (arrows) which is devoid of staining. These areas correspond to the fibrous bodies. ×300

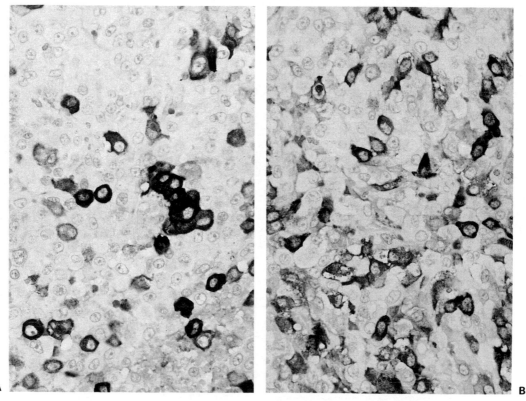

Fig. 8.6 A mixed GH-PRL secreting adenoma. The tumour is stained with antibody to GH in **A**, strongly immunoreactive cells occur in groups and are also widely scattered. In **B** stained with antibody to PRL there are many immunopositive tumour cells. EM examination showed an admixture of GH cells with PRL cells and this tumour is therefore a GH cell-PRL cell adenoma. Peroxidase-antiperoxidase. ×300

immunoreactivity in the bihormonal monomorphous adenomas can be difficult to distinguish from those of a mixed GH cell-PRL cell adenoma and for a definitive diagnosis of these tumours electron microscopy is required. Distinction of the acidophil stem cell adenoma from others in this group is important as it is usually a fast growing aggressive tumour (Horvath & Kovacs 1986).

Corticotroph cell adenomas

The majority of these tumours are well-granulated basophil adenomas and occur in patients with either Cushing's disease or Nelson's syndrome. In adenomas of Cushing's disease, tumour cells frequently contain a perinuclear cytoplasmic vacuole (Fig. 28.7) which resembles Crooke's hyaline change induced in normal corticotrophs in response to increased adrenal glucocorticoid feedback. These so-called 'enigmatic bodies' (Russfield 1968) present in corticotroph cells need to be distinguished from the fibrous bodies present in some chromophobe GH adenoma cells (see above); immunostaining demonstrates that both are immunoreactive for cytokeratin.

Silent corticotroph adenomas are unassociated with signs and symptoms of adrenal hypercorticism (Horvath et al 1980), but, like the functional corticotroph tumours, contain ACTH and other peptides derived from ACTH precursor POMC. Charpin et al (1982) have suggested that the biological activity of ACTH and nature of POMC fragments could be influenced by the site of origin of the tumour; adenomas producing Cushing's disease are thought to arise from corticotrophs in the anterior part of the gland while non-functioning tumours may be derived from the subset of 'intermediate lobe-like' corticotrophs.

Gonadotroph cell adenomas

Tumours secreting gonadotrophins rarely cause endocrine symptoms and it is only since the introduction of antibodies to FSH and LH that this type of chromophobe adenoma has been more readily identifiable. The same cell usually produces both FSH and LH. Because these tumours can be technically difficult to immunostain, their true incidence may be considerably underestimated.

Thyrotroph cell adenomas

This type of adenoma is uncommon, usually chromophobe and with variable immunopositivity for TSH. Cells containing TSH may also occur in the *plurihormonal adenomas*, which are an ill-

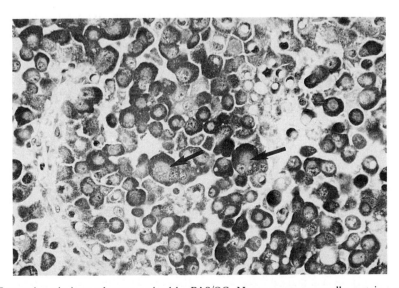

Fig. 28.7 An ACTH-secreting pituitary adenoma stained by PAS/OG. Numerous tumour cells contain a perinuclear hyaline cytoplasmic area (arrows) which represent the so-called 'enigmatic bodies'. ×300

defined group showing production of unexpected combinations of hormones such as GH-TSH, GH-PRL-TSH and PRL-TSH.

Null cell adenomas

Tumours of this class are chromophobe adenomas with the oncocytic variant characterized by an eosinophilic cell cytoplasm packed with mitochondria. Although these tumours are considered to be non-functioning, the use of the avidin-biotin-peroxidase complex method for immunocytochemistry has shown that many of them contain scattered cells positive for various pituitary hormones or α-subunits. It seems likely that this group may well be heterogeneous and include some tumours which are in fact functional.

Non-hormonal immunocytochemical cell markers

Immunostaining for cytokeratin in pituitary adenomas may assist in the often difficult interpretation of a small biopsy and, when dealing with a differential diagnosis between pituitary adenoma and intrasellar paraganglioma or ependymoma, the latter two are usually negative for cytokeratin, in contrast with the strong positivity of the pituitary adenomas. Recently, a study of a nuclear proliferation antigen using monoclonal antibody Ki-67 on smear preparations and frozen sections of pituitary adenomas has been carried out (Burger et al 1986, Landolt et al 1987). Results so far suggest that this antibody may provide useful information concerning the growth potential of the different types of adenoma.

CONCLUSIONS

This chapter reviews the considerable contribution of modern staining techniques, in particular immunocytochemistry, to the understanding of the normal anterior pituitary gland and pituitary adenomas. The advances that are being made in this field are all the more necessary in view of an increasing incidence of pituitary adenomas (Scheithauer 1984a) associated with more sensitive serum hormone assays and improved imaging techniques. Operative intervention is occurring at an earlier stage when the tumour is often quite discrete, and the transsphenoidal approach to pituitary lesions with selective adenomectomy is gaining popularity. This approach has the disadvantage of producing small and fragmented specimens and the application of a panel of histochemical and immunocytochemical stains is an important adjunct in arriving at the correct diagnosis. It is now routine practice to immunostain pituitary adenomas for the presence of hormones and to classify these tumours according to their secretory activity. Preliminary results suggest that tumour size, invasiveness and likelihood for recurrence (Scheithauer et al 1986), as well as response to various pharmacological agents, may all be related to the functional differentiation of an adenoma. There is now a great need for further long-term clinicopathological studies and international agreement on tumour nomenclature. In the future it is likely that certain adenomas will be further subclassified on the basis of their production of hormone fragments or subunits and the non-hormonal immunocytochemical markers, notably nuclear proliferation association antigens, will become of increasing importance.

REFERENCES

Adams J H, Graham D I, Doyle D 1981 Brain biopsy. The smear technique for neurosurgical biopsies. Biopsy pathology series, Chapman and Hall, London
Budka H, Majdic O, Knapp W 1985 Cross-reactivity between human hemopoietic cells and brain tumors as defined by monoclonal antibodies. Journal of Neuro-oncology 3: 173–179
Burger P C, Shibata T, Kleihues P 1986 The use of the monoclonal antibody Ki-67 in the identification of proliferating cells. American Journal of Surgical Pathology 10: 611–617

Charpin C, Hassoun J, Oliver C et al 1982 Immunohistochemical and immunoelectronmicroscopic study of pituitary adenomas associated with Cushing's disease. American Journal of Pathology 109: 1–7
Childs G, Unabia G 1982 Application of a rapid avidin-biotin-peroxidase complex (ABC) technique to the localization of pituitary hormones at the electron microscope level. Journal of Histochemistry and Cytochemistry 30: 1320–1324
Doniach I 1985 Histopathology of the pituitary. Clinics in Endocrinology and Metabolism 14: 765–789

Duello T M, Halmi N S 1980 Immunocytochemistry of prolactin-producing human pituitary adenomas. American Journal of Anatomy 158: 463–469

Höfler H, Denk H, Walter G F 1984 Immunohistochemical demonstration of cytokeratins in endocrine cells of the pituitary gland and in pituitary adenomas. Virchows Arch (A) 404: 359–368

Horvath E, Kovacs K 1986 Identification and classification of pituitary tumours. In: Cavanagh J B (ed) Recent advances in neuropathology 3. Churchill Livingstone, Edinburgh, p 75–93

Horvath E, Kovacs K, Killinger D W, Smyth H S, Platts M E, Singer W 1980 Silent corticotropic adenomas of the human pituitary gland. A histologic, immunocytologic and ultrastructural study. American Journal of Pathology 98: 617–638

Hsu S M, Raine L, Fanger H 1981 Use of avidin-biotin-peroxidase complex (ABC) in immunoperoxidase techniques: a comparison between ABC and unlabeled antibody (PAP) procedures. Journal of Histochemistry and Cytochemistry 29: 577–580

Landolt A M, Krayenbühl H 1972 A modified cytological technique for rapid differentiation of pituitary adenomas. Journal of Neurosurgery 37: 289–293

Landolt A M, Shibata T, Kleihues P 1987 Growth rate of human pituitary adenomas. Journal of Neurosurgery 67: 803–806

McNicol A M 1987 Pituitary adenomas. Histopathology 11: 995–1011

Ordronneau P, Petrusz P 1986 Non-hormonal markers in the pituitary. In: Polak J M, Van Noorden S (eds) Immunocytochemistry modern methods and applications, 2nd edn. Wright, Bristol, p 425–437

Pearse A G E 1949 The cytochemical demonstration of gonadotropic hormone in human anterior hypophysis. Journal of Pathology and Bacteriology 61: 195–202

Robert F 1979 Electron microscopy of human pituitary tumors. In: Tindall G T, Collins W F (eds) Clinical management of pituitary disorders. Raven Press, New York, p 113–131

Rolleston H D 1936 The endocrine organs in health and disease with an historical review. Oxford University Press, Oxford, p 42–67

Russfield A B 1968 Adenohypophysis. In: Bloodworth J M B (ed) Endocrine pathology. Williams and Wilkins, Baltimore, p 75–116

Scheithauer B W 1984a Surgical pathology of the pituitary: the adenomas. Part 1. Pathology Annual 19: (1) 317–374

Scheithauer B W 1984b Surgical pathology of the pituitary: the adenomas. Part 2. Pathology Annual 19: (2) 269–329

Scheithauer B W, Kovacs K, Laws E R, Randall R V 1986 Pathology of invasive pituitary tumors with special reference to functional classification. Journal of Neurosurgery 65: 733–744

Springall D R 1986 Immunocytochemistry in diagnostic cytology. In: Polak J, Van Noorden S (eds) Immunocytochemistry modern methods and applications, 2nd edn. Wright, Bristol, p 547–567

Sternberger L A 1979 Immunocytochemistry, 2nd edn. John Wiley, New York

Van Noorden S, Lewis P D, Polak J M 1985a Pituitary adenomas. In: Polak J M, Bloom S R (eds) Endocrine tumours. Churchill Livingstone, Edinburgh, p 241–263

Van Noorden S, Stuart M C, Cheung A, Adams E F, Polak J M 1985b Localization of pituitary hormones by multiple immunoenzyme staining procedures using monoclonal and polyclonal antibodies. Journal of Histochemistry and Cytochemistry 34: 287–292

Varndell I M, Polak J M 1986 Electron microscopical immunocytochemistry. In: Polak J M, Van Noorden S (eds) Immunocytochemistry modern methods and applications, 2nd edn. Wright, Bristol, p 146–166

Vila-Porcile E, Olivier L 1984 The problem of the folliculo-stellate cells in the pituitary gland. In: Motta P M (ed) Ultrastructure of endocrine cells and tissues. Martinus Nijoff, Boston, p 57–76

29. Neuroendocrine tumours and hyperplasias

E. Solcia   G. Rindi   C. Capella

GENERAL CONCEPTS

A variety of epithelial endocrine cells scattered in different tissues of endodermal (gastrointestinal, bronchial mucosa and thyroid gland) or ectodermal origin (paraganglia, adrenal medulla, pituitary and skin) have been shown to share with nerve cells a series of histochemical markers (Pearse 1969). These include monoamines, regulatory peptides, secretory proteins like chromogranins and secretogranins, membrane proteins of secretory vesicles and granules like synaptophysin, enzymes like neuron specific enolase, cholinesterase, dopa decarboxylase or glutamic acid decarboxylase, calcium binding proteins like calbindin, soluble cytosolic proteins of unknown function like protein gene product (PGP) 9.5, neurofilament proteins, and certain gangliosides of the cytoplasmic membrane (Polak & Bloom 1985). These cells form the so-called neuroendocrine system, either 'diffuse' (gut, pancreas, lung, thyroid C cells, skin, urogenital tract), or 'glandular' (adrenal medulla, paraganglia, pituitary). Related growths are called 'neuroendocrine tumours' or 'hyperplasias' (Gould et al 1983a) (Table 29.1). In this morphofunctional sense, the designation 'neuroendocrine' has no relationship with the embryological derivation of the cells, either from neuroectoderm (paraganglionic cells, adrenal medullary cells and thyroid C cells), the epiblast (Merkel cells, pituitary cells), or the endoderm (gut, pancreas, respiratory tract and at least most urogenital tract cells). Although the earlier concept of APUD cells (Pearse 1969) suggested neural crest origins for some, if not all, this concept, useful initially, has been modified with recent knowledge.

Tumour markers

The diagnosis of neuroendocrine tumours is based firstly on the recognition of its neuroendocrine nature, and secondly on the identification of the neuroendocrine cell type.

The more widely used general markers are the Grimelius silver or lead haematoxylin methods for secretory granules, and immunohistochemical methods for granular secretory proteins like chromogranins (Wilson & Lloyd 1984, Solcia et al 1986a), or granule membrane proteins like synaptophysin (Wiedenmann et al 1986, Buffa et al 1987). A close relationship between chromogranins and silver-binding sites of endocrine granules has been suggested. As a general screening method, chromogranin is one of the more useful markers and may be positive even in the absence of hormone secretion (Sobol et al 1989). Neuron specific enolase, despite its questionable specificity, and PGP 9.5 are useful for agranular or poorly granular cells, especially of neuroendocrine carcinomas.

In both normal and neoplastic tissues, neuroendocrine cell typing is based mainly on the identification of specialized hormonal or modulatory substances produced and stored by the cells (Solcia et al 1987) (Table 29.1). As a rule, in different cell types distinct genes are expressed, coding for different propeptides from which different regulatory peptides are obtained, thus providing a sound basis for cell and tumour characterization and correlation with clinical hyperfunctional syndromes.

This approach, however, suffers from some limitations. The same peptide may be produced by different types of tumours having different origin, cytology, evolution and prognosis. This is illustrated by the dual nature of vasoactive intes-

Table 29.1   Cells and tumours of the human neuroendocrine system

Site	Normal cell types	Main products	Main tumours
Pituitary	GH, PRL, ACTH, TSH, FSH/LH	GH, PRL, ACTH, TSH, FSH, LH	Adenomas·
Non-chromaffin paraganglia	Type I	Enkephalins, NAD, DA, 5HT	Non-chr. paragangliomas
Chromaffin paraganglia	Main cells	NAD, DA, enkephalins	Chromaffin paragangliomas
Sympathetic ganglia	SIF cells	DA, 5HT, enkephalins	Ganglioneuromas/paragangliomas
Adrenal medulla	AD cell NAD cell	AD, enkephalins, NPY NAD, dynorphins, bombesin	Phaeochromocytomas and neuroblastomas
Skin	Merkel cell	Unknown	Merkel cell tumour or NE Ca
Nose, ear, salivary glands	—		NE Ca, carcinoids
Thyroid	C cell	Calcitonin, CGRP	Medullary Ca
Thymus	—		NE Ca, carcinoids
Lung, larynx	P cells	Bombesin (GRP)	NE Ca, P cell carcinoids
Oesophagus	EC, non-EC	5HT	NE Ca, carcinoids
Stomach	ECL, EC, P/D1, G, D, X	H, 5HT, gastrin, somatostatin	Carcinoids: ECL, EC, P cells G/D cell tumour; NE Ca
Pancreas	A, B, D, PP, P, EC	Glucagon, insulin, somatostatin, PP	Islet cell tumours gastrinoma, VIPoma, etc.
Biliary tract	EC, D	5HT, somatostatin	Carcinoids, G/PP cell tumour
Small bowel	EC, P, M, L, N, D, G, CCK, S, GIP	See Table 29.2	Carcinoids: EC, non-EC cells G and/or D cell tumour
Appendix Large bowel	EC, L, D EC, L, D	5HT, subst P, entero-glucagons, somatostatin	Carcinoids: EC, L cells Carcinoids: EC, L cells
Prostate, urethra	EC, non-EC	5HT	Mixed endocrine exocrine Ca, carcinoids
Uterus	—		NE Ca, carcinoids
Gonads	—		Carcinoids: EC, L, C cells
Breast	—		Mixed endocrine/exocrine Ca, NE Ca

ABBREVIATIONS: see also Tables 29.2, 29.3 & 29.4

A	= Glucagon producing cell		GRP	= Gastrin releasing peptide
ACTH	= Adrenocorticotrophic Hormone		H	= Histamine
AD	= Adrenaline		5-HT	= 5-hydroxytryptamine
ADH	= Antidiuretic Hormone		L	= GLI and PYY producing cell
B	= Insulin producing cell		LH	= Luteinizing Hormone
C	= Calcitonin producing cell		M	= Motilin
CCK	= Cholecystokinin		N	= Neurotensin
CGRP	= Calcitonin gene related peptide		NAD	= Noradrenaline
D	= Somatostatin producing cell		NPY	= Neuropeptide Tyrosine
DA	= Dopamine		P	= Cells with very small round haloed granules
D1	= Ultrastructural variety of PP cell		PP	= Pancreatic polypeptide
EC	= Enterochromaffin cell		PRL	= Prolactin
ECL	= Enterochromaffin cell-like		PTH	= Parathyroid Hormone
F	= Ultrastructural variety of PP cell		PYY	= PP-like peptide with N-Terminal Tyrosine and C-Terminal Tyrosine amide
FSH	= Follicle-stimulating hormone			
G	= Gastrin-producing cell		S	= Somatostatin
GH	= Growth Hormone		SIF	= Small intensely fluorescent cell
GIP	= Gastric Inhibitory Polypeptide		TSH	= Thyroid stimulating hormone
GLI	= Glucagon-like immunoreactants: glicentin, glucagon-37, glucagon-29		VIP	= Vasoactive intestinal polypeptide
			NE Ca	= Neuroendocrine carcinoma
GRF	= Growth Hormone releasing factor		Ca	= Carcinoma

tinal peptide (VIP)-producing tumours, which may be either pancreatic epithelial endocrine tumours or retroperitoneal ganglioneuromas and ganglioneuroblastomas (Solcia et al 1988b). Moreover, alternative processing of the same m-RNA may result in distinct propeptides and peptides. This is exemplified by calcitonin and calcitonin gene-related peptide (CGRP), coded by the same gene through different propeptides showing partly overlapping tissue and tumour specific distribution (Sabate et al 1984, Tschopp et al 1984). Two or more regulatory peptides may be found in the same cell, either as products of the same propeptide coded by a single gene, as in the case of ACTH and endorphins — both arising from pro-opiomelanocortin, or as products of different genes and prohormones, as with glucagon and PP-related peptides in intestinal L cells and rectal carcinoids (Böttcher et al 1984, Fiocca et al 1987).

Tumour diagnosis

In addition to hormone histochemistry, parallel histopathological and ultrastructural investigations are needed to establish the correct diagnosis of neuroendocrine tumours. Ultrastructural immunocytochemistry, combining the distinctive structure of secretory granules with the immunochemical identification of their stored products, can be helpful in tumour diagnosis. The ultrastructural characterization of tumour cells is especially needed when no hormonal products have been identified and the function of related endocrine cells has not yet been established, as in the case of gastric argyrophil carcinoids (Capella et al 1980). However, the use of chromogranin may obviate the need for ultrastructural study (Sobol et al 1989).

The coexistence of multiple endocrine cell types in an endocrine tumour or of different hormonal products within the same tumour cell may make it difficult to classify neuroendocrine tumours on purely cytological grounds. However, most difficulties are overcome when both the cytological and clinical findings are taken into account. This is especially true for cases developing a clinically functional syndrome, which, as a rule, expresses a single and distinctive pattern, even in those patients with immunocytochemically heterogeneous tumours (Solcia et al 1984).

Tumours with appropriate and inappropriate secretion

From Table 29.1 it may be concluded that many neuroendocrine tumours arise in tissues normally showing related endocrine cells. The behaviour of these 'appropriate' tumours differs markedly according to their site of origin, with prevalence of benign tumours in the pancreas, appendix, rectum and lung — especially for tumours less than 2–3 cm in size — and malignant tumours in the skin and thyroid.

Endocrine tumours may arise in human tissues substantially lacking endocrine cells of the peptide/amine type, as, for instance, thymus, salivary glands, larynx, oesophagus, cervix and ovary. As a rule, these 'inappropriate' tumours arising in 'non-endocrine' tissues are malignant.

Other 'inappropriate', often malignant, tumours may arise in tissues normally having endocrine cells of a different type. Some tumours, such as ACTH and calcitonin cell tumours (Deftos & Burton 1980), arise in many different tissues. In contrast, gastrin cell tumours are prevalent in the pancreas. This behaviour could be explained as a latent potential of the pancreatic endocrine stem cell to express gastrin secretion and gastrin cell morphology, since gastrin cells have been found in the fetal neonatal pancreas of the rat and cat.

The origin of some endocrine tumours from metaplastic epithelia (e.g. intestinal-type tissue occurring in the stomach, gall-bladder, oesophagus, nose and ovary; pyloric-type tissue in the pancreas, biliary tract and ovary) may explain some 'ectopic' endocrine tumours and mixed endocrine exocrine tumours (Cocco & Conway 1978, Solcia et al 1979, Margolis & Jang 1984).

Hyperplasias

Neuroendocrine cells scattered in the mucosa of the gastrointestinal, biliary, bronchial and urogenital tracts may increase in number and become hyperplastic while remaining:

1. Intraepithelial or intraglandular
2. As single scattered elements (simple or diffuse hyperplasia)

3. Form chains and lines at the base of the epithelium (chain-forming or linear hyperplasia)
4. Form minute micronodules up to 100–150 μm in size (micronodular hyperplasia) (Solcia et al 1979).

Sometimes, budding of small clusters of neuroendocrine cells from the base of the epithelium is found, a process called 'burgeonnement'.

Extraepithelial (hyperplastic) micronodules of neuroendocrine cells may be found scattered in the lamina propria, especially of the gastric and appendiceal mucosa. Micronodules are usually surrounded by basal membrane — better seen by electron microscopy — and appear benign. They are often found in areas of chronic inflammation with fibrosis and glandular atrophy, which may induce reactive proliferation of neuroendocrine cells, sometimes intimately contacting hyperplastic nerve endings and Schwann cells.

In the pancreatic islets, adrenal medulla and paraganglia, the enlarged size, slight histological rearrangement and cell crowding of the endocrine gland or unit are signs of diffuse hyperplasia. Focal concentration of the hyperplastic growth may lead to micronodular or nodular hyperplasia.

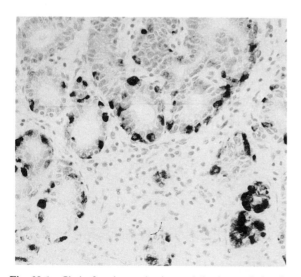

Fig. 29.1 Chain-forming and micronodular hyperplasia of neuroendocrine cells in type A chronic atrophic gastritis, stained by chromogranin A monoclonal antibody PhE5 and streptavidin-biotinylated peroxidase complex (SABC)-Haematoxylin counterstain. ×280

Disappearance of the original stroma and histological pattern, as well as profound changes in the proportion and distribution of the various cell types, usually mark the initiation of tumour growth, either benign or malignant.

Preneoplastic changes

In the case of diffuse neuroendocrine cells preneoplastic changes may include:

1. *Adenomatoid hyperplasia*, a collection of micronodules closely adherent to each other, with interposition of basal membranes, mimicking the back-to-back pattern typical of intestinal adenomas.
2. *Enlarging micronodules*, averaging from 150 to 500 μm, in size.
3. *Fusing micronodules* resulting from disappearance of the intervening basal membranes between micronodules.
4. *Microinfiltrative lesions* formed by irregular cords and nests of relatively large, moderately atypical cells filling the lamina propria in between glands or below the epithelium (Solcia et al 1979, 1986b).

All these lesions may contribute to the so-called 'tumourlets' so far reported in the lung and gut (Bonikos et al 1967, Bhagavan et al 1974).

Neoplastic changes

Any neuroendocrine growth confined to the mucosa is considered to be a neoplasia when showing:

1. Trabecular, microlobular, solid nests, rosette-like or diffuse structures with newly-formed stroma, or
2. At least one infiltrating or expansive lesion exceeding 0.5 mm in size.

Infiltration of the muscularis mucosae or the muscular wall and/or angioinvasion are distinctive marks of invasive tumours.

Whenever possible, benign neuroendocrine tumours or adenomas (for instance, islet cell adenomas) should be distinguished from well differentiated, low-grade carcinomas. For some tumours (for instance many carcinoids and pancreatic tumours), separation of benign from low-grade malignancy on conventional histological grounds may be difficult or even impossible. Characterization of tumour cell lineage and correlation with clinical syndromes may be helpful in these cases.

CYTOCHEMICAL CHARACTERIZATION OF SOME NEUROENDOCRINE GROWTHS

Gastrointestinal tract

The neuroendocrine cells scattered in normal human gastroenteropancreatic tissues are shown in Table 29.2. At least 14 cell types have been identified in the gastrointestinal mucosa, 12 of which have known hormonal products. For some endocrine cells of the upper small intestine (secretin, cholecystokinin, GIP and motilin cells), no tumour counterpart has so far been identified in man.

Presently characterized endocrine tumours of the gastrointestinal tract are outlined in Table 29.3. Most gastric argyrophil carcinoids have been

shown ultrastructurally to be ECL cell tumours or to have a prominent ECL cell component (Capella et al 1980). Argyrophilia and positivity for chromogranin A and histamine in the absence of gastrin and somatostatin immunoreactivity is diagnostic (Håkanson et al 1986, Solcia et al 1986b). Gastric argyrophil carcinoids and ultrastructurally proven ECL cell tumours may arise in a background of chronic atrophic gastritis of the body-fundus, usually with achlorhydria and secondary hypergastrinaemia, and often with pernicious anaemia. Rarely, they may arise in hypergastrinaemic patients with Zollinger-Ellison syndrome, especially when coupled with multiple endocrine neoplasia type 1 (MEN1) syndrome. Tumour regression has been reported in patients whose hypergastrinaemia was abolished by antrectomy, thus confirming the gastrin-dependance of such tumours (Solcia et al 1986, Bordi et al 1988). Gastric argyrophil carcinoids are often multiple and associated with argyrophil cell hyperplasia.

Gastrin cell hyperplasia of the pyloric mucosa may be secondary to achlorhydria in patients with

Table 29.2 Human gastroenteropancreatic endocrine cells

| Cell | Main product | Pancreas | Stomach | | Intestine | | | | | |
| | | | Oxyntic | Antral | Small | | | Appendix | Large | |
					Duod.	Jej.	Ileum		Colon	Rectum
P	Unknown	f	+	+	+	f	f		f	f
EC	5HT+Peptides	f	+	+	+	+	+	+	+	+
EC₁ subtype	5HT+Substance P				+	+	+	+	+	+
D	Somatostatin	+	+	+	+	+	f	+	f	+
L	GLI + PYY				f	+	+	+	+	+
A	Glucagon	+	a							
PP (F/D1)	Pancr. Peptide	+								
B	Insulin	+								
X	Unknown		+							
ECL	Histamine		+							
G	Gastrin			+	+					
CCK	CCK				+	+	f			
S	Secretin				+	+				
GIP	GIP				+	+	f			
M	Motilin				+	+	f			
N	Neurotensin				f	+	+			

f = few; a = fetus and newborn; Abbreviations: see Table 29.1

Table 29.3   Cytologic and clinicopathologic characterization of gut endocrine tumours

Tumour	Preferred site	Prevalent cell type	Main hormonal products	Associated syndrome or pathologic condition
Well differentiated				
1. *Gastric argyrophil carcinoid*	Body/fundus	ECL	Histamine, 5HT/5HTP	CAG, ZES/MEN1 syndrome Atypical carcinoid syndrome G cell hyperplasia
2. *Gastrin cell tumour*	Duodenum, antrum, jej.	G	Gastrin	ZES; ECL cell growth
3. *Somatostatin cell tumour*	Duodenum, antrum, jej.	D	Somatostatin	Cholelithiasis
4. *Gangliocytic 'paraganglioma'*	Duodenum, antrum, jej.	PP, D	Somatostain, PP	Neurofibromatosis
5. *Argentaffin carcinoid*	Appendix, ileum, jej, cecum	EC	5HT, subst. P	Carcinoid syndrome
6. *Hindgut trabecular carcinoid*	Rectum, colon	L	Glicentin/ glucagon-37, PP, PYY	
7. *'Inappropriate' tumours*	Stomach, intestine		ACTH/MSH, VIP, calcitonin	Cushing syndrome VIPoma syndrome
Poorly differentiated				
Neuroendocrine carcinoma	Esophagus, stomach, intestine	Protoendocrine	Variable	Variable

CAG = Chronic atrophic gastritis; ZES = Zollinger-Ellison syndrome; MEN1 = Multiple Endocrine Neoplasia, type 1.

severe chronic atrophic gastritis of the body-fundus (type A gastritis) or it may appear as a primary lesion causing peptic ulcer disease (Fig. 29.2). Gastrin cell tumours, seldom found in the stomach and jejunum, are much more frequent in the duodenum, and may occur with or without associated overt hypergastrinemia and peptic ulcer disease (gastrinoma syndrome). Gastrin immunohistochemistry is diagnostic (Fig. 29.3) for such tumours, which may also show varying numbers of somatostatin cells (Solcia et al 1988a).

Pure somatostatin cell tumours have been observed in the duodenum (Dayal et al 1983). Cholelithiasis, but no overt 'somatostatinoma' syndrome, has been reported in association with these tumours. Increased numbers of somatostatin cells (Sjölund et al 1979) and even a small somatostatin cell tumour have been found in the

small intestine in celiac disease. Extreme somatostatin cell hyperplasia coupled with dwarfism, obesity and goitre, apparently due to severe, lifelong hypersomatostatinism, has also been described (Holle et al 1986).

A tumour growth resulting from variable admixtures of:

1. Epithelial neuroendocrine cells, often immunoreactive for somatostatin, pancreatic polypeptide or other hormones
2. Mature ganglion cells
3. Schwann-like spindle cells enveloping nerve cells, axons and epithelial cells, or forming small bundles

has been called 'ganglioneuromatous', or 'gangliocytic' paraganglioma. Most of these tumours arise in the periampullary region of the

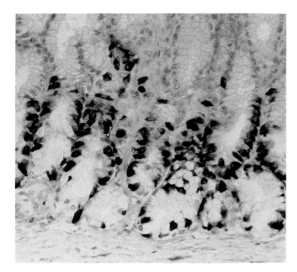

Fig. 29.2 Palisade-forming gastrin cell hyperplasia in the pyloric mucosa of a patient with Stempien disease (hypertrophic gastropathy of the acidopeptic mucosa coupled with gastric hypersecretion and protein dispersion); C-terminus gastrin antibodies; SABC-Haematoxylin. ×280

duodenum and some of them may be associated with von Recklinghausen's multiple neurofibromatosis. This may represent the earlier described 'neurocarcinoid'.

The EC cell argentaffin carcinoid producing serotonin and, in most cases, substance P may be diagnosed with classic histochemical tests for 5HT or by peptide immunohistochemistry. Those arising in the lower small bowel and caecum are frequently metastatic and associated with the 'carcinoid' syndrome.

Carcinoid tumours arising in the hindgut (i.e. rectum and colon, but excluding caecum) have a characteristic trabecular structure and staining pattern. Cells producing various glucagon- and PP/PYY-related peptides and mimicking intestinal L cells, are the dominant cell type of trabecular hindgut carcinoid (Fiocca et al 1987) (Fig. 29.4).

Poorly differentiated neuroendocrine carcinomas are relatively rare in the gastrointestinal tract,

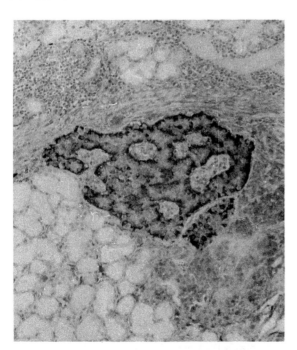

Fig. 29.3 Submucosal microgastrinoma in the duodenum of a patient with multiple duodenal microgastrinomas and the Zollinger Ellison syndrome. Note the close topographical relationship of the tumour with Brünner glands, where a few gastrin cells are normally found. C-terminus gastrin antibodies; ABC-technique-Haematoxylin. ×160

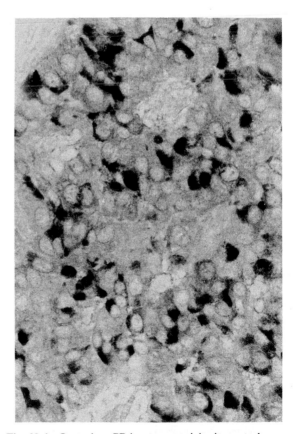

Fig. 29.4 C-terminus PP immunoreactivity in a rectal trabecular carcinoid, ABC-Haematoxylin. ×450

though not exceptional in the stomach. They resemble corresponding lung tumours as regards morphology, cytochemistry and clinicopathological behaviour.

Urogenital tract

Both serotonin-producing EC cells and various types of non-EC neuroendocrine cells have been described in human non-neoplastic urethra, prostate and uterus (Abrahamsson et al 1986). Scattered serotinin, ACTH, endorphin, somatostatin, and glucagon-like immunoreactive cells have been frequently reported in prostatic carcinomas, ovarian mucinous cystadenocarcinomas and, less frequently, in cervical or endometrial adenocarcinomas (Bosman 1984). Mixed endocrine-exocrine carcinomas with intimate admixtures of abundant endocrine as well as exocrine cells have been observed with some frequency in the prostate (Capella et al 1981). A few neuroendocrine carcinomas, either poorly or moderately differentiated, have been described in the prostate and uterus (Wenk et al 1977, Albores-Saavedra et al 1979, Johannessen et al 1980).

Well differentiated endocrine tumours or carcinoids are rare in the prostate, urethra, bladder and kidney. Both argentaffin EC cells carcinoids and trabecular L cell carcinoids closely resembling those arising in the rectum are not infrequent in the ovary (Solcia et al 1984).

Pancreas

Four main types of endocrine cells (A, B, D and PP cells) have been recognized in the human pancreas. Cells with very small, round, haloed granules (P cells) and rare enterochromaffin EC cells have also been identified (Table 29.2).

The classification of pancreatic endocrine tumours reported in Table 29.4 is based on extensive cytological and clinicopathological investigations (Capella et al 1977, Bordi & Tardini 1980, Heitz et al 1982).

The origin of pancreatic endocrine tumours from ductular stem cells is supported by many histopathological, embryological, and phylogenetic studies. However, a possible origin from transformation of mature islet cells is suggested by recent experiments in transgenic mice (Rindi et al 1988).

Nesidiodysplasia is the term used to describe the pathological changes in persistent neonatal hyperinsulinaemic hypoglycaemia, in which there is an increase in size and DNA content of B-cells (Ariel et al 1988). Earlier it had been considered that nesidioblastosis (budding of endocrine cells from the pancreatic ducts) was the hallmark of persistent neonatal hyperinsulinaemic hypoglycaemia before it was recognized that nesidioblastosis is a normal phenomenon of the neonatal pancreas (Gould et al 1983).

Unexpected, clinically silent islet cell tumours, usually of small size, are occasionally found in the pancreas. A, PP and D cells are often detected in

Table 29.4 Pancreatic endocrine tumours

A. Islet cell tumours	
1. Clinically silent, A, B, C, D, PP and P cell tumours	Mostly adenomas
2. Insulinoma: B cells with or without other islet cells	
3. Glucagonoma: A cells with or without other islet cells	
4. Somatostatinoma: D cells with or without other islet cells	Mainly low grade carcinomas
5. Non-functioning A, B, D, PP and P cell tumours with local symptoms	
B. 'Inappropriate' tumours	
1. Gastrinoma: gastrin cells with or without other cells	
2. Vipoma: VIP cells with or without other cells	Mostly low grade carcinomas
3. Other functioning tumours with inappropriate (ACTH, ADH, calcitonin, neurotensin, PTH, etc.) syndromes and cells, with or without islet cells	
C. Poorly differentiated carcinomas	Highly malignant

these tumours, most of which can be interpreted as harmless adenomas (Grimelius et al 1975, Ruttman et al 1980).

Of those pancreatic endocrine tumours associated with insulinoma, about 85% have been shown to be benign. Most insulinomas are diagnosed when rather small (less than 3 cm), while tumours associated with the glucagonoma or the somatostatinoma syndrome are generally much larger. A mean size of 7.4 cm can be calculated from published cases of 32 glucagonomas. It is likely that a larger tumour mass or a longer tumour history is required for development of the hyperfunctional syndrome. This may select tumours of larger growing potential and lower degree of differentiation, thus explaining their higher (62%) malignancy rate (Rutmann et al 1980). Similarly, the higher incidence (around 50%) of malignancies among those non-functioning tumours in which local symptoms develop may depend on their larger size (Heitz et al 1982).

So far, there is no evidence that the appearance in islet cell tumours of a few inappropriate cells, e.g. ACTH or gastrin cells, without pertinent hyperfunction increases their malignancy rate. There is no doubt, however, that tumours with inappropriate cells (e.g. gastrin, VIP, GRF, ACTH, PTH, calcitonin cells, etc.), coupled with a pertinent hyperfunctional syndrome, are mostly malignant (Heitz et al 1982, Capella et al 1983). Immunohistochemical reactivity for the α-chain of human chorionic gonadotropin (α-HCG) seems to represent a marker of malignancy for all types of pancreatic endocrine tumours (Heitz et al 1983).

Poorly differentiated neuroendocrine carcinomas are very rare.

Lung and other upper gut derivatives

Neuroendocrine cells producing bombesin/GRP-like peptides and serotonin are well represented in human fetal and neonatal lung, while being rare in non-pathological adult lung. These cells may form linear or micronodular hyperplasia in lungs with chronic inflammation, bronchiectasia and scar (Gould et al 1983a). Occasionally, aggregates of fusing and enlarging micronodules called *tumourlets* are found in lung parenchyma. These are composed of small, round-to-spindle shaped cells which ultrastructurally resemble normal lung P cells and histologically show diffuse immunoreactivity for bombesin-like peptides and more scanty serotonin and ACTH-immunostaining (Tsutsumi et al 1983, Solcia et al 1984a). Bombesin-immunoreactive peripheral carcinoids with spindle cells and paraganglioid structure arise from these tumourlets. Often, S-100 protein immunoreactive sustentacular (supporting) cells surround the solid alveoli formed by tumour cells in paraganglioid carcinoids (Fig. 29.5), thus mimicking the typical structure of paragangliomas (Capella et al 1979, 1988).

Central *carcinoids* show a trabecular or mixed trabecular-paraganglioid structure and, despite their abundance of argyrophil secretory granules, show only scarce cells with bombesin-like peptides. A large proportion of these tumours display abundant immunostaining for α-HCG (Fig. 29.6) or serotonin, while lacking reactivity for Masson's

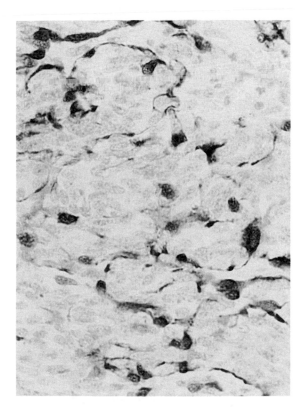

Fig. 29.5 S-100 protein immunoreactive sustentacular cells surrounding small, unstained tumour nests in a peripheral lung carcinoid with paraganglioid structure. ABC-Haematoxylin. ×350

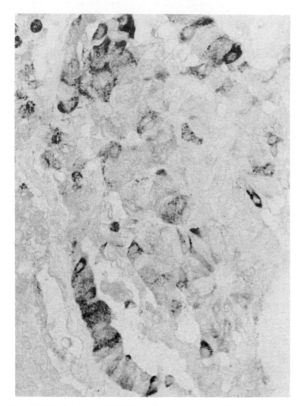

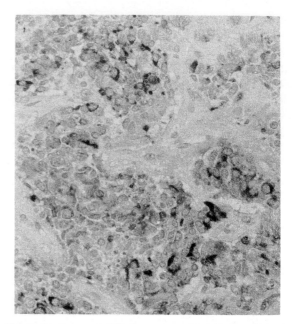

Fig. 29.7 Scattered GRP-immunoreactive cells in a liver metastasis of a moderately differentiated neuroendocrine carcinoma from the lung. C-terminus GRP antibodies; ABC-Haematoxylin. ×140

Fig. 29.6 Diffuse α-HCG immunoreactivity of tumour cells in a trabecular carcinoid arising in a main bronchus. 5E8 monoclonal antibody (from Dr S. Ghielmi, Brescia, Italy); ABC. ×350

argentaffin reaction, and are negative for substance P which is present in intestinal EC cell carcinoids. Ultrastructural investigations confirm the absence of EC cells in most lung carcinoids, which show plenty of small, round, thin-haloed granules resembling those of pulmonary P cells. Pituitary adenomas producing α-HCG (Capella et al 1979), calcitonin, leu-enkephalin, somatostatin, ACTH, GRF and VIP immunoreactive cells have been also detected.

Small cell and intermediate cell *neuroendocrine carcinomas* of high-grade malignancy with a scarce and variable content of neuroendocrine substances arise frequently in the lung. Their prognosis and behaviour do not substantially differ from those of small/intermediate cell carcinomas lacking signs of endocrine differentiation. Probombesin C-flanking peptide antibodies have been shown to be superior to bombesin/GRP antibodies as a marker of neuroendocrine carcinoma, while being less effective for staining lung carcinoids (Hamid et al 1987). Poorly differentiated neuroendocrine carcinomas are rarely associated with overt hyperfunctional syndromes (Fig. 29.7).

Laryngeal endocrine tumours are rare. They comprise the well-differentiated, but always malignant, carcinoid and the clearly more aggressive and relatively more common oat cell carcinoma. Calcitonin and ACTH have been detected in some cases (Cefis et al 1983).

Thymic and *mediastinal* carcinoids resemble lung carcinoids. However, they are more frequently associated with ACTH production and Cushing's syndrome (34%), and have a higher malignancy rate (around 50% with metastases) (Rosai et al 1976).

Very few *oesophageal carcinoids* have been reported (Brenner et al 1969), in contrast with more than 60 published cases of poorly differentiated endocrine carcinomas, in some of which ACTH or calcitonin immunoreactivity was detected (Briggs & Ibrahim 1983). Cells resembling EC or P cells may also be present.

Skin

Despite the reported immunoreactivity of rodent Merkel cells with enkephalin antibodies and that of human Merkel cells with some anti-VIP antibodies (Hartschuh et al 1983), no consistent reactivity of skin neuroendocrine carcinomas (so called Merkelomas) with the same antibodies has been documented. Only sporadic somatostatin, calcitonin and ACTH immunostaining has been obtained, despite their widespread staining with Grimelius' silver, chromogranin A or synaptophysin antibodies and their abundant secretory granules shown by electron microscopy (Gould et al 1980, Frigerio et al 1983, Buffa et al 1987).

Pheochromocytoma, paragangliomas and nerve cell tumours

In addition to catecholamines, various regulatory peptides have been detected immunohistochemically in the adrenal medulla and extra-adrenal paraganglia (Capella et al 1988). Met-enkephalin and leu-enkephalin immunoreactivities have been found in human adrenal medullary cells (Linnoila et al 1980), and carotid body cells (Wharton et al 1980). Leu-enkephalin-like immunoreactivity has been reported in all diffuse medullary hyperplasias, in contrast with hyperplastic medullary nodules, pheochromocytomas, and paragangliomas, where only half were positively immunostained. Somatostatin immunoreactive cells have also been observed in human adrenal medulla and NE-secreting extra-adrenal paragangliomas (Lundberg et al 1979). Coexisting overproduction of both catecholamines and ACTH has been reported in a pheochromocytoma as-

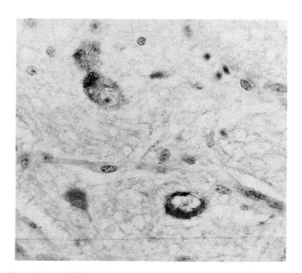

Fig. 29.8   NPY immunoreactive neurons scattered in a retroperitoneal ganglioneuroma. ABC-Haematoxylin. ×300

sociated with Cushing's syndrome (Spark et al 1979). Hypercalcitoninaemia combined with elevated catecholamines has been found in a few patients with sporadic or familial pheochromocytoma, and calcitonin-like immunoreactivity has been extracted from some of these tumours (Kalager et al 1977). A large number of neuropeptide Y (NPY)-producing cells and very high concentrations of NPY-like immunoreactivity have been found in tumour tissue from patients with pheochromocytoma (Adrian et al 1983).

Several neuropeptides can be detected in both nerve cell bodies and fibres of ganglioneuromas and ganglioneuroblastomas, especially NPY, somatostatin and VIP, with or without an associated VIPoma syndrome (Adrian et al 1983, Solcia et al 1988b) (Fig. 29.8).

REFERENCES

Abrahamsson P A, Wadström L B, Alumets J, Falkmer S, Grimelius L 1986 Peptide-hormone- and serotonin-immunoreactive cells in normal and hyperplastic prostate glands. Pathology Research and Practice 181: 675–683

Adrian T E, Allen J M, Terenghi G et al 1983 Neuropeptide Y in pheochromocytomas and ganglioneuroblastomas, Lancet ii: 540–542

Albores-Saavedra J, Rodriguez-Martinez H A, Larraza-Hernandez O 1979 Carcinoid tumors of the cervix. Pathology Annual 14: 273–291

Ariel I, Kerem E, Schwartz-Arad D et al 1988 Nesidiodysplasia — a histological entity? Human Pathology 19: 1215–1218

Bhagavan B S, Hofkin G A, Woel G M, Koss L G 1974 Zollinger-Ellison syndrome. Ultrastructural and histochemical observations in a child with endocrine tumorlets of gastric antrum. Archives of Pathology 98: 217–222

Bonikos D S, Archibald R, Bensch K G 1967 On the origin of the so-called tumorlets of the lung. Human Pathology 7: 461–469

Bordi C, Tardini A 1980 Electron microscopy of islet cell tumours. In: Progress in surgical pathology, vol. I. Masson, New York, p 135–155

Bordi C, D'Adda T, Pilato F P, Ferrari C 1988 Carcinoid (ECL cell) tumour of the oxyntic mucosa of the stomach: a hormone-dependent neoplasm? Progress in Surgical Pathology 9: 177–195

Bosman F T 1984 Neuroendocrine cells in non-neuroendocrine tumours. In: Evolution and tumour pathology of the neuroendocrine system. Elsevier, Amsterdam, p 519–543

Böttcher G, Sjolund K, Ekblad E, Håkanson R, Schwartz T W, Sundler F 1984 Coexistence of peptide YY and glicentin immunoreactivity in endocrine cells of the gut. Regulatory Peptides 8: 261–266

Brenner S, Heimlich H, Widman M 1969 Carcinoid of the esophagus. New York State Journal of Medicine 69: 1337–1339

Briggs J C, Ibrahim N B N 1983 Oat cell carcinoma of the oesophagus: a clinico-pathological study of 23 cases. Histopathology 7: 261–277

Buffa R, Rindi G, Sessa F et al 1987 Synaptophysin immunoreactivy and small clear vesicles in neuroendocrine cells and related tumours. Molecular Cellular Probes 1: 367–381

Capella C, Polak J M, Frigerio B, Solcia E 1980 Gastric carcinoids of argyrophil ECL cells. Ultrastructural Pathology 1: 411–418

Capella C, Usellini L, Buffa R, Frigerio B, Solcia E 1981 The endocrine component of prostatic carcinomas, mixed adenocarcinoma-carcinoid tumours and non-tumour prostate. Histochemical and ultrastructural identification of the endocrine cells. Histopathology 5: 175–192

Capella C, Solcia E, Frigerio B, Buffa R, Usellini L, Fontana P 1977 The endocrine cells of the pancreas tumours. Ultrastructural study and classification. Virchows Archiv A Pathological Anatomy and Histology 373: 327–352

Capella C, Gabrielli M, Polak J M, Buffa R, Solcia E, Bordi C 1979 Ultrastructural and histological study of 11 bronchial carcinoids. Evidence for different cell types. Virchows Archiv A Pathological Anatomy and Histology 381: 313–329

Capella C, Polak J M, Buffa R et al 1983 Morphological patterns and diagnostic criteria of VIP-producing endocrine tumours. A histological, histochemical, ultrastructural and biochemical study of 32 cases. Cancer 52: 1860–1874

Capella C, Riva C, Cornaggia M, Chiaravalli A M, Frigerio B, Solcia E 1988 Histopathology, cytology and cytochemistry of pheochromocytomas and paragangliomas including chemodectomas. Pathology Research and Practice 183: 176–187

Cefis F, Cattaneo M, Carnevale-Ricci P M, Frigerio B, Usellini L, Capella C 1983 Primary polypeptide hormones and mucin-producing malignant carcinoid of the larynx. Ultrastructural Pathology 5: 459–467

Cocco A E, Conway S J 1978 Zollinger-Ellison syndrome associated with ovarian macinous cystadenocarcinoma. New England Journal Medicine 298: 144–146

Dayal Y, Nunnemacher G, Doos W G, De Lellis R A, O'Brien M J, Wolfe H J 1983 Psammomatous somatostatinomas of the duodenum. American Journal of Surgical Pathology 7: 653–665

Deftos L J, Burton D W 1980 Immunohistological studies of non-thyroidal calcitonin-producing tumors. Journal of Clinical Endocrinology Metabolism 50: 1042–1045

De Lellis R, Blount M, Tischler A S, Wolfe H, Lee A K 1983 Leu-enkephalin-like immuno-reactivity in proliferative lesions of the human adrenal medulla and extra-adrenal paraganglia. American Journal of Surgical Pathology 7: 24–37

Fiocca R, Rindi G, Capella C et al 1987 Glucagon, glicentin, proglucagon, PYY, PP and proPP-icosa-peptide immunoreactivities of rectal carcinoid tumors and related non-tumor cells. Regulatory Peptides 17: 9–29

Frigerio B, Capella C, Eusebi V, Tenti P, Azzopardi J G 1983 Merkel cell carcinoma of the skin: the structure and origin of normal Merkel cells. Histopathology 7: 229–249

Gould V E, Dardi L E, Memoli V A 1980 Neuroendocrine carcinomas of the skin: light microscopic, ultrastructural, and immunohistochemical analysis. Ultrastructural Pathology 1: 499–509

Gould V E, Linnoila I, Memoli V A, Warren W H 1983a Neuroendocrine components of the bronchopulmonary tract: hyperplasias, dysplasias, and neoplasms. Laboratory Investigation 49: 519–537

Gould V E, Memoli V A, Dardi L E, Gould N S 1983b Nesidiodysplasia and nesidioblastosis of infancy. Structural and functional correlations with the syndrome of hyperinsulinaemic hypoglycaemia. Pediatric Pathology 1: 7–31

Grimelius L, Hultquist G T, Stenkvist B 1975 Cytological differentia of asymptomatic pancreatic islet cell tumours in autopsy material. Virchows Archiv A Pathological Anatomy and Histology 365: 275–288

Håkanson R, Böttcher G, Ekblad E et al 1986 Histamine in endocrine cells in the stomach. A survey of several species using a panel of histamine antibodies. Histochemistry 86: 5–17

Hamid Q A, Addis B J, Springall D R et al 1987 Expression of the C-terminal peptide of human probombesin in 361 lung endocrine tumours, a reliable marker and possible prognostic indicator for small cell carcinoma. Virchows Archiv A Pathological Anatomy and Histology 411: 185–192

Hartschuh W, Weihe E, Yanaihara N, Reinecke M 1983 Immunohistochemical localization of vasoactive intestinal polypeptide (VIP) in Merkel cells of various mammals: evidence for a neuromodulator function of the Merkel cell. Journal Investigative Dermatology 81: 361–364

Heitz P H U, Kasper M, Polak J M, Klöppel G 1982 Pancreatic endocrine tumours: immunocytochemical analysis of 125 tumours. Human Pathology 13: 263–271

Heitz P H U, Kasper M, Klöppel G, Polak J M, Vaitukaitis J L 1983 Glycoprotein-hormone alpha-chain production of pancreatic endocrine tumours: a specific marker for malignancy. Cancer 51: 277–282

Holle G E, Spann W, Eisenmenger W, Riedel J, Pradayrol L 1986 Diffuse somatostatin-immunoreactive D-cell hyperplasia in the stomach and duodenum. Gastroenterology 91: 733–739

Johanessen J V, Capella C, Solcia E, Davy M, Sobrinho-Simões M 1980 Endocrine cell carcinoma of the uterine cervix. Diagnostic Gynecology and Obstetrics 2: 127–134

Kalager T, Gluck E, Heimann P, Myking O 1977 Pheochromocytoma with ectopic calcitonin production and parathyroid cyst. British Medical Journal ii: 21–22

Linnoila R I, Diaugustine R P, Hervonen A, Miller R J
1980 Distribution of met 5 and leu 5 enkephalin, VIP-
and substance P-like immunoreactivities in human adrenal
glands. Neurosciences 5: 2247–2259

Lundberg J M, Hamberger B, Schultzberg M et al 1979
Enkephalin- and somatostatin-like immunoreactivities in
human adrenal medulla and pheochromocytoma.
Proceedings of the National Academy of Sciences of the
USA 76: 4079–4083

Margolis R M, Jang N 1984 Zollinger-Ellison syndrome
associated with pancreatic cystadenocarcinoma. New
England Journal of Medicine 311: 1380–1381

Pearse A G E 1969 The cytochemistry and ultrastructure of
polypeptide hormone-producing cells of the APUD series,
and the embryologic, physiologic and pathologic
implications of the concept. Journal of Histochemistry and
Cytochemistry 17: 303–313

Polak J M, Bloom S R (eds) 1985 Endocrine tumours. The
pathology of regulatory peptide-producing tumours.
Churchill Livingstone, Edinburgh

Rindi G, Bishop A E, Murphy D, Solcia E, Hogan B, Polak
J M 1988 A morphological analysis of endocrine tumour
genesis in pancreas and anterior pituitary of AVP/SV40
transgenic mice. Virchows Archiv A Pathological Anatomy
and Histology 412: 255–266

Rosai J, Levine O, Weber W R, Higa E 1976 Carcinoid
tumors and oat cell carcinomas of the thymus. Pathology
Annual 11: 201–226

Ruttman E, Klöppel G, Bommer G, Kiehn M, Heitz P H U
1980 Pancreatic glucanoma with and without syndrome.
Virchows Archiv A Pathological and Histology 388: 51–67

Sabate M I, Carpani M, Varndell I M et al 1984 Calcitonin
gene-related peptide in normal thyroid, and medullary
carcinoma of thyroid. Journal of Pathology 142: A29

Sjölund K, Alumets J, Berg N-O, Håkanson R, Sundler F
1979 Duodenal endocrine cells in adult coeliac disease.
Gut 20: 547–552

Sobol R E, Memoli V, Deftos L J 1989 Hormone-negative,
chromogranin A-positive endocrine tumours. New
England Journal of Medicine 320: 444–447

Solcia E, Capella C, Riva C, Rindi G, Polak J M 1988b The
morphology and neuroendocrine profile of pancreatic
epithelial VIPomas and extrapancreatic, VIP-producing,
neurogenic tumors. Annals of the New York Academy of
Sciences, in press

Solcia E, Capella C, Buffa R, Usellini L, Frigerio B
Fontana P 1979 Endocrine cells of the gastrointestinal
tract and related tumors. Pathobiology Annual 9: 163–203

Solcia E, Capella C, Buffa R et al 1984 The contribution of
immunohistochemistry to the diagnosis of neuroendocrine
tumors. Seminars in Diagnostic Pathology 1: 285–296

Solcia E, Capella C, Buffa R, Tenti P, Rindi G, Cornaggia
M 1986a Antigenic markers of neuroendocrine tumors:
their diagnostic and prognostic value. In: New concepts in
neoplasia as applied to diagnostic pathology. IAP
Monograph. Williams and Wilkins, Baltimore,
p 242–261

Solcia E, Capella C, Sessa F, Rindi G, Cornaggia M, Riva C
1986b Gastric carcinoids and related endocrine growths.
Digestion 35: suppl 1: 3–22

Solcia E, Usellini L, Buffa R et al 1987 Endocrine cells
producing regulatory peptides. Experimentia 43: 839–850

Solcia E, Capella C, Fiocca R, Tenti P, Sessa F, Riva C
1988a Disorders of endocrine system. In: Pathology of the
gastrointestinal tract, ch 13. W B Saunders, Philadelphia,
in press

Spark R F, Connolly P B, Gluckin D S, White R, Sacks B,
Landsberg L 1979 ACTH secretion from a functioning
pheochromocytoma. New England Journal of Medicine
301: 416–418

Tschopp F A, Tobler P H, Fischer J A 1984 Calcitonin
gene-related peptide in the human thyroid, pituitary and
brain. Molecular and Cellular Endocrinology 36: 53–57

Tsutsumi Y, Osamura Y, Watanabe K, Yanaihara N 1983
Immunohistochemical studies on gastrin-releasing peptide
and adrenocorticotropic hormone-containing cells in the
human lung. Laboratory Investigation 48: 623–632

Wenk R E, Bhagavan B S, Levy R, Miller D, Weisburger
W 1977 Ectopic ACTH, prostatic oat cell carcinoma, and
marked hypernatremia. Cancer 40: 773–778

Wharton J, Polak J M, Pearse A G E et al 1980
Enkephalin-, VIP- and substance P-like immunoreactivity
in the carotid body. Nature 284: 269–271

Wiedenmann B, Franke W W, Kuhn C, Moll R, Gould
V E 1986 Synaptophysin. A marker protein for
neuroendocrine cells and neoplasms. Proceedings of the
National Academy of Sciences of the USA 83: 3500–3504

Wilson B S, Lloyd R V 1984 Detection of chromogranin in
neuroendocrine cells with a monoclonal antibody.
American Journal of Pathology 115: 458–468

30. Fine needle aspiration cytology

S. R. Orell   J. M. Skinner

FINE NEEDLE ASPIRATION CYTOLOGY

The diagnostic accuracy of fine needle aspiration cytology (FNAC) depends, not only on the skill and experience of the cytopathologist who examines the smears, but to an equal extent on the use of correct techniques when sampling and processing.

FNAC was developed as a simple, quick, inexpensive and minimally traumatic means to obtain a preoperative morphological assessment of localized lesions in cases of suspected malignancy. As immunohistochemistry, electron microscopy and other advanced techniques have become available to most laboratories, it is now more often possible to make definitive diagnoses based on FNAC samples. However, the object of FNAC is to provide information necessary for the preoperative planning of appropriate management, not a definitive diagnosis. If surgical intervention is likely to follow, it is more rational to save these costly investigations for the larger tissue specimen.

THE BIOPSY PROCEDURE

The sampling mechanism has been described by Thompson (1982). The sharp edge of the oblique needle tip scrapes cells from along the track as the needle is advanced. The negative pressure holds the tissue against the edge of the needle and keeps the detached cells inside its lumen. A strong negative pressure is not essential, in fact the capillary pressure within a fine needle is usually sufficient, as shown by Zajdela et al (1987). Sampling is selective; friable tissue components such as epithelial and lymphoid cells are sampled more easily than stromal structures. Most samples contain, not only single cells and small clusters of cells, but also tiny tissue fragments which can display diagnostically useful microarchitectural patterns. The aspirate must therefore be smeared gently so that this valuable diagnostic feature is not destroyed.

The size of the needles used in FNAC varies between 25 and 21 gauge. Standard, disposable injection needles of suitable lengths are adequate to obtain samples for the preparation of cytological smears. 22-gauge needles with trocar (9 cm disposable lumbar puncture needles, or longer Chiba needles) are suitable for biopsy of deep organs. Fine calibre needles of 22 gauge, designed to obtain tissue cores for paraffin embedding, are also available. Such core biopsies are preferred by some pathologists and can be of great value in selected cases, particularly if electron microscopy is indicated. We have mainly used the Rotex screw needle (Ursus Konsult AB, Stockholm, Sweden) (Nordenstrom 1984), but other makes are available. However, the usual FNAC biopsy has the advantage of speed, simplicity and low cost, and we can see no reason to use core needles as a routine.

Accurate positioning of the biopsy needle is obviously all-important. In superficial, palpable lesions, this is achieved by feeling and fixing the target with one hand, operating the needle, syringe or pistol grip with the other, and by feeling the consistency of the tissues with the probing needle tip. If aspiration is used, the syringe should be mounted in a pistol grip (Cameco, Taby, Sweden) to leave one hand free to feel the target. The feel of the tissues is of course more direct if the needle is used alone without a syringe, held between the fingertips (see below). Biopsy of deep, non-palpable lesions must be guided by some form of imaging technique, such as fluoroscopy, ultrasonography or computerized tomography (Langlois 1986).

The sample is usually obtained by aspiration through the biopsy needle with a disposable 10–20 cc syringe. The needle is introduced into the target tissue as described, a negative pressure is created, the needle is moved back and forth within the target tissue, the negative pressure is released and the needle is withdrawn. The range and duration of the needle movements must be adjusted to the cellularity and the vascularity of the tissue to maximize cell yield and minimize admixture with blood. The sample should remain inside the needle lumen and not be drawn into the syringe. It is then expelled onto clean glass slides. Sometimes a portion of the aspirate, often the most cellular, is trapped in the hub of the biopsy needle. It cannot be expelled in the usual way, but can easily be aspirated from the hub with another needle. When lesions in deep organs and tissues are biopsied, the needle is advanced to the periphery of the target before the trocar is removed, in order to avoid contamination from adjacent tissues. The activities of radiologist and pathologist need to be well co-ordinated so that the samples can be assessed immediately as they are obtained. This allows for adequacy of material and need for supplementary investigations, such as microbiology, immunocytochemistry and EM, to be determined in one sampling session.

Recently, Zajdela et al (1987) described a modified technique of fine needle biopsy without aspiration. The needle, usually 23 or 25 gauge, is held directly with the fingers, is advanced into the target tissue and is moved back and forth in various directions before it is withdrawn. The capillary pressure is sufficient to keep the cells scraped from the track inside the lumen of the needle. The main advantage of this technique is that the operator has a much better feel of the consistency of the tissues than via a syringe. An additional advantage is that admixture with blood tends to be less. The cell yield is similar to that obtained with aspiration, except perhaps from fibrous tissues. Recently we have routinely used needle biopsy without aspiration for lesions in the thyroid, salivary glands and lymph nodes, and also

for some highly vascular deep lesions, for example tumours in the liver, the kidneys and in bone. Aspiration biopsy is still preferred for fibrous breast lumps, benign and malignant.

Rinsing the needle with sterile saline after smears have been prepared seldom gives sufficient material for microbiological investigation and culture. If frank pus is obtained, this should be submitted immediately to the bacteriology laboratory within the aspiration needle. Smaller amounts of purulent material need to be washed into a sterile container with a small amount of saline to prevent desiccation. The usefulness depends to a large extent on the amount of material submitted, and it is therefore advisable to use the whole of one or several samples. Consultation with the microbiologist is recommended whenever unusual infections, for example in immunosuppressed patients, are suspected.

THE PROCESSING OF SAMPLES

The quickest, simplest and least expensive technique is to make direct smears. If possible, the sample should be divided onto several slides in case special stains are needed. The experienced operator can usually tell if the aspirate has a high cell content when it is expelled from the needle. A highly cellular aspirate is thick, creamy or putty-like. Such a sample (a 'dry' sample) is best smeared with the flat of another slide (or a 0.4 mm coverslip) in a single, sliding movement. If the sample is more voluminous due to an abundance of blood or other fluid (a 'wet' sample), a two-step smearing technique should be adopted. In the first step, the smearing slide is held at a blunt angle to the specimen slide, and the fluid is allowed to fill the angle between the slides. The smearing slide is then moved quickly to the mid of the specimen slide. Most of the cells follow the smearing slide like a buffy coat, while the fluid is left behind. The cells concentrated at the middle of the slide are smeared with the flat of the slide like a 'dry' sample. The smearing pressure must be finely balanced to give a thin, even, quickly drying film of cells without crush artefacts. This is essential when a Romanowsky type of stain is to be used. Immediate fixation to avoid drying artefacts is more important than a thin film if smears are to be stained according to Papanicolaou or by haematoxylin/eosin (Orell et al 1986).

Various techniques to increase the cell yield from fine needle biopsies by rinsing the needles and the syringe with saline or with a fixative followed by filtration or centrifugation, have been described (Boon & Lykles 1908, Coleman et al 1975, Smith et al 1980). Several laboratories have increased their proportion of satisfactory and diagnostic aspirates by using such methods. However, they are more time consuming and expensive than direct smears and are rarely necessary if biopsies are taken correctly. Occasionally, it may prove impossible to avoid admixture with a lot of blood. If the volume is too large to handle by two-step smearing, a cell separation technique, e.g. a Ficoll-Hypaque gradient (Spalsbury et al 1973, Spriggs 1975), is recommended.

We do not find it necessary to make cell blocks from FNAC sample as a routine, but paraffin sections of cell blocks or tissue fragments are certainly useful in selected cases (Kern & Haber 1986). Tissue fragments visible in the aspirate should be processed and sectioned. Tissue fragments can easily be recovered from a voluminous, bloody aspirate if this is expelled and spread on a watch glass. A simple agar technique (Keebler et al 1976) is most commonly used for the preparation of cell blocks.

The preparation of a cell suspension from the needle biopsy sample is useful in the investigation of suspected malignant lymphoma. The sample is ejected into a small volume (1 ml) of Hank's balanced salt solution with the addition of 20–30% fetal calf serum. A number of slides are then prepared in the Cytocentrifuge, without delay, using a low speed (300 r.p.m.) to avoid cell rupture. These slides can be used for routine stains as well as for immune marker studies.

Fine needle biopsies can also be processed for electron microscopic examination if the aspirate has a high cell content.

If a 'dry' aspirate cannot be obtained from a fibrous or vascular lesion, the best alternative is to extract a tissue fragment with a fine-calibre core needle. Various techniques to concentrate cells from voluminous, bloody samples obtained by aspiration have been reported (Lazaro 1983, Lorenzetti et al 1986, Odselius et al 1987).

FIXATION

Standard fixation for FNAC smears are either air drying or fixation in 70-95% ethanol. In air dried smears, cell detail is best preserved if smears are thin and if drying is hastened by warming the slide (hair dryer, back of hand). Once dried, smears keep well over several days at room temperature. Alcohol fixation, conveniently done in Coplin jars, should be immediate to avoid drying artefacts. Carnoy's fixative may be used to lyse red blood cells. We do not recommend spray fixatives for FNAC smears.

STAINING

Romanowsky staining of air dried smears, and staining either according to Papanicolaou, or with hamatoxylin-eosin of ethanol fixed smears, should be used in parallel whenever sufficient material is available. The two fixation/staining methods are supplementary: certain information is better provided by one method than by the other (Orell et al 1986).

Romanowsky stains

All commonly used Romanowsky type stains are equally suitable for FNAC smears: May-Grunwald-Giemsa, Jenner-Giemsa, Wright's stain, etc. Staining routines developed for blood films need to be slightly modified, e.g. staining with Giemsa should be prolonged to enhance nuclear chromatin. The 'Diff-Quick' stain (Harleco, Philadelphia) is used for urgent cases, particularly for immediate assessment in the clinic. Many laboratories find it suitable also for routine use, but the staining is less uniform than with the usual Romanowsky stains. Some modification of the times recommended by the manufacturer is suggested: a minimum of 1 minute fixation, 15 seconds in solution 1, and 25 seconds in solution 2. The staining solutions keep for months in room temperature.

FNAC smears can be stained according to Papanicolaou with the routine gynaecological and other wet fixed smears. The staining method basically follows that given in the Compendium on Cytopreparatory Techniques (Keebler et al 1976).

Modifications of the Papanicolaou method for rapid staining have been developed (Pak et al 1981). Haematoxylin-eosin staining is carried out as for histological sections.

Histochemical stains

The commonest reason for a histochemical stain in FNAC is for the demonstration of mucosubstances. PAS/diastase is a satisfactory method for epithelial mucin, but mucicarmine and Alcian blue can also be used. Only distinct, intracytoplasmic, positively staining vacuoles should be accepted as clear evidence of mucin secretion. Other intracytoplasmic products sometimes need to be histochemically confirmed, such as fat (Oil-red-O) or glycogen (PAS) (Sachdeva & Kline 1981). Air dried smears are adequate for this purpose. Although now largely replaced by immunocytochemical methods, silver stains for argyrophil and argentaffin cells may be useful in the cytological diagnosis of carcinoid tumours (Wilander et al 1985). We use the Grimelius technique with air dried smears postfixed in formalin.

The specific identification of pigment is a frequent problem in FNAC. Melanin, haemosiderin and lipofuscin all stain a similar green-blue-black colour with Romanowsky stains, and the only clue to the nature of the pigment is its distribution and its physical character: melanin is usually fine, dust like, haemosiderin coarser and lipofuscin very coarse. In Papanicolaou or haematoxylin-eosin stained smears, melanin appears brown, haemosiderin golden. However, histochemical methods are often required to identify the pigment with certainty. Formalin-induced fluorescence is a simple and quick method to identify melanin precursors and some other amines in smears. Masson-Fontana is more reliable, but also more time consuming. Perls' iron is suitable for the identification of haemosiderin. The demonstration of bile pigment within tumour cells is diagnostic of hepatocellular carcinoma. For this purpose, Fouchet's reagent counterstained with haematoxylin/Sirius red gives excellent results.

Enzyme histochemistry is now less commonly used than immunocytochemical staining. However, excellent results can be achieved staining air dried smears for tartrate-resistant acid phosphatase in suspected metastatic carcinoma of the prostate, for alkaline phosphatase in some bone tumours, for naphthylamidase in hepatocellular carcinoma and for esterase and peroxidase in the typing of lymphoma (Lennert 1981, Stoward & Pearse 1990). Enzyme activity does not deteriorate in air dried smears left for a few days at room temperature.

IMMUNOCYTOCHEMISTRY

A full and definitive immunocytochemical investigation is usually frustrated by the limited amount of material available in FNAC. Immunocytochemical investigation is most helpful when the differential diagnosis is limited to two or three alternatives, or to confirm a diagnosis made on routine smears. It is of less value in poorly differentiated tumours when routine smears give no clue to diagnosis (Domagala et al 1986, Chess & Hajdu 1986). If the need for immunocytochemistry is anticipated by the pathologist at the time of FNAC biopsy on the basis of a rapidly stained smear, the sample can either be divided onto several slides, or a cell block or a cell suspension can be prepared.

Two or three tests can be made on the same slide by simply drawing circles 3 mm apart with a diamond pencil and wiping off the smear between the circles. Time and expensive antisera will not be wasted if slides are checked for adequacy of cell content before staining, using a phase contrast microscope, preferably with a Nomarsky attachment.

Most antigens of value in cytological diagnosis are well preserved in air dried smears, at least for a couple of days. If longer delays occur, smears are fixed in acetone and kept refrigerated. Initial wet fixation in ethanol, formalin or acetone offers no advantages. Cell blocks have the obvious advantage of allowing multiple sections for a battery of antisera, but not all antigens are preserved in formalin fixed tissue. It is possible to use routinely Papanicolaou-stained smears for immunoperoxidase staining in some cases (Travis & Wold 1987).

It is rare to make a definitive diagnosis by immunocytochemistry with the limited material usually available. Thus it is necessary to have a number of different antibody panels at hand to be used on the basis of the cytomorphological or clinical differential diagnosis. The aim is to provide information useful in the formulation of a management plan.

We use the ABC (avidin-biotin-complex) peroxidase method as a routine, whether monoclonal or polyclonal primary antibodies are used (Hsu et al 1981). Diaminobenzidine (DAB) is used as the chromogen. Especially where intrinsic pigments are a potential problem, or in double labelling techniques, the monoclonal alkaline phosphatase anti-alkaline phosphatase (monoAPAAP) method is used (To et al 1983, Hohmann et al 1988) with either Fast Red TR or Fast Blue BB as coupler. Whichever method is adopted it is mandatory to include appropriate positive and negative controls.

The choice of an antibody panel depends on the circumstances of each case. Monoclonal antibodies are preferred for their precision, but good quality polyclonal antibodies are still of value and often more sensitive.

Common cytological problems which can be analysed using reasonably standardized protocols are given below. An exhaustive listing is not given here and is more appropriate for surgical biopsy material.

The small round cell tumour

This tumour is often a problem in lymph node aspirates. A minimum panel of three antibodies will provide useful information:

LCA(CD45) leucocyte common antigen. If positive proceed to lymphoma panel.

Cytokeratin (CAM 5.2). Positive in carcinoma.

Chromogranin (or NSE). Positive in neuroendocrine tumours. Chromogranin is useful in the assessment of other small cell tumours such as Ewing's sarcoma, primitive neuroectodermal tumours and alveolar rhabdomyosarcoma. In the latter case, chromogranin is negative and a positive result with desmin will be seen.

The large cell anaplastic tumour

Again, often a problem in lymph nodes, but more so in aspirates of abdominal masses.

LCA(CD45) leucocyte common antigen. If positive proceed to lymphoma panel.

Cytokeratin (CAM 5.2). Positive in carcinoma.

EMA. Positive particularly in adenocarcinomas.

Germ cell tumours and melanoma have to be considered and the correct antibodies used if smears are available (see below). A judicious choice has to be made based on probabilities.

The spindle cell tumour

Here the question is of carcinoma, sarcoma or melanoma.

Cytokeratin (CAM 5.2). Positive in carcinoma (squamous cell).

Vimentin. Positive in melanoma or fibrosarcoma.

S100-protein. Positive in melanoma.

Desmin. Positive in smooth muscle tumours.

Factor VIII related antigen. Positive in Kaposi's sarcoma and other endothelial tumours. A positive diagnosis in these cases avoids unnecessary surgery.

If a lymphoid proliferation is detected, the first question may be whether the proliferation is benign or malignant? This is followed by what subtype of malignant cell is present? These questions can be answered to a limited degree in FNAC samples but are in almost all cases best studied in lymph node sections where more material is available and sampling errors can be avoided.

If LCA +ve:

Pan T (CD3). All T cells. Note that there is no antibody marker of monoclonality, but disparity in T3/Ti staining may prove to be useful (Lanier et al 1987).

Pan B (CD19 or 24). B cells. We have also found MB2 (Immunotech) to be a reliable marker which can be used after formalin fixation. Monoclonality can be determined using kappa and lambda light chain antibodies.

Specific differential diagnoses can be pursued depending on the initial cytomorphology or anatomical location of the lesion. For example, an undifferentiated tumour in the thyroid can be typed using hormone antibodies to calcitonin or thyroglobulin. If negative for both, the possibility of a lymphoma could be explored using anti-leucocyte common antigen (CD45). Large cell anaplastic tumours of possible germ cell origin can be typed using alpha-fetoprotein or beta-human gonadotrophin. In the prostate or bladder neck region, prostate specific antigen and prostatic acid phosphatase are useful in distinguishing between transitional cell and adenocarcinomas. This combination is also useful in the situation of metastatic adenocarcinoma in the male. All these require only a few smears. An exhaustive listing is unsuited in this chapter, but see Taylor (1986) and the relevant chapters of this book.

FNA smears of breast carcinomas have been successfully used for biochemical hormone receptor assays (Silversward et al 1980, Magdelenat et al 1986), although the cell volume obtained in aspirates is not always sufficient. Immunohistochemical methods using anti-oestrogen receptor antibody are now being increasingly used (Flowers et al 1986). This method is not limited by the cell volume obtained, and it allows an evaluation of heterogeneity of receptor content within a given tumour. Routine air dried smears cannot be used as the cells wash off in the procedure. The aspirate is ejected and suspended in 1/2 ml tissue culture medium (RPMI 1640, Gibco, NY) with 10% fetal calf serum added, and kept on ice until processed. The cells are deposited on glass slides coated with a high molecular weight poly-L-lysine, in a Cytocentrifuge. Staining is carried out using a commercial oestrogen receptor immunoperoxidase kit (ERICA, Abbott) (Horsfall et al 1988).

More precise diagnose, say in the case of presumed solitary primaries, will require a resected specimen. Too much should not be expected of FNAC, or tried.

REFERENCES

Boon M E, Lykles C 1980 Imaginative approach to fine needle aspiration cytology. Lancet ii: 1031–1032

Chess Q, Hajdu S I 1986 The role of immunoperoxidase staining in diagnostic cytology. Acta Cytologica 30: 1–7

Coleman D, Desai S, Dudley H, Hollowell S, Hulbert M 1975 Needle aspiration of palpable breast lesions: A new application of the membrane filter technique and its results. Clinical Oncology 1: 27–32

Domagala W, Lubinski J, Weber K, Osborn M 1986 Intermediate filament typing of tumor cells in fine needle aspirates by means of monoclonal antibodies. Acta Cytologica 30: 214–224

Flowers J L, Cox E B, Geisinger K R et al 1986 Use of monoclonal antiestrogen receptor antibody to evaluate estrogen receptor content in fine needle aspiration breast biopsies. Annals of Surgery 203: 250–254

Hohmann A, Hodgson A J, Wang Di, Skinner J M, Bradley J, Zola H 1988 Monoclonal alkaline phosphatase-anti-alkaline phosphatase (APAAP) complex: Production of antibody, optimisation of activity and use in immunostaining. Journal of Histochemistry and Cytochemistry 36: 137–143

Horsfall D, Jarvis L, Grimbaldeston M, Tilley W, Orell S 1988 Immunocytochemical assay for oestrogen receptor in fine needle aspirates of breast cancer by video. Image analysis: Relationship to biochemical determination. Submitted for publication

Hsu S M, Raine L, Fanger H 1981 Use of avidin-biotin-peroxidase complex (ABC) in immunoperoxidase techniques: A comparison between ABC and unlabelled antibody (PAP) procedures. Journal of Histochemistry and Cytochemistry 29: 577–580

Keebler C M, Reagan J W, Wied G L 1976 Compendium on cytopreparatory techniques, 4th edn. Tutorials of cytology. Chicago, Illinois

Kern W H, Haber H 1986 Fine needle aspiration minibiopsies. Acta Cytologica 30: 403–408

Langlois S LeP 1986 Organ imaging for guidance of biopsy needles. In: Orell S R, Sterrett G F, Walters M N-I, Whitaker D (eds) Manual and atlas of fine needle aspiration cytology. Churchill Livingstone, Edinburgh

Lanier L L, Ruitenberg J, Allison J P, Weiss A 1987 Biochemical and flow cytometric analysis of CD3 and Ti expression on normal and malignant T cells. In: McMichael A J, Beverly P C L, Cobbold S et al (eds) Leucocyte typing III. Oxford University Press, Oxford, p 175–178

Lazaro A V 1983 Technical note: improved preparation of fine needle aspiration biopsies for transmission electron microscopy. Pathology 15: 399–402

Lennert K 1981 Histopathology of non-Hodgkin's lymphoma (based on the Kiel classification). Springer, Berlin

Lorenzetti E, Albedi F M, Nardi F 1986 Use of the Cytocentrifuge for electron microscopy investigations. Acta Cytologica 30: 70–74

Magdelenat H, Merle S, Zajdela A 1986 Enzyme immunoassay of estrogen receptors in fine needle aspirates of breast tumors. Cancer Research 46 (suppl 8): 4265–4267

Nordenstrom B 1984 Technical aspects of obtaining cellular material from lesions deep in the lung. A radiologist's view and description of the screw-needle sampling technique. Acta Cytologica 28: 233–242

Odselius R, Falt K, Sandell L 1987 A simple method for processing cytologic samples obtained from body cavity fluids and by fine needle aspiration biopsy for ultrastructural studies. Acta Cytologica 31: 194–198

Orell S R., Sterrett G F, Walters M N-I, Whitaker D 1986 Manual and atlas of fine needle aspiration cytology. Churchill Livingstone, Edinburgh.

Pak H Y, Yokota S, Teplitz R L, Shaw S L, Werner J L 1981 Rapid staining techniques employed in fine needle aspirations of the lung. Acta Cytologica 25: 178–184

Sachdeva R, Kline T S 1981 Aspiration biopsy cytology and special stains. Acta Cytologica 25: 678–683

Silfversward C, Gustafsson J A, Gustafsson S A, Nordenskjold B, Wallgren A, Wrange O 1980 Estrogen receptor analysis on fine needle aspirates and on histologic biopsies from human breast cancer. European Journal of Cancer 16: 1351–1357

Smith M J, Kini S R, Watsojn E 1980 Fine needle aspiration and endoscopic brush cytology. Comparison of direct smears and rinsing. Acta Cytologica 24: 456–459

Spalsbury C, Brodetsky A M, Teplitz R L 1973 Discontinuous Ficoll gradient separation of normal and malignant cells: cytologic applicability. Acta Cytologica 17: 522–532

Spriggs A T 1975 A simple density gradient method for removing red cells from hemorrhagic serous fluids. Acta Cytologica 19: 470–472

Stoward P J, Pearse A G E 1990 Histochemistry. Theoretical and applied. Churchill Livingstone, Edinburgh

Taylor C R 1986 Immunomicroscopy: A diagnostic tool for the surgical pathologist. In: Major problems in pathology, vol 19. Saunders, Philadelphia

Thompson P 1982 Thin needle aspiration biopsy (letter). Acta Cytologica 26: 262–263

To A, Dearnaley D P, Ormerod M G, Canti G, Coleman D V 1983 Indirect immunoalkaline phosphatase staining of cytologic smears of serous effusions for tumour marker studies. Acta Cytologica 27: 109–113

Travis W D, Wold L E 1987 Immunoperoxidase staining of fine needle aspiration specimens previously stained by the Papanicolaou technique. Acta Cytologica 31: 517–520

Wilander E, Norheim I, Oberg K 1985 Application of silver stains to cytologic specimens of neuroendocrine tumors metastatic to the liver. Acta Cytologica 29: 1053–1057

Zajdela A, Zillhardt P, Voillemot N 1987 Cytological diagnosis by fine needle sampling without aspiration. Cancer 59: 1201–1205

31. Minerals, pigments and inclusion bodies

M. Wolman

INTRODUCTION

Some minerals and all pigments encountered in pathological specimens can be recognized as such by their colour, although their nature cannot be determined in most cases by simple microscopic examination. The presence of others, which are transparent, can sometimes be suspected because of the tissue reaction to them, or clinical history. In many instances transparent minerals might be missed, when the possibility of their occurrence is not considered by the pathologist, and special procedures are not instituted.

Mineral and pigment deposits belong to two main categories: endogenous, derived from the body itself in course of metabolic processes, and exogenous, introduced into the body from the outside world. The two will be discussed here together. Endogenous minerals, which are essential and normal constituents of tissues, cells or active proteins, such as calcium in bones, iron in erythrocytes, copper in caeruloplasmin and zinc in insulin and carbonic anhydrase, are outside the scope of the present volume.

Reactions to various minerals and pigments differ. Some do not elicit any cellular or tissue reactions. Deposition of iron in cells and tissues in haemosiderosis is an indication of increased breakdown of erythrocytes without apparent damage to the storing cells. In haemochromatosis, on the other hand, deposition of iron in tissues is associated with necrosis of some cells and intense reactive fibrosis. It follows that neither the presence, absence, or nature of the reaction can be used at present as a reliable criterion for determining the nature of the iron compound deposited.

This is not the only example in which the type of reaction to a mineral cannot help the pathologist

to diagnose the nature of the mineral. It has been shown that the dermal reactions to zirconium lactate and beryllium oxide in humans (whether granulomatous or absent) depend on the presence or absence of hypersensitivity to these compounds (Elias & Epstein 1968). With pure carbon particles, however, as in anthracosis, the mineral does not seem to be able to function as a hapten and never elicits a granulomatous reaction.

It is clear, therefore, that a definitive diagnosis regarding the nature of the deposit must rely mainly on histochemical and histophysical tests.

The present chapter will not deal with a number of useful techniques, electron microscope microanalysis, direct (Chandler 1975) or after low temperature oxygen incineration (Henderson et al 1975), secondary ion mass spectrometry and laser microprobe mass analysis, which can serve to identify elements and minerals in tissue sections. These techniques require sophisticated means which lie outside the reach of most pathology laboratories and are discussed by Verbueken et al (1985).

CARBON PARTICLES AND TATTOOING MATERIALS

Indian ink, a suspension of carbon particles, is often used in tattooing. Carbon particles are also the main constituent of the pigmented material found in the lungs and lymph nodes in cases of pneumoconiosis. Pure carbon is believed not to elicit any reaction and it can be diagnosed in these organs by a mixture of positive and negative findings. Direct light microscopic examination reveals that it is black and not brown or yellow. As iron-containing pigments, bile and lipid pigments, as well as melanins, are not black, the blackness of the deposit rules them out. Black inorganic deposits (for example, pieces of shrapnel or dental amalgam) can be recognized as they resist ashing. Ashing is a useful procedure for distinguishing inorganic deposits which remain, from organic deposits which are lost.

Graphite

This is a special form of carbon which may be found in pneumoconiosis of miners, or as a result of professional or chance exposure (for example, in printers). It may apparently elicit a granulomatous and fibrous reaction. Graphite can be recognized by the birefringence of the periphery of its particles and by its resistance to ashing at 600°C for 5 minutes, while carbon particles disappear under this treatment (Johnson 1980). A number of other compounds, many of which elicit a granulomatous reaction, are used in polychromatic tattoos.

PROSTHETIC AND DENTAL MATERIALS AND TRAUMATICALLY INTRODUCED FOREIGN BODIES

Macrophages containing ferric ions are often found around iron pins and pieces of shrapnel. A granulomatous foreign body reaction may be found around silver clips, mercury, gold and mercury amalgam. Tiny amalgam granules, liquid mercury and metallic material in soft tissue surrounding the solid masses can be recognized by their black appearance in transmitted light and their persistence in spodograms.

In injuries caused by explosives, pieces of clothing and building materials are often found in the tissues. While clothing can mostly be recognized by ordinary and polarized light microscopy, plaster and cement can be recognized by their calcium content.

Different plastic materials and silicones are used in prostheses. Fine granules are mostly endocytosed by macrophages. Large prostheses often elicit a foreign body reaction with giant cells and occasionally granulomata. The giant cells sometimes contain asteroid inclusions. The polymers are mostly refractile, but they may be birefringent if the compound has an orderly molecular ar-

rangement and may be stained by Congo red. It should be noted that foreign inclusions residing for long periods of time in the human body often become encrusted with endogenous constituents and often contain ferric compounds and lipids. PVP and PVA deposits can be demonstrated by the chlorazol fast pink method.

SILICA, SILICATES, STARCH AND ASBESTOS

Silica and starch

Deposits of silica are mainly found in the lungs and lymph nodes of patients who have inhaled silica dust in the course of their work as miners and millers. Silicates, now replaced by starch, were used in surgical glove powders for many years and produced granulomatous reactions. A granulomatous reaction with necrosis around identifiable starch granules might be indicative of contamination by silicon compounds rather than of an allergic reaction.

Silicates, silica and starch may also be found as arterial emboli in drug addicts who have injected themselves intravenously with dissolved drug tablets. Silica is intensely birefringent and can be diagnosed by its persistent birefringence in spodograms heated to 650°C for 1 hour. Starch is also birefringent with a typical Maltese cross appearance and is intensely PAS-positive.

Asbestos

Exposure to asbestos dust is of great importance as it seems to be causally related not only to mesothelioma, but also to a number of other malignant tumours. Light microscopic study of sections reveals in many cases the so-called 'asbestos bodies', mainly in macrophages or giant cells. It has been found, however, that the 'asbestos bodies' may contain a core of other fibrous materials, such as fibreglass, rather than asbestos. On such a core of asbestos or non-asbestos fibres, protein, calcium and iron are often deposited giving the bodies a yellowish tinge and mostly bulbous ends. 'Asbestos bodies' are often birefringent, give positive reaction for iron, and withstand incineration. It should be stressed, however, that a positive identification of asbestos bodies in bronchoalveolar lavage fluid or in the tissues does not necessarily indicate asbestosis. A minimum of two asbestos bodies per cm^2 of the section, and characteristic pathological findings are required for positive diagnosis (Mark 1986).

CALCIUM

Deposition of calcium salts is a common occurrence in two pathological processes. The first is metastatic calcification in hypercalcaemia, where calcium is deposited in normal tissues where conditions (e.g. the low pH) favour precipitation of calcium salts. In the second, pathological calcification, calcium salts are deposited in damaged or dead tissues because local conditions favour precipitation. In both these instances and in normal calcified tissues the deposit consists mainly of apatite. Many pathologists regard basophilia as evidence for the presence of calcium. This is wrong: basophilia denotes the presence of anionic groups which are free to react with the basic dye. These are chemical moieties which are likely to bind calcium or any other cation present in their neighbourhood. Calcium is sometimes admixed with considerable amounts of iron, as for example in the Gandy-Gamna bodies in the spleen.

Another procedure erroneously believed to indicate calcification is the von Kossa procedure in which silver ions are bound to fixed tissue anionic groups (mainly carbonates and phosphates, which are likely to be sites of calcification) and are then demonstrated by reduction through light.

Insoluble calcium salts can be demonstrated by the alizarin S procedure for light microscopy and the Morin procedure for fluorescence microscopy. These procedures demonstrate the deposits but do not prove their nature as numerous other metals are stained by them. The more sensitive GBHA procedure (Appendix 7) is reported to be specific for calcium.

SILVER, MERCURY AND GOLD

Silver

Silver deposits occur locally in tissues of patients exposed to, or treated with, silver nitrate, protargol or other silver preparations.

Mercury

Poisoning by mercuric salts and organic mercury are mostly related to industrial exposure and, in the case of Minamata disease, by industrial environmental pollution. Mercury often persists in the brain and kidneys for long periods after exposure and can be detected there. Mercuric sulphide and mixtures of mercuric and cadmium sulphide are included in red tattoos and persist there indefinitely.

Gold

Deposition of gold in tissues is mostly due to intra-articular injections of colloidal gold in patients suffering from joint disease.

To localize metals in tissues, techniques such as the Timm silver sulphide procedure, or the autometallography technique of Danscher (1984 — Appendix 7) are sufficient. Differentiating various metals can be best accomplished by electron diffraction or other complex methods or by their solubilities. Treatment for 2 hours in Lugol's iodine followed by a 5% sodium thiosulphate rinse will abolish staining of silver and gold, while treatment for 2 hours of sections with 1% KCN abolishes the staining of silver and mercury. Thus, abolition of staining by both treatments indicates that the deposit consists of silver, while absence of staining only after Lugol treatment indicates mercury, and absence after KCN only indicates gold.

COPPER, ALUMINIUM, ZINC AND TIN

Copper

Copper is deposited in the cornea and the liver in patients suffering from hepatolenticular degeneration (Wilson's disease) and its presence plays a role in the pathogenesis of this disease. However, in spite of the high copper content of hepatic tissue in the disease, histochemical tests yield often negative results, as the metal is in a soluble form. Thus, absence of staining for copper does not rule out the diagnosis of Wilson's disease. Although copper may be detected in pigment granules of copper-associated protein of normal livers, staining for copper can be of help in the differential diagnosis of liver diseases. Positive staining for copper-associated protein is also found in primary biliary cirrhosis, Indian childhood cirrhosis, biliary atresia, extrahepatic chronic biliary obstruction and sclerosing cholangitis (Goldfischer et al 1980) (see also Ch 15).

In Menkes disease copper is increased in the brush border of the intestinal epithelium, but is so soluble that histochemical detection is virtually impossible.

The best and most commonly used methods for demonstrating copper include rubeanic acid, rhodanine and Shikata's orcein method.

Aluminium

In many patients with Alzheimer's dementia, neurofibrillary tangles contain demonstrable aluminium. In other Alzheimer patients aluminium is not increased in the tangles.

Aluminium-poisoning became a major clinical problem in patients suffering from chronic renal diseases treated by dialysis. Aluminium deposition can be demonstrated in bones at the borderline between osteoid and mineralized tissue, in bone marrow cells and occasionally in the renal glomerular basement membrane by aluminon or by fluorescence after staining with Morin (De Boni et al 1974).

Zinc

This is a constituent of some active proteins (insulin, carbonic anhydrase). It is also a constituent of filling material in dentistry, wound dressing material and smoke bombs. It may be found in wounds, in the gums as well as in the lungs of military personnel exposed to explosion of smoke bombs at close quarters. It can be histochemically demonstrated by the dithizone procedure or less specifically by Timm's method (Appendix 7).

Tin

Tin is occasionally found in the lungs of miners and, like aluminium, can be demonstrated by fluorochroming with Morin.

LEAD

This metal is found in tissues, mainly in cases of lead poisoning which occurs mostly as an occu-pational hazard. Deposits are found in various organs, mainly in the gums and bones. In the kidneys, chronic lead poisoning is often associated with striking changes which may almost be considered as pathognomonic. Some renal tubular cells are enormously hypertrophied and their nuclei contain large inclusion bodies which are acid-fast (Goyer & Rhyne 1973). Similar, but less striking, inclusions are produced by bismuth salts. Lead deposits are difficult to diagnose accurately by histochemical means and most authors rely on the sulphide-silver procedure of Timm (see Appendix 7) which demonstrates all the cations which yield black sulphides. For electron microscopy the Timm procedure has been usefully modified by Danscher & Zimmer (1978). However, the light-microscopic appearance of the inclusions and their acid fastness are sufficient for a definite diagnosis.

IRON

Haemosiderin has to be differentiated from other brownish pigments. Haemosiderin is deposited locally in phagocytic cells wherever haemoglobin breaks down, and diffusely in different cells of various tissues in haemochromatosis and haemosiderosis, as well as in patients treated with high doses of iron-containing drugs. Localization of iron in the tissues does not allow differentiation between primary (genetically determined) and secondary (acquired) haemochromatosis.

Haemosiderin can easily be detected by Perls' method for ferric ions (Fig. 31.1). In haemachromatosis the liver shows intensely stained pigment granules in macrophages within fibrous septa, but significant amounts of iron are also present in hepatocytes and especially in Kupffer cells.

Iron-containing substances found in the body often do not contain ionized iron. Haemoglobin and other haemoproteins, such as myoglobin, cannot be demonstrated by Perls' method. They, as well as compounds containing ionized iron, can be shown by microincineration followed by examination with either polarized light or dark field microscopy, where they appear red. Figure 31.2 shows a spodogram of a liver of a patient who suffered from transfusion haemosiderosis, photographed under crossed polars. In contrast to Figure 31.1, iron and other inorganic materials are

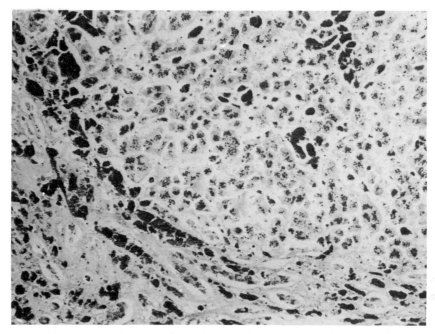

Fig. 31.1 Section of liver in haemochromatosis stained for iron by Perls' method. At the left and lower margins are large amounts of intensely stained pigment situated in macrophages lying in a fibrous band delimiting a pseudolobule. In the centre and right-upper part of the figure small discrete granules are situated in hepatocytes, while the solid-stained masses are Kupffer cells filled with haemosiderin. ×320

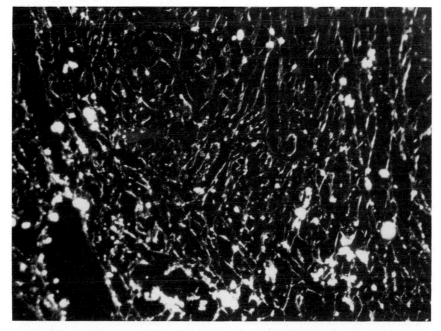

Fig. 31.2 Spodogram of a liver in haemosiderosis photographed under polarized light. Birefringent crystals are seen mainly in the perilobular macrophages at the bottom part of the figure. Few crystals can be seen within the lobule. ×255

found almost only at the periphery of the lobule. In the lobule birefringent crystals are few and occur practically only in Kupffer cells. Alternatively, in unfixed tissues and often also in tissues fixed for short periods of time in formalin or ethanol, the haemoproteins can be demonstrated by their intrinsic peroxidase activity (Appendix 5).

IRON-FREE HAEMOGLOBIN DERIVATIVES AND PRECURSORS

Porphyrins

Porphyrins, presumed to be precursors of haem, occur in various organs in porphyrias of different types (acquired and congenital). They can be detected by examination of frozen (preferably cryostat) sections under ultraviolet light, where they emit red or orange-red fluorescence (see Ch. 15).

Haemoglobin breakdown results in the formation of two pigmented substances: haemosiderin, which has been discussed above, and haematoidin which is identical with bilirubin.

Bilirubin

This can be demonstrated in paraffin wax-embedded or preferably fresh frozen sections by the van den Bergh reaction (Appendix 7), which can demonstrate both the direct (conjugated) and the total (direct and indirect) bilirubin. The procedure is specific and, in cases of brownish granules — for example, in the liver, may be aided by reactions for iron to exclude haemosiderin and by one of the reactions for lipid pigments (e.g. autofluorescence) to exclude chromolipids. A third pigment, often found together with granules of haemosiderin and haematoidin (for example, in haemosiderosis) and called haemofuscin by some authors, is in reality a chromolipid (lipofuscin).

Formalin and malaria pigments

The first is an artefact which forms in tissues congested with blood which were fixed in formalin at low pH levels for long periods. The pigment is dark brown, is not situated within cells, but rather on them, and its distribution and abundance together with its dark colour mostly allow easy recognition. Its formation can be inhibited by the inclusion of 2% phenol in the fixative (see Ch. 1). Confirmation of the nature of this and of the malaria pigment (which is very similar in appearance, but is found mainly in erythrocytes, endothelial cells and phagocytes) can be obtained by their birefringence and by their easy bleaching (within minutes) by concentrated formic acid.

MELANINS, NEUROMELANINS AND LIPOMELANINS

Definite demonstration of melanin granules is often of cardinal importance in diagnostic pathology. While demonstration of 'melanin' in melanosis coli suggests that the patient has probably been using a laxative based on anthraquinone, the finding of melanin in anaplastic tumour cells can clinch the diagnosis of malignant melanoma.

Melanins in the skin and melanocytic tumours are the products of oxidation of aromatic aminoacids. They are yellowish-brown in colour and cannot be distinguished from other brownish pigments in routinely stained sections. The most sensitive method for their detection is the Masson-Fontana procedure. As reducing activity is present in some lipid pigment granules, the use of complementary tests such as bleaching is advisable. A most useful procedure for diagnosing melanin and premelanin granules is staining for DOPA oxidase (Fig. 31.3) (Appendix 5). There are two drawbacks in the use of this method: 1. it can be used only on unfixed or lightly fixed tissue; 2. it stains mast cell granules well, which might lead to errors in diagnosis. Mast cell granules can easily be distinguished from melanin granules, however, as they are intensely metachromatic. The pigment of ochronosis is a melanin and gives the same histochemical reactions as other melanins.

The 'melanin' pigment of melanosis coli exhibits the same histochemical characteristics as skin melanin. However, many of the pigment granules observed in this condition consist of the lipid pigment lipofuscin.

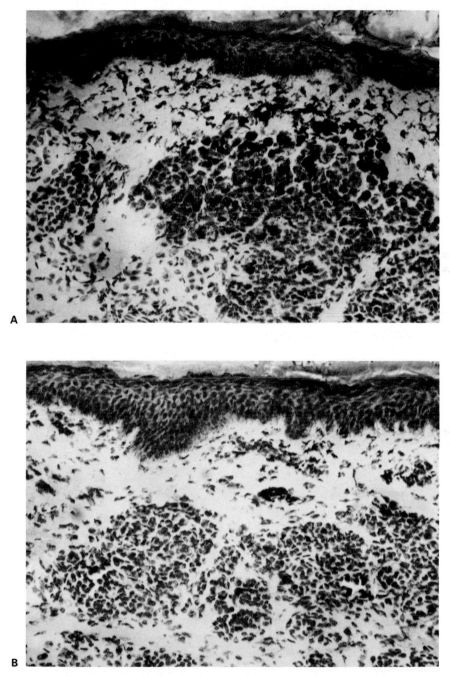

Fig. 31.3 Two unfixed cryostat sections of an intradermal naevus. **A** is stained by the DOPA-oxidase procedure counterstained with haemalum. **B** is a control section, incubated in buffer only, showing pre-existent melanin. Pigment present in cell aggregates in **A** which is absent in **B** is the product of the enzymic reaction. The cells in **A** appear larger than those of **B**, although the magnification in both figures is identical. As in **B** the melanin-containing cytoplasm shows around the nucleus. ×200

Neuromelanins

These are produced mainly at the expense of catecholamines which serve as synaptic transmitters. The different pathogenesis from cutaneous melanins is reflected in some histochemical differences between neuromelanin and skin melanin. Both reduce silver diammine and can be bleached by oxidizing agents (although the times required might be slightly different in the two cases). However, neuromelanin is a lipomelanin and not a true melanin and can be identified by frank sudanophilia and autofluorescence demonstrated after bleaching. Without bleaching, fluorescence is quenched and sudanophilia is masked by the melanin moiety. The hepatic pigment of the Dubin-Johnson syndrome and of some similar syndromes is also a lipomelanin. As such it often exhibits autofluorescence (in addition to the typical melanin reactions), the intensity of which can be increased by bleaching.

CHROMOLIPIDS (LIPOFUSCINS, CEROIDS)

Lipid pigments occur in many organs and the intensity of pigmentation is in most instances directly related to age. The pigments are formed by autoxidation of unsaturated lipids with inclusion of different agents, including oxidation catalysts, antioxidants and various proteins. They are therefore increased in different organs also in relation to functional activity, degree of unsaturation of the lipids, presence of oxidation catalysts and relative lack of antioxidants. Chromolipids are a highly heterogeneous group of substances: they differ in their building blocks, in the degree of polymerization and in the nature and amount of non-lipid materials included in the polymerizing mass (Wolman 1980).

Intense pigmentation of the brain, heart and liver is mainly age-associated. In humans, smooth muscle lipid pigmentation is mostly an indication of malabsorption which does not allow ready absorption of fats and fat-soluble vitamins (vitamin E). In analgesic (mainly phenacetin) abuse, lipid pigment granules occur in the kidneys and liver. Chapter 9 deals with the identification of lipid storage disorders in which lipofuscin occurs.

Chromolipids being mainly lipid polymers can be stained by the Sudan dyes (e.g. Sudan black B, oil red O), while most of them are not dissolved by the lipid solvents used in paraffin wax embedding. Thus, sudanophilia in paraffin wax-embedded sections is helpful.

Another constant feature of lipid pigments is their yellow (or orange) autofluorescence. Reducing activity as in the Schmorl reaction or the Masson-Fontana procedure serves well for staining chromolipid granules, but is also present in melanin. Chromolipids are often acid-fast, PAS-positive and basophilic. These characteristics, like their insolubility in solvents, depend mainly on the extent of peroxidation and polymerization. Other characteristics, such as presence of reactions for unsaturated sterols or phospholipids, depend on the nature of the building blocks. Thus, the lipid pigment granules occurring in steroid-secreting organs (adrenal, ovary, testis) often give positive cholesterol reactions and are birefringent.

As has been stated in the preceding sections, sudanophilia and autofluorescence cannot always be demonstrated in pigment granules of mixed character. In lipomelanins both these characteristics can best be demonstrated after bleaching of the melanin moiety.

PIGMENTATION CAUSED BY DRUGS

Antimalarial and other drugs and some antibiotics cause skin pigmentation and often pathological changes in the eyes. The pigment granules may contain the dye or drug, which can often be recognized by its fluorescence. In many cases, however, the pigment is a melanin-type deposit often containing iron and other constituents, or else haemosiderin.

URATES AND OXALATES

In gout, articular tophi and deposits in the kidneys mainly consist of monosodium urate. This acid salt, although only slightly soluble in water, is easily lost from sections treated with aqueous solutions except when bound to proteins. Absolute ethanol or methanol, or Carnoy's fixative, are therefore recommended. Sodium urate crystals can easily be visualized by polarization microscopy or alternatively by Gomori's silver methenamine procedure.

Positive identification of sodium urate crystals is of importance in the differential diagnosis of gout versus pseudogout. Calcium pyrophosphate dihydrate crystals (which occur in pseudogout) exhibit weak birefringence positive in respect to the crystals' length. Sodium urate exhibits negative birefringence in respect to the length. A simple way to differentiate the two consists of examining the crystals under polarized light with a gypsum plate. Crystals lying in the same direction as collagen fibres, which assume the same colour as the collagen, exhibit positive birefringence. If they are yellow when collagen is blue or vice versa, they exhibit negative birefringence.

Calcium oxalate

Calcium oxalate is deposited in the kidneys in conditions associated with uraemia, in ileal disease, in poisoning by different substances, in *Aspergillus niger* infection, in patients infused with xylitol and in primary oxalosis. In the last-mentioned process oxalates can also be found in other organs. Calcium oxalate is not stained by the commonly used von Kossa method for 'calcium' and can be recognized as such by its birefringence, von Kossa negativity and preferably by staining with Alizarin red S following the procedure of Proia and Brinn (described in Appendix 7).

INCLUSION BODIES

Inclusion bodies can be divided into two classes: cytoplasmic and intranuclear. The intranuclear inclusion bodies can be of type A, occupying most of the nucleus, often surrounded by a halo, with clumped chromatin pushed against the nuclear envelope. Type B intranuclear inclusions are smaller, often multiple, and do not disrupt the normal architecture of the nucleus and its chromatin.

The presence of inclusion bodies in tissues and in cytology specimens poses the question whether they are indicative of viral infection. This question can be answered in many cases by the combined use of histochemical and morphological approaches.

Viruses are living organisms containing DNA or RNA. Non-viral inclusions do not contain nucleic acids except: 1. inclusions of remnants of nuclei of dead cells in phagocytes (in necrosis) or in neighbouring parenchymatous cells (in apoptosis); or 2. inclusion of chromatin clumps in malignant cells. In both instances the irregular form of the inclusions and their pyknotic texture allows their differentiation from viral inclusion bodies. The presence of nucleic acids can be demonstrated by methyl green-pyronin staining, acridine orange fluorescence, or toluidine blue, directly and after DNse or RNse digestions. DNA can also be shown by the Feulgen reaction.

Viral inclusion bodies

These bodies are formed as a result of the effects of virions on the host cell. In an early stage of cellular infection the cells mostly contain viral nucleic acids and the inclusion bodies are mostly basophilic and stain for DNA and/or RNA in accordance with the nature of the virus (whether a DNA or RNA virus). In DNA viruses, e.g. the various pox diseases, herpes zoster, herpes simplex, cytomegalic inclusion disease, and the conditions caused by the papova virus family, inclusion bodies can be stained by the Feulgen or the other histochemical DNA procedures. The typical type A eosinophilic inclusion bodies, which mostly do not contain, or contain little, demonstrable nucleic acid, occur in later stages of cellular infection. These inclusion bodies often contain viral proteins, but they are composed mainly of the host cell proteins.

Pathologists are sometimes confronted with a difficulty which occurs mainly in neonates and in immunologically compromised patients. In various organs intranuclear inclusion bodies are found in swollen cells. Differentiation between *herpes* and *cytomegalovirus* rests mostly on morphological criteria: cells with cytomegalic inclusions are mostly larger and the inclusions are found in both the nucleus and the cytoplasm. Clearcut PAS-positivity of the cytoplasmic inclusions indicate cytomegalovirus rather than herpes.

The histochemical demonstration of the Australia antigen in *hepatitis B* is discussed in Chapter 15.

Correct diagnosis of presence or absence of infection by the *rabies* virus is of primary clinical and epidemiological importance. Traditionally, and in most laboratories today, the diagnosis of rabies is based on purely morphological criteria by the finding of Negri and lyssa bodies in neurons (Fig. 31.4). However, morphology alone can lead to diagnostic errors (Derakshan et al 1978) and immunohistochemical procedures are essential for correct diagnoses.

A definite diagnosis of the nature of the virus responsible for the presence of inclusion bodies can be achieved by immunohistochemical procedures (Fig. 31.5), or by in situ hybridization techniques to demonstrate viral DNA or RNA (see

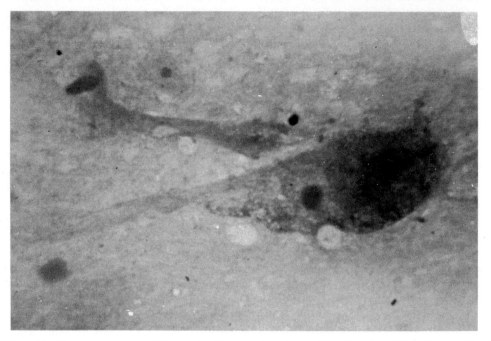

Fig. 31.4 Rabies. Negri body in a neuron: Mann stain. Photograph courtesy of Drs Orgad and Perl, Kimron Veterinary Institute, Beth Dagan, Israel.

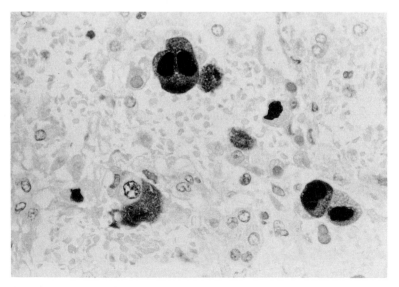

Fig. 31.5 Cytomegalovirus. Intranuclear and cytoplasmic staining is present in this paraffin section of lung. ABC peroxidase method with Dako monoclonal CMV antibody; 5 min protease pretreatment.

Ch. 3). Absence of demonstrable viral products does not necessarily indicate that the process was not caused by the tested virus.

Non-viral inclusions

In the nervous system the most common inclusions are the *corpora amylacea*, which increase with age in the CNS of all humans. The inclusions are located in fibrillar astrocytes and are PAS-positive and mostly acidophilic. Occasionally they are amphiphilic or basophilic and can then be stained by Alcian blue or colloidal iron. In these bodies, and more so in the *myoclonus (Lafora) bodies*, the structure can be concentric-lamellar. The myoclonus bodies appear as targets inside the cytoplasm of neurons with a basophilic centre surrounded by a pale amphiphilic outer zone. The centre stains intensely with PAS and Alcian blue and gives a Maltese cross appearance under polarized light. The centre often contains a granule of protein which is stainable by the ninhydrin-, or alloxan-Schiff procedures while the rest of the body remains unstained. The periphery stains weakly with PAS and Alcian blue and is less intensely birefringent. The myoclonus bodies are pathognomonic for Lafora body myoclonus epilepsy. They may also be found in sweat gland epithelial cells and skin biopsy may be used for diagnosis (Berkovic et al 1986).

A rare hyaline inclusion, *Lewy body*, is found in the cytoplasm of pigmented neurons in Parkinson's disease, occasionally in other diseases and also in aged brains. These bodies are mostly concentric, PAS-negative, isotropic and acidophilic, and contain protein. These characteristics and morphology differ from those of Lafora bodies and corpora amylacea. Their presence in the substantia nigra has been recently proposed as a pathognomonic sign of Parkinson's disease.

Many of the different inclusion bodies are composed of specific intermediate filaments to which the protein ubiquitin is linked (Lowe et al 1988). Using a polyclonal antibody to ubiquitin on paraffin sections, Lewy bodies, neurofibrillary tangles, Pick bodies, Rosenthal fibres, cytoplasmic bodies in muscle and Mallory bodies in liver can be demonstrated (Figs 31.6 and 31.7).

In the Hallervorden-Spatz disease and in neuroaxonal dystrophies *spheroids* representing localized dilatations of axons can be found in selected sites. The spheroids, as with the Lewy bodies, can also be found in senescent and normal brains. Immunohistochemical staining for neurofilaments can be used to establish their identity (see also Ch.9).

In the liver, inclusions with diagnostic significance can be found. The *Councilman bodies*, or acidophilic necrosis, represent apoptotic dead hepatocytes and are mostly extracellular. Bodies of a similar nature (*Civatte bodies*) are found in the epidermis of grade II graft vs host disease and in lichen planus.

Mallory bodies, common in alcoholic cirrhosis, may also be found in chronic hepatic disease. They do not contain DNA, but react positively for RNA and proteins. The protein is basic, rich in tyrosine and tryptophan and mostly rather poor in -SH and -SS-groups. The bodies are weakly PAS-positive and contain ubiquitin (Fig. 31.7).

Alpha 1-antitrypsin globules are pathognomonic for the genetically determined alpha-l-antitrypsin deficiency which causes chronic liver disease and emphysema. Their identification is discussed in Chapter 15.

Cytoplasmic needle-like inclusions can be found in numerous hepatocytes of patients with porphyria cutanea tarda. These inclusions (which might represent a 'useful artefact') are stained by the ferric ferricyanide reduction test and are weakly birefringent (see also Ch. 15).

In malakoplakia, which affects mostly the urinary system, the *Michaelis-Gutmann* bodies are diagnostic. The bodies invariably contain calcium and protein, are PAS and Alcian-blue-positive, often containing stainable iron, and may be slightly sudanophilic and Luxol fast blue-positive.

In the blood and hemopoietic system a number of diagnostically important inclusions can be identified. *Auer bodies* can be found in acute promyelocytic leukaemia and in monocytic leukaemias. They are PAS-positive, peroxidase and acid phosphatase-positive, metachromatic, contain proteins and appear to contain RNA but no DNA. They are occasionally sudanophilic.

Inclusions containing immunoglobulins or their constituents occur in leukaemias and other

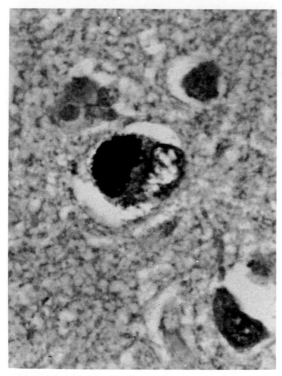

Fig. 31.6 Parkinson's disease. Lewy body in a cortical neuron in a paraffin section stained with ubiquitin antibody. Indirect immunoperoxidase and haematoxylin. Photograph courtesy of Dr J Lowe.

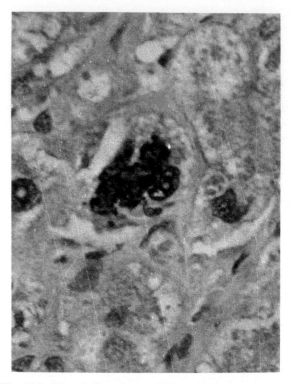

Fig. 31.7 Chronic liver disease. Mallory bodies contain ubiquitin as shown by an indirect immunoperoxidase method with ubiquitin antibody. Photograph courtesy of Dr J Lowe.

neoplastic processes of the B lymphocytic series. Best known among these are the *Russell bodies*. In primary macroglobulinaemia (Waldenstrom) few cells contain small (3–5μm in diameter) intranuclear inclusions which stain variably with PAS. These inclusions are not pathognomonic for the disease.

In the Chediak-Higashi syndrome, leucocytes and sometimes cells of parenchymatous organs contain eosinophilic inclusions. These large inclusions appear to represent in part fused

lysosomes and in part fused secretory granules. They are sudanophilic and PAS-positive after diastase digestion. They may contain acid phosphatase, and/or peroxidase and/or alkaline phosphatase.

In red blood cells the *Howell-Jolly* bodies represent nuclear fragments and are accordingly positive for DNA, RNA and proteins. *Heinz bodies* are also found in red blood cells, consist of altered haemoglobin and are stained by protein tests.

REFERENCES

Berkovic S F, Andermann F, Carpenter S, Wolfe L S 1986 Progressive myoclonus epilepsies. Specific causes and diagnosis. New England Journal of Medicine 315: 296–305

Chandler J A 1975 Electron probe X-ray microanalysis in cytochemistry. In: Glick D, Rosenbaum R M (eds) Techniques of biochemical and biophysical morphology, vol 2. Wiley-Interscience, New York

Danscher G 1984 Autometallography. A new technique for light and electron microscopic visualization of metals in biological tissues (gold, silver, metal sulphides and metal selenides). Histochemistry 81: 331–335

Danscher G, Zimmer J 1978 An improved Timm sulphide silver method for light and electron microscopic localization of heavy metals in biological tissues. Histochemistry 55: 27–40

De Boni U, Scott J W, Crapper D R 1974 Intracellular aluminium binding: a histochemical study. Histochemistry 40: 31–37

Derakshan I, Bahmanyar M, Fayaz A, Noorsalehi S, Mohammad M, Ahouraii P 1978 Light microscopical diagnosis of rabies. A reappraisal. Lancet i: 302–303

Elias P M, Epstein W L 1968 Ultrastructural observations on experimentally induced foreign body and organized epithelioid-cell granulomas. American Journal of Pathology 52: 1207–1223

Golfischer S, Popper H, Sternlieb I 1980 The significance of variations in the distribution of copper in liver diseases. American Journal of Pathology 99: 715–730

Goyer R A, Rhyne B C 1973 Pathological effects of lead. International Reviews of Experimental Pathology 12: 1–77

Henderson W J, Melville-Jones C, Barr W T, Griffiths K 1975 Identification of talc on surgeons' gloves and in tissue from starch granulomas. British Journal of Surgery 62: 941–944

Johnson F B 1980 Identification of graphite in tissue sections. Archives of Pathology and Laboratory Medicine 104: 491–492

Lowe J, Blanchard A, Morrell K et al 1988 Ubiquitin is a common factor in intermediate filament inclusion bodies of diverse type in man, including those of Parkinson's disease, Pick's disease and Alzheimer's disease, as well as Rosenthal fibres in cerebellar astrocytomas, cytoplasmic bodies in muscle, and Mallory bodies in alcoholic liver disease. Journal of Pathology 155: 9–15

Mark E J 1986 Pathological discussion. In: Case records of the Massachusetts General Hospital. New England Journal of Medicine 315: 437–449

Verbueken A H, Van de Vyver F L, Van Grieken R E, De Broe M E 1985 Microanalysis in biology and medicine: ultrastructural localization of aluminium. Clinical Neurology 24 suppl 1: S58-S57

Wolman M 1980 Lipid pigments (chromolipids): their origin, nature and significance. Pathobiology Annual 10: 253–267

Appendices

The appendices contain methods mentioned in the text which the authors and editors have found to be useful and reliable in the situations described. Methods and techniques not included in the appendices can be found in any of the standard histochemical and histological texts.

Paraffin sections — sections of fixed, wax-embedded tissue. May also be called routine sections

Frozen sections — sections of fixed tissue cut on the freezing microtome or cryostat

Cryostat sections — sections of unfixed tissue cut in a cryostat.

Plastic sections — sections of fixed tissue embedded in resin (methacrylate, Epon, Araldite, etc.)

Appendix 1 General preparative methods, fixatives and general stains

2 Proteins, neurosecretory granules, biogenic amines, silver staining and nucleoproteins

3 Carbohydrates, mucins, mucopolysaccharides

4 Lipids

5 Enzymes

6 Immunohistochemical methods

7 Pigments, metals etc.

Note

The quality of water used for making solutions, referred to as distilled water in the methods, is an important factor in achieving successful results. Wherever possible, HPLC quality water should be used.

Appendix 1

General preparative methods, fixatives and general stains

METHODS FOR FREEZING TISSUE

1. Place tissue in container and put into deep freeze. Suitable only for some biochemical examinations. *Useless for histochemistry and morphology.*

2. Freeze tissue on solid carbon dioxide (first place aluminium cooking foil on the carbon dioxide). Suitable for some tissues if small enough samples are frozen (no thicker than 3–4 mm). Useless for muscle or brain.

3. Place tissue on cryostat chuck and freeze with CO_2 jet or in tissue freezing attachments of some cryostats. (Remarks as for 2.)

4. Drop tissue directly into liquid nitrogen. Suitable for some tissues. Layer of gaseous nitrogen around the tissue prevents rapid freezing.

5. *Dust tissue (no larger than 1 cm^3) with starch powder and drop into liquid nitrogen. Suitable for all tissues. Good histochemistry and good morphology can be obtained.

6. Drop tissue (1 cm^3 maximum) into isopentane at — 170°C in a liquid nitrogen bath. Suitable for all tissues. The temperature of the isopentane is critical and if too cold or too warm, ice crystal artefacts may occur.

7. *Drop tissue (1 cm^3 maximum) into hexane maintained at the temperature of solid carbon dioxide in a methanol/CO_2 bath, or acetone/CO_2 bath. Suitable for all tissues.

Note

The coating of the tissue with starch powder (as provided with surgical gloves) gives good heat conducting conditions and allows very rapid freezing.

  * Methods 5 and 7 are recommended because they give consistently good results even in inexperienced hands.

  Method 6 requires skill and experience but when mastered will give the best results.

  Method 2 is just acceptable if no other method is available (except for muscle where unacceptable results are obtained).

  In extreme circumstances tissue may be frozen on aluminium cooking foil on the freezing surface of the freezer compartment of a domestic fridge. Results will be variable but may be preferable to no frozen tissue.

ORIENTATION AND SUPPORT OF SMALL SPECIMENS FOR CRYOSTAT SECTIONING

Small specimens (portions of needle or punch biopsies, suction rectal biopsies etc.) can be supported and oriented:

  a. in OCT compound (Tissue Tek), or
  b. on small blocks of gelatine (12%), or
  c. on small blocks of animal liver or kidney.

The specimen and supporting medium is then frozen as described above, and can be sectioned without damage to the knife.

OCT compound is inert but should not be used for tissue which might be needed later for biochemical analysis.

GENERAL FIXATIVES

Buffered formaldehyde

40% formaldehyde	100 ml
$NaH_2PO_4.H_2O$	4 g
Na_2HPO_4 (anhydrous)	6.5 g
Distilled water	900 ml

Formol-calcium fixative

40% formaldehyde	100 ml
Dried calcium acetate	15.8 g
Water to 1 litre	

The pH is approximately 6.8 as made, and requires no adjustment and no marble chips are needed. This formula is more convenient than using calcium chloride.

Formol-saline fixative

40% formaldehyde	100 ml
Distilled water	900 ml
Sodium chloride	8.5 g

Paraformaldehyde fixative

Paraformaldehyde	4 g
PBS (see p 476)	100 ml

Dissolve the paraformaldehyde by heating (to 56°C) and stirring in a fume cupboard. A few drops of 0.1 mmol/l NaOH may be necessary to clear the solution. Paraformaldehyde is a solid which, under the influence of heat, depolymerizes to produce pure formaldehyde without the methanol present in commercial formalin.

FIXATIVE TREATMENT FOR LYSOSOMAL ENZYME DEMONSTRATION

Formol-calcium: gum-sucrose

1. Fix small (2 mm thick) blocks of tissue in formol-calcium for 6–18 h at 4°C.
2. Transfer to gum sucrose solution (1 g gum acacia, 30 g sucrose, water to 100 ml) at 4°C with several changes.
3. Keep at 4°C in gum sucrose until needed *or* snap-freeze the tissue and keep frozen.

Note: For some lysosomal enzymes this procedure affords the best means for their preservation and subsequent demonstration.

FIXATIVE TREATMENT FOR BIOGENIC AMINES

Chromaffin reaction

Formaldehyde (40%)	12 ml
5% aqueous potassium dichromate	50 ml
1 mol/l sodium acetate	20 ml
Distilled water	18 ml

Make up immediately before use. The pH should be 5.8.

FIXATIVE TREATMENT FOR POLYPEPTIDE HORMONES

Benzoquinone fixative

This should be prepared immediately before use:

Phosphate buffered saline pH 7.1–7.4	99.6 ml
Benzoquinone (recrystallized)	0.4 g

Bubble nitrogen through container for at least 5 min per 100 ml solution. Do not shake the container. Keep the fixative in the dark at 4°C. It is rendered useless by oxidation. Fixation time 30 min to 4 h for intestine or overnight for 1 cm cubes of brain. After fixation, wash tissue in PBS containing 15% sucrose and keep at 4°C. For long-term storage add 0.1% sodium azide.

FIXATIVES FOR HAEMATOLOGY

Formaldehyde vapour fixation for lipid staining

Fresh 40% formaldehyde: a few drops are added to a piece of filter paper, which is placed in a Coplin jar. Fixation in the vapour is for 10 min and slides are washed briefly in 70% alcohol.

Buffered-formalin-acetone for non-specific esterases

Na_2HPO_4	100 mg
KH_2PO_4	500 mg
Acetone	225 ml
40% formaldehyde	125 ml
Distilled water	150 ml
(store at 4°C)	pH 6.6

Fix fresh films 30 sec at 4–10°C and wash in distilled water × 3. Allow to dry 10–30 min.

STORAGE AND TRANSPORT OF TISSUE SPECIMENS FOR IMMUNOCYTOCHEMISTRY

After snap-freezing, specimens can be preserved in a $-20°C$ ($-70°C$ is better) deep freezer for several days before transport to the laboratory.

Transport medium (Michael) is used in some centres as a means of transfer of unfrozen tissue to the laboratory at ambient temperature. On reception, wash in the buffer through three changes, allow to dry by draining, snap-freeze and then store deep frozen until the tests are performed. Test results comparable with those using frozen material may be obtained by the use of the transport medium so long as the buffer is carefully prepared *with special attention to its pH*.

Fixative transport medium

55 g $(NH_4)_2SO_4$ in 100 ml buffer.

Buffer   2.5 ml of 1 mol/l potasssium citrate buffer pH 7.0

5 ml of 0.1 mol/l magnesium sulphate

5 ml of 0.1 mol/l N-ethyl maleimide

87.5 ml distilled water

Adjust final solution with 1 mol/l KOH to pH 7.0

NUCLEAR COUNTERSTAINS

Carrazzi's haematoxylin

This is a light progressive haematoxylin which can be used for many counterstaining procedures.

Haematoxylin	1 g
Sodium iodate	0.2 g
Potassium alum	50 g
Glycerol	200 ml
Distilled water	800 ml

Dissolve haematoxylin and alum separately in water. Mix and leave overnight. Add iodate and glycerol and leave overnight, when it will be ready to use.

Staining time 30 seconds to 2 minutes. Wash to blue. No differentiation necessary.

Kernechtrot

This red nuclear counterstain is delicate and does not leach. Mounting can be in glycerine jelly or DPX.

Bring 100 ml 5% aluminium sulphate to the boil and add 0.5 g Kernechtrot (nuclear fast red, calcium red). Boil for 5–15 min until no further colour change occurs. Cool and filter. Staining time 3 min, wash and mount.

HAEMATOXYLIN AND EOSIN STAINING OF CRYOSTAT SECTIONS

This simple staining procedure is so often done badly with poor results. The following is a recommended method which gives consistent, clear, clean and crisp results.

Method

1. Cut cryostat sections at 5–6 μm and air dry
2. Flood with alcohol for 5–10 s
3. Wash briefly and fix in formol-calcium for 30 s
4. Wash briefly
5. Flood with Harris's haematoxylin for 30 s to 1 min
6. Wash to blue. DO NOT DIFFERENTIATE
7. Flood with eosin for 30 s
8. Rapidly rinse, and dehydrate, clear and mount.

GOMORI TRICHROME FOR MUSCLE BIOPSIES

Preparation of tissue

Unfixed cryostat sections (8 μm).

Solutions

A. Harris's haematoxylin — beware of the pH; it should be pH 2.3

B. 0.6 g chromotrope 2R

0.3 g fast green FCF

0.6 g phosphotungstic acid

1 ml glacial acetic acid

Make up to 100 ml with distilled water and adjust pH with 1 mol/l NaOH to pH 3.4

Method

1. Stain nuclei with solution A for 5 min
2. Rinse briefly with distilled water
3. Stain in solution B for 10 min
4. Wash in 0.2% acetic acid
5. Dehydrate, clear and mount in DPX.

Results

Mitochondria, nemaline rods— red
Nuclei— purple
Connective tissue— green
Muscle fibres— blue.

FLOW CYTOMETRY (For paraffin sections)

Solutions

A. Physiological saline

0.9 g sodium chloride

100 ml distilled water

B. Phosphate buffered saline
  1 g PBS (FA buffer) powder
  100 ml distilled water
C. Pepsin
  75 mg pepsin
  15 ml physiological saline (A)
  Bring solution to pH 1.5 using 2 mol/l HCl
D. RNA solution
  10 mg RNse
  10 ml PBS
  Mix together and put on a heater/stirer to simmer
  for 3 min
E. Propidium iodide (PI)
  100 ml PBS
  10 mg PI
  10 mg magnesium chloride hexahydrate
  0.1 ml Triton X100

Vortex together 10 ml of PBS and Triton X100. Add the rest of ingredients, mix well, then filter and store in the dark at 4°C.

Note

Wear gloves and mask when weighing.

Method

1. Cut 40 μm paraffin sections. Sections will roll up. If an area needs to be selected from the block, stick a piece of sellotape on the block, the section will adhere and the appropriate area can be cut out using a dissecting microscope.
Store these sections in small test tubes or sterilin pots
2. Put each section into a fine mesh cassette with a paper label. The cassettes can be collected in a jar of xylene. Using occasional agitation, dewax sections for 1 h.
3. Replace xylene with alcohol, agitate again and leave for 30 min.
4. Replace with 50% alcohol. The tissues can be left in this overnight.
5. Replace alcohol with distilled water and leave for 10 min.
6. Open the cassette carefully; remove the scraps of tissue (the adhesive and the sellotape will have dissolved) and put into a labelled Sterilin pot containing 1 ml of pepsin solution.
7. Agitate the pots on a vortex stirer.
8. Incubate in a 37°C waterbath for 45 min, agitate again on the vortex.

9. Cool under running tap water, add 5 ml of PBS to each pot to stop digestion.
10. Filter the sample. Fit a swinnex filter onto a labelled sterilin pot, using a 10 ml hypodermic syringe with a 19 g needle; extract the sample, remove the needle and fix the syringe onto the filter. Slowly push the sample through the filter.
11. Centrifuge the samples at 3000 r.p.m. for 10 min at room temperature.
12. Using a clean pasteur pipette for each sample, remove the supernatant leaving 1 ml of solution in the pot.
13. The pellet must be broken up using a 5 ml syringe and a 25 g needle. Needle sample at least three times (this is very important) to avoid clumping when sample is passing through the beam.
14. Add 0.5 ml of RNse to each sample and incubate in a waterbath for 15 min 37°C.
15. Add 1 ml of propidium iodide to each sample and leave for 20 min. (If they need to be kept for longer, store on ice).

Suppliers

Sigma Chemical Co., Fancy Rd, Poole, Dorset

  Pepsin — Sigma (P7012), store at −20°C
  RNse — Sigma (R5000), store at −20°C
  Propidium iodide (P4170) store desiccated at 4°C
  Triton X100

Difco Laboratories Ltd, PO Box 14B, Central Avenue, East Molesey, Surrey

  F.A. buffer dried 2314–15–0

Millipore (UK) Ltd, The Boulevard, Ascot Road, Croxley Green, Watford

  Swinnex W/O filter 12/PK (cat. no. 5X00250)

Henry Simon, PO Box 31, Stockport, Cheshire

  35 μm Gauze.

FLOW MICRODENSITOMETRY

Nuclear smears for microdensitometry

1. Follow the flow cytometry method up to step 13.
2. Remove 100 μm of the flow cytometry sample and keep in a LP3 tube.
3. These samples are then spun in a Cytospin machine (Shandon) to concentrate the sample. Allow to dry.

4. The slides are stained using a Feulgen technique:
   a. Hydrolyse slides in 1 mol/l HCl (preheated) at 60°C for 8 min.
   b. Transfer to Schiff's reagent, in a coplin jar, for 60 min at room temperature.
   c. Wash well in running tap water.
   d. Dehydrate, clear and mount.

Sections for microdensitometry

1. Cut 6 μm sections and dry overnight.
2. Take section to water.
3. The sections are Feulgen stained as in 4 above.

Quantify in any microdensitometer.

MICROWAVE FIXATION OR STABILIZATION

1. Blocks of tissue up to 3 cm thick may be used.
2. Put container with blocks (in plastic cassettes if necessary) and buffer up to a total of 250–300 ml in microwave oven. Bring the temperature to 55–60°, checking with a rapid reading thermometer. The following solutions give satisfactory results: phosphate buffered saline pH 7.4, cacodylate buffer 0.1 mol/l pH 7.4, 1% calcium chloride. The blocks are quite firm at this point and thinner slices may be cut. 3 cm slices of brain may be hardened by this technique and finer cuts subsequently made.
3. The stabilized tissue is dehydrated and embedded in paraffin wax as normal.

Appendix 2

Proteins, neurosecretory granules, biogenic amines, silver staining and nucleoproteins

AMYLOID
SHIKATA ORCEIN METHOD
ALCOHOLIC BASIC FUCHSIN FOR CYSTINE
WOOLLASTON TEST
ARGYROPHILIA (GRIMELIUS)
MASKED METACHROMASIA
LEAD HAEMATOXYLIN
FORMALDEHYDE-INDUCED FLUORESCENCE (FIF)
GLYOXYLIC ACID METHOD FOR FIF (TISSUE
  BLOCKS)
GLYOXYLIC ACID METHOD FOR FIF (SECTIONS)
PICROSIRIUS FOR CONNECTIVE TISSUE
SILVER STAINING FOR ARGYROPHILIC PLEXUS
AgNORs (NUCLEOLAR ORGANIZER REGIONS)
BrdU UPTAKE

AMYLOID. DIFFERENTIATION OF AL FROM AA TYPE

Preparation of tissue

Paraffin sections of tissue fixed in any routine fixative.

Method

1. Bring sections to water
2. Treat with acidified potassium permanganate (equal volumes of 5% potassium permanganate and 0.3% sulphuric acid) for 2.5–3 min
3. Drain and replace with 5% oxalic acid until section becomes colourless
4. Rinse twice in distilled water
5. Stain with a standard Congo red method
6. Examine by ordinary and polarizing microscopy
7. Treat duplicate section from step 5.

Result

Secondary amyloid (AA type) loses its affinity for Congo red (staining and birefringence) after acidified potassium permanganate treatment. Primary amyloid (AL type) Congo red affinity is resistant to this treatment.

ORCEIN STAIN FOR HEPATITIS B SURFACE ANTIGEN AND COPPER-ASSOCIATED PROTEIN (SHIKATA)

Formalin-fixed paraffin sections.

Solutions required

Acidified potassium permanganate (oxidizing solution)

Potassium permanganate	0.15 g
Dist. water	100 ml
Conc. sulphuric acid	0.15 ml

Orcein solution

Orcein	1.0 g
70% ethanol	100 ml
Conc. HCl	2.0 ml

The pH of the solution should be 1.0 to 2.0

Technique

1. Bring sections to water
2. Treat with acidified potassium permanganate for 15 min
3. Rinse in water and decolorise in 2% oxalic acid for 10 min
4. Rinse in dist. water, then wash in tap water for 3 min
5. Stain in orcein solution for 2–4 h at room temperature
6. Rinse in water, then differentiate in 1% HCl in 70% ethanol
7. Dehydrate, clear and mount.

Results

HBsAg, copper associated protein and elastic tissue stain brown.

Note

The concentrations recommended are for orcein from BDH. Other orceins may require twice the conc. and twice the amount of HCl.

The orcein solution should be freshly prepared every two weeks.

Source

Orcein (natural). BDH (British Drug Houses).

ALCOHOLIC BASIC FUCHSIN (FOR CYSTINE)

Preparation of tissue

Air-dried cryostat sections (5–8 μm) of snap-frozen tissue; air-dried bone marrow films.

Method

1. Flood slides with alcohol
2. Stain for 2–5 min in 0.7% basic fuchsin in 70% alcohol (7 ml 1% basic fuchsin in ethanol, 3 ml water)
3. Rinse in alcohol, clear and mount.

Result

Nuclei: red; cystine: unstained.

Note

View in polarized light (partial or full). The cystine crystals are birefringent and can be seen without the nuclear stain. However, the presence of nuclei is a help in orientation.

WOOLLASTON TEST FOR CYSTINE

Preparation of tissue

Air-dried cryostat sections (10 μm) of snap-frozen liver, spleen or lymph node containing high concentrations of cystine. Bone marrow films have insufficient cystine for this test.

Method

Place a coverslip over the dry section. Choose an area with abundant crystals and observe with polarized light while concentrated hydrochloric acid is introduced under the coverslip. The 'brick'-shaped crystal aggregates of cystine will dissolve and recrystallize in a fan-shaped cluster of needle-shaped crystals of cystine hydrochloride.

Notes

Beware of fumes and corrosion.
Low concentrations of cystine will fail to recrystallize. Inorganic crystals (phosphates mainly) will disappear completely.

ARGYROPHILIA FOR NEUROSECRETORY GRANULES IN THE DNES

Preparation of tissue

Fix in Bouin or formalin. Tissues fixed otherwise can be postfixed in Bouin for several hours. Paraffin sections.

Preparation of solutions

Buffered silver

Dissolve 50 mg $AgNO_3$ in 100 ml of 0.02 mol/l acetate buffer at pH 5.6. Make up fresh buffer with freshly distilled water. Discard if solution is cloudy.

Reducing solution

Hydroquinone (quinol) 1 g; anhydrous sodium sulphite 5 g; glass distilled water 100 ml.

Method

1. Bring sections to distilled water
2. Treat for 3 h at 60°C
3. Transfer directly to reducing solution at 60°C
4. If reaction is weak rinse slides in distilled water and return them to fresh silver solution at room temperature for 10–15 min
5. Reduce again as required
6. Wash in distilled water
7. Dehydrate, clear, mount in Canada balsam.

Results

Brown to black staining indicates argyrophilia.

MASKED METACHROMASIA FOR NEUROSECRETORY GRANULES

Preparation of tissue

Fix preferably in glutaraldehyde-picric acid (25% glutaraldehyde 25 ml, saturated aqueous picric 75 ml, and anhydrous sodium acetate 1 g) or in 6% glutaraldehyde (NaH_2PO_4, 0.2 mol/l 48.75 ml; Na_2HPO_4, 0.2 mol/l, 76.25 ml; 25% glutaraldehyde 72.5 ml). Check pH to 7.0 and make up to 250 ml with distilled water. Paraffin sections.

Method

1. Bring sections to water
2. Hydrolyse for 5–20 min in 5 mol/l HCl at 60°C
3. Wash in distilled water
4. Stain 2–3 min in 0.1% toluidine blue in 0.1 mol/l acetate buffer at pH 5.0
5. Examine wet, under coverslip.

Results

A reddish stain shows proteins with high levels of side chain carboxyls or carboxamides.

LEAD-HAEMATOXYLIN FOR NEUROSECRETORY GRANULES

Preparation of tissue

Glutaraldehyde or formalin-fixed paraffin wax-embedded tissue; fixed frozen sections; cryostat sections.

Preparation of solutions

Stabilized lead solution

Add equal parts of 5% lead nitrate in distilled water and saturated aqueous ammonium acetate; filter. To each 100 ml of filtrate add 2 ml of concentrated (40%) formalin. This stock solution keeps indefinitely in a refrigerator at 0–4°C.

Haematoxylin solution

Dissolve 0.2 g haematoxylin in 1.5 ml of 95% ethanol. Add 10 ml stock lead solution and dilute with 10 ml of distilled water. Stir repeatedly for 30 min, filter and make up to 50–75 ml with distilled water.

Method

1. Bring sections to water
2. Stain in the lead-haematoxylin solution for 1–2 h at 45°C or 2–3 h at 37°C
3. Rinse in distilled water, examine and stain further if necessary
4. Rinse in distilled water
5. Dehydrate, clear and mount.

Results

Blue-black staining of neurosecretory granules, nuclear chromatin, nucleoli, nerve fibres, muscle fibres, keratohyalin granules and calcium deposits.

FORMALDEHYDE-INDUCED FLUORESCENCE (FIF) (FREEZE-DRIED FORMALDEHYDE VAPOUR (FDFV) METHOD)

Preparation of tissue

Quench small pieces of tissue in isopentane precooled in liquid nitrogen. Transfer with cold forceps to freeze dryer and dry at −40°C for 8 h. Transfer to closed paraformaldehyde chamber or vessel and incubate for 3 h at 60–80°C. Impregnate in vacuo with wax containing a high proportion of plastic polymers, and embed in the same wax.

Method

1. Cut sections 2–5 μm and pick up on dry albuminized prewarmed slides.
2. Mount in Styrolite or Fluorolite and examine by fluorescence microscopy.

Results

Catecholamines and indolamines fluoresce in distinct colours if correct filter combinations are used (dopamine and noradrenalin, greenish; 5-HT yellow).

Notes

For catecholamines, excitation using the 405 nm mercury line is convenient. For 5-HT excitation may be with either the 365 or 405 nm mercury lines. Excitation at 365 nm requires excitation filters UG1 and BG38.

GLYOXYLIC ACID METHOD FOR CATECHOLAMINES (FOR TISSUE BLOCKS)

Preparation of fixation medium

Make a 2% solution of glyoxylic acid monohydrate in 0.1 mol/l phosphate buffer, pH 7.2. Adjust pH to 7.0 to 7.2 with 1 mol/l NaOH.

Method

1. Immerse small pieces of tissue in the fixative for up to 20 min at 4°C.
2. Quench tissues in liquid nitrogen-cooled isopentane, and mount on chucks.
3. Cut sections 5–10 μm thick and mount on glass slides.
4. Immerse in fixative medium at 4°C for 2–10 min.
5. Blot dry and expose to 37°C heat for 3–4 h.
6. Mount in Entellan, or other non-fluorescent medium.
7. Examine by fluorescence microscopy. Use HBO 200 lamp, primary filters BG38 and BG3, plus interference filter (405 nm). Secondary filter K470 (Leitz).

GLYOXYLIC ACID TECHNIQUE FOR SECTIONS

1. Cut cryostat sections, pick up on slides
2. Dip × 3 in the following solution at room temperature:

Sucrose	10.2 g
Potassium dihydrogen phosphate	4.8 g
Glyoxylic acid monohydrate	1.5 g
Distilled water	100 ml

Bring solution to pH 7.4 with 1 mol/l NaOH (about 35 ml). Make up to 150 ml with distilled water. Prepare and use at room temperature the same day. Remove excess fluid quickly with absorbent paper.

3. Dry with hair dryer (cool).
4. Heat dry sections in oven at 80 ± 1°C for exactly 5 min.
5. Cover slip with mineral oil (BP) and place on hot plate at 80°C for 1½ min to remove air bubbles.
6. Examine with a fluorescence microscope where catecholamines fluoresce.

PICROSIRIUS FOR CONNECTIVE TISSUE

Preparation of tissue

10 μm cryostat sections; paraffin sections.

Reagents

Picrosirius: 0.1% sirius red in saturated aqueous picric acid.

Method

1. Fix cryostat sections 5 min–2 h in formol-calcium.
2. Wash.
3. Stain in picrosirius solution 10–60 min.
4. Rinse in water.
5. Dehydrate, clear and mount in DPX.

Results

Connective tissue (collagens I, II and III) — red.

Notes

Length of fixation not critical.
Length of staining not critical.
Nuclear counterstaining reduces the intensity of the colour.

SILVER METHOD FOR THE INTESTINAL ARGYROPHILIC PLEXUS

Preparation of tissues

Open the portion of bowel along one of the taeniae coli and place mucosal side down onto a piece of cork. Insert pins into the four corners taking a care not to stretch the specimen. Invert the cork into a large volume of fixative. After 48 hours the tissue can be unpinned and transferred into a smaller container for storage. It is important to transfer the same 'dirty' fixative into the storage container as prolonged fixation in the same fluid is necessary. Fixation in excess of 14 days is needed.

Fixative

10% formol saline

Sodium chloride	8.5 g
Formalin	100 ml
Distilled water	900 ml

Cutting of the specimen

Cut 50 μm thick frozen sections on a freezing microtome, in a plane parallel to the myenteric plexus. The specimen should be placed mucosal surface on the freezing stage. Take sections when the outer muscle coat is sufficiently removed. With specimens from young children sections can be taken immediately. Generally 6–9 sections can be taken (before the plexus is all taken). Collect the sections in distilled water and stain all free floating.

Reagents

Cajal's formol ammonium bromide:

Formalin	15 ml
Ammonium Bromide	2 g
Distilled water	85 ml

Alcoholic pyridine:

Abs. alcohol	50 ml
Distilled water	40 ml
Pyridine	10 ml

Ammoniacal silver:
20% silver nitrate into which is added 0.880 sg. ammonia solution until the precipitate first formed is just dissolved.

10% Formalin:

Concentrated formalin	100 ml
Tap water	900 ml
Marble chips	

Carbol Xylene:

Molten anhydrous phenol	25 ml
Xylene	75 ml

Method

1. Cut 50 μm sections of approximately 2×2 cm of tissue, after trimming the outer muscle coat.
2. Select sections with greatest area of plexus.
3. Postfix in Cajal's formol ammonium bromide for 24 h.
4. Rinse in two changes of distilled water, and transfer to alcoholic pyridine for 1 h at 37°C.
5. Rinse in two changes of distilled water and transfer sections to 20% silver nitrate in a glass container, at 37°C for 1 h. Sections must be flat.
6. Take sections through three changes of 10% formalin without washing.
7. Treat sections in 2% formalin for 30 s.
8. Collect sections in distilled water — two changes.
9. Treat sections with ammonical silver solution with constant agitation for 1 min.

10. Take sections through three changes of 1% formalin with rapid agitation.
11. Collect sections in distilled water.
12. Fix in 5% sodium thiosulphate for 5 minutes.
13. Wash thoroughly.
14. Sections are picked up singly on a slide and all the creases are teased out. When nearly dry blot the sections several times. Dehydrate in absolute alcohol and clear in carbol xylene then xylene. Mount in natural resin, e.g. balsam.

Results

Argyrophil cells and nerves — black
Argyrophobic cells — golden

Notes

It is not necessary to treat sections singly in any step during the method. All sections can be transferred together provided at least 50 ml of the formalin solutions and ammonical silver solution are used. Make sure that all the sections are freely floating in these solutions.

When making the ammonical silver solution it is important that no residual smell of ammonia is present. If too much ammonia is added the solution becomes too alkaline and sensitivity is lost. It is advisable to stop adding ammonia whilst the solution is still slightly turbid and allow it to stand for at least 10 minutes prior to use.

Synthetic resin should not be used as too much shrinkage occurs. The carbol xylene softens the sections and prevents distortion when the mountant has dried.

Fixation in the same 'dirty' fixative is necessary because progressive degradation of blood and tissue constituents slowly lowers the pH of the fixative.

Alcoholic pyridine blocks the staining of reticulin fibres. 10% formalin should be made several days before it is to be used.

A SILVER TECHNIQUE TO DEMONSTRATE NUCLEOLAR ORGANIZER REGIONS (AgNORs)

NB: All glassware to be used must be cleaned for 1 h in potassium dichromate cleaning solution:

Potassium dichromate	100 g
Distilled water	850 ml
Conc. sulphuric acid	100 ml

Dissolve dichromate in the water. Stand beaker in a sink of running cold water whilst *carefully* adding the acid.

Staining solutions

A. 0.3 g gelatine
15 ml distilled water
0.15 ml formic acid

Add gelatine to water and stand in hot water until dissolved. Add formic acid.

B. 15 g silver nitrate
30 ml distilled water

Mix together and add to gelatine solution just before use.

1. Take sections to distilled water

2. Put slides in a coplin jar, add silver solution quickly.

3. Cover coplin jar with a box for 35 min at room temperature (in the dark). NB: Time may vary (try 25–35 min).

4. Rinse in distilled water.

5. Dehydrate, clear and mount.

BROMODEOXYURIDINE BrdU UPTAKE

Solutions and equipment

1. 10 mg BrdU (for a final concentration of
50 μmol/l)
67 ml distilled water
340 mg HEPES buffer

Aliquot this solution into 1000 μl samples and freeze at −20°C.

2. Fetal calf serum (FCS)

Aliquot into 1500 μl samples and freeze at −20°C.

3. Modified Eagles Medium (MEM) — tissue culture medium

4. HEPES buffer

5. Gelatine capsules containing a few mgs of catalase

6. Polypropylene LP3 tubes and stoppers.

7. 'Bomb' capsules and rubber washers (Fig. A)

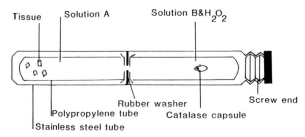

Fig A

Method

1. Samples of fresh tissue must be cut into 1 mm pieces.

2. Set up 3 'bombs' as follows:

Take one tube of BrdU and allow to thaw.

Take one tube of FCS and allow to thaw.

Solution A. Put 9000 μl of MEM, 52 mg HEPES buffer, 1000 μl FCS and 1000 μl BrdU in a sterilin pot

Solution B. In another sterilin pot put 4500 μl of MEM, 26 mg HEPES buffer and 500 μl FCS.

3. Place tissue into polypropylene tubes and add 1500 μl BrdU solution (A), put a stopper on with a hole in it. In another tube put 1500 μl solution (B), add a gelatine capsule and 2 drops of 30 volume hydrogen peroxide. Assemble apparatus with a washer between the two tubes.

Incubate at 37°C for 1 h on a shaker.

4. Gently release the screw under water to allow the build-up of oxygen to escape. Fix tissue overnight, process as for routine paraffin embedding.

Demonstration of BrdU in tissue sections

1. Cut 5 μm sections. Pick up on poly-L-lysine coated slides and dry overnight at 37°C.

2. Take sections through xylene, 100% alcohol, 90% alcohol to 70% alcohol.

3. Block endogenous peroxidase with 0.1% H_2O_2 in methanol for 30 min at room temperature. Wash in tap water, then distilled water.

4. Transfer to freshly prepared trypsin solution (see Appendix 6) preheated to 37°C for 13 min. Wash in cold water.

5. Transfer to preheated 1 mol/l HCl at 60°C for 15 min. Wash well in tap water.

6. Flood slides with PBS.

7. Incubate in Dako anti-BrdU diluted 1:30 in 1% human serum in PBS for 1 h. Wash 3 times in PBS.

8. Add Dako biotinylated rabbit-anti-mouse IgG antibody diluted 1:300 in 1% human serum in PBS for 1 h. Make up the ABC complex. Wash 3 times in PBS.

9. Add ABC complex for 1 h. Wash 3 times in PBS.

10. DAB solution — 5 mg DAB in 10 ml PBS and 2 drops of H_2O_2 for 5 min. Wash in tap water.

11. Counterstain in Mayer's haemalum and eosin.

12. Dehydrate, clear and mount.

Material and Suppliers

1. MEM solution. 500 ml (12–312–54). Fetal calf serum 100 ml (29–101–49)

Flow Laboratories Ltd, Woodcock Hill Industrial Estate, Harefield Rd, Richmansworth, Herts, WD3 IPQ

2. LP3 tubes-polypropylene and stoppers BHLP3P

Luckham Ltd, Victoria Gardens, Burgess Hill, Sussex, UK

3. 5-Bromo-2-deoxyuridine 100 mg (B5002), HEPES sodium 100 g (H7006), Catalase 5 g (C-10)

Sigma Chemicals Ltd, Fancy Rd, Poole, Dorset, UK

4. BrdU Monoclonal (MO74401), rabbit anti-mouse IgG biotinylated. (EO35401), ABC complex (K035500)

Dako Ltd, 22 The Arcade, The Octagon, High Wycombe, Bucks, UK

5. 'Bombs' Hospital Workshop made: Denekamp J, Kallman R H 1973 Cell Tissue Kinetics 6: 217–227

Appendix 3

Carbohydrates, mucins, mucopolysaccharides

PAS
PAS FOR SIALIC ACID
PERIODIC ACID THIONIN SCHIFF/KOH/PAS
ALCIAN BLUE — PAS
ALCIAN BLUE — (CEC METHOD)
HIGH IRON DIAMINE - ALCIAN BLUE
COLLOIDAL IRON
FEYRTER'S THIONIN
TOLUIDINE BLUE (FOR LYMPHOCYTES)
TOLUIDINE BLUE (HAUST & LANDING) FOR
  SOLUBLE MUCOPOLYSACCHARIDES
HYALURONIDASE DIGESTION
CHONDROITINASE DIGESTION

PERIODIC ACID-SCHIFF (PAS)

Preparation of tissue

Air-dried films of blood or marrow fixed in formol-calcium; fresh films preferred, but satisfactory staining has been obtaned with material methanol-fixed or Romanowsky-stained months to years before. Sections of fixed embedded tissue, postfixed cryostat sections, fixed frozen sections, celloidin protected cryostat sections (cell PAS).

Celloidin protection for cell PAS technique

Air-dried cryostat sections, blood films or cytology smears are dipped in 0.25% celloidin in absolute alcohol and rapidly dried. Follow with the PAS method from step 1.

Preparation of solutions

Schiff's reagent

Add 1 g pararosaniline (or purified basic fuchsin) to 100 ml of distilled water at 95°C. Stir well to dissolve and cool to 60°C. Add 2 g potassium metabisulphite and 20 ml 1 mol/l HC1. Stopper closely, and leave over-night. Add 1 teaspoonful (1–2 g) of activated charcoal, shake and filter. Store at 4°C in stoppered bottles to which xylene is added to maintain the sulphur dioxide concentration.

Method

1. Treat sections with 0.5% aqueous periodic acid, 8 min
2. Wash in running tap water 2 min
3. Rinse in distilled water
4. Cover with Schiff's reagent 15 min
5. Rinse in distilled water and wash in tap water
6. Counterstain with suitable nuclear stain (Mayer H; Carrazzi H. *NOT* Harris H)
7. Wash, dehydrate, clear and mount.

Results

Carbohydrates containing 1:2 glycol groups (*vic*-glycols) are stained red. This includes glycogen, glycolipids, neutral mucosubstances and some non-sulphated acid mucosubstances.

Notes

Glycogen can be removed from sections by treating with salivary amylase (10 min) or 0.1% diastase in 0.02 mol/l phosphate buffer pH 6 containing 0.9% sodium chloride for 1 h at room temperature. Salivary amylase will digest glycogen in celloidin protected sections *through* the celloidin protective layer. Celloidin protection retains glycogen and other water-soluble oligosaccharides.

A delicate nuclear counterstain is required. Harris' haematoxylin overlies the PAS reaction and gives a 'muddy' appearance. Carrazzi's haematoxylin is preferred.

PAS FOR SIALIC ACID

Preparation of tissue

Unfixed cryostat sections (5 μm) on polylysine coated slides, celloidin protected.

Method

1. Treat sections with 0.03% sodium metaperiodate, 30 min.
2. Steps 2–7 of standard PAS method.

Notes

The dilute sodium metaperiodate will only attack the sialic acid side chain under these conditions. Glycogen and other carbohydrates are not stained while gangliosides and other sialic acid compounds are demonstrated.

THE PERIODIC ACID-THIONIN SCHIFF-POTASSIUM HYDROXIDE-PERIODIC ACID SCHIFF SEQUENCE FOR THE DEMONSTRATION OF NORMAL PERIODATE REACTIVE AND ACETYL SIALOMUCINS

Solutions required

1% Aq. *freshly prepared* periodic acid
0.5% Potassium hydroxide in 70% alcohol
Thionin-Schiff reagent

Preparation of Thionin-Schiff reagent

Dissolve 1 g thionin in 200 ml boiling distilled water, removing from the heat before adding the dye to avoid frothing. Allow the solution to cool to 50°C and add 2 g potassium metabisulphite with mixing. Allow to cool to room temperature and add 2 ml conc. HCl, mix and add 2 g activated charcoal. Mix and leave in a stoppered container in the dark overnight at room temperature. Filter through a Whatman no. 1 filter paper before use. The solution should be pale blue. Store at 4°C and use within 2 days.

Method

1. Dewax and bring to water
2. Place in 1% periodic acid (freshly prepared) for 2 h
3. Wash in running water for 10 min
4. Place in thionin-Schiff reagent for 10 min

5. Rinse in distilled water
6. Wash in running tap water for 10 min
7. Treat with 1% periodic acid for 10 min
8. Treat with freshly prepared 0.1% sodium borohydride in 1% aq. disodium hydrogen phosphate for 30 min
9. Rinse in 70% alcohol
10. Treat with 0.5% KOH in 70% alcohol for 5–10 min
11. Wash *gently* in running tap water for 10 min
12. Rinse in distilled water for 5 min
13. Treat with 1% periodic acid for 15 min
14. Wash in running tap water for 5 min
15. Rinse in distilled water
16. Treat with Schiff's reagent (p. 449) for 15 min
17. Rinse in distilled water
18. Wash in running tap water for 15 min
19. Dehydrate, clear and mount.

Results

Normal periodate reactive substances, including glycogen, will stain blue. KOH/PAS positive substances will stain red. Mixtures will stain purple.

For sialomucins of the gastrointestinal tract the C0, C7 and C9 classes of O-acylated sialic acids will stain blue. The C8 (C7, 8, C8, 9 and C7, 8, 9) classes are stained red. Mixtures stain purple. Biochemical studies have shown that C9 O-acylated sialomucin are not present in the colonic mucins of man.

Notes

Controls should include steps 1–6 (PAT stage), and steps 1–3, 8, 9 and 13–19 to check the PA/PB/KOH/PAS stage.

Slide cleaning solution

Potassium dichromate	100 g
Distilled water	850 ml
Concentrated sulphuric acid	80 ml

Carefully add the concentrated sulphuric acid very slowly with constant stirring. (Use a pair of goggles while preparing chromic/sulphuric acid mixture.)

One by one add new glass slides to the chromic/sulphuric acid mixture and stir them with a glass rod. Leave them there for at least 12 h.

The next day, clean the slides in running hot water for 10–15 min. Wash in distilled water and then in absolute alcohol. Wipe dry with a clean piece of cloth.

Coat slides with poly-L-lysine (see Appendix 6).

ALCIAN BLUE — PAS

Preparation of tissue

Paraffin sections

Method

1. Dewax and bring sections to water.
2. Stain in freshly filtered 1% alcian blue 8GX in 3% aqueous acetic acid at room temperature for 30 min.
3. Rinse briefly in distilled water.
4. Proceed with PAS method.

Result

Acid mucosubstances — blue
Neutral mucosubstances — red

Notes

Some acid mucosubstances containing *vic*-glycol groups may stain purple.

Alcian blue solutions should be freshly prepared, or kept for a maximum of 1 week.

ALCIAN BLUE — CEC METHOD

Preparation of tissue

Postfixed cryostat sections; frozen sections; paraffin sections.

Solutions required

0.1% alcian blue 8GX in 0.025 mol/l sodium acetate buffer at pH 5.8, containing 0.1 mol/l, 0.2 mol/l, 0.4 mol/l, 0.6 mol/l, 0.8 mol/l and 1.0 mol/l $MgCl_2$.

Method

1. Bring to water.
2. Stain overnight in each solution, at room temperature.
3. Rinse each section individually in distilled water and then transfer to a fresh distilled water bath for 5 min.
4. Dehydrate quickly in alcohols, clear and mount.

Results

Carboxyl groups lose basophilia at low salt concentrations; basophilia of sulphated groups persists at higher salt concentrations.

HIGH IRON DIAMINE/ALCIAN BLUE (HID + AB)

Preparation of tissue

Paraffin sections.

Solutions required

1. Diamine solution: Dissolve 120 mg N-N'-dimethyl-*m*-phenylene diamine dihydrochloride and 20 mg N-N'-dimethyl-*p*-phenylene diamine dihydrochloride simultaneously in 50 ml of distilled water. Add 1.4 ml of 60% ferric chloride. Prepare the diamine solution fresh each time.
2. 1% alcian blue in 3% acetic acid (freshly prepared).

Method

1. Dewax and bring sections to water.
2. Treat sections with diamine solution — at room temperature — 18 h.
3. Rinse very briefly in distilled water.
4. Treat with 1% alcian blue in 3% acetic acid — 30 min.
5. Wash well in 80% alcohol.
6. Dehydrate, clear and mount in DPX.

Result

Carboxylic groups in mucosubstances stain blue.
Sulphated groups stain black or dark brown.

Note

The phenylene diamines are carcinogenic. *Handle with care.* Eastman-Kodak reagents are preferred.

Be sure to use the dihydrochloride salts of the phenylene diamines. The original published method gave 1.4 ml of 10% ferric chloride in error and this has led to many bad results. If used as above, i.e. 1.4 ml of 60% ferric chloride, good results will be obtained.

COLLOIDAL IRON

Prepation of tissue

Paraffin sections.

Preparation of reagents

Add 2 ml of 60% ferric chloride solution (AR quality) slowly, drop by drop to 250 ml boiling distilled water.

Allow to cool. This solution of colloidal iron is stable for years.

Method

1. Bring sections to water.
2. Flood with 3% acetic acid for 2 min.
3. Stain for 1 h at room temperature in:
   4 parts colloidal iron solution
   3 parts distilled water
   1 part acetic acid.
4. Wash in several changes of 3% acetic acid.
5. Flood with freshly prepared mixture of equal volumes of 2% potassium ferrocyanide and 2% hydrochloric acid (6 ml conc. HC1, 94 ml water) for 10 min.
6. Wash well in tap water.
7. Counterstain with neutral red.
8. Wash, dehydrate, clear and mount in DPX.

Result

Acid groups (carboxyl, phosphate, sulphate, sialic acid residues) stain blue.

Note

This is a relatively non-specific method which can be controlled with hyaluronidase, chondroitin sulphatase and sialidase digestions. Pre-existing ferric iron deposits will also be shown. A control Perls' reaction is necessary for critical studies.

FEYRTER'S THIONIN FOR SIALIC ACID AND OTHER ACIDIC SUBSTANCES

Preparation of tissue

Cryostat sections cut at 5–8 μm from snap-frozen tissue, lightly fixed in formol-calcium for 5–10 min, frozen sections of formalin-fixed tissue mounted on slides.

Method

1. Briefly rinse sections in tap water, wipe around section and allow to come to the point of drying
2. Cover section with 2–3 drops of 1% thionin in 0.5% aqueous tartaric acid
3. Add coverslip and remove excess dye
4. Seal edges with nail varnish, glyceel etc.

Result

Nuclei–blue; acidic substances — red.

Notes

Observe at intervals of up to 24 h after mounting. Although this is a non-specific method it appears to be remarkably selective for sialic acid residues. Thus gangliosides and other sialic acid-containing substances show a delicate rose-red metachromasia very soon after mounting.

TOLUIDINE BLUE (MUIR, MITTWOCH AND BITTER) FOR LYMPHOCYTIC INCLUSIONS

Preparation of tissue

Air-dried blood films.

Method

1. Fix in methanol for 10 min at room temperature
2. Stain in 0.1% toluidine blue in 30% methanol (1 ml 1% toluidine blue, 3 ml methanol, 6 ml water) for 30 min at room temperature
3. Rinse in acetone, clear and mount in DPX.

Result

Nuclei — blue
Inclusions in lymphocytes — red
Basophils — red granules.

Note

Under oil immersion objective ($\times$100) count 100 lymphocytes and record the number of them containing metachromatic inclusions.

TOLUIDINE BLUE (HAUST AND LANDING) FOR SOLUBLE MUCOPOLYSACCHARIDES

Preparation of tissue

Cryostat sections (5–8 μm) of snap-frozen tissue; air-dried blood films; air-dried bone marrow films.

Method

1. Fix sections or films for 20 min in an acetone:tetrahydrofuran mixture (1:1 by volume) at room temperature

2. Rinse in acetone

3. Immerse and agitate in: 0.5% toluidine blue in 25% acetone (5 ml 1% toluidine blue, 2.5 ml acetone, 2.5 ml water) for 2 min

4. Rinse in acetone, clear in xylene and mount DPX.

Result

Nuclei — blue

Mucopolysaccharide — red to purple (can appear almost black).

HYALURONIDASE DIGESTION

1. Bring dewaxed sections to water

2. Incubate in 0.1 mol/l phosphate buffer pH 6.0 containing 0.3–0.6 mg/ml testicular hyaluronidase (bovine testicular, Sigma type IV, 750 units/mg) for 2 h at 37°C *or* 5 h at room temperature

3. Wash in distilled water

4. Stain with the appropriate technique (alcian blue, colloidal iron etc.).

Appendix 4

Lipids

OIL RED O
SUDAN BLACK B
BROMINE SUDAN BLACK
SUDAN BLACK FOR HAEMATOLOGY
CHOLESTEROL (SCHULTZ AND PAN)
NILE BLUE SULPHATE
LUXOL FAST BLUE — NEUTRAL RED
FERRIC HAEMATOXYLIN
TOLUIDINE BLUE FOR SULPHATIDES

OIL RED O METHOD

Preparation of tissue

10 μm Cryostat sections unfixed or postfixed in formol calcium; frozen sections of fixed tissue; air-dried films of blood, bone marrow or touch preparations unfixed or fixed in formalin vapour for 10 min.

Method

1. Rinse sections in 70% alcohol
2. Stain in oil red O solution (saturated in 70% alcohol) for up to 2 h
3. Rinse in 70% alcohol
4. Wash and counterstain nuclei with Mayer's or Carrazzi's haematoxylin for 3 min
5. Blue sections in tap water, rinse in distilled and mount in glycerol gelatin.

Results

Cholesteryl esters and triglycerides—red
Some phospholipids—pale pink
Nuclei—blue

Notes

1. Unstained crystalline cholesterol and its esters can be distinguished from fats when the stained section is viewed in polarized light with partly crossed polars. Crystals appear birefringent.

2. For gross staining of arteries with oil red O, the unfixed tissue is cleaned of frak adventitial fat and processed as described for sections, followed by differentiation.

3. Alternatively, any of the usual oil red O methods with isopropanol, triethyl phosphate or propylene glycol as solvent may be used with identical results.

SUDAN BLACK B (SBB) METHOD FOR GENERAL USE

Preparation of tissue

Cryostat sections unfixed or postfixed in formol-calcium; frozen sections of fixed tissue.

Method

1. Rinse sections in 70% ethanol.
2. Stain in saturated Sudan Black B in 70% ethanol, for up to 2 h.
3. Differentiate with 70% ethanol.
4. Counterstain nuclei with Kernecht-rot.
5. Wash in water and mount sections in glycerol-gelatin.

Result

Unsaturated cholesterol and triglyceride esters, —blue-black.
Some phospholipids appear grey.
Nuclei are red.

Note

If sections are viewed in polarized light, unstained free and ester cholesterol will appear birefringent whilst the

455

stained phospholipids in myelin exhibit a bronze dichroism. Propylene glycol may be used as solvent for the Sudan black.

The staining is clean unless the Sudan black solution is oversaturated. If the tissue section and lipids are stained brown, something is wrong!

BROMINE SUDAN BLACK B (BSBB) METHOD

Preparation of tissue

Cryostat sections postfixed in formol-calcium 1 h; short fixed frozen sections.

Method

1. Treat sections with 2.5% aqueous bromine at room temperature, inside a fume cupboard, 30 min
2. Wash in distilled water and remove excess bromine with 0.5% sodium metabisulphite, 2 min
3. Wash thoroughly in distilled water and proceed with the Sudan black B method outlined above.

Results

Bromination enhances the reaction of all sudanophilic lipids. In addition, lecithin, free fatty acids and free cholesterol are stained.

SUDAN BLACK B FOR HAEMATOLOGY

Preparation of tissue

Air-dried films of blood or marrow.

Preparation of solutions

Formol-ethanol: 1 part to 9.
Stock solution–mix together:
Phenol 12 g in ethanol 25 ml,
Disodium hydrogen phosphate ($Na_2HPO_4.12H_2O$) 0.225 g (or anhydrous 0.089 g) in 75 ml water
Sudan black B (Gurr or BDH) 0.45 g in ethanol 150 ml.
Filter just before use.

Method

1. Fix slides in formol-ethanol for 30–45 s
2. Rinse in distilled water
3. Rinse in 70% ethanol
4. Stain in stock solution for 1 h
5. Rinse in 70% ethanol
6. Counterstain with May-Grünwald-Giemsa or Leishman
7. Rinse in running tap water 1 min
8. Mount with aqueous mounting medium (Gurr).

Positive result

Dark brown to black staining of granulocytes; occasionally, especially in pathological states, red metachromatic staining occurs.

Note

This method does not stain lipids.

CHOLESTEROL (AND ITS ESTERS)

Preparation of tissue

Postfixed cryostat sections of snap-frozen tissue; frozen sections of formalin-fixed tissue.

Schultz method

1. Treat sections in 2.5% ferric ammonium sulphate in 0.2 mol/l acetate buffer pH 3.0 for up to 7 days at 37°C
2. Wash 3 × 1 h each in 0.2 mol/l acetate buffer pH 3.0
3. Rinse in distilled water
4. Treat with 5% formalin in tap water for 10 min
5. Drain and allow sections to almost dry
6. Apply reaction medium.

Reaction medium

Add 0.5 ml acetic acid drop by drop slowly to concentrated sulphuric acid with much agitation and cooling. The solution becomes hot and viscous but must remain colourless. Add 1–2 drops of this mixture to the section and cover with a coverslip. Observe immediately. A blue-green colour indicates cholesterol.

Adams PAN method

1. Treat sections in 1% ferric chloride for 4–18 h
2. Wash well in distilled water 10 min
3. Air dry
4. Paint the sections sparingly with the following solution using a soft camel-hair brush:

1:2-Naphthoquinone-4-sulphonic acid	10 mg
Ethanol	5 ml

60% Perchloric acid	2.5 ml
40% Formaldehyde	0.25 ml
Distilled water	2.25 ml

Heat painted sections on a hot plate (60–70°C) for 1–2 min. The sections are kept moist by gently repainting. Mount in 60% perchloric acid. A blue colour indicates cholesterol.

Notes

Oxidation is an essential requirement for the successful demonstration of cholesterol.

NILE BLUE SULPHATE (Cain's method)

Preparation of tissue

Postfixed cryostat sections of snap-frozen tissue; frozen sections of formalin-fixed tissue; air-dried bone marrow films postfixed in formol-calcium.

Preparation of solution

10 g Nile blue
100 ml 1% sulphuric acid
Reflux gently for 4 h. Cool. Check suitability by adding xylene to 1 ml of dye solution with shaking. A red colour in the xylene layer shows the solution is suitable for staining.

Method

1. Stain sections at 60°C for 5 min
2. Wash (dip only) in distilled water at 60°C
3. Differentiate for 30 s. only, in 1% acetic acid at 60°C
4. Wash in cold water
5. Mount in glycerine jelly.

Result

Neutral fat—red
Free fatty acids—blue
Nuclei—pale blue.

Notes

Many authors refer to 'positive with Nile blue' without specifying the colour — or the method used. Both are important in interpretation. The method is only of pathological significance in Wolman's disease where the storage cells stain a deep purple (the red of neutral fat plus the blue of free fatty acids).

LUXOL FAST BLUE — NEUTRAL RED

Preparation of tissue

Postfixed cryostat sections of snap-frozen tissue; frozen sections of formalin-fixed tissue; paraffin sections.

Method

1. Bring sections to 95% alcohol
2. Stain for 16–18 h at 60°C in 0.1% luxol fast blue in 95% alcohol (in a closed container)
3. Rinse in 70% alcohol to remove excess surface dye
4. Wash in tap water
5. Differentiate in 0.05% aqueous lithium carbonate until no further dye comes away (overdifferentiation is not possible)
6. Wash well
7. Counterstain with 1% aqueous neutral red
8. Rinse, dehydrate, clear and mount.

Result

Nuclei—red
Myelin and several substances in various neuronal storage disorders are stained a deep blue-black
Connective tissue and muscle are pale blue/green.

Note

The differentiation step 5 is not critical as in some other methods for luxol fast blue, and over-differentiation is not a problem.

FERRIC HAEMATOXYLIN FOR PHOSPHOLIPIDS

Preparation of tissues

10 μm Cryostat sections on poly-L-lysine subbed slides; frozen sections.

Solutions

Ferric haematoxylin solution:

A)	Distilled water	596 ml
	Concentrated HCl	4 ml
	FeCl$_3$ 6H$_2$O	5 g
	FeSO$_4$ 7H$_2$O	9 g

This solution keeps well

| B) | 1% aqueous haematoxylin |
| | Must be made fresh |

Mix three parts A with one part B and use at once (preferably) or within an hour.

Method

I *To stain all phospholipids*

1. Fix in formol-calcium 20–30 min
2. Wash
3. Stain in ferric haematoxylin 7–8 min
4. Rinse in distilled water
5. Mount in glycerin jelly.

II *Cholesterol extraction*

1. Immerse sections in dry acetone (dried with anhydrous $CaCl_2$ overnight) 10–15 min
2. Fix in formol-calcium 30 min
3. Stain as for I.

III *Extraction of most lipids*

1. Immerse section in chloroform:methanol (2:1) for 30 min
2. Fix in formol-calcium
3. Stain as for I.

IV *Alkaline hydrolysis*

1. Fix in formol-calcium for 30 min
2. Wash
3. Immerse in 1 mol/l NaOH for 1 h at room temperature, *gently* remove the slide, wipe and lower into 1 mol/l HCl for 5–10 s — *this is where sections may fall off!*
4. Wash
5. Stain as for I.

Results

I   Ferric haematoxylin—stains all phospholipids
II   Dry acetone—extracts cholesterol
III Chloroform methanol—extracts most lipids
IV   Alkaline hydrolysis—hydrolyses ester linkages; amide linkages are not hydrolysed. Therefore sphingomyelin is stained while other phospholipids are removed.

TOLUIDINE BLUE FOR SULPHATIDES IN METACHROMATIC LEUCODYSTROPHY

Preparation of tissue

Cryostat sections cut at 5–8 μm of snap-frozen tissue, fixed for 10 min in formal-calcium; frozen sections of formalin-fixed tissue; smears of urinary deposits fixed in formaldehyde vapour.

Method

1. Rinse fixed sections and smears in tap water
2. Stain overnight at room temperature in 0.01% toluidine blue in McIlvaine buffer pH 5.0
3. Rinse in tap water, dehydrate in acetone, clear and mount in DPX.

Result

Sulphatides are stained yellow, brown or purple. Nuclei are blue.

Note

Dehydration must be in acetone. Alcohol will destroy the metachromasia. Mast cells are also stained.

Appendix 5

Enzymes

HEXAZOTIZED PARAROSANILINE
ACID PHOSPHATASE (LEAD PRECIPITATION
  METHOD) GOMORI
ACID PHOSPHATASE (SIMULTANEOUS COUPLING
  TECHNIQUE)
ACID PHOSPHATASE FOR HAEMATOLOGY
ALKALINE PHOSPHATASE (AZO DYE METHOD)
ALKALINE PHOSPHATASE (BCIP-NBT METHOD)
PHOSPHORYLCHOLINE PHOSPHATASE (PROSTATIC)
LEUCOCYTE ALKALINE PHOSPHATASE
MYOFIBRILLAR ATPASE
ATPase AFTER ACID PREINCUBATION
GLUCOSE-6-PHOSPHATASE
ACETYLCHOLINESTERASE (STANDARD METHOD)
ACETYLCHOLINESTERASE (RAPID METHOD)
NADH DEHYDROGENASE
SUCCINATE DEHYDROGENASE
CYTOCHROME OXIDASE
MYOPHOSPHORYLASE
PHOSPHOFRUCTOKINASE
MYOADENYLATE DEAMINASE
LACTASE
DISACCHARIDASES WITH NATURAL SUBSTRATES
BRUSH BORDER α-GLUCOSIDASES
ENTEROPEPTIDASE (ENTEROKINASE)
DIPEPTIDYL AMINO PEPTIDASE IV
TRYPTASE OF MAST CELLS
NON-SPECIFIC ESTERASE FOR GENERAL USE
α-NAPHTHYL ACETATE ESTERASE FOR
  HAEMATOLOGY
ACID ESTERASE IN BLOOD FILMS
ACID ESTERASE IN TISSUES
CHLOROACETATE ESTERASE (CAE)
DUAL ESTERASE (ANAE-CAE) FOR HAEMATOLOGY
β-GALACTOSIDASE
HEXOSAMINIDASE
β GLUCURONIDASE
α-MANNOSIDASE
DOPA-OXIDASE
CATALASE (PEROXIDASE) FOR HUMAN
  PEROXISOMES

HEXAZOTIZED PARAROSANILINE (HPR)

Stock solution

4% pararosaniline hydrochloride or basic fuchsin in
2 mol/l HCl (heat gently, cool and filter).

Add equal volumes of stock 4% pararosaniline to
freshly prepared 4% aqueous sodium nitrite; wait
1-2 min until the solution becomes a pale straw colour.
For critical studies add about 5 mg solid sulphamic acid
per ml of hexazotized pararosaniline to destroy excess
nitrous acid.

ACID PHOSPHATASE (GOMORI) for general use

Preparation of tissue

Postfixed cryostat sections; fixed frozen sections; touch
preparations; bone marrow films.

Preparation of solutions

Incubating medium

Sodium β-glycerophosphate	31.5 mg
dissolved in:	
0.1 mol/l acetate buffer pH 5.0	5 ml
0.008 mol/l lead acetate	5 ml
(add dropwise with gentle agitation)	

Dilute ammonium sulphide

1 ml of 10% ammonium sulphide diluted to 200 ml with
distilled water.

Method

1. Incubate at 37°C for 30 min
2. Wash in tap water, 30 s
3. Immerse in dilute ammonium sulphide
4. Wash well, counterstain nuclei, mount in glycerine
   jelly.

Result

Activity is indicated by a brown/black reaction product.

Notes

Use medium immediately, do not filter, do not adjust pH. Good quality water (HPLC grade) is essential for clean reproducible results.

ACID PHOSPHATASE (simultaneous coupling technique) for general use

Preparation of tissue

Cryostat or frozen sections of tissue fixed in formaldehyde, or less preferably, glutaraldehyde; or freeze-dried cryostat sections coated with celloidin. Fixed tissue must be rinsed in buffer thoroughly before sections are cut to remove excess fixative; touch preparations; bone marrow films.

Preparation of incubation medium:

Naphthol AS-BI phosphate   5 mg
  dissolved in 0.5 ml N, N-dimethylformamide
  Add to
0.1 mol/1 acetate buffer, pH 5.2,   50 ml
  containing:
Hexazotized pararosaniline (p. 459)   2 ml
  or
Fast Blue B, BB, RR or Fast Red Violet, Fast
Ponceau salt L   30 mg
Mix thoroughly, adjust pH to 5.2 and filter before use.

Method

1. Incubate sections for 30–120 min at 37°C
2. Wash thoroughly in tap water
3. Place in 4% formaldehyde for several hours (to prevent the formation of gas bubbles in the sections)
4. Rinse in tap water
5. If desired, counterstain nuclei with haematoxylin or nuclear fast red or methyl green
6. Mount in glycerine jelly or dehydrate clear and mount in DPX.

Results

Enzyme-active sites are red (if hexazotized pararosaniline is used) or blue or violet (if fast blue salts are used).

Note

The formaldehyde postincubation (step 3) can be replaced with 70% ethanol for 10–30 min at room temperature.

ALKALINE PHOSPHATASE for smears, touch preparations, cryostat sections unfixed or postfixed

Preparation of incubating medium

0.05 mol/1 Tris buffer pH 9.2   10 ml
Naphthol AS-TR phosphate
  in a drop of N, N-dimethylformamide   2 mg
Fast Red salt TR   6 mg

Method

1. Incubate 10–60 min at room temperature
2. Transfer to 10% formalin
3. Carrazzi haematoxylin, 2–3 min
4. Wash in running water till blue
5. Mount in glycerin jelly.

Result

Sites of alkaline phosphatase activity — red.

ALKALINE PHOSPHATASE (BCIP-NBT METHOD)

Preparation of tissues

Paraffin sections (5 μm) dried at 37°C; cryostat sections postfixed in formol-calcium 5 min.

Incubation medium

5-Bromo-4-chloro-3-indolyl phosphate (p-toluidine salt; Sigma) 2.5 mg dissolved in 0.5 ml dimethylformamide
0.2 mol/1 veronal acetate buffer pH 9.5   10 ml
Nitro BT   5 mg
1 mol/1 magnesium chloride   0.08 ml

Method

1. Dewax and rehydrate paraffin sections; wash cryostat sections.
2. Incubate at 37°C for 10 min (jejunal biopsies), up to 30 min for cryostat sections and for immunocytochemistry.
3. Wash.
4. Counterstain with Kernechtrot 2 min.
5. Wash, dehydrate clear and mount.

Results

Alkaline phosphatase activity is dark blue to black. Nuclei are red.

Notes

The intestinal enzyme is resistant to formalin fixation and this method will demonstrate the brush border in paraffin sections of small bowel.

BCIP = Bromo *c*hloro *i*ndolyl *p*hosphate
NBT = Nitro *b*lue *t*etrazolium

PHOSPHORYL CHOLINE PROCEDURE FOR PROSTATIC ACID PHOSPHATASE (PRAP) ACTIVITY

Preparation of tissue

Cryostat of frozen sections of lightly fixed tissue.

Method

As for the general acid phosphatase (Gomori) method (p. 459) using 32 mg phosphoryl choline chloride in place of β-glycerophosphate.

Three sets of sections are treated as follows:

1. No fixation
2. Fixation in formol-calcium 20 30 min at 4°C
3. Fixation as in 2. but with the addition of 2.3 mg sodium-L(+)-tartrate to the incubating medium (final concentration 1 mmol/l).

Incubation is for 4–8 min; post treatment is as for the general method.

Note

Prostatic acid phosphatase is active against phosphoryl-choline, shows no inhibition on fixation and is tartrate sensitive. At 1 mmol/l inhibition is virtually complete; at the usually recommended 10 mmol/l lead precipitation occurs. Acid phosphatases of other tissues are only weakly active against phosphorylcholine and show almost no reaction in the short incubation time given. Prostatic acid phosphatase will hydrolyse β-glycerophosphate. Both benign and malignant prostatic epithelial cells are positive. The method should be controlled with liver and/or kidney sections.

LEUCOCYTE ALKALINE PHOSPHATASE

Preparation of tissue

Air-dried films of blood, preferably unanticoagulated, or marrow; fix without delay; if staining cannot be done within 6 hours, fixed dried slides may be stored at −20°C.

Preparation of solutions

Fixative

Formol-ethanol: neutral formalin (40% formaldehyde) to ethanol, 1 part to 9 parts. Keep refrigerated; may be used for up to 3 weeks.

Substrate

Naphthol AS phosphate: dissolve 30 mg in 0.5 ml N, N-dimethyl formamide; add 100 ml 0.05 mol/l tris buffer pH 9.0. Freeze in 10 ml aliquots; thaw at room temperature and check pH just before use.

Incubation medium

To a thawed aliquot of Naphthol AS phosphate solution add 10 mg Fast Blue BB and agitate.

Method

1. Fix for 30 sec at 0–5°C
2. Wash in running tap water for 10–15 sec and allow to dry
3. Filter incubation medium immediately onto slide and leave for 10 min
4. Rinse gently in running tap water and air-dry
5. Counterstain with 0.1% aqueous neutral red 10 min
6. Rinse in tap water
7. Air-dry, do not coverslip, and examine immediately under oil-immersion.

Positive result

Blue granules; reaction may be scored:
   0: no granules
   1: few granules
   2: few to moderate number granules
   3: numerous granules
   4: cytoplasm crowded with granules.
Normal range for 100 neutrophils 35–100.

MYOFIBRILLAR ATPase

Preparation of tissue

Cryostat sections (10 μm) postfixed in methanol-free formalin.

Preparation of solutions

Methanol-free formalin (MFF) fixative

Dissolve 20 g paraformaldehyde in 100 ml distilled water by heating to 60°C. Add 10% NaOH gradually until formalin solution clears. Add sodium cacodylate 10.7 g, and make total volume up to 500 ml. Adjust pH to 7.4; store at 4°C.

Incubating medium

ATP (disodium salt)	30 mg
0.1 mol/l Tris-HCl buffer	
containing 0.018 mol/l calcium chloride	20 ml
Adjust pH to 9.5.	

Method

1. Air dry sections
2. Fix in MFF fixative 4°C for 5 min
3. Wash in running tap water 2 min; rinse in distilled water
4. Immerse sections in 0.1 mol/l Tris-HCl buffer containing 0.018 mol/l calcium chloride pH 10.2, 30 min
5. Wash in running tap water and rinse in distilled
6. Incubate at 37°C for 1 h
7. Wash in two changes of 1% calcium chloride
8. Immerse in 2% cobalt nitrate 3 min
9. Rinse rapidly in 3 changes tap water, 3 changes distilled water
10. Immerse in 0.5% ammonium sulphide
11. Wash in tap water; rinse in distilled water
12. Counterstain nuclei in methyl green or Carrazzi haematoxylin
13. Mount in aqueous mountant.

Results

Type 1 fibres — pale
Type 2A fibres — intermediate
Type 2B fibres — dark

Note

Mounting in synthetic mountant after clearing in xylene will result in rapid fading of the reaction product.

ATPase AFTER ACID PREINCUBATION

Preparation of tissue

Air-dried cryostat sections (10 μm)

Preparation of solutions

Preincubating solutions

0.2 mol/l acetate buffer at pH 4.3 and pH 4.6.

Incubating medium

As for myofibrillar ATPase method (1)

Method

1. Air-dry sections
2. Immerse sections in 0.2 mol/l acetate buffers at pH 4.3 and pH 4.6 at room temperature for 30 min
3. Wash in running tap water 2 min
4. Rinse 3 times in distilled water
5. Incubate at 37°C for 40 min
6. Follow steps 7–13 in myofibrillar ATPase method.

Results

Preincubation at pH 4.3

Type 1 fibres — dark
Type 2A fibres — pale
Type 2B fibres — pale
Type '2C' fibres — intermediate

Preincubation at pH 4.6

Type 1 fibres — dark
Type 2A fibres — pale
Type 2B fibres — intermediate
Type '2C' fibres — intermediate

Note

The exact conditions for acid preincubation may vary from one laboratory to another. Some may find pH 4.4 or pH 4.5 gives better results. Each laboratory should define the conditions which give the best (most easily interpretable) results for them.

GLUCOSE-6-PHOSPHATASE

Preparation of tissue

Air-dried cryostat sections (5-8 μm) of snap-frozen tissue.

Incubation medium

Glucose-6-phosphate (dipotassium salt)	5 mg
Distilled water	5.5 ml

0.05 mol/l Tris buffer pH 7.0	4 0 ml
2.29% lead acetate	0.6 ml

Mix well and filter.

Method

1. Incubate sections for 15 min at 37°C
2. Wash in tap water
3. Treat with dilute ammonium sulphide (1 ml of 10% solution diluted to 200 ml with tap water)
4. Wash well in tap water
5. Lightly counterstain with Carrazzi's haematoxylin
6. Wash and mount in glycerine jelly.

Result

Sites of enzyme activity are stained brown.

Notes

Sections of very fatty tissue may float off.

The dilute ammonium sulphide should be only a pale yellow.

Specificity of the reaction may be checked by pre-incubation of the sections in 0.05 mol/l acetate buffer pH 5.0 for 10 min at 37°C. This procedure inactivates glucose-6-phosphatase but has no effect on other phosphatases.

The tissue should be *fresh* and the delay between excision and freezing should ideally be less than 5 min. Postmortem tissue may be suitable, but negative results cannot be considered conclusive.

Tris buffer at pH 7.0 is used because the coefficient of temperature variation for this buffer reduces the pH during incubation to 6.5, which is optimal for glucose-6-phosphatase.

ACETYLCHOLINESTERASE (STANDARD METHOD)

Preparation of tissue

Cryostat sections of snap-frozen tissue cut at 10 μm are air-dried and fixed for 30 s in 4% formaldehyde in 0.1 mol/l calcium acetate (formol-calcium). Frozen sections of formol-calcium-gum sucrose treated blocks of tissue.

Incubation medium

A. Acetylthiocholine iodide	5 mg
0.1 mol/l acetate buffer pH 6.0	6.5 ml
0.1 mol/l sodium citrate	0.5 ml

30 mmol/l copper sulphate	1.0 ml
Distilled water	1.0 ml
4 mmol/l iso-octamethylpyrophosphoramide (iso OMPA) *or* 4 mmol/l ethopropazine.	0.2 ml

B. Add 1.0 ml 5 mmol/l potassium ferricyanide just before use.

A and B can be made in bulk and stored in aliquots at −20°C.

Method

1. Rinse the fixed sections for 10 s in tap water
2. Incubate at 37°C for 1 h in the above medium
3. Wash briefly in tap water
4. Treat with 0.05% *p*-phenylene diamine dihydrochloride in 0.05 mol/l phosphate buffer pH 6.8 for 45 min at room temperature
5. Wash in tap water
6. Treat with 1% osmium tetroxide for 10 min at room temperature
7. Wash well in tap water, counterstain lightly (10 s) in Carrazzi haematoxylin (or Mayer's haemalum), wash, dehydrate, clear and mount in DPX.

Result

Nerve fibres and cells containing acetylcholinesterase are stained dark brown to black.

Notes

Do not counterstain too heavily.

Thoroughly dehydrate because areas with strong activity are resistant to dehydration.

Either inhibitor is adequate; ethoproprazine is less toxic and causes less alarm among safety officers.

Beware of areas of RBCs which appear to have nerve fibres lying over. This is due to the acetylcholinesterase of the RBC membrane.

ACETYLCHOLINESTERASE (RAPID METHOD)

Preparation of tissue — as for standard method

Incubation medium — as for standard method

Method

1. Rinse the fixed sections in tap water
2. Incubate at 37°C for 20–30 min
3. Wash in distilled water
4. Treat with dilute ammonium sulphide (1 ml of 10% ammonium sulphide diluted to 200 ml with tap water) 30 s

5. Wash in tap water then in distilled water
6. Treat with 0.1% silver nitrate for 1 min
7. Wash in distilled water
8. Counterstain nuclei lightly with Carazzi's haematoxylin
9. Dehydrate, clear and mount in DPX.

Results

Nerve fibres containing acetylcholinesterase are stained dark brown to black.
Nuclei stain blue.

NADH DEHYDROGENASE

Preparation of tissue

Fresh cyostat sections (10 μm).

Preparation of solutions

Incubating medium
MTT (1 mg/ml)   2.5 ml
0.1 mol/l Tris-HCl buffer pH 7.4   2.5 ml
0.5 mol/l cobalt chloride   0.5 ml
Distilled water   3.5 ml
Adjust pH to 7.0; store at $-20°$C. NADH added just before use, 2 mg per 1 ml of stock incubating solution.

Method

1. Incubate at 37°C for 40 min
2. Drain off incubating medium and postfix in formol-calcium at room temperature for 15 min
3. Wash in tap water 2 min
4. Rinse in distilled water
5. Counterstain nuclei in methyl green
6. Wash in distilled water
7. Mount in glycerine jelly.

Results

Black formazan deposits indicate sites of activity (mainly mitochondrial). In muscle higher activity in type 1 fibres is due to larger number of mitochondria per unit area, and the presence of a sarcoplasmic enzyme.

Note

MTT = (3–4), [5-dimethylthiazol-2-yl]-2,5-diphenyl-tetrazolium bromide).

SUCCINATE DEHYDROGENASE

Preparation of tissue

Fresh cryostat sections (10 μm).

Preparation of solutions

Prepare stock solution containing MTT and cobalt chloride in Tris-HCl buffer as in method for NADH-dehydrogenase.

Succinate solution (0.6 mol/l)

Sodium succinate   1.62 g
Distilled water   8.0 ml
1 mol/lHCl   0.05 ml
Adjust pH to 7.0 and make up total volume to 10 ml.
Store at $-20°$C.

Incubating medium

Add 0.1 ml succinate solution per 0.9 ml of stock incubating solution just before use.

Method

As for NADH dehydrogenase.
Incubation 30 min for muscle, 2 h for liver.

Note

For the demonstration of dehydrogenase activity in fatty liver (in cases of ?Reye's syndrome) the lipid should be extracted with cold (4°C) acetone for 5 min before incubation.

CYTOCHROME OXIDASE

Preparation of tissue

Cryostat sections (10 μm), air-dried.

Preparation of solutions

3, 3′ diaminobenzidine
   tetrahydrochloride (DAB)   2.5 mg
Catalase (20 μg/ml)   0.5 ml
Cytochrome c (type 2)   5 mg
Sucrose (optional)   375 mg
0.1 mol/l phosphate buffer pH 7.4   4.5 ml
   Adjust pH to 7.4 before use.

Method

1. Incubate at 22°C for up to 1 h
2. Rinse in distilled water
3. Postfix in formol-calcium 15 min
4. Dehydrate, clear xylene and mount in DPX.

Results

Brown reaction product indicates sites of cytochrome oxidase activity.

Distribution is mitochondrial with higher activity in Type 1 muscle fibres.

Note

DAB may be carcinogenic — **handle with care**.

A more intense reaction can be obtained by treating with 1% aq. osmium tetroxide for 10 min at step 3. Neutral fat does not interfere. Other methods for cytochrome oxidase may not be reliable.

MYOPHOSPHORYLASE

Preparation of tissue

Cryostat sections (10 μm) of snap-frozen skeletal muscle. Air-dried.

Preparation of incubation medium

0.1 mol/l acetate buffer pH 5.9	10 ml

To this add in the order given with stirring between additions:

Glucose-1-phosphate (dipotassium salt)	100 mg
Sodium fluoride	180 mg
Glycogen (rabbit liver *or* oyster)	2 mg
AMP	5 mg
ATP	5 mg
0.1 mol/l $MgCl_2$	1 ml
Ethanol	2 ml
PVP (MW 44 000)	900 mg

Store at −20°C in 2 ml aliquots.

Method

1. Incubate for 30 min at 37°C
2. Rinse (2 s) in 40% ethanol and rapidly air-dry
3. Fix in ethanol 3 min
4. Rapidly air-dry
5. Treat sections in large volume (10 ml) of Lugol's iodine diluted 1–10 with water, for 5 min
6. Mount in glycero: Lugol's iodine (9 vol: 1 vol).

Result

Newly synthesized glycogen in skeletal and smooth muscle stains brown — purple — blue.

Note

Skeletal muscle shows no reaction (i.e. stains yellow) in McArdle's disease, while the smooth muscle enzyme in blood vessel walls is unaffected and acts as an internal control for the method.

Lugol's iodine as used here is 1 g iodine, 2 g potassium iodide, 100 ml distilled water.

PHOSPHOFRUCTOKINASE

Preparation of tissues

Cryostat sections (10 μm) of air-dried.

Incubation media

Sodium arsenate in 5 ml water	31.2 mg
Nitro BT	4 mg
ATP	5.5 mg
NAD	7.4 mg
Magnesium chloride 1 mol/l	0.01 ml

Make up to 10 ml with distilled water. Adjust to pH 7.0 with 3% acetic acid, and add 1 drop glyceraldehyde-3-phosphate dehydrogenase (Sigma).

Reserve two drops of this for the negative control section. Divide the remainder into two equal amounts of 5 ml each. To one add 3 mg fructose-6-phosphate (for PFK, medium A). To the other add 4.3 mg fructose-1:6-diphosphate (for aldolase, medium B).

The reserved control medium without substrate is medium C.

Method

1. Incubate sections in A, B and C at 37°C for at least 1 h
2. Wash and mount in glycerine jelly.

Results

Enzyme activity is blue. Absent activity in A, with activity in B indicates a deficiency of phosphofructokinase. The medium containing fructose-1:6-diphosphate (medium B) acts as a control for the integrity of the pathway beyond phosphofructokinase.

This is a multistep method in which the final reaction product reflects the localization of the final enzyme in the pathway (NADH-dehydrogenase).

MYOADENYLATE DEAMINASE

Preparation of tissues

Cryostat sections (10 μm), air-dried.

Preparation of solutions

Incubating medium

Adenosine-5'-monophosphate	4 mg
Distilled water	7.0 ml
Nitro blue tetrazolium (5 mg/ml)	2.0 ml
Potassium chloride 3 mol/l	0.7 ml

Add potassium chloride slowly while stirring. Adjust pH to 6.1 and add dropwise 5 mg dithiothreitol dissolved in 0.3 ml distilled water just before using.

Method

1. Incubate at room temperature for 1 h
2. Rinse briefly in distilled water
3. Mount in glycerine jelly.

Results

Deep blue reaction end-product more intense in type 1 muscle fibres.

Note

For control use 4 mg inosine-5'-monophosphate instead of AMP.

LACTASE

Preparation of tissue

Cryostat sections (5 μm) unfixed or fixed in cold (4°C) chloroform-acetone (1 : 1) 5 min.

A. Indigogenic method

Incubation medium

5-Bromo-4H chloro-3-indolyl-β-D-fucoside	3 mg
Dissolve in N,N-dimethylformamide	0.3 ml
0.1 mol/l citric acid phosphate buffer, pH 6	6 ml
1.650% potassium ferricyanide (50 mol/l)	0.5 ml
2.11% potassium ferrocyanide (50 mmol/1)	0.5 ml

Mix thoroughly.

Method

1. Incubate sections in the medium at 37°C 2 h
2. Rinse in distilled water
3. Fix in 4% formaldehyde for 5 min at room temperature
4. Rinse in distilled water
5. Counterstain with Kernechtrot
6. Wash, dehydrate, clear and mount.

Results

Enzyme-active sites are stained turquoise (blue)

Notes

The medium can be used repeatedly. After incubation it is filtered and stored frozen in a closed vessel. This can be repeated at least 10 times and is sufficient for 50 biopsies.

B. Azo-coupling method

Incubation medium

1-naphthyl-β-D-glucoside	5 mg
Dissolve in N, N-dimethylformamide	0.25 ml
Buffered hexazotized pararosaniline	10 ml

(made up of 0.3 ml hexazotized pararosaniline and 9.7 ml 0.1 mol/l citric acid-phosphate buffer, pH 6.5; if the pH is between 5.5 and 6 correction is unnecessary, otherwise adjust with 1 mol/l and 0.1 mol/l NaOH)
Mix well and filter.

Method

1. Incubate sections in the medium in a refrigerator 15 h
2. Rinse in distilled water
3. Place in 4% formaldehyde for several hours at room temperature
4. Rinse in tap water
5. Nuclei can be counterstained with haematoxylin
6. Wash, dehydrate, clear and mount.

Results

Enzyme-active sites are stained brown.

Notes

This method is recommended only if indolyl fucoside is not available.

DISACCHARIDASES USING NATURAL SUBSTRATES (lactase, trehalase, isomaltase, sucrase and maltase)

Preparation of tissue

Unfixed cryostat sections at 10 μm.

Glucose oxidase-peroxidase-diaminobenzidine (GO-PO-DAB) method

Preparation of solutions

1. 1% aqueous solution of low temperature gelling agarose (Sigma type VII); check pH with indicator paper and adjust to 6.0. Store at 4°C
2. 4% solutions of sucrose, trehalose, lactose, palatinose (for isomaltase) and maltose (chromatographically pure). Store at 4°C
3. Diaminobenzidine solution: dissolve 2–4 mg diaminobenzidine tetrahydrochloride in a few drops of N,N-dimethylformamide, add 0.3 ml distilled water and 2.1 ml 0.1 mol/l citric acid-phosphate buffer, pH 6.5. Prepare freshly
4. Glucose oxidase — peroxidase solution: dissolve 0.5–1 mg glucose oxidase (degree of purity I, Boehringer, about 70–140 units) and 0.1 mg horseradish peroxidase (type VI, Sigma, about 30 purpurogallin units) in 0.9 ml 0.1 mol/l citric acid — phosphate buffer, pH 6.5. Prepare freshly.

Incubation medium

4% disaccharide	0.5 ml
Diaminobenzidine solution	0.8 ml
Glucose oxidase-peroxidase solution	0.3 ml
Mix	
1% aqueous low temperature gelling agarose at 37°C	1.6 ml
Mix thoroughly.	

Method

1. Cover sections mounted on slides with gel medium and allow to set; the solution is enough for three slides containing several sections each
2. Incubate in a moist chamber at 37°C for 2 h (sucrase, trehalase, maltase) or 18 h (lactase, isomaltase)

3. Wash slides carefully (substrate gel must not peel off) in 5% acetic acid for 5 min
4. Rinse in distilled water
5. Cover slides with wet blotting paper and allow to dry at room temperature.

Results

When active disaccharidase is present a brown staining develops in sections and in the gel overlay. Its intensity is the measure of enzyme activity.

Notes

If the activity of disaccharidases is low, controls are carried out in which the substrates are replaced by an aliquot of distilled water. Most of the reaction product (diaminobenzidine brown) is formed in the gel overlay. Therefore the gel must not be removed. It is left to dry into a film and an assessment of total activity is possible.

If six different biopsies are placed on the slide, each acts as a control for the other, and an immediate impression of relative activities can be gained.

Low temperature gelling agarose is preferred since its gel point is below 30°C and it is fluid above this, thus allowing the heat-sensitive reagents and enzymes in the tissue sections less chance to degrade.

BRUSH BORDER α-GLUCOSIDASES (SUCRASE, ISOMALTASE AND GLUCOAMYLASE)

Preparation of tissue

Cryostat sections 5 μm fixed in cold 4°C chloroform-acetone (1:1) 5 min.

A. Indigogenic procedure

Incubation medium

5-Bromo-4-chloro-3-indolyl-α-D-glucoside	3 mg
Dissolve in N, N-dimethylformamide	0.3 ml
0.1 mol/l citric acid-phosphate buffer, pH 6	6 ml
1.65% potassium ferricyanide (50 mmol/l)	0.5 ml
2.11% potassium ferrocyanide (50 mmol/l)	0.5 ml
Mix thoroughly.	

Method

1. Incubate sections in the medium at 37°C 45 min
2. Rinse in distilled water
3. Fix in 4% formaldehyde at room temperature for 5 min

4. Counterstain with Kernechtrot
5. Wash, dehydrate, clear and mount.

Results

Enzyme-active sites are stained turquoise.

Notes

The medium can be used repeatedly. After incubation it is filtered and stored frozen in a closed vessel. This can be repeated at least 10 times. The medium is sufficient for 50 biopsies. Although the indoxyl substrate is cleaved by glucoamylase more efficiently than 6 Br-2-naphthyl-α-D-glucoside it is well suited for the asessment of sucrase-isomaltase in the human jejunal mucosa because sucrase-isomaltase is the main α-glucosidase of the brush border.

B. Azo-coupling method

Incubation medium

6-Br-2-naphthyl-α-D-glucoside	2 mg
or	
2-naphthyl-α-D-glucoside	5 mg
Dissolve in N, N-dimethylformamide	0.5 ml
Buffered hexazotized pararosaniline	10 ml

(made up of 9.7 ml 0.1 mol/l citric acid-phosphate buffer, pH 7, and 0.6 ml hexazotized pararosaniline; adjust pH to 6.5 with 1 mol/l and 0.1 mol/l NaOH)
Mix well with filter

Method

1. Incubate sections in the medium 2 h at room temperature or 15 h at 4°C
2. Rinse in distilled water
3. Place in 4% formaldehyde for several hours at room temperature
4. Rinse in tap water
5. If desired, counterstain nuclei with haematoxylin
6. Mount in glycerin jelly.

Results

Enzyme-active sites are stained orange-red.

Notes

The 2-naphthyl-derivative is cleaved more rapidly than the bromonaphthyl-derivative, is less expensive and more soluble. However, the bromonaphthyl-derivative is more easily obtainable.

In human jejunal biopsies both substrates are suitable for the assessment of sucrase/isomaltase deficiencies, although they are cleaved to some extent by glucoamylase.

ENTEROPEPTIDASE (ENTEROKINASE)

Preparation of tissue

Unfixed cryostat sections at 10 μm.

Indirect procedure

Preparation of solutions

1. 1% aqueous solution of low temperature gelling agarose (Sigma type VII), check pH with indicator paper and adjust to 6.5. Store at 4°C.
2. Incubation medium:

Benzyloxycarbonyl-Glycyl-Glycyl-Arginine-4-methoxy-2-naphthylamide	2 mg
or	
Benzoyl-L-arginine-2-naphthylamide (BANA)	4 mg
Dissolve in N, N-dimethylformamide	0.15 ml
0.1 mol/l Tris-maleate or Tris-HCl buffer, pH 6.5, with 0.1% calcium chloride	2 ml
Mix well	
Fast Blue B	5 mg

Mix thoroughly and filter
Add 2 mg trypsinogen and 2 ml 1% warm (37°C) agarose
Mix well.

Method

1. Cover sections mounted on slides with gel medium and allow to set; the solution is enough for 3 slides containing sections of several biopsies each
2. Incubate 2 h in a moist chamber at 37°C
3. Carefully place the gel-covered slides bearing the sections for 5 min in Petri dishes with 2% copper sulphate at room temperature (optional)
4. Wash 5 min in distilled water
5. Dry the underside of each slide, examine under the microscope with 10 × magnification, and where appropriate take pictures
6. For storage let gels dry into a film.

Results

Sites of reaction product in the section and in the overlying gel appear reddish or blue-violet.

Notes

It is necessary to perform controls with media without trypsinogen. Because generated trypsin can activate trypsinogen it is impossible to assess the enteropeptidase activity correctly. However, a similar condition is also in the gut 'in vivo'. The method reflects the activation of trypsinogen.

For better localization but less sensitivity use Gly-(Asp)$_4$-Lys-2-naphthylamide as substrate. This can be used in the presence of trypsin.

DIPEPTIDYL (AMINO) PEPTIDASE IV

Preparation of tissue

Cryostat sections (5 μm) fixed in cold (4°C) chloroform-acetone (1:1) 5 min

Azo-coupling method

Incubation medium

Glycyl-Proline-4-methoxy-2-naphthylamine	4 mg
Dissolve in N, N-dimethylformamide	0.5 ml
Buffered Fast Blue B	10 ml

   (made up of 10 mg Fast Blue B dissolved in 10 ml 0.1 mol/l phosphate or cacodylate buffer, pH 7.2–7.4) Mix well and filter.

Method

1. Incubate sections in the medium at room temperature 2 h
2. Rinse in distilled water
3. Place in 2% copper sulphate for 5 min
4. Rinse in distilled water
5. Place in 4% formaldehyde for several hours
6. Rinse in distilled water
7. Mount in glycerin jelly, Apathy's gum syrup or similar medium.

Results

Enzyme-active sites, capillaries, brush borders and T-lymphocytes (mainly T-helper) are stained deep red.

Notes

It is advisable to evaluate the reaction as soon as possible because of fading of the reaction product after several days. Sites with low activity become less discernible or cannot be seen at all.

TRYPTASE OF HUMAN MAST CELLS

Preparation of tissue

Cryostat sections of snap frozen samples; frozen sections of tissue fixed in cold formaldehyde; paraffin sections after short embedding of samples fixed in cold aldehyde fixatives or in Carnoy solution. Sections of unfixed samples are fixed in cold chloroform-acetone (1:1) 5 min.

Incubation medium

Benzyloxycarbonyl-Alanyl-Alanyl-Lysine-4-methoxy-2-naphthylmide	3 mg
or	
D-Valyl-Leucyl-Arg-4-methoxy-2-naphthylamide	3 mg
Dissolve in N,N-dimethylformamide	0.5 ml
Buffered Fast Blue B	10 ml

   (made up of 10 mg Fast Blue B supplied as double zinc salt or of 4 mg of Fast Blue B supplied as fluoroborate dissolved in 0.1 mol/l Tris-HCl or phosphate buffer pH 6.5)
Mix well and filter.

Method

1. Incubate sections in the medium 30–120 min at 37°C (depending on the pretreatment)
2. Rinse in distilled water
3. Place in 2% copper sulphate for 5 min
4. Rinse in distilled water
5. Place in 4% formaldehyde for several hours
6. Wash in tap water
7. Mount in glycerin jelly, Apathy's gum syrup or similar medium.

Results

Enzyme active-sites are stained deep red.

Notes

It is advisable to evaluate the reaction within several days because of fading. Cells with low activity become then less discernible.

NON-SPECIFIC ESTERASE (α-NAPHTHYL ACETATE METHOD) FOR GENERAL USE

Preparation of tissue

Fresh-frozen cryostat sections (5-10 μm) postfixed in acetone or formol-calcium for 2–5 min. Frozen sections of formol-calcium/gum sucrose-treated small blocks.

Incubation medium

α-Naphthyl acetate	2.5 mg
(dissolved in 2-methoxyethanol)	0.2 ml
0.2 mol/l disodium hydrogen phosphate	7.5 ml
Distilled water	2.5 ml
Hexazotized pararosaniline	0.8 ml

Adjust pH to 6.5 with 0.2 mol/l disodium hydrogen phosphate or 1 mol/l HCl.

Method

1. Incubate sections at room temperature for 1–20 min (15 min is a general average)
2. Wash well in tap water
3. Counterstain nuclei with methyl green or Carrazzi's haematoxylin
4. Wash
5. Mount in glycerine jelly or dehydrate, clear and mount in DPX.

Results

Suites of esterase activity are coloured brown.

α-NAPHTHYL ACETATE ESTERASE (ANAE) (FOR HAEMATOLOGY)

Preparation of tissue

Air-dried films of blood or marrow; unfixed films may be stored at room temperature for up to 2 weeks without appreciable loss of enzyme activity.

Incubation medium

α-Naphthyl acetate dissolved in 2.5 ml 2-methoxyethanol	50 mg

67 mmol/l phosphate buffer pH 7.6	44.5 ml
Hexazotized pararosaniline	3.0 ml

Adjust pH to 6.1, filter.

Method

1. Fix in buffered formalin acetone (p. 436) at 4–10°C for 30 s
2. Wash in distilled water 3 times
3. Incubate for 45 min at room temperature
4. Rinse with distilled water 3 times
5. Counterstain with methyl green for 15 min
6. Wash with tap water, dry and mount with DPX.

Positive result

Dark brown-red reaction product.

Note

For testing fluoride sensitivity, add 1.5 mg sodium fluoride per ml of phosphate buffer.

ACID ESTERASE IN BLOOD FILMS (FOR WOLMAN'S DISEASE)

Preparation of tissue

Air-dried blood films. Fix for 10 min in formol-calcium.

Incubation medium

α-Naphthyl acetate (dissolved in 0.2 ml 2-methoxy ethanol)	12.5 mg
0.05 mol/l phosphate buffer pH 7.4	50 ml
Hexazotized pararosaniline	3 ml

Adjust to pH 5.8 with 1 mol/l NaOH.

Method

1. Rinse fixed films in tap water
2. Incubate for up to 4 h at 37°C
3. Wash in tap water
4. Counterstain with Carrazzi's haematoxylin
5. Wash, dehydrate, clear and mount in DPX.

Result

Enzyme activity is shown by a dark reddish-brown colour.

Notes

Lymphocytes (T lymphocytes) usually have one or two small dense spots of activity. No activity is seen in neutrophils. Monocytes (non-specific esterase) are densely stained.

The inhibitor E600 is not necessary for blood film studies because no non-specific esterase activity is present in lymphocytes, and monocytes are easily recognizable by their intense activity.

ACID ESTERASE IN TISSUES

Preparation of tissue

Frozen sections (5-10 μm) of formal-calcium gum-sucrose treated tissue give the most reliable results.

Incubation medium

α-Naphthyl acetate (in 0.2 ml
   2-methoxyethanol)   2.5 mg
0.05 MOL/L phosphate buffer pH 5.0   10 ml
Hexazotized pararosaniline   0.6 ml

Adjust to pH 5.0 with 1 mol/l NaOH and add 0.1 ml 10^{-2} mol/l E600 in 0.1 mol/l phosphate buffer pH 5.0.

Method

1. Treat free-floating sections in 10^{-4} mol/l E600 in 0.1 mol/l phosphate buffer pH 5.0 at 37°C for 30 min
2. Transfer sections to the incubating medium for 30 min at 37°C
3. Wash, mount on slides, counterstain with Carrazzi's haematoxylin, wash, dehydrate, clear and mount in DPX.

Result

Acid esterase activity is shown by a reddish-brown stain.

Notes

The inhibitor E600 is used to inhibit the non-specific esterases which are very active in liver and other tissues and which would mask acid esterase activity unless inhibited. Fixed tissue is essential because acid esterase is labile and its activity cannot be demonstrated reliably in cryostat sections of snap-frozen tissue.

It only snap-frozen tissue is available, the use of a semipermeable membrane technique (using 5 ml 1% low temperature gelling agarose — Sigma type VII — and 5 ml 0.1 mol/l phosphate buffer pH 5.0 in the medium instead of 10 ml 0.5 mol/l phosphate buffer pH 5.0) can give informative results.

CHLOROACETATE ESTERASE (CAE) FOR HEAMATOLOGY AND GENERAL USE

Prepation of tissue

Aid-dried films of blood or marrow; unfixed films may be stored at room temperature for up to 2 weeks without appreciable loss of enzyme activity. Paraffin sections; frozen sections of fixed tissue; postfixed cryostat sections.

Incubation medium

67 mmol/l phosphate buffer pH 7.6   47.5 ml
Hexazotized pararosaniline   0.24 ml
Naphthol AS-D chloroacetate   5 mg
   (dissolved in 2.5 ml N,N-dimethylformamide)
   Adjust pH to 7.6.

Method

1. Fix films in buffered formalin acetone (p. 436) at 4–10°C for 30 s
2. Wash in distilled water 3 times
3. Incubate away from direct light, at room temperature for 10 min
4. Wash with tap water
5. Counterstain with methyl green for 15 min
6. Wash with tap water, dry and mount with DPX.

Positive result

Bright orange-red reaction product.

Notes

This medium can also be used on routine sections of fixed, wax-embedded tissue to show mast cells (including mucosal mast cells) and granulocytes. Incubate for up to 2 h; counterstain with Carrazzi's haematoxylin.

DUAL ESTERASE (α-NAPHTHYL ACETATE ESTERASE ANAE: CHLOROACETATE ESTERASE, CAE) FOR HAEMATOLOGY

Preparation of tissue

Air-dried films of blood or marrow; unfixed films may be stored at room temperature for up to 2 weeks without appreciable loss of enzyme activity.

Preparation of solutions

CAE medium:

67 mmol/l phosphate buffer pH 7.6	9.5 ml
Naphthol AS-D chloroacetate	1 mg
in N,N-dimethylformamide	0.5 ml
Fast Blue BB	5 mg

Method

1. Fix, wash and incubate in ANAE medium for haematology (p. 470)
2. Wash with distilled water 3 times
3. Filter CAE medium (above) onto slides and leave 20 min at room temperature
4. Wash with distilled water 3 times
5. Counterstain with methyl green 15 min
6. Wash with tap water, dry and mount with DPX.

Positive result

ANAE activity dark red-brown; CAE activity blue.

β-GALACTOSIDASE

Preparation of tissue

Air-dried cryostat sections (5–8 μm) of snap-frozen tissue; frozen sections of formol-calcium gum-sucrose treated tissue; air dried bone marrow films.

Incubation medium

5-Bromo-4-chloro-3-indolyl-β-galactoside	3 mg
dissolved in 2 drops 2-methoxyethanol	
McIlvaine buffer pH4.0	7 ml
50 mmol/l (2.11%) Potassium ferrocyanide	0.5 ml
50 mmol/l (1.65%) Potassium ferricyanide	0.5 ml
Sodium chloride	47 mg

Method

1. Cover sections with medium (use small volumes and prevent drying out) and incubate at 37°C for 5–18 h
2. Rinse in water
3. Counterstain with Kernechtrot
4. Rinse, dehydrate, clear and mount in DPX.

Result

Enzyme activity is blue. Nuclei are red.

Notes

Blood films often have a layer of uncoloured precipitate over them after incubation. This can be removed, without damaging the film, by a jet of water from the wash bottle.

The enzyme is stable in dried films and its activity can be shown in films several years old. Neutrophils have diffuse cytoplasmic activity; lymphocytes have one or two dense spots of activity; eosinophils have granular activity; platelets are also positive.

HEXOSAMINIDASE (β-GLUCOSAMINIDASE)

Preparation of tissue

Air-dried blood films; postfixed cryostat sections of snap-frozen tissue; frozen sections of formol calcium-gum sucrose treated tissue; air-dried monolayers of cultured fibroblasts on cover slips.

Incubation medium

2.5 mg Naphthol AS-BI-β-D-glucosaminide dissolved in 0.1 ml 2-methoxyethanol
5 ml Distilled water
5 ml McIlvaine buffer pH 5.0
0.5 ml Hexazotized pararosaniline
   Adjust pH to 4.1 (for maximum activity)
   Adjust pH to 5.0 (for best localization).

Method

1. Incubate lightly fixed films (5 min in formol-calcium) or sections for 1–4 h at 37°C
2. Wash in tap water
3. Counterstain in Carrazzi's haematoxylin
4. Wash (mount), dehydrate, clear and mount in DPX.

Result

A red colour indicates enzyme activity.

Notes

This method is of use only in testing for the total deficiency in Sandhoff's disease and cannot be used to detect Tay–Sachs disease.

β-GLUCURONIDASE

Preparation of tissue

Air-dried films of blood or marrow; treat preferably within 4 h. Heparin and citrate are inhibitory. EDTA should be less than 5 mg/ml.

Cryostat sections (10 μm) unfixed; frozen sections of formol-calcium gum sucrose treated tissue.

Incubation medium

Naphthol AS-BI β-D-glucuronic acid (dissolved in 1–2 drops N, N-dimethyl formamide)	2.5 mg
0.1 mol/l Acetate buffer pH 5.0	10 ml
Hexazotized pararosaniline (p. 459)	0.5 ml

Adjust to pH 5.2; filter if necessary.

Method

1. Fix blood and bone marrow films in buffered formalin acetone (p. 459) at 4–10°C 30 s; or in formol-calcium for 1–2 min
2. Wash in distilled water 3 times
3. Incubate for 90 min at 37°C (2–3 h for tissue sections)
4. Wash well in distilled water
5. Counterstain with methyl green for 15 min or Carrazzi haematoxylin for 2 min
6. Rinse in tap water, air-dry and mount with DPX.

Positive result

Red reaction product.

α-MANNOSIDASE FOR LANGERHANS CELLS, LANGERHANS CELL HISTIOCYTOSIS AND MELANOMA

Preparation of tissue

Cryostat sections (10 μm) mounted on a semipermeable membrane.

Preparation of medium

2.5 mg α-Naphthyl — α-mannoside dissolved in 0.1 ml 2-methoxyethanol
5 ml 1% Low temperature gelling agarose (Sigma type VII)

5 ml McIlvaine (citrate-phosphate) buffer pH 5.0
0.5 ml Hexazotized pararosaniline
Adjust pH to 5.5.

Incubate for 4 hours at 37°C or overnight at room temperature, cut out membrane, counterstain with Carrazzi's haematoxylin, dehydrate, clear and mount in DPX.

Activity is shown by a reddish brown colour with yellow background.

Note

Similar, but less intense activity can be shown in unfixed cryostat sections on cover slips incubated in the above medium, substituting distilled water for the agarose and adding 500 mg polyethylene glycol 6000. This latter method is ideal for marrow aspirates.

DOPA-OXIDASE REACTION

Preparation of tissue

Cryostat or frozen sections of unfixed tissue or of tissue fixed in cold formalin for 12–24 h.

Method

1. Incubate 10 μm serial sections for 3 h at 37°C in the following:
   a) 0.1 mol/l phosphate buffer, pH 7.4
   b) 0.1 mol/l phosphate buffer, pH 7.4 10 ml containing 2 mg of tyrosine and 0.2 mg DL-DOPA
   c) As in b), but also containing 1 mmol/l diethyldithiocarbamate (=38.7 mg/100 ml)
2. Rinse in water, counterstain with Mayer's haemalum or Kernechtrot, dehydrate, clear and mount.

Results

Pigment formed as a result of the enzymic activity appears brown-black in section b). By comparing with sections a) and c) this pigment can be distinguished from pre-existing melanin.

Note

Presence of the reaction indicates melanogenetic capacity of cells and is useful, when present, in identifying naevus and melanoma cells. Mast cells are also stained by this procedure and are often difficult to distinguish from melanocytes. The perivascular location of mast cells and metachromasia of their granules can be used for definitive diagnosis.

Diethyldithiocarbamate is added (lc) to chelate the copper on which DOPA-oxidase is dependent.

CATALASE FOR THE DEMONSTRATION OF HUMAN PEROXISOMES

Preparation of tissue

Fix thin (1 mm) slices or portions of needle biopsy specimens of liver at room temperature for 24 h in 4% formaldehyde in 0.1 mol/l cacodylate buffer pH 7.3 containing 1% calcium chloride, dihydrate. Without rinsing, snap-freeze tissue. Cut 5–7 μm sections and wash (free floating) in 0.1 mol/l cacodylate buffer pH 7.3 containing 1% clacium chloride, for light microscopy. For EM cut 40 μm sections into the same buffer

Incubation medium

Solution A

Diaminobenzidine dihydrochloride 20 mg in a few drops of water

Add 10 ml 0.1 mol/l 2-amino-2-methyl-3-propandiol containing 7.5% sucrose and 1 mg KCN
The pH is 9.3–9.4 and should need no adjustment

Solution B

3% hydrogen peroxide

Method

1. Transfer sections into solution A preheated to 45°C, with continuous agitation (rotary mixer in the incubator)
2. Add 0.2 ml solution B
3. Incubate for 1 h at 45°C with continuous agitation
4. Wash in water
5. Treat with 1% osmium tetroxide
6. Dehydrate, clear and mount for light microscopy, or process into resin for electron microscopy.

Results

Peroxisomes, red blood cells and granulocytes are dark brown to black (and electron dense).

Appendix 6

Immunohistochemical methods

TISSUE PREPARATION

Fresh frozen tissue sections

Mount on chrome-gelatine- or poly-L-lysine-coated slides. Air-dry at room temperature for at least 1 h (better overnight).

Storage

Wrap slides in foil or cling-film and seal in a plastic bag or container with desiccant (e.g. silica gel). Store at $-20°C$ indefinitely. Allow entire bag/container to reach room temperature before opening to avoid condensation of atmospheric water on the sections.

Fixation

Fix as required, e.g. 10–30 min in acetone at room temperature. After fixation in non-aqueous fixatives, sections may be allowed to dry or immersed in buffer. Once sections have become wet they must not be allowed to dry at all throughout the procedure or poor structure and staining will result.

Cytospin preparations, cell cultures and smears

Treat as frozen sections. Culture medium should be well washed off the preparation before fixation.

Permeabilization

Alcohol, acetone or acetone/chloroform fixation is usually adequate, but after formalin fixation whole cell preparations may require permeabilization of cell membranes if internal antigens are to be immunostained. If cell surface antigens are to be localized, permeabilization is not necessary.

After fixation the preparation may be treated for 30 min with buffer containing detergent (e.g. 0.2% Triton X-100), then rinsed in buffer without detergent to avoid dispersal of subsequently applied immune reagents over the entire surface of the slide. Alternative ways of permeabilizing the preparation are to freeze and thaw it several times (if insoluble antigens are to be localized), or to treat with acetone after the formalin fixation.

Prefixed frozen sections

After formalin or other fixation (e.g. *p*-benzoquinone for immunostaining of peptides), soak the tissue blocks in buffer containing 15% sucrose. For long-term storage add 0.1% sodium azide. The sucrose acts as a cryoprotectant and makes the tissue easier to section. Mount 5–40 μm sections on coated slides and allow to dry well before use, or use free-floating (40–80 μm sections). These preparations may need permeabilization (see below).

Prefixed Vibratome sections

Use free-floating (40–80) μm. These preparations may need permeabilization. A useful method for these and other whole-mount preparations, provided the antigens withstand the treatment, is to dehydrate through graded alcohols to a solvent such as xylene, and then rehydrate. In addition, detergent (0.1% Triton X-100) may be added to the buffer rinses in the immunostaining schedule and/or to the antibody diluent.

Paraffin sections

Most routine fixation procedures will give adequate results, although for some antigens particular fixatives are preferred. For details see main text.

Cut at 2–4 μm. Pick up from warm water in the usual way on poly-L-lysine-coated slides and dry (preferably overnight) in an oven at 37°–45°C. Sections should not be baked on a hot-plate as this destroys some antigens. Sections can usually be stored indefinitely at room temperature although it is possible that some antigens will deteriorate with time. If this occurs, sections freshly cut from the block should be used.

Poly-L-lysine coating of slides

Poly-L-lysine hydrobromide, mol.wt 150 000–300 000 (Sigma P-1399) is used in a 1 g/l aqueous solution. Aliquots of 1 ml are stored frozen at -20°C. For use, thaw an aliquot and mix the solution thoroughly. Place a small drop (about 10 μl) at one end of a clean microscope slide. Spread over the surface with another slide in the same way as making a blood smear. The layer of poly-L-lysine should be even and thin enough for interference colours to be seen as it is spread. It will dry almost at once and the slide is then ready for use. The coated side should be marked as the film is invisible. A batch of slides can be prepared in advance and stored for at least a week before being used. Coating solution remaining in the vial may be refrozen.

BUFFERS

0.01 mol/l phosphate-buffered saline, pH 7.0–7.2 or 0.05 mol/l Tris-buffered saline, pH 7.6 may be used.

Phosphate-buffered saline (PBS)

It is convenient to make a large quantity of 0.1 mol/l PBS and dilute it as required. The salts are dissolved separately (disodium hydrogen orthophosphate may require gentle heat) then mixed together and made up to the correct volume. The pH of the stock 0.1 mol/l solution is checked after dilution to 0.01 mol/l.

For 10 l of 0.1 mol/l PBS:

Sodium chloride	870.0 g
Potassium dihydrogen phosphate	27.2 g
Disodium hydrogen orthophosphate	113.6 g
(141.0 g of the dihydrate)	

After dilution to 0.01 mol/l the pH should be 7.0–7.2

Tris-buffered saline (TBS)

This is conveniently made as 0.5 mol/l and diluted to 0.05 mol/l for use. The pH does not change on dilution. For 10 l of 0.2 mol/l TBS

Sodium chloride	870 g
[Tris (hydroxymethyl) aminomethane]	605 g

Dissolve in distilled water. Adjust pH to 7.6 with concentrated hydrochloric acid and make up to 10 l with distilled water.

Antibody diluent for general purposes

PBS (TBS) containing 0.1% bovine serum albumin and 0.1% sodium azide. Do not use azide when diluting peroxidase-linked reagents as it inhibits the enzyme action. Poly-L-lysine, 2 mg/ml, and 2.5% sodium chloride (see above) may be included as a further precaution against non-specific binding.

BLOCKING ENDOGENOUS ENZYMES

Peroxidase

Blocking is usually carried out before application of the primary antibody, but if the procedure is found to damage the antigen, the blocking treatment can be carried out at any stage prior to application of a peroxidase-linked reagent.

Paraffin sections

After dewaxing and hydrating through graded alcohols to water, immerse sections for 30 min in 0.3% hydrogen peroxide in water (or buffer or methanol) (1 ml of 30% hydrogen peroxide in 100 ml of water freshly prepared from stock 30% solution). Rinse in buffer before immunostaining.

Cryostat sections, cell cultures, smears

After fixation, immerse preparations for 1 h at 37°C in PBS containing:

10 mmol/l D-glucose
1 mmol/l sodium azide
100 units of glucose oxidase (Sigma, G-6891) per 100 ml added just before use.

Alkaline phosphatase

Blocking is not necessary for paraffin sections. For cryostat sections or other fresh material 1 mmol/l levamisole (Sigma L-9756) (10 mg in 4 ml) is added to the enzyme incubation medium. Note that this does not inhibit the brush border alkaline phosphatase of small intestine.

PROTEASE DIGESTION

Trypsin is a good general protease, but pronase or protease XXIV (Sigma) may also be useful. Protease treatment may be carried out before or after endogenous peroxidase is blocked.

Trypsin digestion

Each batch of enzyme must be tested to establish a standard time of incubation, usually between 5 and 15 min. The solution must be freshly made.

0.1% trypsin (Sigma T-8128) is added to 0.1% calcium chloride (stored at 37°C). The enzyme will not be completely dissolved and the pH will be about 5.0. Sodium hydroxide (0.1 mol/l) is added dropwise to bring the pH to 7.8 and clear the solution. Sections are incubated at 37°C for the established time, then washed well with distilled water and rinsed in buffer.

Protease XXIV (formerly protease VII) digestion

Protease (Sigma P-8038) is used as an 0.05% solution in PBS. As this enzyme is considerably more expensive than trypsin it is suitably used as drops applied to the preparations. Incubation is for 5–20 minutes at 37°C.

Neuraminidase

1 unit neuraminidase (type V, Sigma N-2876) in 10 ml 0.1 mol/l sodium acetate buffer, pH 5.5, containing 0.01% calcium chloride.

Incubate preparations at 37°C for 5–20 min.

BACKGROUND BLOCKING

Non-specific binding of immunoglobulin

Before applying the primary antibody incubate the preparation for a minimum of 10 minutes with non-immune serum a) from the species providing the second antibody in an indirect or three-layer, or b) the primary antibody in a direct (one layer) technique. The serum is diluted 3–20% in buffer and is then drained off, but the preparation is not rinsed before the primary antibody is applied.

If labelled Protein A, which binds to the Fc portion of immunoglobulins, is the detecting reagent, blocking with serum should not be carried out as it will provide extra non-specific binding sites.

For paraffin sections in which Fc receptors have been destroyed in the processing, non-immune serum may be replaced by chicken egg albumen (5.0%) or casein to block non-specific protein attachment sites. Albumen should not be used with biotinylated reagents as it contains avidin and thus may increase the background binding.

Eliminating high levels of background staining

Aldehyde groups and strong endogenous peroxidase (catalase). The following schedule is often helpful, but may destroy some antigens (e.g. leucocyte common antigen):

1. *Bleach acid haematin* with 6% hydrogen peroxide 5 min
2. Rinse in water
3. *Block endogenous peroxidase* with 2.5% periodic acid 5 min
4. Rinse in water
5. *Block aldehyde groups* with freshly made 0.2% sodium or potassium borohydride 2 min
6. Rinse in water.

Charged sites

Dilute the reagents in buffer containing 2 mg/ml poly-L-lysine, mol.wt 3800, or with a high content (2.5%) of sodium chloride, raise the pH of the buffer to 9.0 and/or add 0.2% of detergent to the rinsing buffer.

Endogenous biotin

Apply unlabelled avidin for 30 min
Rinse, and apply unlabelled biotin for 30 min.

Crossreacting anti-immunoglobulins

Add to the anti-immunoglobulin (second layer) 5% of serum from the species to be immunostained.

ENZYME DEVELOPMENT

Peroxidase

Diaminobenzidine method

25–50 mg 3,3'-diaminobenzidine tetrahydrochloride 2 H_2O (DAB) (Aldrich Chemical Co.) in 100 ml PBS (TBS). Just before use, add 50–100 μl of 30% hydrogen peroxide solution (0.015–0.03% final concentration).

Develop preparations for 5–10 minutes, checking microscopically. The reaction product should be dark brown and the background clear. When the reaction is complete, rinse the preparations well in running water, then counterstain lightly in Carrazzi or Meyer haematoxylin, and blue in running tap water. Dehydrate, clear and mount in permanent medium.

Safe storage of DAB. Buy DAB in 5 g quantities. Make an aqueous solution of the entire amount in 200 ml of distilled water (25 mg per ml). Stir in a fume cupboard with a magnetic stirrer for about 15 min to dissolve the powder. Dispense into vials in suitable aliquots for one development (e.g. 1 ml of concentrate will be adequate for 100 ml of peroxidase development medium). Freeze and store at -20°C. The solution may be thawed and refrozen without deteriorating.

Safe use of DAB. Wear gloves when using DAB. Restrict its use to one area of the laboratory, preferably a fume cupboard, and wipe up spills immediately using bleach (see below).

Safe disposal of DAB. Add a few ml of household bleach (sodium hypochlorite) to the DAB then pour it down the sink, preferably in a fume cupboard, with plenty of running water. Wash the development dish well with running water to remove all traces of bleach which might contaminate the next DAB solution to be made in it.

Amino ethyl carbazole method

Care should be taken as for DAB when using this potential carcinogen.

Stock solution:	3-amino-9-ethylcarbazole, 4% in dimethyl formamide (stable at room temperature)
Incubating solution:	0.5 ml stock solution in 9.5 ml 0.05 mol/l acetate buffer, pH 5.0 10 μl 30% hydrogen peroxide.

Filter onto preparations. Incubate for 5–10 minutes at room temperature. The reaction-product is red to reddish brown and is alcohol-soluble. Counterstain with Carrazzi of Meyer haematoxylin, wash in water and mount in glycerine jelly or Hydromount (National Diagnostics).

Alkaline phosphatase

To inhibit endogenous alkaline phosphatase add 1 mmol/l levamisole to the incubating medium (see above). Naphthol phosphate solutions may be made up in a concentrated form (e.g. 1 mg per ml of buffer) and stored frozen in aliquots until needed.

For a blue end-product (alcohol-soluble)

2 mg Naphthol AS-MX phosphate (Sigma N-5000 or N-4875) dissolved in 0.2 ml dimethyl formamide. Add 9.8 ml 0.1 mol/l Tris/HCl buffer, pH 8.2. Just before use add Fast Blue BB (Sigma F-3378) at 1 mg/ml and filter the yellow solution onto the preparations. Incubate 10–15 minutes at room temperature or 37°C, checking microscopically for a bright blue reaction product. Wash in water, counterstain in Kernechtrot, mount in an aqueous medium.

For a red end-product (alcohol-soluble)

Substitute Fast Red TR (Sigma F-1500) for the Fast Blue BB.

For a red end-product (less soluble in alcohol)

2 mg Naphthol AS-TR phosphate (Sigma N-6125 N-or 6000) dissolved in 40 μl dimethylformamide. Add 8 ml 0.2 mol/l Tris/HCl buffer, pH 9.2. Just before use add 100 μl hexazotized New Fuchsin solution, freshly made as for HPR (p. 459).

Filter onto the preparations. Incubate at room temperature or 37°C for 5–15 min. Wash sections in water. Counterstain with Carrazzi or Meyer haematoxylin. Mount in an aqueous medium (fix the preparations in formalin for a few hours before counterstaining to reduce bubble formation in the mountant), or dry the preparations thoroughly at 37°C then dip briefly into xylene and mount in a permanent medium.

For a permanent blue-brown end-product

Use the BCIP-NBT method (Appendix 5).

IMMUNOSTAINING PROCEDURE — GENERAL NOTES

Incubating chamber

Slide-mounted preparations are placed on a rack in a covered Petri-dish or other container. The atmosphere is kept moist by including wet tissue or cottonwool.

Application of reagents

Immune reagents in antibody diluent or in simple PBS or TBS are applied as drops to cover the preparation. Between layers the slides are rinsed as below and wiped dry except for the area of the tissue preparation which must be kept moist with buffer. The drop of reagent will then spread easily over the tissue and remain confined to that area and drying artefact will be avoided.

Rinsing

Preparations are rinsed from a wash bottle or immersed in three changes of buffer of 5 min each. It is not necessary to agitate the buffer.

IMMUNOSTAINING METHODS

Paraffin sections are dewaxed and brought to water. Treatment with a protease is carried out if required. Cryostat sections, cytospins, cultures or other types of preparation are suitably fixed. Endogenous peroxidase is blocked at the start of the immunoperoxidase procedure (see above), the preparations are rinsed in buffer and the slides are then dried, except for the area of the tissue.

Direct method

1. Blocking serum (from species supplying labelled antibody) 10 minutes (or longer)
2. Drain off serum and apply suitably diluted, labelled primary antibody 1 h or longer at room temperature
3. Rinse in three changes of buffer
4. Fluorescence preparations: mount in aqueous non-fluorescent mountant, e.g. Hydromount or buffered glycerine (9 parts glycerine to 1 part PBS).

Enzyme-labelled preparations: develop enzyme, counterstain nuclei, mount in aqueous medium or dehydrate, clear and mount in permanent medium.

Indirect method

1. Blocking serum from species supplying second (labelled) antibody, 10 min (or longer)
2. Drain off and apply suitably diluted, unlabelled primary antibody, 1 h or longer (preferably overnight 4°C) for polyclonal antibodies
3. Rinse in three changes of buffer
4. Apply second (labelled) antibody, suitably diluted, 1 h at room temperature
5. Rinse in 3 changes of buffer
6. Fluorescence preparations: mount in aqueous medium.

Enzyme-labelled preparations: develop enzyme, counterstain nuclei, and mount appropriately.

Peroxidase antiperoxidase (PAP) method

1. Blocking serum from species providing second (unlabelled) antibody, 10 min (or longer)
2. Drain off and apply primary antibody, suitably diluted, overnight at 4°C
3. Rinse in three changes of buffer
4. Apply second (unlabelled) antibody, suitably diluted, 30 min at room temperature
5. Rinse in three changes of buffer
6. Apply PAP complex suitably diluted, 30 min at room temperature
7. Rinse in three changes of buffer
8. Develop enzyme, counterstain nuclei, and mount.

The procedure for alkaline phosphatase anti-alkaline phosphatase (APAAP) is similar to the above. If the primary antibody is a mouse monoclonal antibody, the second antibody is an unlabelled anti-mouse immunoglobulin followed by mouse PAP or APAAP.

A more intense reaction product can be obtained by repeating steps 4–7 for 10 minutes each layer before developing the enzyme, but background staining will also be accentuated.

Avidin-biotin methods

1. Background blocking: do not block with albumen. If necessary, use serum or irrelevant immunoglobulins from the species providing the biotinylated second layer antibody (or primary antibody in a direct method). Avidin binding may be blocked with avidin 1 mg/ml for 20 min followed by rinsing, then biotin 0.1 mg/ml for 20 min and rinsing
2. Primary antibody, biotinylated in a direct method, unlabelled in an indirect method
3. Rinse in three changes of buffer
4. Direct method: labelled avidin (fluorescent or enzyme label)
   Indirect method: biotinylated second antibody, 30 min
5. ABC method: make the avidin-biotin complex by mixing biotinylated label (e.g. biotinylated peroxidase) with avidin in the proportion recommended by the supplier. The complex should be made about 30 min before use, but should not stand too long or the complex will become too large and steric hindrance will prevent full binding to the biotinylated second antibody
6. Rinse in three changes of buffer
7. Labelled ABC or labelled avidin, 30 min
8. Rinse in three changes of buffer
9. Develop enzyme label and process to mountant.

ENHANCEMENT OF IMMUNOPEROXIDASE DIAMINOBENZIDINE REACTION PRODUCT

Simultaneous intensification

Imidazole

To darken the end product of reaction, a solution (0.1 mol/1) of imidazole (Sigma, I-0250) is brought to the buffer pH with 1 mol/l HCl and added to the DAB development solution to a concentration of 0.01 mol/l.

Heavy metal intensification

Cobalt chloride method

Make the DAB solution using 50 mg DAB per 100 ml of buffer. Add 0.5 ml of 1% cobalt chloride with stirring.

Incubate the preparations in this solution for 3–4 min then mix in 10 µl of 30% hydrogen peroxide and continue the incubation for a further 3–5 min, checking microscopically. The reaction should be dark blue-black and the background clear.

Nickel ammonium sulphate and nascent hydrogen peroxide method

This method provides a gentle and continuous release of hydrogen peroxide. Rinse preparations in 0.1 mol/1 acetate buffer, pH 6.0. Incubate in the following for about 20 min:

A. Nickel ammonium sulphate		2.5 g
0.2 mol/1 Acetate buffer, pH 6.0		50 ml
B. DAB		30–50 mg
Distilled water		50 ml

Just before use, mix A and B and add

β-D-Glucose 200 mg	200 mg
Ammonium chloride	40 mg
Glucose oxidase	0.5–1 mg
(Sigma, type V S, G-6891)	

Rinse in acetate buffer. The end-product is blue-black and the background clear.

Postdevelopment intensification

Osmium tetroxide-potassium ferrocyanide method

After development with DAB treat preparations with 5% osmium tetroxide for 30 min (in a fume cupboard — osmium vapour is toxic). Reaction product will become grey-brown.

Wash well in distilled water

Treat with the following for 15 min:

2% potassium ferrocyanide	
(freshly prepared)	20 ml
1% hydrochloric acid	80 ml

Wash well in distilled water.

The osmicated reaction product will become dark brown-black.

Silver intensification

After development with DAB rinse the preparations well in distilled water, then immerse in the following solution at 60°C for 5–10 min, checking microscopically.

To 25 ml of stock solution (0.125% silver nitrate in 1.5% hexarnine stored at 4°C add 25 ml of distilled water and 2 ml of 5% sodium tetraborate (borax). Heat to 60°C in a water bath, protected from light.

The reaction product will become dark brown-black. Wash in distilled water. The preparation may be 'toned' with 0.2% gold chloride to reduce intensity if desired.

Fix in 5% sodium thiosulphate for 30 s wash, dehydrate, clear and mount.

Suppression of tissue argyrophilia

All methods using silver salts, including the IGSS method, run the risk of silver deposition on argyrophilic elements of the tissue. This can be avoided by suppressing the argyrophilic reaction before treating the preparations with silver solutions. The simplest method uses copper sulphate and hydrogen peroxide.

1. After the DAB reaction, rinse the preparations in two changes of 1% acetic acid, 1 min each
2. Treat with 10 mmol/l copper sulphate for 10 min
3. Wash with two changes of distilled water, 1 min each
4. Treat with 3% hydrogen peroxide in 1% sodium acetate. 3 H$_2$O, 10–20 min
5. Rinse in two changes of 1% sodium acetate, 1 min each.

Rinse well in several changes of distilled water before proceeding to silver intensification.

IMMUNOGOLD STAINING WITH SILVER ENHANCEMENT (IGSS)

1. Treat preparations with Lugol's iodine (1% iodine in 2% potassium iodide) for 5 min. *This oxidation step*

is essential, regardless of whether the fixative contained mercury.

2. Rinse in tap water
3. Bleach in sodium thiosulphate (2.5–5%)
4. Wash well in tap water
5. Wash in IGSS buffer I, 2 changes of 5 min each. IGSS buffer I: 0.05 mol/l Tris/HCl buffer, pH 7.4, containing 2.5% sodium chloride and 0.5% Tween 80 or 0.2% Triton X-100
6. Treat with undiluted goat serum (or serum from other species providing second layer antibody) for 15 min
7. Drain off serum but do not rinse preparations
8. Apply primary antibody suitably diluted for 90 min. The diluent is TBS containing 0.1% bovine serum albumin and 0.1% sodium azide. The dilution must be determined by titration but is usually about that used overnight for a PAP method.
9. Rinse in 2 ×10 min changes of IGSS buffer I
10. Immerse in IGSS buffer II for 10 min. IGSS buffer II: 0.05 mol/l Tris/HCl buffer, pH8.2, containing 0.9% sodium chloride
11. Apply undiluted goat serum for 10 min
12. Incubate in gold-adsorbed second antibody suitably diluted for 60 min. The diluent is IGSS buffer II containing 0.8% bovine serum albumin. Dilution is determined by titration.
13. Rinse in 3×5 min changes of IGSS buffer II
14. Rinse briefly in several changes of distilled water then in 4 ×5 min changes of distilled water. Thorough washing in very pure distilled water is essential to remove all traces of halide before the silver development.
15. The silver development solution is prepared shortly before use. Preparations are incubated in the dark (container wrapped in foil or put in a cupboard) or under photographic safelight for 4–8 min. Microscopic checks can be made on well rinsed preparations.

Silver development solution
Use scrupulously clean glassware (as for all silver methods) and highest purity water. Protect the silver lactate solution from light and add it just before use.

A. Citrate buffer:

Trisodium citrate dihydrate	2.35 g
Citric acid	2.55 g
Distlled water	85.0 ml

The pH should be about 3.5

B. Hydroquinone   0.85 g
Dissolve in solution A

C. Silver lactate	0.11 g
Distilled water	15 ml

Gum arabic may be included to slow down the reaction and help prevent non-specific deposition of silver grains. Use 60 ml of a 500 g/l solution in place of 60 ml of water.

Alternative method of enhancement with silver acetate

This method has the advantage of being light-insensitive and may produce less background staining than the silver lactate method. It is important to use gold particles of the smallest possible size.

Enhancement medium

A. Hydroquinone	0.250 g
Citrate buffer (as above), adjusted to pH 3.8 approx.	50.0 ml
B. Just before use add Silver acetate (from FLUKA — a powdered form; others are hard to dissolve) freshly dissolved in:	0.1 g
Distilled water	50.0 ml

Incubate for a few minutes, then check microscopically. Continue to incubate (under the microscope if you wish) until the desired intensity is reached.

If slower development is required, solution A should consist of

Hydroquinone	0.500 g
Citrate buffer	50.0 ml
Gum arabic (500 g/l)	10.0–40.0 ml

Kit developers are also available (e.g. Janssen's Inten-SE), which are not light-sensitive.

16. Rinse briefly in distilled water then fix in photographic fixer (diluted 1:4) or 5% sodium thiosulphate for 2 min
17. Wash in running tap water for 10 min. (Use warm water [40°c] to remove gum arabic). If staining is too weak, repeat steps 14–16 with fresh silver development solution with a shorter development time to allow further deposition of silver
18. Counterstain as required (e.g. haematoxylin and eosin)
19. Dehydrate, clear and mount.

MULTIPLE IMMUNOSTAINING

Double immunoenzymatic staining

Two separate antigens are localized simultaneously using primary antibodies raised in different species (e.g. mouse and rabbit). They are developed with species-specific non-crossreacting second antibodies (e.g. goat anti-mouse and goat anti-rabbit immunoglobulins), which may be conjugated to peroxidase and alkaline phosphatase respectively or used unconjugated and followed by a third layer mixture of PAP and APAAP. The enzymes are then developed separately.

Sample schedule (PAP and APAAP)

1. Block background using albumen or serum from same species as second layer antibodies
2. Apply mixture of first layer antibodies (mouse anti-x and rabbit anti-y) at optimal dilutions. Negative control: mixture of inappropriate mouse and rabbit antibodies or normal serum at similar dilutions
   Incubate overnight
3. Rinse in three changes of buffer
4. Apply mixture of second layer antibodies at optimal dilutions (goat anti-mouse and goat anti-rabbit immunoglobulins) for 30 min
4. Rinse in three changes of buffer.
5. Apply mixture of mouse APAAP and rabbit PAP at optimal dilutions for 30 min.
6. Rinse in three changes of buffer
7. Develop alkaline phosphatase in blue with naphthol AS-MX phosphate and Fast Blue BB (see above)
8. Rinse in water and buffer
9. Develop peroxidase with DAB to a light brown colour. Too intense a reaction will detract from the contrast with the blue and prevent a mixture of colours from being seen
10. Rinse in water. Counterstain lightly if desired and mount in an aqueous mountant or dry in an oven and mount in synthetic medium.

Antigen x will be stained in blue and antigen y in brown. If the two are present in the same cell, the colour will be greyish-purple.

HYBRIDIZATION HISTOCHEMISTRY FOR mRNA USING RADIOLABELLED cRNA PROBE

The fixation method chosen will depend on the tissue to be analysed and the quantity of available RNA. The schedule supplied here applies to frozen or Vibratome sections of paraformaldehyde-fixed tissue.

Gloves should be worn throughout the procedure to avoid contaminating the material with RNase from the fingers. Instruments, containers and solutions should be sterile if possible.

Tissue preparation

Tissue must be collected as fresh as possible to avoid degradation of RNA. Small ($1 \times 1 \times 0.5$ cm) pieces of tissue are immersed in freshly made 4% paraformaldehyde in PBS, pH 7.2. Tissue is fixed for 4 h then rinsed in 3×1 h changes of PBS containing 15% sucrose. It may be left in the rinsing buffer overnight or longer at 4°C, but it is best to make frozen blocks as soon as possible and store them at −40°C or lower in clean containers.

Slide preparation

Glass microscope slides are used:

1. Wash in 0.1 mol/l HCl for 20 min
2. Rinse in deionized water
3. Immerse in 100% ethanol for 20 min
4. Allow to dry at room temperature
5. Autoclave to remove any traces of RNase
6. Coat with poly-L-lysine (see above), chrome gelatine or 3-aminopropyltriethoxysilane.

Sections

Cryostat or Vibratome sections cut at 15–30 μm are mounted on the prepared slides and placed in a clean slide rack. This is covered with foil and kept overnight at 37°C to ensure adherence of the sections. They may then be stored at -20°C until used.

Pretreatment of sections (permeabilization)

1. Rehydrate sections in PBS, pH 7.2, for 5 min
2. Immerse in 0.3% Triton X-100 in PBS for 15 min
3. Wash with PBS, 2×5 min
4. Incubate with 1 μg/ml Proteinase K in 0.1 mol/l Tris containing 50 mmol/l ethylene diamine tetra-acetic acid (EDTA), pH 8.0 at 37°C for 20 min
5. Immerse in 0.1 mol/l glycine, in PBS for 5 min (to stop protease reaction)
6. Fix in 4% paraformaldehyde in PBS for 5 min (to prevent diffusion of target nucleic acid)
7. Wash with PBS, 2×5 min
8. Immerse in 0.25% acetic anhydride containing 0.1 mol/l triethanolamine, pH 8.0 for 10 min (to reduce background in the autoradiographs)
9. Immerse in 50% formamide in double strength (2 ×) standard sodium citrate (SSC) at 37°C for at least 15 min

Standard sodium citrate (SSC) 1 × SSC:

0.15 mol/l sodium chloride
0.015 mol/l sodium citrate

Note: different strengths of SSC are indicated as 1×, 2×, 4× and 0.1×

Hybridization

1. Drain slides briefly—do not dry
2. Apply 10–20 μl of preheated (37°C) hybridization mixture per section.

Hybridization mixture

0.5–1×10^6 c.p.m ^{32}P-labelled cRNA probe in 10 μl of solution consisting of (final concentration):

    0.5% sodium pyrophosphate
    2 × SSC
    50% formamide
    10% dextran sulphate
    0.25% bovine serum albumin
    0.25% Ficoll 400
    0.25% polyvinyl pyrrolidone 360
    250 mmol/l Tris/HCl buffer, pH 7.5
    0.5% sodium dodecyl sulphate
    250 μg/ml denatured salmon sperm DNA (this decreases non-specific binding by competing with the labelled RNA for tissue sites)

3. Cover sections with dimethyldichlorosilane-coated coverslips and incubate at 43°C for 16 h
4. Wash with 4 × SSC (coverslips float off) at 37°C, 3 × 20 min with gentle shaking
5. Immerse for 30 min at 37°C in 10 mmol/l Tris/HCl buffer, pH 8.0 containing:

20 μg/ml RNase A
0.5 mol/l sodium chloride,
1.0 mmol/l EDTA

6. Wash with 2 × SSC at 37°C for 30 min with gentle shaking
7. Wash with 0.1 ×SSC at 37°C for 30 min
8. Dehydrate through graded alcohols (70%, 95%, 2 × 100%) containing 0.3 mol/l ammonium acetate (pH 8.0), 10 min each, at room temperature, and allow to air dry.

Autoradiography

1. Dip slides in emulsion (Kodak NTB-2 or Ilford K5) and expose for an appropriate time (3–5 days for ^{32}P, 5–7 days for ^{35}S, 2 weeks for ^{125})
2. Develop with Kodak D19 developer, stop and fix
3. Wash in water, counterstain, dehydrate, clear and mount. View under both light- and dark-field illumination.

HYBRIDIZATION HISTOCHEMISTRY FOR DNA USING BIOTINYLATED DNA PROBE
(generalized method — many variations may be found in the literature)

As for mRNA hybridization, fixation and tissue preparation should be optimized for the system under investigation. Crosslinking (e.g. paraformaldehyde, glutaraldehyde), or precipitant (e.g. alcohol) fixatives may be suitable. Cell preparations or frozen or paraffin sections should be mounted on slides prepared as for mRNA hybridization. All solutions should be sterilized by autoclaving. Glassware, slide racks, etc. should be baked for 4 h at 250°C or soaked in 1% aqueous diethylpyrocarbonate and then autoclaved.

Pretreatment of preparations

Bring paraffin sections to water.

1. If detection of the biotinylated probe is to be carried out by peroxidase-linked avidin, endogenous peroxidase in the preparations must be blocked by immersion in 0.3% H_2O_2 in methanol for 20 min
2. Permeabilize as for mRNA, steps 1–8. Proteinase treatment is particularly necessary when a crosslinking fixative has been used.
3. Wash preparations in distilled water, 1 min
4. Digest mRNA derived from target DNA. Treat with 100 μg/ml RNase in 2 × SSC at 37°C for 1 h
5. Wash in three changes of 2 × SSC, 5 min each
6. Wash in distilled water, 1 min
7. Dry at 37°C for 10 min (to prevent dilution of probe).

Hybridization

Add biotinylated DNA probe to warm (37°C) hybridization mixture, freshly prepared as for RNA hybridization. The concentration of the probe must be determined by experiment. It should be in the order of 1 ng DNA / ml of solution.

1. Apply about 20 μl of probe/hybridization mixture per cm^2 of preparation
2. Cover with dimethyldichlorosilane-coated coverslip
3. Denature native and probe DNA by placing preparation on a hot plate at 95°C for 5 min
4. Place preparation in damp chamber at hybridization temperature for 1–2 h. The temperature (melting temperature of DNA) depends on the amount of formamide in the solution. At 50% formamide the optimum temperature for DNA probes is 42°C.

Post-hybridization

Wash the preparations as for RNA hybridization. The RNase step is not necessary.

Detection of the biotinylated probe

Apply avidin-biotinylated enzyme complex, incubate, wash and develop with a suitable substrate and chromogen. Counterstain, dehydrate, clear and mount as appropriate.

Negative controls

Use a non-homologous probe.
Omit the denaturation step.
Digest the target DNA with DNase before hybridization.
Use tissue that does not contain the target DNA.

Appendix 7

Pigments, metals, etc.

BILIRUBIN
SILVER SULPHIDE METHOD FOR HEAVY METALS
  (TIMM-DANSCHER)
AUTOMETALLOGRAPHY METHOD FOR GOLD,
  SILVER AND MERCURY
ALIZARIN RED S METHOD FOR OXALATE
CALCIUM (GLYOXAL-BIS-2-HYDROXYANIL)
  METHOD
COPPER (RUBEANIC ACID)
COPPER (RHODANINE)
SCHMORL REACTION
AUTOFLUORESCENT PIGMENTS
CHLORAZOL FAST PINK FOR PVA AND PVP

BILIRUBIN REACTION

Preparation of tissue

For staining all bilirubin, routine sections of formalin-fixed tissue may also be used. For differential staining of conjugated bilirubin fresh cryostat sections obtained after 18–24 h fixation in cold $CaCl_2$ (1%)-formalin should be used. The fixed tissue should be immersed for 1–2 days in 0.88 mol/l sucrose containing 1% gum acacia at 4°C.

Preparation of solution

Stock diazo solution

To 100 ml of distilled water add 2 ml of concentrated HCl, then 200 mg of 2,4 dichloroaniline, shake and filter. Keeps for 2 weeks at 4°C.

Sodium nitrite

1%. Keeps for 2 weeks at 4°C.

Accelerator solution

Dissolve 6 g caffeine, 10 g sodium benzoate and 10 g urea in 100 ml water. To 35 ml of this solution add 25 ml concentrated (40%) formaldehyde and 30 ml water.

Method for bilirubin (direct procedure)

1. Immerse sections for 30 min in the staining solution (prepared by adding 0.5 ml of the sodium nitrite to 25 ml of the stock diazo solution and leaving in ice water for 20 min before use)
2. Wash for 3 min in running water
3. Counterstain with haematoxylin
4. Dehydrate rapidly in ethanol, clear and mount.

Method for all bilirubin (the indirect procedure)

1. Prepare the solution as above (0.5 ml of sodium nitrite and 25 ml of the stock diazo solution, left in ice water for 20 min), add to 50 ml of the accelerator solution. If a precipitate forms, dissolve by heating to 38–40°C. Immerse sections in this for 30 min
2. Wash, counterstain, dehydrate and mount as in the direct bilirubin procedure.

Results

The method for all bilirubin stains bilirubin everywhere yellow-brown. The direct procedure stains hepatic and renal bilirubin but not that found in neurons in kernicterus.

SILVER SULPHIDE METHOD FOR HEAVY METALS (Timm-Danscher)

Preparation of tissue

For paraffin wax embedding

Fix small pieces of tissue for 10–24 h in 70% ethanol containing 1–2 ml of concentrated (about 10%)

ammonium sulphide per 100 ml. For cryostat sections: suspend thin (less than 2 mm) pieces of tissue in a covered container containing 1–2 ml of concentrated ammonium sulphide at room temperature for 20 min, then snap freeze.

Preparation of solution

To 100 ml of 20% gum arabic, prepared with daily stirring 14 days ahead of time, add 1 ml of 10% $AgNO_3$ and stir well. Then add 10 ml of a solution containing 0.5 g citric acid and 0.2 g hydroquinone. All reagents should be dissolved in double distilled water and the solution used within seconds of its preparation.

Method

1. Dewax and bring to water. Cryostat sections should be mounted on slides or coverslips
2. Treat in the solution for variable lengths of time (20 min to 4–6 h), until sections are light brown
3. Wash well in distilled water
4. Counterstain with a suitable nuclear stain (e.g. haemalum, or Kernechtrot)
5. Rinse, dehydrate, clear and mount.

Results

Heavy metals (Pb, Au, Ag, Fe, Cd, Cu, Co, Ni, Zn, Hg) brown-black.

Notes

The procedure does not allow identification of the metal. Metals present in protein complexes will be demonstrated only if they are ionized. Thus, haemoglobin is not stained while haemosiderin (ferritin) is easily demonstrable by the technique.

AUTOMETALLOGRAPHY PROCEDURE FOR GOLD, SILVER AND MERCURY

Method

1. Cryostat, paraffin, or Epon sections on glass slides are dried and dipped in a suitable photographic emulsion (e.g. Ilford 95) in the dark
2. Dry for 30 min
3. Develop section in a photographic developer for different lengths of time, to find the optimal time for the preparation

4. Treat for 1 min in 1% acetic acid
5. Treat for 3 min in a photographic fixation bath
6. Take out of the dark room, rinse, counterstain by toluidine blue, dehydrate, clear and mount. Epon sections can be left uncovered.

ALIZARIN RED S METHOD FOR OXALATE

Preparation of tissues

Paraffin sections: postfixed cryostat sections: frozen sections.

Preparation of solutions

Solution A: 2% aq. Alizarin Red S adjusted to pH 4.2 with NaOH
Solution B: 2% aq. Alizarin Red S adjusted to pH 7.0 with NaOH

Method

Section 1: treat with solution A for 3 min, rinse quickly twice with distilled water, dehydrate, clear and mount.

Section 2: treat with solution B for 3 min, rinse quickly twice with distilled water, dehydrate, clear and mount.

Section 3: treat with 2 mol/l acetic acid for 30 min and stain as for section 2.

Result

Section 1: treat with solution A for 3 min, rinse quickly twice with distilled water, dehydrate, clear and mount.

Section 2: treat with solution B for 3 min, rinse quickly twice with distilled water, dehydrate, clear and mount.

Section 3: treat with 2 mol/l acetic acid for 30 min and stain as for section 2.

CALCIUM: GBHA METHOD

Preparation of tissue

Cryostat sections of fresh tissue or sections of unfixed freeze-dried tissue preferred. Routine sections of fixed tissue can also be used, although much of the original calcium content may not be present in such tissue.

Preparation of solutions

GBHA solution

Glyoxal-bis-(2-hydroxyanil) (GBHA) dissolved in:	2.5 g
75% ethanol containing:	50 ml
NaOH	1.7 g

Na₂CO₃-KCN solution

Add solid anhydrous Na_2CO_3 to 50 ml 96% ethanol and shake until no more carbonate appears to dissolve. Filter, and add solid KCN. Shake again until the solution is saturated. Filter, and store solution in stoppered bottle.

Method

1. Mount sections on albuminized slides
2. Dewax sections in chloroform
3. Immerse sections in 0.1% celloidin (dissolved in 50:50 v/v absolute ethanol:diethyl ether) for about 30 s, and dry in air
4. Flood sections with GBHA solution at room temperature and decant after 10 min
5. Rinse in 75% ethanol and subsequently in 96% ethanol
6. Immerse for 15 min in the ethanolic Na_2CO_3-KCN solution at room temperature
7. Rinse in 96% ethanol and subsequently in three changes of absolute ethanol
8. Clear in xylene and mount in a synthetic resin (e.g. Entallan or DPX)
9. Examine the sections with and without a green filter in the substage of microscope.

Result

Calcium deposits are coloured red, and black when viewed with a green filter.

Note

Both ionic and insoluble calcium appears to be visualized by this method. In early versions of the method, only some forms of calcium, probably insoluble salts, are stained. The carbonate-KCN solution decolourizes the GBHA complexes of Ba, Sr, Cd, Co and Ni increasing the specificity for Ca.

RUBEANIC ACID METHOD FOR COPPER

Preparation of tissue

Thin (1 mm or less) slices of fresh or formalin-fixed tissue; fixed or unfixed cryostat or frozen sections, or formalin-fixed routine sections after dewaxing and hydration. Metal-containing fixatives, such as Zenker or Susa and fixatives kept in metal containers should be avoided.

Method

1. Immerse for 10 min in 0.1% rubeanic acid in 70% ethanol
2. Add to the alcoholic solution sodium acetate to reach a concentration of 0.2%, stir and leave at room temperature for 2 days
3. Rinse in 70% ethanol. Leave in 70% ethanol for 1 day
4. Slices of tissue: dehydrate, clear, embed in paraffin wax, cut sections, deparaffinize, hydrate. Cryostat and other sections: hydrate
5. Counterstain with Mayer's haemalum or Kernechtrot
6. Wash, dehydrate, clear and mount.

Results

Copper salts are stained greenish-black.

Notes

Other metal salts (for example, iron and lead) also react with rubeanic acid, but they do not exhibit the greenish tinge.

RHODANINE STAIN FOR COPPER

Preparation of tissue

Postfixed cryostat sections; frozen sections; paraffin sections.

Solutions required

Rhodanine stock solution

p-Dimethylaminobenzylidene rhodanine	0.2 g
Ethanol	100 ml

To prepare working solution, dilute 3 ml of stock solution (well shaken) with 47 ml distilled water.

Borax solution

Disodium tetraborate	0.5 g
Distilled water	100 ml

Technique

1. Bring sections to water
2. Incubate in rhodanine working solution at 37°C (or 56 °C) 18 h (or 3 h)
3. Rinse in several changes of distilled water and stain with Carrazzi's haematoxylin, 1 min
4 Rinse in distilled water and then quickly in borax solution. Rinse well in distilled water
5. Dehydrate, clear and mount.

Results

Copper deposits stain bright red. Bile stains green.

Note

If fading occurs, it can be reduced by staining at 56°C and by using certain mounting media (e.g. DPX).

SCHMÖRL REACTION

Preparation of tissues

Cryostat sections (10 μm); frozen sections; paraffin sections.

Preparation of solutions

Schmörl reagent

Add 18 ml of 1% ferric chloride to 6 ml of 1% potassium ferricyanide. Mix and use immediately.

Method

1. Air-dry sections 15 min
2. Cover with reagent for 15 min
3. Rinse in 1% acetic acid
4. Wash in distilled water and blot gently
5. Allow to air-dry completely
6. Clear in xylene and mount in DPX.

Results

Blue granular reaction indicates the presence of lipo-fuscin pigment, copper associated protein, etc.

AUTOFLUORESCENT PIGMENTS

Preparation of tissue

Air-dried cryostat sections (5–8 μm) of snap-frozen tissue; paraffin sections; frozen sections.

Method

1. Mount cryostat sections in DPX (dewax and mount paraffin sections in DPX)
2. Examine by fluorescence microscopy with dark-ground illumination (excitation filter 300-370 nm (UG5); barrier filter 410 nm).

Result

Lipofuscin (wear and tear pigment) has orange-yellow fluorescence. Pigment in Batten's disease has a yellowish fluorescence.

Notes

The background tissue is deep blue on initial mounting and becomes brighter on storage.
Cryostat sections have much less background than paraffin sections.

CHLORAZOL FAST PINK FOR PVA AND PVP

Preparation of tissue

Routine sections of formalin-fixed tissue.

Method

1. Dewax and bring sections to water
2. Stain in 1% chlorazol fast pink in 50% ethanol at 60°C for 10-30 min
3. Rinse in water
4. Counterstain with Carrazzi's haematoxylin
5. Wash, dehydrate, clear and mount.

Results

PVP and PVA stain pink-red.

Index